Lecture Notes in Artificial Intelligence 10576

Subseries of Lecture Notes in Computer Science

More information about this series at http://www.springer.com/series/1244

Jörg Rothe (Ed.)

Algorithmic Decision Theory

5th International Conference, ADT 2017
Luxembourg, Luxembourg, October 25–27, 2017
Proceedings

 Springer

Editor
Jörg Rothe
Heinrich Heine University
Düsseldorf
Germany

ISSN 0302-9743 ISSN 1611-3349 (electronic)
Lecture Notes in Artificial Intelligence
ISBN 978-3-319-67503-9 ISBN 978-3-319-67504-6 (eBook)
DOI 10.1007/978-3-319-67504-6

Library of Congress Control Number: 2017954897

LNCS Sublibrary: SL7 – Artificial Intelligence

Printed on acid-free paper

This Springer imprint is published by Springer Nature
The registered company is Springer International Publishing AG
The registered company address is: Gewerbestrasse 11, 6330 Cham, Switzerland

Preface

The 5th International Conference on Algorithmic Decision Theory (ADT 2017), held in October 2017 in Luxembourg, brought together researchers and practitioners coming from areas as diverse as artificial intelligence, operations research, discrete mathematics, theoretical computer science, decision theory, game theory, multiagent systems, computational social choice, multi-criteria decision aiding, resource allocation, matching, and argumentation theory. Their joint aim is to improve the theory and practice of modern decision support. Previous ADT conferences were held in Venice, Italy (2009); Piscataway, NJ, USA (2011); Brussels, Belgium (2013); and Lexington, KY, USA (2015).

Among the scientific challenges the ADT community faces are difficult questions such as how to elicit and aggregate big preference data, how to deal with combinatorial structures and with partial or uncertain information, and how to manage distributed decision making. In real-world decision making, such challenges are encountered in various domains, for example in electronic commerce, recommender systems, argumentation tools, network optimization (communication, transport, energy), risk assessment and management, and e-government.

The papers in this volume were presented at ADT 2017. Each submission to ADT 2017 was peer-reviewed by at least three Program Committee (PC) members in a double-blind fashion. Out of 45 submissions the PC decided to accept 25 papers for oral presentation, giving an acceptance rate of about 55%. This volume contains 22 of these 25 accepted full papers. The other three accepted papers were submitted only for presentation at the conference and are not contained in the proceedings:

- Khaled Belahcène, Christophe Labreuche, Nicolas Maudet, Vincent Mousseau, and Wassila Ouerdane: *An Efficient SAT Formulation for Learning Multicriteria Non-compensatory Sorting Models*
- Giuseppe De Marco: *On Ambiguous Games under Imprecise Probabilities*
- Dominik Peters: *Single-Peakedness and Total Unimodularity for Multiwinner Elections*

In addition, the PC selected six short papers for poster presentation and the program also included six peer-reviewed short papers written for the associated doctoral consortium. Last but not least, this volume also contains the abstracts of the three invited keynote speeches and the abstract of a tutorial presented at the doctoral consortium.

I am very grateful to Raymond Bisdorff for his unwearying efforts as local chair; to Eleni Pratsini, Carmine Ventre, and Toby Walsh for accepting my invitation and giving great keynote speeches at the conference; to Serge Gaspers for holding a wonderful tutorial at the doctoral consortium; to Anja Rey for organizing the doctoral consortium; to Alexis Tsoukiàs, Gerhard J. Woeginger, and Nick Mattei for their advice; to the Program Committee and their additional reviewers for their help with selecting and preparing the conference program; and to the many volunteers who helped in some way or another. I also thank the authors for submitting and presenting their interesting recent research results. And I thank our sponsors for their financial support: the Fonds National de la Recherche, Luxembourg (FNR), the University of Luxembourg (FSTC/CSC Research Unit), the CSC Interdisciplinary Lab for Intelligent and Adaptive Systems (ILIAS), the ADT 2015 organizers (in particular, Judy Goldsmith from the University of Kentucky, Lexington, KY, USA), the Advances in Preference Handling Multidisciplinary Working Group affiliated to EURO, CNRS (France), and Springer. I also very much appreciated using EasyChair for organizing the reviewing process and preparing the proceedings.

Most notably, I thank my wife, Irene, and my daughters, Ella and Paula, for their loving support during the months I was working as a program chair, having less time to spend with them, and for still making my life such a joy.

July 2017   Jörg Rothe

Organization

Program Committee

Raymond Bisdorff	University of Luxembourg, Luxembourg
Sylvain Bouveret	LIG - Grenoble INP, Université Grenoble-Alpes, France
Robert Bredereck	TU Berlin, Germany, and University of Oxford, UK
Simina Brânzei	The Hebrew University of Jerusalem, Israel
Katarína Cechlárová	P.J. Šafárik University Kosice, Slovakia
Jiehua Chen	Ben-Gurion University of the Negev, Israel
Gabrielle Demange	EHESS, Paris School of Economics, France
Paul Goldberg	University of Oxford, UK
Judy Goldsmith	University of Kentucky, Lexington, KY, USA
Edith Hemaspaandra	Rochester Institute of Technology, USA
Sébastien Konieczny	CNRS, CRIL, Université d'Artois, France
Jérôme Lang	CNRS, LAMSADE, Université Paris-Dauphine, France
Nicholas Mattei	IBM T.J. Watson Research Center, NY, USA
Brice Mayag	LAMSADE, Université Paris-Dauphine, France
Joana Pais Ribeiro	ISEG, Technical University of Lisbon, Portugal
Hans Peters	Maastricht University, The Netherlands
Gabriella Pigozzi	LAMSADE, Université Paris-Dauphine, France
Eve Ramaekers	Université catholique de Louvain, Belgium
Anja Rey	TU Dortmund, Germany
Federica Ricca	Università di Roma La Sapienza, Italy
Francesca Rossi	IBM T.J. Watson Research Center, NY, USA, and University of Padova, Italy
Jörg Rothe	Heinrich-Heine-Universität Düsseldorf, Germany
Ildikó Schlotter	Budapest University of Technology and Economics, Hungary
Claudia Schulz	TU Darmstadt, Germany
Piotr Skowron	University of Oxford, UK, and TU Berlin, Germany
Marija Slavkovik	University of Bergen, Norway
Alexis Tsoukiàs	CNRS, LAMSADE, Université Paris-Dauphine, France
Leon van der Torre	University of Luxembourg, Luxembourg
Kristen Brent Venable	Tulane University and IHMC, New Orleans, LA, USA
Angelina Vidali	De Montfort University, Leicester, UK
Toby Walsh	Data61 and UNSW, Sydney, Australia, and TU Berlin, Germany
Gerhard J. Woeginger	RWTH Aachen, Germany
Stefan Woltran	TU Wien, Austria
Ying Zhu	University of Kentucky, Lexington, KY, USA

Additional Reviewers

Barrot, Nathanaël
Bliem, Bernhard
Brill, Markus
Chalki, Aggeliki
Cornelio, Cristina
Fitzsimmons, Zack
Hosseini, Hadi
Kaczmarczyk, Andrzej
Kesselheim, Thomas
Komusiewicz, Christian

Kuckuck, Benno
Lonca, Emmanuel
Maly, Jan
Miles, Luke Harold
Narváez, David
Schwind, Nicolas
Siler, Cory
Tauer, Björn
Xia, Lirong

Abstracts of Invited Talks

Algorithmic Decision Theory and Cognitive IoT

Eleni Pratsini

IBM Research - Zurich, 8803 Rüschlikon, Switzerland
pra@zurich.ibm.com

Abstract. The Internet of Things is changing the way we interact with our surroundings. It is estimated that the number of connected devices will grow to 21 billion by 2020. The IoT is generating an explosion of sensor data, most of it is unstructured and multi-modal, requiring sophisticated techniques to analyze and interpret. We can only keep up with the complexity and unpredictability of this information through the use of cognitive computing systems that self-learn, reason and automatically adjust to the surroundings. The enterprise that can analyze the data on the fly and generate transformational decisions to constantly adapt to the changing environment will be a leader in its field. Decision Theory plays a key role in determining these transformational decisions, and our techniques need to address the effects of the solution cycle, separating the operator actions from the model. Through the use of examples, we will highlight the differentiation and power of cognitive computing and decision theory in IoT, and point to new research directions.

Keywords: Decision theory · Cognitive computing · IoT

Novel Mechanism Design Paradigms

Carmine Ventre

University of Essex, Colchester, UK
c.ventre@essex.ac.uk
http://csee.essex.ac.uk/acstaff/carven/

Abstract. The main difficulty of algorithm design is classically linked with understanding the combinatorial structure of the optimization problem at hand. However, this picture is by now outdated. The emergence of the Internet as computing platform has, in fact, highlighted the presence of agents that selfishly evaluate the outcome of computation and might not "follow the rules" if this benefits them. Algorithms have now also to work in the presence of these selfish interests, which are often in contrast with the ultimate objective of the computation (e.g., optimality). The field of Algorithmic Mechanism Design has as its main scope the realignment of the objective of the designer with those of the selfish agents. "Good" mechanisms set the rules "properly" and guarantee that agents have no interest in misguiding the mechanism towards the computation of "wrong" (e.g., suboptimal) outcomes.

The great theoretical findings in the field have not been accompanied so far by notable technological advances. The reasons behind this apparent failure can be found in the assumptions that are often underlying either the modelling of the problem at hand (e.g., full rationality of agents involved in the computation) or the hypothesis needed for the theorems (e.g., the possibility to exchange money at the end of the computation).

In this talk, I focus on some recent work aimed at building the theoretical foundations for a more applied use of Algorithmic Mechanism Design. I reconsider the role of money as 'necessary evil' and discuss ways to incorporate real-world hypotheses in the mechanism design setting in order to reconcile computation and incentives without the use of monetary transfers. I show how these hypotheses are also sufficient to obtain "good" mechanisms for agents that are not fully rational. Finally, as a complementary avenue towards pragmatic mechanism design, I explore the definition of "subject model" needed for the success of lab experiments devised by neuroscientists to measure awareness.

This work was partially supported by EPSRC under grant EP/M018113/1.

Deceased Organ Matching in Australia

Toby Walsh[1,2]

[1] UNSW Sydney, Data61
tw@cse.unsw.edu.au
[2] TU Berlin

Abstract. Despite efforts to increase the supply of organs from living donors, most kidney transplants performed in Australia still come from deceased donors. The age of these donated organs has increased substantially in recent decades as the rate of fatal accidents on roads has fallen. The Organ and Tissue Authority in Australia is therefore looking to design a new mechanism that better matches the age of the organ to the age of the patient. I discuss the design, axiomatics and performance of several candidate mechanisms that respect the special online nature of this fair division problem.

Introduction

Kidney disease is a major problem in Australia. Thousands of people are on dialysis. Many spend years waiting for a transplant, each costing the health care budget hundreds of thousands of dollars. In addition, as dialysis takes up several days each week, many are unable to work and depend on support from the state. The total cost to the Australian economy runs into billions of dollars annually. In 2016, 85% of transplants involved a kidney coming from a deceased person, whilst only 15% of transplants came from a living donor. Whilst there has been considerable focus in the literature of late on increasing the supply of organs by developing mechanisms for paired exchange, only 2.5% of these living donations came from paired exchange. Most living donors were a spouse, family member or friend of the recipient.

Organs coming from deceased people still provide the majority of all transplanted kidneys. Many come from people killed in road traffic accidents. Matching such organs to patients on the waiting list is becoming more challenging as roads become safer. The mean age of donated organs has increased from 32 years in 1989 to 46 years in 2014. Advances in medicine mean that doctors are also now willing to transplant older kidneys. In 2014, the oldest organ transplanted came from a person who was 80 years old. This compares to 1989, the first year for which records are available, when the oldest organ transplated came from a person aged just 69. The Organ and Tissue Authority of Australia, the government body charged with the task of allocating organs to patients fairly and efficiently, is therefore looking to develop a new matching mechanism. Their goal is to develop a new procedure which matches the age of the patients and organs.

Organ Matching Mechanisms

The mechanism used at present in Australia does not explicitly take age of the patients or organs into account. As a result, young organs will be offered to old patients, and old organs to young patients. Neither are very desirable. Even if an old patient would like a young organ, from a societal perspective, this is not a very good outcome. The old patient will die from natural causes with an organ inside them that could have continued to function in a younger patient. And transplanting an old organ into a young patient is not a good outcome for both the individual or society. The graft will likely fail after a few years, meaning the patient will need a new transplant. In addition, the patient's immune system will now be highly sensitized, so that a new match will be more difficult.

The Organ and Tissue Authority is looking therefore to develop a new mechanism in which organs are ranked by the Kidney Donor Patient Index (KDPI). This is an integer from 0 to 100 that is calculated from the age of the donor, and a number of other factors like diabetic status. A donated organ with a KDPI of X% has an expected risk of graft failure greater than X% of all donated organs. Similarly the Organ and Tissue Authority wish to rank patients waiting transplant with the Expected Post-Transplant Survival (EPTS) score. This is also an integer from 0 to 100 that is calculated from the age of the recipient, and a number of other factors like diabetic status, and time on dialysis. A patient on the waiting list with a lower EPTS is expected to have more years of graft function from high-longevity kidneys compared to candidates with higher EPTS scores. Our goal is to provide the Organ and Tissue Authority with a new mechanism that is fair and efficient, matching organs so that the KDPI of an arriving organ is as close as possible to the EPTS score of their allocated patient.

Other Applications

This work fits into a broader research programme to design mechanisms for resource allocation problems that better reflect the complexity and richness of the real world [1, 2]. Unlike traditional resource allocation problems [3], one of the fundamental features of the deceased organ matching problem is that it is online. We do not know when organs will arrive to be match. And we must match and transplant them shortly after they arrive, before we know what organs or patients will arrive in the future. At the end of the year, we could find an optimal allocation. However, we do not have the luxury of waiting till the end of the year as organs must be transplanted immediately. There are many other domains where resources are allocated in a similar online manner. A food bank might start allocating and distributing food to charities soon after it is donated [4]. An airport must start allocating landing slots before all demands are known. A particle accelerator might start allocating beam time before all requests have come in. A university might allocate rooms to students for the current term, not knowing what rooms might be wanted in future terms. This work offers a case study in how we can efficiently and fairly solve such *online* allocation problems. We study axiomatic properties of such online fair division problems, as well as run experiments on real world organ data [5]. Axiomatic analysis covers such properties as fairness and efficiency (e.g. [6–11]), as well as strategic behaviour and manipulation (e.g. [12–18]). Insights from this research may prove valuable in a range of other domains. In future, we plan to identify and study phase transition behaviour [19–24] which has proved valuable in a wide range of computational domains [25–33] including social choice [34–36].

References

1. Walsh, T.: Allocation in practice. In: Lutz, C., Thielscher, M. (eds.) KI 2014. LNCS, vol. 8736, pp. 13–24. Springer, Switzerland (2014)
2. Walsh, T.: Challenges in resource and cost allocation. In: Proceedings of the 29th AAAI Conference on AI. Association for Advancement of Artificial Intelligence, pp. 25–30 (2015)
3. Chevaleyre, Y., Dunne, P., Endriss, U., Lang, J., Lemaitre, M., Maudet, N., Padget, J., Phelps, S., Rodriguez-Aguilar, J., Sousa, P.: Issues in multiagent resource allocation. Informatica (Slovenia) 30(1), 3–31 (2006)
4. Aleksandrov, M., Aziz, H., Gaspers, S., Walsh, T.: Online fair division: analysing a food bank problem. In: Yang, Q., Wooldridge, M. (eds.) Proceedings of the Twenty-Fourth International Joint Conference on Artificial Intelligence, IJCAI 2015, pp. 2540–2546 (2015)
5. Mattei, N., Saffidine, A., Walsh, T.: Mechanisms for online organ matching. In: Proceedings of the 26th International Joint Conference on Artificial Intelligence (2017)
6. Bouveret, S., Lang, J.: A general elicitation-free protocol for allocating indivisible goods. In: Walsh, T. (ed.) Proceedings of the 22nd International Joint Conference on Artificial Intelligence, IJCAI 2011, IJCAI/AAAI, pp. 73–78 (2011)
7. Kalinowski, T., Narodytska, N., Walsh, T.: A social welfare optimal sequential allocation procedure. In: Proceedings of the 23rd International Joint Conference on Artificial Intelligence, IJCAI-2013 (2013)
8. Baumeister, D., Bouveret, S., Lang, J., Nguyen, N., Nguyen, T., Rothe, J., Saffidine, A.: Axiomatic and computational aspects of scoring allocation rules for indivisible goods. In: 5th International Workshop on Computational Social Choice, COMSOC 2014 (2014)
9. Aziz, H., Gaspers, S., Mackenzie, S., Walsh, T.: Fair assignment of indivisible objects under ordinal preferences. In: Bazzan, A., Huhns, M., Lomuscio, A., Scerri, P. (eds.) International conference on Autonomous Agents and Multi-Agent Systems, AAMAS 2014, pp. 1305–1312 (2014)
10. Aziz, H., Walsh, T., Xia, L.: Possible and necessary allocations via sequential mechanisms. In: Yang, Q., Wooldridge, M. (eds.) Proceedings of the Twenty-Fourth International Joint Conference on Artificial Intelligence, IJCAI 2015, pp. 468–474 (2015)
11. Aziz, H., Kalinowski, T., Walsh, T., Xia, L.: Welfare of sequential allocation mechanisms for indivisible goods. In: Kaminka, G., Fox, M., Bouquet, P., Hüllermeier, E., Dignum, V., Dignum, F., van Harmelen, F. (eds.) ECAI 2016 - 22nd European Conference on Artificial Intelligence. Frontiers in Artificial Intelligence and Applications, pp. 787–794. IOS Press (2016)
12. Roos, M., Rothe, J.: Complexity of social welfare optimization in multiagent resource allocation. In: Proceedings of the 9th International Conference on Autonomous Agents and Multiagent Systems, AAMAS 2010, pp. 641–648 (2010)
13. Kalinowski, T., Narodytska, N., Walsh, T., Xia, L.: Strategic behavior when allocating indivisible goods sequentially. In: Proceedings of the Twenty-Seventh AAAI Conference on Artificial Intelligence, AAAI 2013. AAAI Press (2013)

14. Aziz, H., Walsh, T., Xia, L.: Possible and necessary allocations via sequential mechanisms. In: Yang, Q., Wooldridge, M. (eds.) Proceedings of the Twenty-Fourth International Joint Conference on Artificial Intelligence, IJCAI 2015 (2015)

15. Aziz, H., Gaspers, S., Mackenzie, S., Mattei, N., Narodytska, N., Walsh, T.: Manipulating the probabilistic serial rule. In: Weiss, G., Yolum, P., Bordini, R., Elkind, E. (eds.) Proceedings of International Conference on Autonomous Agents and Multiagent Systems, AAMAS 2015 (2015)

16. Nguyen, N., Baumeister, D., Rothe, J.: Strategy-proofness of scoring allocation correspondences for indivisible goods. In: Proceedings of the 24th International Joint Conference on Artificial Intelligence, IJCAI 2015. IJCAI (2015)

17. Aziz, H., Schlotter, I., Walsh, T.: Control of fair division. In: Kambhampati, S. (ed.) Proceedings of the Twenty-Fifth International Joint Conference on Artificial Intelligence, IJCAI 2016, pp. 67–73. IJCAI/AAAI Press (2016)

18. Walsh, T.: Strategic behaviour when allocating indivisible goods. In: Schuurmans, D., Wellman, M. (eds.) Proceedings of the Thirtieth AAAI Conference on Artificial Intelligence, pp. 4177–4183. AAAI Press (2016)

19. Huberman, B., Hogg, T.: Phase transitions in artificial intelligence systems. Artif. Intell. **33**, 155–171 (1987)

20. Hogg, T.: Refining the phase transition in combinatorial search. Artif. Intell. **81**(1–2), 127–154 (1996)

21. Cheeseman, P., Kanefsky, B., Taylor, W.: Where the really hard problems are. In: Proceedings of the 12th IJCAI, International Joint Conference on Artificial Intelligence, pp. 331–337 (1991)

22. Mitchell, D., Selman, B., Levesque, H.: Hard and easy distributions of sat problems. In: Proceedings of the 10th National Conference on AI, pp. 459–465. Association for Advancement of Artificial Intelligence (1992)

23. Gent, I., Walsh, T.: The SAT phase transition. In: Cohn, A.G. (ed.) Proceedings of 11th ECAI, pp. 105–109. John Wiley & Sons (1994)

24. Gent, I., Walsh, T.: Easy problems are sometimes hard. Artif. Intell. 335–345 (1994)

25. Gent, I., Walsh, T.: Phase transitions from real computational problems. In: Proceedings of the 8th International Symposium on Artificial Intelligence, pp. 356–364 (1995)

26. Gent, I., Walsh, T.: Phase transitions and annealed theories: Number partitioning as a case study. In: Proceedings of 12th ECAI (1996)

27. Gomes, C., Selman, B.: Problem structure in the presence of perturbations. In: Proceedings of the 14th National Conference on AI, pp. 221–226. Association for Advancement of Artificial Intelligence (1997)

28. Gomes, C., Selman, B., Crato, N.: Heavy-tailed distributions in combinatorial search. In: Smolka, G. (ed.) CP 1997, pp. 121–135. Springer, Berlin (1997)

29. Gomes, C., Selman, B., McAloon, K., Tretkoff, C.: Randomization in backtrack search: exploiting heavy-tailed profiles for solving hard scheduling problems. In: The Fourth International Conference on Artificial Intelligence Planning Systems, AIPS 1998 (1998)

30. Gent, I., Walsh, T.: The TSP phase transition. Artif. Intell. **88**, 349–358 (1996)

31. Gent, I., Walsh, T.: Beyond NP: the QSAT phase transition. In: Proceedings of the 16th National Conference on AI. Association for Advancement of Artificial Intelligence (1999)

32. Bailey, D., Dalmau, V., Kolaitis, P.: Phase transitions of PP-complete satisfiability problems. In: Proceedings of the 17th IJCAI, International Joint Conference on Artificial Intelligence, pp. 183–189 (2001)

33. Walsh, T.: From P to NP: COL, XOR, NAE, 1-in-k, and Horn SAT. In: Proceedings of the 17th National Conference on AI, Association for Advancement of Artificial Intelligence (2002)

34. Walsh, T.: Where are the really hard manipulation problems? The phase transition in manipulating the veto rule. In: Proceedings of 21st IJCAI, International Joint Conference on Artificial Intelligence, pp. 324–329 (2009)
35. Walsh, T.: An empirical study of the manipulability of single transferable voting. In: Coelho, H., Studer, R., Wooldridge, M. (eds.) Proceedings of the 19th European Conference on Artificial Intelligence (ECAI-2010). Volume 215 of Frontiers in Artificial Intelligence and Applications, pp. 257–262. IOS Press (2010)
36. Walsh, T.: Where are the hard manipulation problems? J. Artif. Intell. Res. **42**, 1–39 (2011)

An Introduction to Parameterized Complexity with Applications in Algorithmic Decision Theory

Serge Gaspers (iD)

UNSW Sydney, Sydney, Australia
sergeg@cse.unsw.edu.au
Data61, CSIRO, Canberra

Abstract. The main motivation for parameterized complexity is that in many applied settings, the difficulty of solving a problem does not merely depend on the size of the instance, but on other parameters of the instance as well. It enables a much more fine-grained complexity analysis than the classical theory around NP-hardness. In this talk, we will see some of the basic algorithmic techniques for taking advantage of small parameters of the input and briefly discuss fixed-parameter intractability. The concepts will be illustrated with studies of computational problems in game theory, combinatorial game theory, voting, resource allocation, and matching markets.

In parameterized complexity [3, 4], also known as multivariate complexity theory [6], we equip the instances of a computational problem with a parameter k that depends on the instance. The parameter could be the number of agents or candidates, the treewidth of the input graph, the largest length of a preference list, etc. We say that a parameterized problem is *fixed-parameter tractable* if there is an algorithm solving it in time $f(k)N^{O(1)}$, where N is the size of the input instance and f is an arbitrary computable function. Note that the superpolynomial part of the running time depends only on the parameter and not on N. This is in contrast to undesirable running times of the form $N^{f(k)}$, which are polynomial for every fixed value of the parameter, but where the degree of this polynomial grows with increasing k.

We will focus on algorithmic techniques that are particularly important in algorithmic decision theory, including branching algorithms, kernelization [10], integer linear programming [8], and parameterizing by the number of numbers [7]. They will be illustrated with applications in computational social choice [1, 9], game theory [5], combinatorial games [2], and stable matchings [11].

References

1. Betzler, N., Bredereck, R., Chen, J., Niedermeier, R.: Studies in computational aspects of voting. In: Bodlaender, H.L., Downey, R., Fomin, F.V., Marx, D. (eds.) The Multivariate Algorithmic Revolution and Beyond. LNCS, vol. 7370, pp. 318–363. Springer, Heidelberg (2012)

2. Bonnet, É., Gaspers, S., Lambilliotte, A., Rümmele, S., Saffidine, S.: The parameterized complexity of positional games. In: Proceedings of the 44th International Colloquium on Automata, Languages and Programming (ICALP 2017), Track A, Schloss Dagstuhl – Leibniz-Zentrum für Informatik, LIPIcs 80, pp. 90:1–90:14 (2017)
3. Cygan, M., Fomin, F.V., Kowalik, L., Lokshtanov, D., Marx, D., Pilipczuk, M., Pilipczuk, M., Saurabh. S.: Parameterized Algorithms. Springer (2015)
4. Downey, R.G., Fellows, M.R.: Fundamentals of Parameterized Complexity. Springer (2013)
5. Estivill-Castro, V., Parsa, M.: Computing Nash equilibria gets harder: New results show hardness even for parameterized complexity. In: Downey, R., Manyem, P. (eds.) Proceedings of the 15th Computing: The Australasian Theory Symposium (CATS 2009). CRPIT, vol. 94, pp. 81–87. Australian Computer Society (2009)
6. Fellows, M.R., Gaspers, S., Rosamond, F.: Multivariate complexity theory. In: Blum, E.K., Aho, A.V. (eds.) Computer Science: The Hardware, Software and Heart of It, Chapter 13, pp. 269–293. Springer (2011)
7. Fellows, M.R., Gaspers, S., Rosamond, F.A.: Parameterizing by the number of numbers. Theory Comput. Syst. **50**(4), 675–693 (2012)
8. Lenstra, H.W.: Integer programming with a fixed number of variables. Math. Oper. Res. **8**(4), 538–548 (1983)
9. Lindner, C., Rothe, J.: Fixed-parameter tractability and parameterized complexity, applied to problems from computational social choice. In: Holder, A. (ed.) Mathematical Programming Glossary. INFORMS Computing Society (2008)
10. Lokshtanov, D., Misra, N., Saurabh, S.: Kernelization – preprocessing with a guarantee. In: Bodlaender, H.L., Downey, R., Fomin, F.V., Marx, D. (eds.) The Multivariate Algorithmic Revolution and Beyond. LNCS, vol. 7370, pp. 129–161. Springer, Heidelberg (2012)
11. Schlotter, I.: Parameterized complexity of graph modification and stable matching problems. Ph.D. thesis, Budapest University of Technology and Economics (2010)

Contents

Short Papers (Poster Presentations)

Doctoral Consortium

Full Papers (Oral Presentations)

Constructive Preference Elicitation for Multiple Users with Setwise Max-margin

Stefano Teso[1], Andrea Passerini[2], and Paolo Viappiani[3(✉)]

[1] KU Leuven, Leuven, Belgium
stefano.teso@cs.kuleuven.be
[2] University of Trento, Trento, Italy
andrea.passerini@unitn.it
[3] Sorbonne Universités, UPMC Univ Paris 06 CNRS, LIP6 UMR 7606, Paris, France
paolo.viappiani@lip6.fr

Abstract. In this paper we consider the problem of simultaneously eliciting the preferences of a group of users in an interactive way. We focus on *constructive* recommendation tasks, where the instance to be recommended should be synthesized by searching in a constrained configuration space rather than choosing among a set of pre-determined options. We adopt a setwise max-margin optimization method, that can be viewed as a generalization of max-margin learning to sets, supporting the identification of informative questions and encouraging sparsity in the parameter space. We extend setwise max-margin to multiple users and we provide strategies for choosing the user to be queried next and identifying an informative query to ask. At each stage of the interaction, each user is associated with a set of parameter weights (a sort of alternative options for the unknown user utility) that can be used to identify "similar" users and to propagate preference information between them. We present simulation results evaluating the effectiveness of our procedure, showing that our approach compares favorably with respect to straightforward adaptations in a multi-user setting of elicitation methods conceived for single users.

1 Introduction

Preferences are a widely studied concept in artificial intelligence [17]; the design of effective methods for preference elicitation is a particularly important topic in order to support the development of personalized systems such as recommender systems, electronic commerce applications and personal agents.

Recently, a number of techniques have been proposed allowing to incrementally elicit the preferences of a user by asking specifically chosen questions. These methods include Bayesian elicitation techniques [4,12,23] and regret-based methods [5,24]. The advantage of the Bayesian approaches is that they can identify informative queries in a principled manner and as well handle inconsistencies in preference feedback, but they require computationally intensive Bayesian

Part of this research was done while ST was at University of Trento, partially supported by CARITRO Foundation grant 2014.0372.

J. Rothe (Ed.): ADT 2017, LNAI 10576, pp. 3–17, 2017.
DOI: 10.1007/978-3-319-67504-6_1

updates; on the other hand regret-based methods can efficiently deal with larger configuration spaces but assume that all preference information is "noiseless".

Recently, setwise max-margin optimization has been proposed [22] as a paradigm for elicitation that has the following distinctive characteristics: (1) it allows to determine informative queries, (2) it can efficiently deal with large configuration spaces, (3) it is robust to user inconsistencies in preference feedback, and (4) it can be coupled with regularization terms if sparsity is required.

The focus of works in preference elicitation has been so far on acquiring the preferences of a single user. However, we claim that real systems - such as electronic commerce websites - do not usually interact with a user in isolation, but may be accessed by several users at the same time. Moreover, typical users of a web application may only provide very little information to the system. This means that it is crucial to exploit as much as possible the available preference information and to leverage the knowledge about the preferences of similar users.

In this paper we consider the problem of preference elicitation in the case that a number of users are simultaneously present. We focus on *constructive* recommendation problems, where the task is that of arranging novel configurations subject to feasibility constraints and user preferences, rather than selecting an item among a set of candidates. This setting rules out standard collaborative filtering techniques [21], where recommendations are propagated between users based on shared ratings over the same or similar objects. We instead rely on a notion of similarity in model space, i.e. similar users have similar utility functions, and propagate information while simultaneously learning user utilities. We extend setwise max-margin preference elicitation to the multi-user setting, by: (i) defining a user similarity as a kernel over user utility models; (ii) measuring the reliability of the learned model for each user; (iii) defining for each user an aggregate utility function combining her utility model with those of the other users, weighted by their respective reliability and similarity to the user being recommended. We show how to incorporate these aspects in the setwise max-margin optimization problem, while retaining the formulation as mixed integer-linear problem (MILP) which allows for efficient computation.

Our experimental evaluation on both syntetic and real datasets shows how a simple procedure iteratively querying the least elicited user succeeds in improving recommendation quality with respect to independent elicitation of users.

2   Related Work

The basic idea of max-margin is as follows. The utility (often assumed to be linear) is determined by a set of parameter weights (unknown to the system). The currently known preferences of the user are encoded by a set of inequalities (typically stating the preference for an alternative over another one) on the feasible parameters; a shared non-negative margin is introduced as a decision variable that is maximized in the objective function. Noisy feedback can be addressed by relaxing the constraints using slack variables and adding a penalty term in the objective for violated constraints.

This intuition has been adopted for preference learning by different authors in essentially the same way. In particular, Gajos and Weld [10] proposed the use of maxmargin optimization for learning preferences in the context of personalized user interfaces using a volumetric heuristic to choose the next question to ask. More recently, maxmargin methods have been used to assess preferences expressed in terms of Choquet capacities [1]. These works are limited by the lack of a principled way to determine informative queries and offer low scalability.

In our work we follow the ideas of [22] that extends maxmargin to produce a set of solutions (instead of a single one); in this way we can use such a set to devise a query to ask to user. *Setwise maxmargin* can then provide an efficient method for interactive preference elicitation handling user inconsistencies (the preference reported by the user may not be always true) and particularly suited to large configuration spaces. One main advantage is that determination of the next question (that is conflated into the problem of generating a set of diverse recommendations as in [23,24] in different paradigms) is the output of an optimization problem and therefore much more scalable that ad-hoc heuristic that need to iterate over available items.

While preference elicitation is a well studied topic in the community of artificial intelligence and algorithmic decision theory, the elicitation of preferences in a multi user setting is still underexplored, with the exception of [13]. However some recent works in the computational social choice community [3,8,15,25] have considered the problem of eliciting interactively the preferences of several users (agents) in order to determine a choice for the group. The difference with our work is that these approaches aim at establishing a best choice for the whole community (according to a voting rule that is fixed in advance); instead we wish to make recommendations that are personalized to each user while exploiting similarity between users' preferences. Some authors have instead considered how to combine interactive elicitation with collaborative filtering for predicting ratings given to items [9]; however this differs form our setting as we consider multi attribute utilities.

The idea of pooling together information about related learning tasks is not new. Our work is related to multi-task approaches (see for instance [2]), where information (data or parameters) is transferred beteween similar tasks to reduce the labelling effort required to achieve good generalization. Like in our work, task similarities are often expressed with a kernel function [20]. The multi-task active learning approach of Saha et al. [19] is perhaps the most closely related: in both methods the task similarity is estimated during the learning process. However, Saha et al. assume that labelled examples are received from some external source, and therefore do not propose any query selection strategy. On the contrary, we rely on the proven setwise max-margin approach for selecting informative queries especially designed for interaction with human decision makers.

3   Background

Notation. We indicate scalars x in italics, column vectors $\mathbf{x}$ in bold, and sets $\mathcal{X}$ in calligraphic letters. Important scalar constants N, M, K are upper-case. The

Table 1. Notation used throughout the paper.

$N, M \in \mathbb{N}$	Number of attributes and users, respectively
$K \in \mathbb{N}$	Cardinality of the query sets
$\mathcal{X} \subseteq \{0,1\}^N$	Set of feasible configurations
$\boldsymbol{w}_*^u \in \mathbb{R}_+^N$	True preferences of user $u \in [M]$
$\boldsymbol{x}_*^u \in \mathcal{X}$	One of the configurations most preferred by user u
$\boldsymbol{w}_1^u, \ldots, \boldsymbol{w}_K^u \in \mathbb{R}_+^N$	Estimated preferences of user u
$\boldsymbol{x}_1^u, \ldots, \boldsymbol{x}_K^u \in \mathcal{X}$	Query set made to user u
$\boldsymbol{x}^u \in \mathcal{X}$	Recommendation made to user u
$v(u) \geq 0$	Variability within $\{\boldsymbol{w}_i^u\}$
$k(u,y) \geq 0$	Similarity between $\{\boldsymbol{w}_i^u\}$ and $\{\boldsymbol{w}_i^y\}$
$\boldsymbol{\alpha} := (\alpha, \beta, \gamma) \in \mathbb{R}_+^3$	Hyperparameters of the MU-SWMM algorithm

inner product between vectors is written as $\langle \boldsymbol{w}, \boldsymbol{x} \rangle = \sum_i w_i x_i$, the Euclidean ($\ell_2$) norm as $\|\boldsymbol{x}\| := \sqrt{\sum_i x_i^2}$, and the ℓ_1 norm as $\|\boldsymbol{x}\|_1 := \sum_i |x_i|$. We abbreviate the set $\{\boldsymbol{w}_i^u\}_{i=1}^K$ as $\{\boldsymbol{w}_i^u\}$ whenever the range for index i is clear from the context, and the set $\{1, \ldots, n\}$ as $[n]$. Table 1 summarizes the most frequently used symbols.

Constructive setting. We consider a feasible set of products $\mathcal{X}$ populated by multi-attribute configurations $\boldsymbol{x} = (x_1, \ldots, x_N)$ over N attributes. In this presentation we will concentrate on 0–1 attributes only, a common choice in the preference elicitation literature [12,23]. Categorical attributes can be handled by using a one-hot encoding. Linearly dependent numerical attributes can be dealt with too; we refer the reader to the detailed discussion in [22] for space constraints. In contrast to standard recommendation, the set of products $\mathcal{X}$ is not explicitly provided, but rather defined by a set of hard (feasibility) constraints. These are assumed to be linear in the attributes. This setup is rather general, and naturally allows to encode both arithmetical and logical constraints. For instance, under the usual mapping *true* $\mapsto 1$ and *false* $\mapsto 0$, the logical disjunction between two 0–1 attributes $x_1 \vee x_2$ can be written as $x_1 + x_2 \geq 1$. Similarly, logical implication $x_1 \Rightarrow x_2$ can be translated to $(1 - x_1) + x_2 \geq 1$.

Following previous work on preference elicitation [12,23], user preferences are modeled as additive utility functions [14]. The *true* user preferences are represented by a non-negative weight vector $\boldsymbol{w}_* \in \mathbb{R}_+^N$, and the utility (i.e. subjective quality) of a feasible configuration $\boldsymbol{x}$ is given by the inner product $\langle \boldsymbol{w}_*, \boldsymbol{x} \rangle$. The *true* most preferred configurations, i.e. with maximal true utility, are analogously indicated as $\boldsymbol{x}_*$. In the remainder all weight vectors (both true and estimated) will be assumed to be non-negative and bounded, i.e. for every attribute $z \in [N]$ there exist two finite non-negative constants $w_z^\top$ and $w_z^\perp$ that bound w_z from above and below, respectively. Bipolar preferences, i.e. user dislikes, can be modeled by negated attributes $1 - x_z$, $z \in [N]$, if needed.

Algorithm 1. The SWMM single-user algorithm. T is the maximum number of iterations, K the query set size, and α the hyperparameters.

1: **procedure** SWMM (M, K, T, α)
2:     $\mathcal{D} \leftarrow \emptyset$
3:     **for** $t = 1, \ldots, T$ **do**
4:         $\{w_i\}, \{x_i\} \leftarrow$ SOLVE_OP1$(\mathcal{D}, K, \alpha)$
5:         *the user selects x^+ from $\{x_i\}$*
6:         $\mathcal{D} \leftarrow \mathcal{D} \cup \{x^+ \succcurlyeq x^- : x^-$ was not selected$\}$
7:         $w, x \leftarrow$ SOLVE_OP1$(\mathcal{D}, 1, \alpha)$

The single-user SWMM *algorithm.* The true user preferences w_* are not directly observed, and must be estimated. The SWMM algorithm tracks an estimate of the preferences at all times, iteratively improving it through user interaction.

The pseudocode of SWMM is reported in Algorithm 1. At every iteration, the algorithm selects K query configurations $x_1, \ldots, x_K \in \mathcal{X}$ based on the previously collected user feedback $\mathcal{D}$ (line 4). The query set $\{x_i\}$ is presented to the user, who is invited to select a most preferred configuration x^+ from the K alternatives (line 5). The cases where the user selects a sub-par item are accounted for in the mathematical formulation, as discussed later on. The user choice is interpreted as a set of pairwise ranking constraints $\{x^+ \succcurlyeq x^- : x^-$ was not selected$\}$, and added to $\mathcal{D}^1$ (line 6). At this point, a recommendation x is computed by leveraging all user feedback and presented to the user (line 7). If the user is satisfied with the suggested product, the procedure ends. Otherwise it is repeated, up to a maximum number of rounds T.

The primary goal of any preference elicitation system is to recover a satisfactory recommendation with minimal user effort. The choice of queries is crucial for reaching this goal [4]: the number of queries that can be afforded is small, thus every query should be chosen to be as "informative" as possible. Bayesian approaches to preference elicitation [12,23] model uncertainty about user preferences as a probability distribution over utility weights w, and select queries that maximize expected informativeness, as measured by expected value of information (EVOI) [7] or its approximations. However, even the approximate strategies for EVOI maximization [23] are extremely time consuming and cannot scale to fully constructive scenarios as the ones we are dealing with here [22].

The SWMM query selection strategy addresses this problem by taking a space decomposition perspective inspired by max-margin ideas. The algorithm jointly learns a set of weight vectors, each representing a candidate utility function, and a set of candidate configurations, one for each weight vector, maximizing diversity between the vectors, consistency with the available feedback, and quality of each configuration according to its corresponding weight vector.

---

[1] In [22], the authors convert user choices to pairwise ranking constraints using a custom procedure. Here we opted for a straightforward winner-vs-others representation, as described in the main text. This modification did not appear to significantly alter the performance of the SWMM algorithm in our simulations (data not show).

More formally, user preferences are estimated by a *set* of K weight vectors $\boldsymbol{w}_1, \ldots, \boldsymbol{w}_K \in \mathbb{R}_+^N$. Each weight vector $\boldsymbol{w}_i$ is required to agree with all collected feedback $\mathcal{D}$. More precisely, the weights $\{\boldsymbol{w}_i\}$ are chosen as to provide the largest possible separation margin, i.e. for all $i \in [K]$ and $(\boldsymbol{x}^+ \succcurlyeq \boldsymbol{x}^-) \in \mathcal{D}$ the utility difference $\langle \boldsymbol{w}_i, \boldsymbol{x}^+ - \boldsymbol{x}^- \rangle$ should be as large as possible. Mistakes are absorbed by slack variables ε, as customary. Query configurations $\boldsymbol{x}_1, \ldots, \boldsymbol{x}_K$ are chosen according to two criteria: each $\boldsymbol{x}_i$ should have maximal utility with respect to the associated weight vector $\boldsymbol{w}_i$, and the K products should be as diverse as possible. Diversity is encouraged by requiring that each weight vector $\boldsymbol{w}_i$ separates its associated configuration $\boldsymbol{x}_i$ from all the others configurations in the query set with a high margin, i.e. for all $i, j \in [K]$ with $i \neq j$, $\langle \boldsymbol{w}_i, \boldsymbol{x}_i - \boldsymbol{x}_j \rangle$ should be larger than the margin.

The previous discussion leads directly to the *quadratic* version of the SWMM optimization problem over the variables $\mu, \{\boldsymbol{w}_i, \boldsymbol{x}_i, \varepsilon_i\}$:

$$\max \quad \mu - \alpha \sum_{i=1}^k \|\varepsilon_i\|_1 - \beta \sum_{i=1}^k \|\boldsymbol{w}_i\|_1 + \gamma \sum_{i=1}^k \langle \boldsymbol{w}_i, \boldsymbol{x}_i \rangle \tag{1}$$

$$\text{s.t.} \quad \langle \boldsymbol{w}_i, \boldsymbol{x}_s^+ - \boldsymbol{x}_s^- \rangle \geq \mu - \varepsilon_{is} \qquad \forall i \in [k], \boldsymbol{x}_s^+ \succcurlyeq \boldsymbol{x}_s^- \in \mathcal{D} \tag{2}$$

$$\langle \boldsymbol{w}_i, \boldsymbol{x}_i - \boldsymbol{x}_j \rangle \geq \mu \qquad \forall i, j \in [k], i \neq j \tag{3}$$

$$\mu \geq 0, \; \boldsymbol{w}^\perp \leq \boldsymbol{w}_i \leq \boldsymbol{w}^\top, \; \boldsymbol{x}_i \in \mathcal{X}, \; \varepsilon_i \geq 0 \qquad \forall i \in [k] \tag{4}$$

The non-negative variable $\mu \in \mathbb{R}_+$ is the separation margin. The objective has four parts: the first part drives the maximization of the margin μ; the second minimizes the total sum of the ranking errors $\{\varepsilon_i\}$; the third one introduces an ℓ_1 regularizer encouraging sparsity of the learned weights; finally, the last part requires the configurations $\{\boldsymbol{x}_i\}$ to have high utility with respect to the associated $\{\boldsymbol{w}_i\}$. The hyperparameters $\alpha, \beta, \gamma \geq 0$ modulate the contributions of the various parts. We refer to this optimization problem as **OP1**.

Constraint 2 encourages consistency of the learned weights with respect to the collected user feedback; ranking mistakes are absorbed by the slack variables $\{\varepsilon_i\}$. Constraint 3 enforces the generated configurations to be as diverse as possible with respect to the corresponding weight vectors. Finally, Constraint 4 ensures that all variables lie in the corresponding feasible sets.

Unfortunately the above optimization problem is *quadratic* (due to Constraint 3) and difficult to optimize directly. Here we use the tight *mixed-integer linear* formulation proposed in [22], which can be solved using off-the-shelf MILP solvers. In that paper, the MILP formulation was shown to perform well empirically, reaching or outperforming two state-of-the-art Bayesian approaches.

4   Multi-user Setwise Max-margin

Now we generalize the SWMM algorithm to simultaneously elicit the preferences of M users. Our goal is to exploit preferences shared by similar users for computing queries and recommendations. As a consequence, the cognitive effort of query answering can be distributed (fairly) among users, while maintaining the same or better recommendation quality.

Algorithm 2. The MU-SWMM algorithm. M is the number of users; T, K and α are as in Algorithm 1.

1: **procedure** MU-SWMM (M, K, T, α)
2:   $v(u) \leftarrow 1$, $k(u, y) \leftarrow \mathbb{I}(u = y)$
3:   **for** $u = 1, \ldots, K$ **do**
4:     $\{w_i^u\} \leftarrow$ SOLVE_OP2$(u, 0, k, \emptyset, K, \alpha)$
5:     $\mathcal{D}^u \leftarrow \emptyset$
6:   **for** $t = 1, \ldots, T$ **do**
7:     $select\ u \in \mathrm{argmin}_u |\mathcal{D}^u|\ uniformly\ at\ random$
8:     $\{w_i^u\}, \{x_i\} \leftarrow$ SOLVE_OP2$(u, v, k, \mathcal{D}^u, K, \alpha)$
9:     $user\ u\ selects\ x^+\ from\ \{x_i\}$
10:    $\mathcal{D}^u \leftarrow \mathcal{D}^u \cup \{x^+ \succcurlyeq x^- : x^-\ was\ not\ selected\}$
11:    $update\ v(\cdot)\ and\ k(\cdot, \cdot)\ based\ on\ \{w_i^1\}, \ldots, \{w_i^M\}\ using\ Eq.\ 5\ and\ 6$
12:    $w^u, x^u \leftarrow$ SOLVE_OP2$(u, v, k, \mathcal{D}^u, 1, \alpha)$

Our strategy, dubbed MU-SWMM, is outlined in Algorithm 2. At every iteration, MU-SWMM picks a user u to be queried, based on some criterion (discussed later on). Then it proceeds like SWMM, by selecting a query set $\{x_i^u\}$ for user u, adding the feedback to $\mathcal{D}^u$, and suggesting a recommendation x^u. Once a user has received a satisfactory recommendation, it is removed from the pool of selectable users and skipped at later iterations[2]. The algorithm iterates until all users are satisfied, or T iterations are reached.

We remark that only *one* user is queried at every iteration. The MU-SWMM algorithm relies on **OP2**, a modification of **OP1** where the utility of configurations is determined both by the estimated preferences of the selected user u, as well as those of similar (non-selected) users $y \neq u$. This "aggregate utility" takes into consideration both the degree of similarity of u to the other users y and how good their preference estimates are. This avoids interferences due to similar users with unreliable preference estimates.

In order to implement this strategy, we must solve several problems: how to quantify the quality of the estimated preferences of a user; how to measure the similarity between two users; how to appropriately inject the information about other users into SWMM optimization problem **OP1**; and finally how to select the user to be queried. We discuss these points separately.

Measuring how much we know about a user. Ideally, one could quantify how much is known about a user u by using the regret

$$\left(\max_x \langle w_*^u, x \rangle \right) - \langle w_*^u, x^u \rangle$$

i.e. the difference in true utility between a true most preferred recommendation x_*^u and the current recommendation x^u. Unfortunately, computing the regret requires to observe w_*^u, so we must rely on surrogate measures.

---

[2] This detail is omitted from Algorithm 2 for simplicity.

A simple surrogate is given by the number of times a user was queried, i.e. the number of collected user responses $|\mathcal{D}^u|$. This quantity, however, may give a simplistic estimate: in general, two users who replied to the same number of queries may have widely different regrets, depending on how difficult their preferences are to learn. This is especially true when $\mathcal{D}^u$ is redundant, i.e. contains repeated or similar constraints.

Taking care of redundancy requires us to look at the geometry of the problem. In particular, we use the *spread* of the weight vectors $\{w_i^u\}$ computed during the query selection. As informative feedback is added to $\mathcal{D}^u$, the space of optimal weight vectors shrinks, and so does the distance between the vectors $w_1^u, \ldots, w_K^u$. More formally, we define the spread of user u as follows:

$$v(u) := c \sum_{i \neq j} \|w_i^u - w_j^u\|^2 \tag{5}$$

where the constant c is chosen so that $v(u) \in [0, 1]$. In other words, the spread is simply the *empirical variance* of the estimated weights $\{w_i^u\}$. Note that the spread is much less affected by redundant constraints than $|\mathcal{D}^u|$ [3].

Evaluating user similarity. The most general mechanism for defining similarities between objects are *kernels* [20]. In particular, two users are similar when their estimated preferences are. While a variety of kernels could be used, we propose using a Gaussian kernel, similar to [19,26]:

$$k(u, y) := \exp\left(-\tau \sum_{i,j} \|w_i^u - w_j^y\|^2\right) \tag{6}$$

Here $\tau \in \mathbb{R}_+$ is an "inverse temperature" parameter controlling the shape of the kernel. Note that, similarly to [19], the kernel is not fixed: rather, it is adapted dynamically as soon as new user weight estimates $\{w_i^u\}$ are computed.

Transferring preferences across users. Given a user u, we want to alter its estimated utility $\langle w_i^u, x \rangle$, $i = 1, \ldots, K$, based on the preferences of other similar, well-known users. We propose the following aggregate utility:

$$(1 - v(u))\langle w_i^u, x \rangle + v(u) \sum_{y \neq u} (1 - v(y))k(u, y)\langle w^y, x \rangle \tag{7}$$

This is the convex combination of the utility of user u (first term) and a weighted combination of the utilities of the other users. Intuitively, the more u is known (e.g. at the end of the elicitation procedure) the closer $v(u)$ is to zero, and the first term dominates, and vice versa. The contributions of the users $y \neq u$ are decided, again, by how much is known about them $(1 - v(y))$ and by their

---

[3] A constraint repeated l times instantiates l slack variables in **OP1**, thus becoming "harder" by a factor of αl. The effect however is much softer than for $|\mathcal{D}^u|$.

similarity to user u, measured by the kernel $k(u, y)$. This formulation implies that there is little influence from users whose preferences are not well known. It is straightforward to introduce Eq. 7 into **OP1**. By rearranging the terms, the aggregate utility can be written as:

$$\langle a\boldsymbol{w}_i^u + \boldsymbol{b}, \boldsymbol{x} \rangle \text{ with } \begin{cases} a = (1 - v(u)) \\ \boldsymbol{b} = v(u) \sum_{y \neq u} (1 - v(y)) k(u, y) \boldsymbol{w}^y \end{cases}$$

Note that this is a linear transformation of the original, single-user utility. As shown in [22], it is easy to incorporate linear transformations of this kind into the SWMM optimization problem. In our case, we need to rewrite Constraint 4 as:

$$a\boldsymbol{w}^\perp + \boldsymbol{b} \leq \boldsymbol{w}_i \leq a\boldsymbol{w}^\top + \boldsymbol{b} \quad \forall i \in [K]$$

We use **OP2** to refer to the modified optimization problem; in Algorithm 2 we use SOLVE_OP2$(u, v, k, \mathcal{D}^u, K, \boldsymbol{\alpha})$ to denote a solution of the problem with respect to user u, spread v, kernel k, input preferences $\mathcal{D}^u$, set cardinality K, and using $\boldsymbol{\alpha} = (\alpha, \beta, \gamma)$ as coefficients in the objective function.

The spread $v(\cdot)$, the user similarity $k(\cdot, \cdot)$, and the transformation parameters $(a, \boldsymbol{b})$ are adapted whenever new feedback is received. The update is computationally inexpensive: whenever user u provides a response, only $k(u, \cdot)$ (i.e. a single row of the Gram matrix) needs to be recomputed.

Fairness-based user selection. The missing piece is how to choose a user at every iteration. Many strategies may be adopted, depending on the objective. In multi-user preference elicitation we are most concerned about fair distribution of queries among users, so to minimize the individual cognitive effort. Therefore we propose to select the user u that received the least queries so far:

$$u \in \underset{u}{\text{argmin}} |\mathcal{D}^u|$$

with ties broken at random. Although other strategies can be conceived, this simple strategy was observed to work well in our simulations, as shown in Sect. 5. Additionally, we found empirically that it is surprisingly difficult to beat: all of the more sophisticated strategies we tested failed to improve on it (data not shown due to space constraints).

5   Empirical Analysis

We studied the behavior of MU-SWMM on a synthetic and a realistic preference elicitation tasks[4], both taken from [22]. Our goal is to provide empirical answers to these research questions: (**Q1**) Does aggregating the utility of similar users, as per Eq. 7, reduce the cognitive effort required to produce good recommendations?

---

[4] Our experimental setup is available at: https://github.com/stefanoteso/musm-adt17.

(**Q2**) Is the number of queries $|\mathcal{D}^u|$ a reasonable user selection criterion? (**Q3**) How does the algorithm behave when there are no common preferences to be shared between users?

Our experimental setup follow closely the ones of [11,22]. We randomly generated 20 groups of $M = 20$ users each using a hierarchical sampling procedure. Each group was split into C clusters, with $\approx M/C$ users each. The users within a cluster are chosen to have similar preferences, simulating different sub-groups of users. For instance, in a PC recommendation scenario, there may be a cluster of users who prefer energy efficient laptops and a cluster of power users who need more capable machines. This cluster structure enables preference information to be transferred. Note that the clusters are *not known to* MU-SWMM *beforehand*: the algorithm estimates them dynamically from the collected user replies.

For each cluster, we sampled the centroid from a uniform distribution in $[1, 100]$. The true preferences of the users in the cluster were obtained by perturbing the centroid randomly according to a normal distribution of mean 0 and standard deviation $25/6$. As done in [22], we considered users with both sparse and dense weight vectors: for sparse users, 80% of the entries of the true weight vector $\boldsymbol{w}_*^u$ were set to zero.

The user responses were simulated with a Plackett-Luce model [16,18], where the probability of a particular answer is dictated by a Boltzmann distribution:

$$P(\text{user chooses } \boldsymbol{x}_i \text{ from } \{\boldsymbol{x}_1, \ldots, \boldsymbol{x}_K\}) = \frac{\exp(\lambda\langle\boldsymbol{w}_*, \boldsymbol{x}_i\rangle)}{\sum_{j=1}^{K} \exp(\lambda\langle\boldsymbol{w}_*, \boldsymbol{x}_j\rangle)}$$

with λ fixed to 1 as in [11,22]. For $K = 2$, this model reduces to the classical Bradley-Terry model for pairwise ranking feedback [6].

Synthetic setting. The first experiment is performed on the synthetic problem introduced in [22]. In this setting the space of products $\mathcal{X}$ is taken to be the Cartesian product of r categorical attributes, each having r possible values. We use a one-hot encoding to represent products, for a total of r^2 0–1 variables. Here we focus on the $r = 4$ case with $r^2 = 16$ variables and $r^r = 256$ total products. While simple, this problem proved to be non-trivial [22].

We compared the MU-SWMM algorithm against a straightforward multi-user adaptation of SWMM where all users are elicited independently. We also included in the comparison an unrealistic variant of MU-SWMM where the user to be queried is selected according to the maximal *true* regret (i.e. assuming that an oracle gives us this information). We assessed the ability of MU-SWMM to propagate preference information between users by varying the number of clusters C in $\{1, 2, 5\}$. We also varied the query set size $K \in \{2, 3\}$. The kernel inverse temperature parameter τ was fixed to 2 in all experiments.

The results for sparse users can be found in Fig. 1. All algorithms were run for $T = 100$ elicitation rounds, i.e. 5 queries per user on average (x-axis). We report the median performance of the three recommenders across the 20 groups. The performance on a group is the average regret over all M users (y-axis). As above, the regret is simply the difference in true utility between a best product $\boldsymbol{x}_*$ and the actual recommendation $\boldsymbol{x}$, i.e. $\langle\boldsymbol{w}_*^u, \boldsymbol{x}_* - \boldsymbol{x}\rangle$.

The plots show clearly that when users are similar enough, transferring preferences across them (MU-SWMM, red line) is better than no transfer at all (SWMM, gray line). This result is not obvious, since the kernel $k(\cdot, \cdot)$ is estimated *dynamically* from the collected feedback to reflect the hidden cluster structure of the users. The regret-based user selection strategy (blue line) provides an upper bound on the performance of MU-SWMM.

Understandably, the amount of improvement depends on C. The simplistic $C = 1$ case showcases the potential of preference transfer: both multi-user methods converge much faster than SWMM. In the more realistic $C = 2$ case, MU-SWMM takes about *half* the number of queries than SWMM to reach zero median regret. In particular, the number of per-user queries drops from more than 5 to about 3.3 for $K = 2$, and from 4 to less than 2 for $K = 3$. For $C = 5$, i.e. 4 users per cluster, both MU-SWMM and the regret-based strategy are closer to the baseline. The results for dense users in Fig. 2 follow the same trend, despite elicitation being more difficult in this case.

In general, MU-SWMM fares better than or similarly to the no-transfer baseline. Notably, enlarging the query set size K from 2 to 3 further improves the performance of MU-SWMM: more ranking constraints are collected at each iteration, thus improving the estimate of the kernel $k(\cdot, \cdot)$ and preference transfer among similar users. These results validate Eq. 7 and allow us to answer affirmatively to question **Q1**.

They also partially answer **Q2**. Clearly selecting the user with the minimal number of queries does improve on the baseline, as shown in Figs. 1 and 2. In addition, we checked whether it distributes queries fairly among users. Indeed, in the $C = 2$, $K = 2$ case the std. dev. in dataset size $|\mathcal{D}^u|$ between users is rather small (1.47) for the MU-SWMM user selection strategy, and 2.58 for regret-based user selection. The other cases behave similarly (data not shown).

PC recommendation. In the second experiment we consider a realistic PC configuration task, as in [22]. The recommender is required to suggest a fully customized PC. A PC configuration is defined by 7 categorical attributes— type (laptop, desktop or tower), manufacturer, CPU model, monitor size, RAM amount, storage amount—and a linearly dependent numerical attribute, the price. The attributes are mutually constrained via Horn clauses, expressing statements like "manufacturer X does not sell CPUs of brand Y", for a total of 16 Horn constraints. The product space has about 700,000 distinct configurations.

In this experiment we increased the number of average queries per user to 10, due to the very large number of products. We also restricted ourselves to sparse users, as in [22], which are more realistic in this setting: a typical customer will be indifferent about many aspects of a PC configuration.

The performance of the three methods, with $K = 2$, can be found in Fig. 3. Again, MU-SWMM behaves better than the baseline for all values of $C \in \{1, 2, 5\}$, while the degree of improvement depends on C. For instance, for $C = 2$ the regret achieved after 10 queries per user is much closer to the performance upper bound (blue line) than to the baseline (gray). These results highlight that, despite

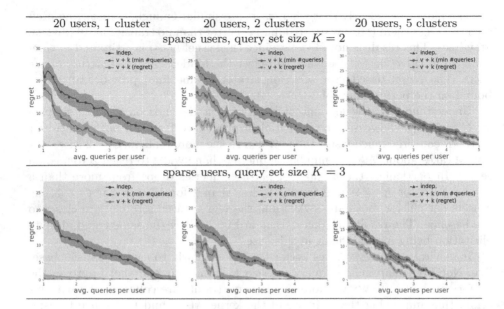

Fig. 1. Results for the synthetic setting with $1, 2, 5$ clusters of *sparse* users, for $K = 2, 3$. The three lines represent the average regret over users of the SWMM baseline (grey line), MU-SWMM with the fair user selection strategy (red), and MU-SWMM with the irrealistic regret-based user selection strategy (blue). The shaded area represents the standard deviation. Best viewed in color. (Color figure online)

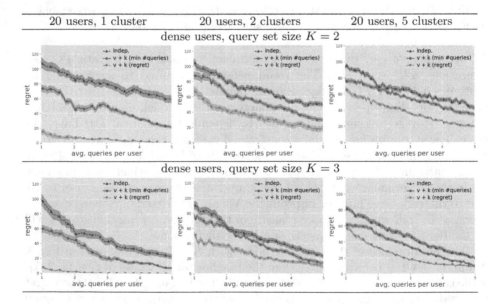

Fig. 2. Results for the synthetic setting with $1, 2, 5$ clusters of *dense* users, for $K = 2, 3$.

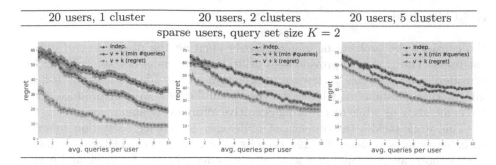

Fig. 3. Results for PC recommendation with $1, 2, 5$ clusters of *sparse* users, for $K = 2$.

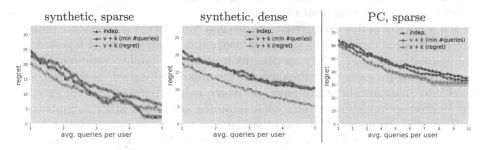

Fig. 4. Results for the $C = M = 20$ case for the synthetic setting with sparse users (left column), synthetic setting with dense users (middle column) and for PC recommendation with sparse users (right column). Here $K = 2$.

the complexity of this recommendation task, user similarity is still estimated sufficiently well for preference transfer to work well.

Worst-case behavior. To answer **Q3**, we studied MU-SWMM when there is only one user in each cluster. This artificial setting is intended to verify the robustness of MU-SWMM against missing user similarities. We ran both the synthetic recommendation (for both sparse and dense users) and the PC recommendation settings (sparse users only) with $C = M = 20$ and $K = 2$; results in Fig. 4.

In the two most difficult settings (i.e. synthetic with dense users and PC), MU-SWMM does not perform worse than the baseline. Unfortunately, in the remaining case (synthetic with sparse users) MU-SWMM performs worse than the baseline, and so does the regret-based user selection strategy. This is probably due to the kernel estimation taking too long to converge to a suitable value, therefore propagating preferences between unrelated users. In order to avoid this kind of interference, it may make sense to reduce the effect of the kernel, especially during the first elicitation rounds, for instance by making the inverse temperature τ increase during the process. We will explore this possibility in a future work.

6   Conclusion

We consider the problem of simultaneously eliciting the preferences of different users. The goal is to provide high utility recommendations to each user by considering all available preference information. A key idea is that, by asking informative questions and leveraging using similarities, we can produce good recommendations while distributing cognitive effort fairly among users.

This work tackles the problem of multi-user preference elicitation by generalizing setwise max-margin [22], thereby inheriting its core features. Namely, our method can gracefully deal with inconsistencies in user feedback and it can work in *constructive* recommendation scenarios, where the product to be recommended is synthesized by searching a large constrained configuration space rather than selected from a set of enumerated options.

A key novelty of this work is that our method can effectively propagate preference information from user to user. Our method estimates a kernel (similarity) function between users utility models, as well as a reliability estimate of the learned model for each user. Preferences of similar, well-known users are combined into an aggregate utility function, which is incorporated into the setwise max-margin optimization problem while retaining an efficient MILP formulation.

We tested our approach on a synthetic and a realistic recommendation task. The experimental results show that our method is able to dynamically recover the similarity between users from their responses, and to exploit it to propagate preference information between more and less known users. When applied to sufficiently similar users, our method often performs much better than a straightforward adaptation of single-user setwise max-margin. On the other hand, it fares well in cases where users are not similar at all. Finally, our simple user selection strategy minimizes the cognitive effort of individual users by distributing queries fairly among them, and was shown to work well in practice.

Acknowledgements. This work is partially supported by the ANR project CoCoRICo-CoDec ANR-14-CE24-0007-01. ST was partially supported by the CARITRO Foundation through grant 2014.0372.

References

1. Ah-Pine, J., Mayag, B., Rolland, A.: Identification of a 2-additive bi-capacity by using mathematical programming. In: Perny, P., Pirlot, M., Tsoukiàs, A. (eds.) ADT 2013. LNCS, vol. 8176, pp. 15–29. Springer, Heidelberg (2013). doi:10.1007/978-3-642-41575-3_2
2. Argyriou, A., Evgeniou, T., Pontil, M.: Multi-task feature learning. Adv. Neural Inf. Process. Syst. **19**, 41 (2007)
3. Benabbou, N., Di Sabatino Di Diodoro, S., Perny, P., Viappiani, P.: Incremental preference elicitation in multi-attribute domains for choice and ranking with the borda count. In: Schockaert, S., Senellart, P. (eds.) SUM 2016. LNCS, vol. 9858, pp. 81–95. Springer, Cham (2016). doi:10.1007/978-3-319-45856-4_6
4. Boutilier, C.: A POMDP formulation of preference elicitation problems. In: Proceedings of AAAI/IAAI, pp. 239–246 (2002)

5. Boutilier, C., Patrascu, R., Poupart, P., Schuurmans, D.: Constraint-based optimization and utility elicitation using the minimax decision criterion. Artif. Intell. **170**(8–9), 686–713 (2006)
6. Bradley, R.A., Terry, M.E.: Rank analysis of incomplete block designs: I. the method of paired comparisons. Biometrika **39**(3/4), 324–345 (1952)
7. Chajewska, U., Koller, D., Parr, R.: Making rational decisions using adaptive utility elicitation. In: Proceedings of AAAI, pp. 363–369 (2000)
8. Dery, L.N., Kalech, M., Rokach, L., Shapira, B.: Reaching a joint decision with minimal elicitation of voter preferences. Inf. Sci. **278**, 466–487 (2014)
9. Elahi, M., Ricci, F., Rubens, N.: Active learning strategies for rating elicitation in collaborative filtering: a system-wide perspective. ACM TIST **5**(1), 13:1–13:33 (2013)
10. Gajos, K., Weld, D.S.: Preference elicitation for interface optimization. In: Proceedings of UIST, pp. 173–182. ACM (2005)
11. Guo, S., Sanner, S.: Real-time multiattribute bayesian preference elicitation with pairwise comparison queries. In: Proceedings of AISTAT, pp. 289–296 (2010)
12. Guo, S., Sanner, S., Bonilla, E.V.: Gaussian process preference elicitation. In: Advances in Neural Information Processing Systems (NIPS), pp. 262–270 (2010)
13. Hines, G., Larson, K.: Efficiently eliciting preferences from a group of users. In: Brafman, R.I., Roberts, F.S., Tsoukiàs, A. (eds.) ADT 2011. LNCS, vol. 6992, pp. 96–107. Springer, Heidelberg (2011). doi:10.1007/978-3-642-24873-3_8
14. Keeney, R.L., Raiffa, H.: Decisions with Multiple Objectives: Preferences and Value Tradeoffs. Wiley, New York (1976)
15. Lu, T., Boutilier, C.: Robust approximation and incremental elicitation in voting protocols. In: Proceedings of IJCAI, pp. 287–293 (2011)
16. Luce, R.D.: Individual Choice Behavior: A Theoretical Analysis. Wiley, New York (1959)
17. Pigozzi, G., Tsoukiàs, A., Viappiani, P.: Preferences in artificial intelligence. Ann. Math. Artif. Intell. **77**(3–4), 361–401 (2016)
18. Plackett, R.L.: The analysis of permutations. Appl. Stat. **24**, 193–202 (1975)
19. Saha, A., Rai, P., Hal, D., Venkatasubramanian, S.: Online learning of multiple tasks and their relationships. In: Proceedings of AISTATS, pp. 643–651 (2011)
20. Scholkopf, B., Smola, A.J.: Learning with Kernels: Support Vector Machines, Regularization, Optimization, and Beyond. MIT press, Cambridge (2001)
21. Shi, Y., Larson, M., Hanjalic, A.: Collaborative filtering beyond the user-item matrix: a survey of the state of the art and future challenges. ACM Comput. Surv. **47**(1), 3:1–3:45 (2014)
22. Teso, S., Passerini, A., Viappiani, P.: Constructive preference elicitation by setwise max-margin learning. In: Proceedings of IJCAI, pp. 2067–2073 (2016)
23. Viappiani, P., Boutilier, C.: Optimal Bayesian recommendation sets and myopically optimal choice query sets. In: Proceedings of NIPS, pp. 2352–2360 (2010)
24. Viappiani, P., Boutilier, C.: Regret-based optimal recommendation sets in conversational recommender systems. In: Proceedings of RecSys, pp. 101–108. ACM (2009)
25. Xia, L., Conitzer, V.: Determining possible and necessary winners given partial orders. J. Artif. Intell. Res. (JAIR) **41**, 25–67 (2011)
26. Zhu, X., Lafferty, J., Ghahramani, Z.: Combining active learning and semi-supervised learning using gaussian fields and harmonic functions. In: ICML 2003 Workshop on the Continuum From Labeled to Unlabeled Data in Machine Learning and Data Mining (2003)

Interactive Thompson Sampling
for Multi-objective Multi-armed Bandits

Diederik M. Roijers$^{(\boxtimes)}$, Luisa M. Zintgraf, and Ann Nowé

Artificial Intelligence Laboratory, Vrije Universiteit Brussel, Brussels, Belgium
{droijers,luisa.zintgraf,ann.nowe}@ai.vub.ac.be

Abstract. In multi-objective reinforcement learning (MORL), much attention is paid to generating optimal *solution sets* for unknown utility functions of users, based on the stochastic reward vectors only. In *online* MORL on the other hand, the agent will often be able to elicit preferences from the user, enabling it to learn about the utility function of its user directly. In this paper, we study online MORL with user interaction employing the multi-objective multi-armed bandit (MOMAB) setting — perhaps the most fundamental MORL setting. We use Bayesian learning algorithms to learn about the environment and the user simultaneously. Specifically, we propose two algorithms: *Utility-MAP UCB (umap-UCB)* and *Interactive Thompson Sampling (ITS)*, and show empirically that the performance of these algorithms in terms of regret closely approximates the regret of UCB and regular Thompson sampling provided with the ground truth utility function of the user from the start, and that ITS outperforms umap-UCB.

1 Introduction

Many real-world decision problems require learning about the outcomes of different alternatives, either by interacting with the real world, or simulations thereof. When the outcomes can be measured in terms of a single scalar objective, such problems can be modelled as a *multi-armed bandit (MAB)* [2]. However, many real-world decision problems are further complicated by the presence of multiple (possibly conflicting) objectives [13]. For example, an agent learning the best strategy to deploy ambulances from a set of alternatives, may want to minimise average response time, while also minimising fuel cost and the stress for the drivers. For such problems, MABs can be extended to *multi-objective multi-armed bandits (MOMABs)* [3,9].

Research on MOMABs has hitherto focussed on settings in which no preference information w.r.t. the available trade-offs between the values in different objectives of the available alternatives is provided by the user during learning [3,18,22]. However, in many situations, this is a limiting assumption. E.g., in the example of ambulance deployment strategies, the ambulances will have to be deployed (i.e., a strategy will have to be executed) while still learning about the expected value of the different strategies in the different objectives. By providing preferences of the responsible human decision makers, i.e., the

© Springer International Publishing AG 2017
J. Rothe (Ed.): ADT 2017, LNAI 10576, pp. 18–34, 2017.
DOI: 10.1007/978-3-319-67504-6_2

management of the ambulance service, w.r.t. the attainable trade-offs between objectives, the agent can focus its learning on those strategies that the user finds most appealing. Other examples of multi-objective learning problems in which interaction with the user can be beneficial are: the exploration of preventive strategies for epidemics using computationally expensive simulations under objectives like minimising morbidity, the number of people infected, and the costs of the preventive strategy; and future household robots that will need to learn the preferences of their users with respect to the outcomes (performance, speed, energy usage) of alternative ways to perform a household task.

In this paper, we focus on such online learning problems, in which the agent can query the user via pairwise comparisons (following, e.g., [6,15,23]). Specifically, we focus on the prevalent case that a user's utility function is linear, in the context of MOMABs.

We propose two new Bayesian learning algorithms for MOMABs that learn about the environment and the user's utility function simultaneously. Specifically, we build upon two popular classes of algorithms for single-objective MABs, UCB1 [2] and Thompson sampling [16], to propose *utility-MAP UCB (umap-UCB)* and *Interactive Thompson Sampling (ITS)*. Both algorithms (umap-UCB and ITS) pose pairwise comparison queries to the user. As a UCB-algorithm, umap-UCB uses explicit exploration bonuses. To decide when to query the user, it computes the MAP of the utility function to determine the arm corresponding to the best mean estimate, and the arm corresponding to the best mean estimate plus exploration bonus. When these two best arms are different, umap-UCB queries the user by asking her to make a pairwise comparison between the estimated mean reward vectors of the two arms. ITS also elicits preferences via pairwise comparisons; it draws two sets of samples from the posteriors of the mean reward vectors of the arms, and the posterior of the utility function. When these two sets of samples have different best arms, ITS queries the user.

We test umap-UCB and ITS empirically, and find that the performance of these algorithms in terms of regret closely approximates the regret of UCB and regular Thompson sampling equipped from the beginning with the ground truth utility function of the user. When comparing umap-UCB and ITS, we find that ITS outperforms umap-UCB both in terms of minimising user regret, and in terms of minimising the number of comparisons that the user is asked to make.

2   Background

Before introducing our setting, we first provide the necessary background on scalar multi-armed bandits and multi-objective decision making in general.

2.1   Multi-armed Bandits

Definition 1. *A scalar* multi-armed bandit (MAB) [1,2,16] *is a tuple* $\langle \mathcal{A}, \mathcal{P} \rangle$ *where*

Algorithm 1. UCB	**Algorithm 2.** Thompson Sampling
$\bar{x}_a \leftarrow$ initialise with single pull, r_a for each a	**Input**: A prior for the reward distributions
$n_a \leftarrow 1$ for each a	$D \leftarrow \emptyset$; // observed data
for $t = |\mathcal{A}|, ..., T$ **do**	**for** $t = 1, ..., T$ **do**
$\quad a(t) \leftarrow \arg\max_a (\bar{x}_a + c(\bar{x}_a, n_a, t))$	$\quad \theta^t \leftarrow$ draw sample from $P(\theta^t|D)$
$\quad r(t) \leftarrow$ play $a(t)$ and observe reward	$\quad a(t) \leftarrow \arg\max_a \mathbb{E}_{P(r|a,\theta^t)}[r]$
$\quad \bar{x}_{a(t)} \leftarrow \frac{n_{a(t)}\bar{x}_{a(t)}+r(t)}{n_{a(t)}+1}$	$\quad r(t) \leftarrow$ play $a(t)$ and observe reward
$\quad n_{a(t)}{+}{+}$	$\quad$ append $(r(t), a(t))$ to D

- $\mathcal{A}$ *is a set of* actions *or* arms, *and*
- $\mathcal{P}$ *is a set of probability distributions,* $P_a(r) : \mathbb{R} \to [0, 1]$ *over scalar rewards,* r, *associated with each arm* $a \in \mathcal{A}$.

We refer to the the mean reward of an arm as $\mu_a = \mathbb{E}_{P_a}[r] = \int_{-\infty}^{\infty} rP_a(r)dr$, *to the optimal reward as the mean reward of the best arm* $\mu^* = \max_a \mu_a$, *and to the expected regret of pulling an arm,* a, *once as* $\Delta_a = \mu^* - \mu_a$.

The goal of an agent interacting with a MAB is to maximise the expected cumulative reward, $E[\sum_{t=1}^{T} \mu_{a(t)}]$, where T is the time horizon, and $a(t)$ is the arm pulled at time t. However, at the start, the agent knows nothing about $\mathcal{P}$, and can only obtain information about the reward distributions by pulling an arm $a(t)$ each timestep, obtaining a sample from the corresponding $P_{a(t)}$. In the MAB literature, this reward maximisation is typically defined via the minimisation of the equivalent measure of expected total regret, i.e., the amount of reward lost due to not playing the optimal arm in each step.

Definition 2. *The expected total* regret *of pulling a sequence of arms for each timestep between* $t = 1$ *and a time horizon* T *(following the definition of* [1]*), is*

$$\mathbb{E}\left[\sum_{t=1}^{T} \mu^* - \mu_{a(t)}\right] = \sum_a \Delta_a \, \mathbb{E}[n_a(T)],$$

where $n_a(T)$ *is the number of times arm* a *is pulled until timestep* T.

In the literature, a popular choice [1,2] for $\mathcal{P}$ is Bernoulli distributions, i.e., distributions with only two possible outcomes: 1 (dubbed 'success'), according to a probability $p_a \in [0, 1]$ or 0 (dubbed 'failure') with a probability $1 - p_a$. The expected reward for a Bernoulli distribution is $\mu_a = p_a$.

The two most popular classes of algorithms for Bernoulli-distributed MABs are UCB and Thompson Sampling. UCB [2,7], provided in a general form in Algorithm 1, keeps estimates of the means of arms $\bar{x}_a$, and uses upper confidence bounds $c(\bar{x}_a, n_a, t)$ for exploration. I.e., at each round the arm is pulled with the highest value for the mean plus exploration bonus, $\bar{x}_a + c(\bar{x}_a, n_a, t)$. This exploration bonus is defined so that it goes down with the number of pulls of an arm, n_a, and up slowly with the total number of pulls of all arms, t. There are many variants of UCB that differ in its definition of $c(\bar{x}_a, n_a, t)$, ranging from the original UCB1 bound [2],

$$c_1(\bar{x}_a, n_a, t) = \sqrt{\frac{2 \ln t}{n_a}}, \tag{1}$$

to the upper confidence bound derived from Chernoff's bound [7]:

$$c_{ch}(\bar{x}_a, n_a, t) = \sqrt{\frac{2 \bar{x}_a \ln \sqrt{t}}{n_a} + \frac{2 \ln \sqrt{t}}{n_a}}. \tag{2}$$

Thompson sampling (Algorithm 2) on the other hand, maintains posterior distributions for the parameters of the reward distributions for each arm $a \in \mathcal{A}$, and pulls arms based on samples from this posterior. Bernoulli distributions have only one parameter, μ_a. The typical prior for these μ_a are beta distributions, with a single count for the number of successes and failures: $\beta(1,1)$ for each arm. The posterior can be calculated by simply counting the number of successes, i.e., the number of times the reward was 1, $s_a(t)$, and the number of failures $f_a(t) = n_a(t) - s_a(t)$, leading to a posterior over the means of each arm, $\beta(s_a(t) + 1, f_a(t) + 1)$. We denote a sample from the joint posterior of all arms as:

$$\theta^t = \langle \theta_1^t ... \theta_{|\mathcal{A}|}^t \rangle \sim P(\theta^t | D) = \prod_{a \in \mathcal{A}} \beta(s_a(t) + 1, f_a(t) + 1).$$

At each iteration, Thompson sampling draws such a sample and pulls the arm corresponding to $\arg\max_a \mathbb{E}_{P(r|a, \theta^t)}[r]$. Because these θ_a^t are samples from the posteriors of the *mean* for each arm, a, the arm corresponding to $\arg\max_a \theta_a^t$ represents the maximum expected reward (for that set of samples).

2.2 Multi-objective Decision Making

In single-objective MABs, an agent must find the alternative a that maximises the expected reward. In multi-objective problems however, there are n objectives, that are all desirable. Hence, the stochastic rewards, $\mathbf{r}(t)$, and the expected rewards for each alternative $\boldsymbol{\mu}_a$ are vector-valued.

Definition 3. *A multi-objective multi-armed bandit (MAB) [3,18,22] is a tuple* $\langle \mathcal{A}, \mathcal{P} \rangle$ *where*

- $\mathcal{A}$ *is a finite set of actions or arms, and*
- $\mathcal{P}$ *is a set of probability distributions, $P_a(\mathbf{r}) : \mathbb{R}^d \to [0,1]$ over vector-valued rewards $\mathbf{r}$ of length d, associated with each arm $a \in \mathcal{A}$.*

As a result, rather than having a single optimal alternative, there can be multiple arms whose value vectors are optimal for different preferences that users may have with respect to the objectives. Such preferences can be expressed using a utility function $u(\boldsymbol{\mu}, \mathbf{w})$ that is parameterised by a parameter vector $\mathbf{w}$ and returns the *scalarized* value of $\boldsymbol{\mu}$. Following the single-objective literature, we make use of Bernoulli distributions (as a worst-case scenario of distributions with high variance). Specifically, we assume that the reward for an arm, a, is a vector of d independent Bernoulli distributions. This probability distribution can thus be compactly described with a vector of means $\boldsymbol{\mu}_a$; for each objective, samples can be drawn independently using the mean for that objective.

When the parameter vector $\mathbf{w}$ is known beforehand, it is possible to *a priori* scalarise the decision problem and apply standard single-objective algorithms like UCB or Thompson sampling. However, often we do not know $\mathbf{w}$ at the start of learning.

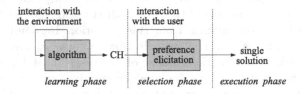

Fig. 1. The offline MORL decision support scenario.

Much *multi-objective reinforcement learning (MORL)* research assumes that $u(\boldsymbol{\mu}, \mathbf{w})$ is unknown throughout the learning phase, and there will only be access to the user in a separate *selection phase*. We refer to this scenario as the *offline MORL decision support scenario*, depicted in Fig. 1, in which an agent provides decision support to the user by presenting her with a set of alternatives at the beginning of the selection phase. In the learning phase, we thus need an algorithm that computes a set of policies containing at least one arm with the maximal scalarised value for *each possible* $\mathbf{w}$. Which alternatives, i.e., arms, should be included in this set depends on what we know about the utility function u. A highly prevalent case is that u is linear.

Definition 4. *A linear utility function is a weighted sum of the values in each objective, $\boldsymbol{\mu}$, of an alternative, i.e.,*

$$u(\boldsymbol{\mu}, \mathbf{w}) = \mathbf{w} \cdot \boldsymbol{\mu}, \tag{3}$$

where $\mathbf{w}$ is a vector of non-negative weights that sum to 1, and express the preferences of the user w.r.t. each objective. Please note that we assume that all objectives are desirable, and thus contribute positively to the utility.

In the offline MORL decision support scenario (Fig. 1) where the utility is an unknown linear utility function, a sufficient solution is the *convex hull (CH)*, the set of all undominated policies under a linear scalarisation:

$$CH(\mathcal{A}) = \{a \in \mathcal{A} \mid \exists \mathbf{w} \forall (a' \in \mathcal{A}) : \mathbf{w} \cdot \boldsymbol{\mu}_a \geq \mathbf{w} \cdot \boldsymbol{\mu}_{a'}\}.$$

When computation time is abundant, and there is enough time until the final decision needs to be made, it can be feasible to do MORL in an offline manner. This has the advantage that the interaction with the user can be done separately after learning, and therefore typically more efficiently (i.e., with fewer interactions with the user). However, as we have indicated in the introduction, there are also situations in which decisions have to be made on a timescale that is relatively short compared to the computation time needed to evaluate arms. Furthermore, it can also be highly important to use the available computation time as efficiently as possible. This is for example the case when evaluating and selecting different alternative preventive strategies against an emerging epidemic using computationally expensive simulations with computational epidemiological models [11], in the presence of multiple objectives like infection ratio, morbidity and economic damage. In other words, offline learning is not possible when there is no time to perform a separate learning phase before acting. In such cases, we need a different approach.

3   Online Interactive Learning with MOMABs

In this paper we focus on the case that we can have interaction with the user during learning, and that the policies executed during learning are important, i.e., accumulate *regret*. This leads to the *online interactive MORL decision support scenario*, which is schematically depicted in Fig. 2. In this scenario, learning and execution of the policy happens simultaneously in the learning phase. Furthermore, we have to interact with the environment as well as the user during learning, in order to maximise the rewards (or minimise the regret). Finally, it can happen that after some amount of time (and/or number of interactions), the learning will stop, and we will move to an *execution only phase*. This happens, e.g., when the computational capacity of simulations is needed for different learning problems, and/or the user becomes unavailable for further input.

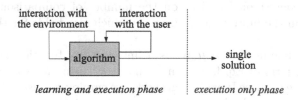

Fig. 2. The online interactive MORL decision support scenario.

In the online interactive MORL decision support scenario, we aim to minimise *user regret*, i.e., the amount of utility that is lost due to playing suboptimal arms. We define the value of the optimal arm as

$$\boldsymbol{\mu}^* = \arg\max_{a} \mathbf{w}^* \cdot \boldsymbol{\mu}_a,$$

where $\mathbf{w}^*$ are the ground truth weights of a linear utility function (Definition 4). Similar to single-objective MABs, we define the expected (in our case vector-valued) regret of pulling an arm, a, once as $\boldsymbol{\Delta}_a = \boldsymbol{\mu}^* - \boldsymbol{\mu}_a$.

Definition 5. *The expected total* user regret *of pulling a sequence of arms for each timestep between $t = 1$ and a time horizon T in a MOMAB is*

$$\mathbb{E}\left[\mathbf{w}^* \cdot (\sum_{t=1}^{T} \boldsymbol{\mu}^* - \boldsymbol{\mu}_{a(t)})\right] = \sum_{a}(\mathbf{w}^* \cdot \boldsymbol{\Delta}_a) \, \mathbb{E}[n_a(T)],$$

where $n_a(T)$ is the number of times arm a is pulled until timestep T.

4   Algorithms

To minimise user regret (Definition 5) in the online interactive MORL setting (Fig. 2), we must interact both with the environment and with the user. Similar to single-objective MABs, we need to learn about the reward, $\mathbf{r}$, but in addition, we must also learn about u and $\mathbf{w}$, as $u(\mathbb{E}[\sum_t \mathbf{r}(t)], \mathbf{w})$ is what we ultimately aim to optimise.

Following Zoghi et al. [23] — who study *relative bandits*; which is an adjacent but different model, in which the reward (vectors) cannot be observed — we assume that we can interact with the user once before (or after) pulling an arm, in the form of a pairwise comparison [6,15,23]. Contrary to [23] however, we present the user with (estimations of) expected reward vectors, rather than (data resulting from) single arm pulls. We thus ask users to compare two vectors, $\mathbf{x}$ and $\mathbf{y}$, and observe whether the user prefers $\mathbf{x}$ to $\mathbf{y}$, denoted $\mathbf{x} \succ \mathbf{y}$. At timestep t, we thus have access to a data set, C, of j of such preference pairs, where $j \leq t$ is the number of comparisons performed until t:

$$C = \{(\mathbf{x}_i \succ \mathbf{y}_i)\}_{i=1}^{j}. \tag{4}$$

There is no predetermined budget on the number of comparisons a user can make, other than the finite-time horizon, T, which also holds for the number of arm pulls.

Because we assume that $u(\boldsymbol{\mu}, \mathbf{w})$ is a linear utility function (Definition 4), and data in the form of Eq. 4, we can estimate $\mathbf{w}^*$ using *logistic regression* [5]. Specifically, as we propose Bayesian methods, we employ *Bayesian logistic regression* [5], enabling us to obtain both a *maximum a posteriori* estimate of the true weights $\mathbf{w}^*$, $\bar{\mathbf{w}}$, as well as a posterior distribution over the true weights.

Along with minimising user regret, we aim to not query the user excessively, as querying the user costs time and can be experienced as bothersome. In other words, we aim to propose algorithms in which both the expected *user regret* per timestep, as well as the expected *number of queries* posed to the user per timestep, goes down steeply as time progresses. In order to achieve this, we build on two state-of-the-art classes of algorithms: UCB (Algorithm 1), and Thompson Sampling (Algorithm 2) for single-objective bandits.

4.1   Utility MAP–UCB

Our first algorithm, that we call *utility-MAP UCB (umap-UCB)* (Algorithm 3), is built upon UCB. In UCB for single-objective MABs (Algorithm 1), actions are chosen based on the estimates of the means for each arm $\bar{x}_a$ plus an exploration bonus, which together form an upper confidence bound on the true means of the arms. When applying the same schema to MOMABs we face the following challenges: (1) that the user has a linear utility function (Eq. 3) with an *unknown* weight parameter $\mathbf{w}^*$; (2) that we must decide how to select which action to play, given the current MAP estimate of the weight vector, $\bar{\mathbf{w}}$; and (3) that we want to estimate $\mathbf{w}^*$, while the *number of queries* posed to the user per timestep goes down steeply as time progresses (without sacrificing too much *user regret*).

First, let us focus on how to estimate $\mathbf{w}^*$. Because we assume a linear utility function (Eq. 3) and pairwise comparisons as data (Eq. 4), we use Bayesian logistic regression to estimate the weights.[1] We thus define a prior on the weights.

Algorithm 3. Utility–MAP UCB

Input: A parameter prior on the distribution of $\mathbf{w}$.

$C \leftarrow \emptyset$;   // previous comparisons
$\bar{x}_a \leftarrow$ initialise with single pull, r_a, for each a
$n_a \leftarrow 1$ for each a
for $t = |\mathcal{A}|, ..., T$ **do**
$\quad \bar{\mathbf{w}} \leftarrow \mathcal{P}_{simplex}(MAP(\mathbf{w}|C))$
$\quad \bar{a}^* \leftarrow \arg\max_a \bar{\mathbf{w}} \cdot \bar{\mathbf{x}}_a$
$\quad a(t) \leftarrow \arg\max_a(\bar{\mathbf{w}} \cdot \bar{\mathbf{x}}_a + c(\bar{\mathbf{w}}, \bar{\mathbf{x}}_a, n_a, t))$
$\quad \mathbf{r}(t) \leftarrow$ play $a(t)$ and observe reward
$\quad \bar{\mathbf{x}}_{a(t)} \leftarrow \frac{n_{a(t)} \bar{\mathbf{x}}_{a(t)} + \mathbf{r}(t)}{n_{a(t)} + 1}$
$\quad n_{a(t)}++$
$\quad$ **if** $\bar{a}^* \neq a(t)$ **then**
$\quad\quad$ perform user comparison for $\bar{\mathbf{x}}_{\bar{a}^*}$ and $\bar{\mathbf{x}}_{a(t)}$ and add result ($(\bar{\mathbf{x}}_{\bar{a}^*} \succ \bar{\mathbf{x}}_{a(t)})$ or $(\bar{\mathbf{x}}_{a(t)} \succ \bar{\mathbf{x}}_{\bar{a}^*})$) to C

Algorithm 4. Interactive Thompson Sampling

Input: Parameter priors on reward distributions, and on $\mathbf{w}$ distribution.

$C \leftarrow \emptyset$;   // previous comparisons
$D \leftarrow \emptyset$;   // observed reward data
for $t = 1, ..., T$ **do**
$\quad \eta_1^t, \eta_2^t \leftarrow$ draw 2 samples from $P(\eta^t|C)$
$\quad \theta_1^t, \theta_2^t \leftarrow$ draw 2 samples from $P(\theta^t|D)$
$\quad a_1(t) \leftarrow \arg\max_a \mathbb{E}_{P(\mathbf{r}, \mathbf{w}|a, \theta_1^t, \eta_1^t)}[\mathbf{w} \cdot \mathbf{r}]$
$\quad a_2(t) \leftarrow \arg\max_a \mathbb{E}_{P(\mathbf{r}, \mathbf{w}|a, \theta_2^t, \eta_2^t)}[\mathbf{w} \cdot \mathbf{r}]$
$\quad \mathbf{r}(t) \leftarrow$ play $a_1(t)$ and observe reward
$\quad$ append $(\mathbf{r}(t), a_1(t))$ to D
$\quad$ **if** $a_1(t) \neq a_2(t)$ **then**
$\quad\quad \tilde{\mu}_{1,a_1(t)} \leftarrow \mathbb{E}_{P(\mathbf{r}|a_1(t), \theta_1^t)}[\mathbf{r}]$
$\quad\quad \tilde{\mu}_{2,a_2(t)} \leftarrow \mathbb{E}_{P(\mathbf{r}|a_2(t), \theta_2^t)}[\mathbf{r}]$
$\quad\quad$ perform user comparison for $\tilde{\mu}_{1,a_1(t)}$ and $\tilde{\mu}_{2,a_2(t)}$ and add result $((\tilde{\mu}_{1,a_1(t)} \succ \tilde{\mu}_{2,a_2(t)})$ or $(\tilde{\mu}_{2,a_2(t)} \succ \tilde{\mu}_{1,a_1(t)}))$ to C

---

[1] We note that logistic regression based on maximum likelihood can lead to problems in earlier iterations of umap-UCB when there is little data available. We observed this empirically. Specifically, in earlier iterations umap-UCB with ML logistic regression instead of Bayesian logistic regression makes an estimate, $\bar{\mathbf{w}}$, with a sheer-infinite weight on one objective, such that no comparison will be asked from the user again. This can be prevented with a reasonable choice of prior in Bayesian logistic regression.

Specifically, we use a multi-variate Gaussian prior $\mathcal{N}(\mathbf{w}|\boldsymbol{\mu}^0, \boldsymbol{\Sigma}^0)$. We use η^0 as a shorthand for $\langle \boldsymbol{\mu}^0, \boldsymbol{\Sigma}^0 \rangle$. Given a prior distribution on $\mathbf{w}$, and user comparisons C as defined in Eq. 4, we obtain a *maximum a posteriori (MAP)* estimate at the beginning of each iteration using Bayesian logistic regression. However, this estimate might not adhere to the simplex constraints. Therefore, we back-project the MAP estimate of the weights onto the simplex for d objectives, leading to the estimate $\bar{\mathbf{w}} = \mathcal{P}_{simplex}(MAP(\mathbf{w}|C))$. The fact that $\bar{\mathbf{w}}$ adheres to the simplex constraints is important for UCB-algorithms, as the exploration bonuses, and the regret-bounds derived from it, use the assumption that the reward samples are within the interval $[0, 1]$ (after applying the utility function).

After umap-UCB obtains a $\bar{\mathbf{w}}$ at the beginning of an iteration, it can proceed to pick actions. To select which arm to play, $a(t)$, umap-UCB follows the standard UCB schema, using the expected scalarised reward, $\bar{\mathbf{w}} \cdot \bar{\mathbf{x}}_a$ plus an exploration bonus c:

$$a(t) \leftarrow \arg\max_a \left(\bar{\mathbf{w}} \cdot \bar{\mathbf{x}}_a + c(\bar{\mathbf{w}}, \bar{\mathbf{x}}_a, n_a, t) \right).$$

We note that we can only use $\bar{\mathbf{w}} \cdot \bar{\mathbf{x}}_a$ as an estimate for the scalarised means because we assume that the estimates of $\mathbf{w}^*$ are independent of the estimates of $\boldsymbol{\mu}_a$.[2] In our setting this assumption holds if the user can objectively compare two vectors, without being influenced by which arms have been pulled in previous iterations, and which comparisons have taken place in previous iterations. We believe this to be a realistic assumption for pairwise comparisons.

The exploration bonus, $c(\bar{\mathbf{w}}, \bar{\mathbf{x}}_a, n_a, t)$, can be implemented in many ways. We note that when $\mathbf{w}^*$ places all the weight on a single objective, the MOMAB becomes a scalar MAB, which is Bernoulli-distributed (the only difference being that the agent does not know $\mathbf{w}^*$). We therefore use exploration bonuses that reduce to those for single-objective MABs in this case. Specifically, we use either $c(\bar{\mathbf{w}}, \bar{\mathbf{x}}_a, n_a, t) = c_1(\bar{\mathbf{w}} \cdot \bar{\mathbf{x}}_a, n_a, t)$ (i.e., UCB1, Eq. 1) or $c(\bar{\mathbf{w}}, \bar{\mathbf{x}}_a, n_a, t) = c_{ch}(\bar{\mathbf{w}} \cdot \bar{\mathbf{x}}_a, n_a, t)$ (Eq. 2). We note that for weights that are more evenly distributed, tighter bounds may hold, e.g., $\mathbf{w}^*$ with equal weights for each objective, and $d \to \infty$, leads to normally distributed scalarised rewards, due to the central limit theorem. However, as the estimation of $\bar{\mathbf{w}}$ is not exact, obtaining a tighter bound is far from trivial, and we leave this open for future work.

Having defined how umap-UCB picks arms to perform, we now define when and which comparison queries umap-UCB poses to the user. We note that we want to decrease the number of queries steeply over time, but never to stop querying (as we may at any point in time, have an estimate $\bar{\mathbf{w}}$ that favours a suboptimal arm). For this reason, we tie in the querying of the user with the exploration mechanism of UCB, which has a similar purpose, i.e., it aims to pull arms so often as to keep on exploring, yet bound the regret of pulling those arms, by rapidly decreasing the number of suboptimal arm pulls over time. To achieve this, umap-UCB explicitly calculates the arm, $\bar{a}^*$, with best estimated *scalarised* mean without exploration bonus, $\bar{\mathbf{w}} \cdot \bar{\mathbf{x}}_a$, at the beginning of each iteration.

---

[2] $\mathbb{E}[\mathbf{x} \cdot \mathbf{y}] = \mathbb{E}[\mathbf{x}] \cdot \mathbb{E}[\mathbf{y}]$, iff $\mathbf{x}$ and $\mathbf{y}$ are independent.

When $\bar{a}^*$ is different from $a(t)$, we query the user for a comparison between the estimates of the means (without exploration bonuses), $\bar{\mathbf{x}}_{\bar{a}^*}$, and $\bar{\mathbf{x}}_{a(t)}$.

Umap-UCB has two important characteristics. Firstly, the algorithm will never stop querying the user. Therefore, if the current estimate of the weights $\bar{\mathbf{w}}$ favours the wrong arm, as time — and with time the accuracy of the estimated means for each a — increases, more and more comparison data will be generated that will eventually lead the MAP estimate of the weights to favour the arm which is best for the ground truth weights, $\mathbf{w}^*$.[3] Furthermore, the expected number of queries is equal to the number of suboptimal arm-pulls, which decreases rapidly over time, and is bounded (in finite time) via the exploration bonuses inherited from the single-objective UCB algorithms that umap-UCB builds upon.

4.2 Interactive Thompson Sampling

For single-objective MABs, UCB algorithms (Algorithm 1) are in practice often outperformed [7] by Thomspon sampling [16] (Algorithm 2). Thompson sampling works according to the following schema: first, it starts with a prior distribution on the parameters of the reward distribution of each arm. Then, it gathers data by drawing samples from the posterior distributions of these parameters, and pulling the arm with the maximal expected rewards according to the sampled parameters for each arm.

We build upon Thompson sampling for MABs by not only sampling from the posteriors of the parameters of the reward distributions, but also those of the user preferences $\mathbf{w}$. We call this algorithm *Interactive Thompson Sampling (ITS)* (Algorithm 4).

ITS starts each iteration by drawing two independent samples both from the posteriors for the parameters of the reward distributions of each arm (θ_1^t and θ_2^t), and from the posterior for the parameters of the utility function (η_1^t and η_2^t). Without loss of generality, we use the first sample to determine the action to play, $a_1(t)$. We note that for our assumptions (independent weights and rewards, Gaussian weights, Bernoulli reward vectors), we can compute which action to select as:

$$\arg\max_a \mathbb{E}_{P(\mathbf{r},\mathbf{w}|a,\theta_1^t,\eta_1^t)}[\mathbf{w} \cdot \mathbf{r}] = \arg\max_a \left(\tilde{\mathbf{w}}_1 \cdot \tilde{\boldsymbol{\mu}}_{1,a}\right),$$

where $\tilde{\mathbf{w}}_1$ (corresponding to η_1^t) is the sampled weights vector, and $\tilde{\boldsymbol{\mu}}_{1,a}$ (corresponding to θ_1^t) is the sampled means vector for the rewards of arm a. Again, we assume that $\tilde{\mathbf{w}}_1$ and $\tilde{\boldsymbol{\mu}}_{1,a}$ can be sampled from their resp. posterior distributions independently.

The second sample is used solely to determine whether and how to interact with the user. ITS determines which actions both samples would select, $a_1(t)$ and $a_2(t)$. If $a_2(t)$ differs from $a_1(t)$, ITS queries the user for a comparison between the expected reward according to the first sample $\tilde{\boldsymbol{\mu}}_{1,a_1(t)}$ to that of the second sample $\tilde{\boldsymbol{\mu}}_{2,a_2(t)}$.

---

[3] Please note that for obtaining 0 regret, it is not necessary that the MAP estimate $\bar{\mathbf{w}}$ is identical to the ground truth $\mathbf{w}^*$, as long as it leads to selecting the same arm.

As the posteriors of both distributions (rewards and weights) become increasingly certain, the number of suboptimal arm-pulls made by ITS goes down. Furthermore, the number of times that two sets of samples from these distributions disagree on which action to take — and thus the number of queries to the user — goes down as well.

5   Experiments

In order to test the performance of umap-UCB and Interactive Thompson Sampling, in terms of user regret (Definition 5) and the number of queries posed to the user, we compare our algorithms on two types of problems: `double circle` MOMABs in Sect. 5.2 and `random` MOMABs in Sect. 5.3. We use two variants of umap-UCB: umap-UCB1, using $c(\bar{\mathbf{w}}, \bar{\mathbf{x}}_a, n_a, t) = c_1(\bar{\mathbf{w}} \cdot \bar{\mathbf{x}}_a, n_a, t)$ (Eq. 1), and umap-UCB-ch using $c(\bar{\mathbf{w}}, \bar{\mathbf{x}}_a, n_a, t) = c_{ch}(\bar{\mathbf{w}} \cdot \bar{\mathbf{x}}_a, n_a, t)$ (Eq. 2).

Besides our own algorithms, we also compare umap-UCB and ITS to single-objective UCB and Thompson sampling provided with the ground truth utility functions of the user. Note that his is an unfair comparison, in the sense that our setting does not actually allow algorithms to know the ground truth utility functions from the beginning. However, it does provide insight into how much utility is lost due to having to estimate the utility function of the user via pairwise comparisons.

5.1   Problems and Experimental Setup

To test the performance of our algorithms, we use two types of MOMABs: the `double circle` and `random`, examples of which are depicted in Fig. 3. Both problems have arms, a, associated with distributions over vector-valued rewards with mean vectors $\boldsymbol{\mu}_a$. We use independent Bernoulli distributions for each objective i, with a mean μ_a^i.

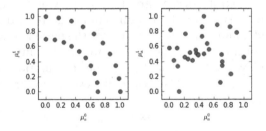

Fig. 3. Examples of `double circle` (left) and `random` (right) MOMABs.

A `double circle` is a two-objective MOMAB that is deterministically generated from two parameters: a number of ticks n_α, and a reduction parameter $p_r \in [0, 1)$. A `double circle` places the mean vectors $\boldsymbol{\mu}_a$ of its arms on two quarter circles. The first is the upper quarter of the unit circle (i.e., the circle with radius 1), and the second that of a circle with radius p_r. On each of the upper quadrants of these circles, n_α arms are evenly distributed. The `double circle` for $n_\alpha = 10$ and $p_r = 0.7$ is depicted in Fig. 3 (left).

A random instance is generated randomly using the number of objectives d, and the number of arms, $|\mathcal{A}|$, as parameters. First, $|\mathcal{A}|$ samples, μ'_a, are drawn from a d-dimensional Gaussian distribution $\mathcal{N}(\mu'_a|\mu_{rnd}, \Sigma_{rnd})$, where $\mu_{rnd} = 1$ (vector of ones), and Σ_{rnd} is a diagonal matrix with $\sigma^2_{rnd} = (\frac{1}{2})^2$ for each element on the diagonal. This set is normalised such that all means fall into the d-dimensional unit hypercube, $\mu_a \in [0,1]^d$. Figure 3 (right) is an example random MOMAB with $d = 2$ and $|\mathcal{A}| = 30$.

Both umap-UCB and ITS require a prior on the weights, $\mathbf{w}$, of the linear user utility function. We employ a multi-variate Gaussian prior, parameterised as $\eta^0 = \langle \mu^0, \Sigma^0 \rangle$. We use the same prior for both algorithms for all experiments. We assume μ is the vector of equal weights, i.e., $\frac{1}{d}$ for each objective, and for the covariance matrix Σ^0 we use a diagonal matrix with $\sigma^2_{cov} = (\frac{1}{3})^2$.

In all experiments, we measure the regret according to Definition 5, i.e., for a single run, when an algorithm pulls an arm a, $\mathbf{w}^* \cdot \Delta_a$, is added to the total accumulated regret.

When querying the user for a comparison between two vectors, $\mathbf{x}$ and $\mathbf{y}$, we first calculate the true utility $u(\mathbf{x})$ and $u(\mathbf{y})$, to which we then add random noise $\varepsilon_{\mathbf{x}}$ and $\varepsilon_{\mathbf{y}}$, independently drawn from a normal distribution $\mathcal{N}(0, \sigma^2_{noise})$. We then compare $u(\mathbf{x}) + \varepsilon_{\mathbf{x}}$ to $u(\mathbf{y}) + \varepsilon_{\mathbf{y}}$. Unless otherwise indicated, $\sigma_{noise} = 0.001$.

5.2 Double Circles

In order to test the performance of our algorithms, we measure their regret, the number of questions they pose to the user, and the L2-norm distance between their estimated weight vectors and the ground truth weights, as a function of time (i.e., the total number of arm pulls) on a double circle with $p_r = 0.7$ and $n_\alpha = 10$ (Fig. 4). We perform ten runs, with different levels of noise. When comparing umap-UCB and ITS, both in terms of regret and in the number of queries, ITS outperforms both UCB algorithms. However, this is not true for the approximation quality of the weights of the linear utility functions. When the noise levels are low ($\sigma_{noise} = 0.001$), ITS quickly focusses on the arms that are close to optimal, and reaches the best estimate of the weights. However, when the noise levels on the comparisons are higher ($\sigma_{noise} \geq 0.01$), the estimates of ITS first improve and then get worse again. We hypothesise that this is the result of ITS quickly focussing on the arms that are close to optimal, whose utility values lay so close that they fall inside the noise interval of the user comparisons, leading to a low signal-to-noise ratio and thus worse estimates. However, the effect on the incurred user regret of these worse estimates is minimal. We thus conclude that ITS outperforms UCB, and that poorer weight estimates are not necessarily detrimental for the incurred user regret.

We also test how much utility is lost by having to estimate the weights of the utility function rather than them being given from the start. In Fig. 4 (left) the regret of the corresponding (cheating) single-objective algorithms to our algorithms are shown as dashed lines. Our best algorithm — ITS — has only little more regret than single-objective Thompson sampling when provided with the ground truth weights from the beginning. Interestingly, umap-UCB1

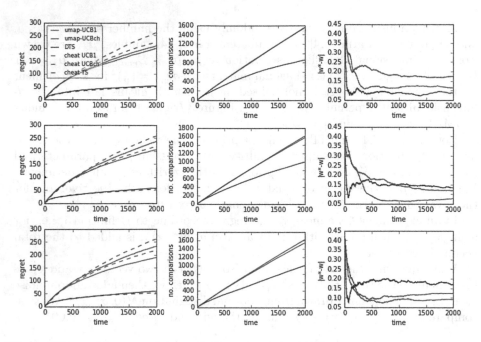

Fig. 4. The performance of umap-UCB and ITS (and UCB and Thompson sampling provided with the ground truth utility function at the start) in terms of regret (left), number of queries asked (middle), and quality of estimation of **w**, on `double circle` instances with $p_r = 0.7$ and $n_\alpha = 10$, averaged over 10 runs. The rows represent the noise level in the user comparisons: $\sigma_{noise} = 0.001$ (top); $\sigma_{noise} = 0.01$ (middle); and $\sigma_{noise} = 0.1$ (bottom). In all runs $\mathbf{w}^* = (0.2, 0.8)$.

and umap-UCB-ch perform even better than their single-objective equivalents (on most individual runs as well as on average). When inspecting the number of pulls per arm of these algorithms, umap-UCB pulls more suboptimal arms in total than single-objective UCB, but the suboptimal arms that are pulled are closer (i.e., have smaller $\mathbf{w}^* \cdot \mathbf{\Delta}_a$) than those of single-objective UCB. We thus conclude that there is only a small increase in regret when having to estimate $\mathbf{w}^*$.

We further note that, if the utility of the arms, $\mathbf{w}^* \cdot \boldsymbol{\mu}_a$, lay close together as in the `double circle`, umap-UCB and ITS keep on querying the user, as a result of not being able to distinguish between the optimal arm and the arms that are just below that in terms of utility (leading to over 40% of the 2000 arm pulls for our best algorithm — ITS — for the lowest noise levels). We expect the number of queries to be less in problems where arms lay further apart, as in `random` MOMAB instances.

5.3   Random MOMABs

In order to test the performance of umap-UCB on MOMABs with arms with mean vectors that lay further apart, we test them on `random` MOMABs with 30 arms each and varying numbers of objectives (Fig. 5). Again, we observe that the user regret for our best performing algorithm, ITS, does not attain a significantly lower regret than single-objective Thompson sampling provided with $\mathbf{w}^*$ from the start. Furthermore, ITS again outperforms both variants of umap-UCB.

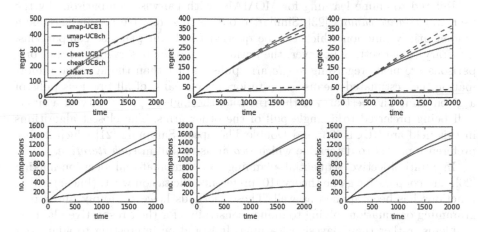

Fig. 5. The performance of umap-UCB and ITS (and UCB and Thompson sampling provided with the ground truth utility function from the start) in terms of regret (top), number of queries asked (bottom) for **random** MOMAB instances with 30 arms and user comparison noise $\varepsilon = 0.001$, randomly drawn $\mathbf{w}^*$ from a uniform distribution, for 2 (left), 4 (middle), and 6 (right) objectives, averaged over 25 runs.

Because the arms lay further apart than on the **double circle**, we observe that ITS — but not umap-UCB — is able to attain significantly lower regret. Furthermore, we observe that for ITS, the number of queries posed to the user is much less than for **double circle**, and often only increases marginally after about 1000 timesteps.

Finally, we tested the effect of the number of objectives of 30-arm **random** MO-MABs on the algorithms' performances. Higher numbers of objectives seem to accrue less regret, while needing about equally many queries. This can be explained by the fact that for higher numbers of objectives, the arms lay further apart (in terms of Euclidean distance), and are thus easier to distinguish. We thus observe a blessing (rather than a curse) of dimensionality for user regret for a fixed number of arms.

In summary, we conclude that ITS outperforms umap-UCB both in terms of regret and in terms of the number of comparison queries posed to the user. Furthermore, when the expected reward vectors of the arms are sufficiently far apart, ITS is able to quickly reduce the number of questions asked per timestep, while umap-UCB cannot. We therefore conclude that ITS is the better algorithm.

6   Related Work

Several papers exist that study MOMABs [3,9,18,22, a.o.]. However, to our knowledge, the previous work on MOMABs all focusses on the *offline* setting rather than the online interactive setting, as this paper does.

Related to online learning for MOMABs with pairwise comparisons by the user, are *relative bandits* [23]. Similar to our setting, relative bandits assume a hidden utility function which can be queried to obtain pairwise comparisons. Contrary to our setting however, the rewards cannot be observed, and the comparisons are made regarding *single* arm pulls, rather than the aggregate information, i.e., estimated means or posteriors over means, of all previous pulls of a given arm. The best arm is the arm with the highest probability of a single pull being preferred to a single pull of the other arms. The closest algorithms in this field are RUCB [23] and Double Thompson Sampling [21] which keep a preference matrix to determine which *two* arms to pull in *each iteration*.

In multi-objective sequential planning [4], combinatorial decision-making [20], and cooperative game theory [10] preference elicitation w.r.t. (linear) utility functions has been applied as well. These methods however, apply linear programming or equation solving to induce constraints for their respective planning problems, rather than Bayesian learning. It would be interesting to adapt our methods to these problem classes to create Bayesian methods for learning in these problems, as well as for *MOMDPs* [13].

7   Conclusions and Future Work

In this paper we proposed *Utility-MAP UCB1 (umap-UCB)* and *Interactive Thompson Sampling (ITS)*, two Bayesian learning methods for *online* multi-objective reinforcement learning in which the agent can interact with its user. Both algorithms build upon state-of-the-art learning algorithms and Bayesian machine learning to learn about the environment and about the utility function of the user simultaneously. Both algorithms pose pairwise comparison queries to the user, and employ Bayesian logistic regression to learn about the linear preferences, representable with a weight vector $\mathbf{w}^*$, of its user. Umap-UCB uses explicit exploration bonuses, and elicits preferences when the best mean estimate for the current MAP estimate of the weight vector, and the best estimated mean plus exploration bonus for the same estimate, recommend different arms. ITS elicits preferences by pulling two sets of samples from both the posteriors of the means of the arms and the posterior for (the estimate of) $\mathbf{w}$, and querying the user when those two sets of samples have a different best arm.

We tested umap-UCB and ITS empirically, and showed that the performance of these algorithms in terms of user regret closely approximates the regret of UCB1 and regular Thompson sampling provided with the ground truth utility functions of the user. Umap-UCB can even perform better than UCB having access to the ground truth. Hence, we conclude that with our algorithms we can come close to the performance of state-of-the-art learning algorithms for single-objective MABs that are provided with the ground truth weights at the start.

Furthermore, we conclude that ITS empirically outperforms umap-UCB both in terms of regret, and in terms of the number of queries posed to the user. We thus conclude that ITS is key for efficient multi-objective learning.

In this paper, we examined the setting of linear utility functions and multi-objective multi-armed bandits. As a next step, we aim to extend ITS to more general utility functions, e.g., additive utility functions [8], or even general-shape monotonically increasing utility functions [14]. Furthermore, we aim extend our methods to multi-objective reinforcement learning in MOMDPs and MOSGs [12,13,17,19].

Acknowledgements. The first author is a postdoctoral fellow of the Research Foundation – Flanders (FWO). This research was in part supported by Innoviris – Brussels Institute for Research and Innovation.

References

1. Agrawal, S., Goyal, N.: Analysis of Thompson sampling for the multi-armed bandit problem. In: COLT, p. 39.1–39.26 (2012)
2. Auer, P., Cesa-Bianchi, N., Fischer, P.: Finite-time analysis of the multiarmed bandit problem. Mach. Learn. **47**(2–3), 235–256 (2002)
3. Auer, P., Chiang, C.-K., Ortner, R., Drugan, M.M.: Pareto front identification from stochastic bandit feedback. In: AISTATS, pp. 939–947 (2016)
4. Benabbou, N., Perny, P.: Combining preference elicitation and search in multiobjective state-space graphs. In: IJCAI, pp. 297–303 (2015)
5. Bishop, C.M.: Pattern Recognition and Machine Learning. Springer, New York (2006)
6. Brochu, E., de Freitas, N., Ghosh, A.: Active preference learning with discrete choice data. In: NIPS, pp. 409–416 (2008)
7. Chapelle, O., Li, L.: An empirical evaluation of Thompson sampling. In: NIPS, pp. 2249–2257 (2011)
8. Clemen, R.T., Decisions, M.H.: An Introduction to Decision Analysis. PWS-Kent, Boston (1997)
9. Drugan, M.M., Nowé, A.: Designing multi-objective multi-armed bandits algorithms: a study. In: IJCNN, pp. 1–8. IEEE (2013)
10. Igarashi, A., Roijers, D.M.: Multi-criteria coalition formation games. In: Rothe, J. (ed.) ADT 2017. LNAI, vol. 10576, pp. 197–213. Springer, Cham (2017)
11. Libin, P., Verstraeten, T., Theys, K., Roijers, D.M., Vrancx, P., Nowé, A.: Efficient evaluation of influenza mitigation strategies using preventive bandits. In: ALA, 9 p. (2017)
12. Mannion, P., Duggan, J., Howley, E.: A theoretical and empirical analysis of reward transformations in multi-objective stochastic games. In: AAMAS, pp. 1625–1627 (2017)
13. Roijers, D.M., Vamplew, P., Whiteson, S., Dazeley, R.: A survey of multi-objective sequential decision-making. JAIR **48**, 67–113 (2013)
14. Roijers, D.M., Whiteson, S.: Multi-objective decision making. Synth. Lect. Artif. Intell. Mach. Learn. **11**(1), 1–129 (2017)
15. Tesauro, G.: Connectionist learning of expert preferences by comparison training. In: NIPS, vol. 1, pp. 99–106 (1988)

16. Thompson, W.R.: On the likelihood that one unknown probability exceeds another in view of the evidence of two samples. Biometrika **25**(3/4), 285–294 (1933)
17. Van Moffaert, K., Nowé, A.: Multi-objective reinforcement learning using sets of Pareto dominating policies. JMLR **15**(1), 3483–3512 (2014)
18. Van Moffaert, K., Van Vaerenbergh, K., Vrancx, P., Nowé, A.: Multi-objective χ-armed bandits. In: IJCNN, pp. 2331–2338 (2014)
19. Wiering, M.A., Withagen, M., Drugan, M.M.: Model-based multi-objective reinforcement learning. In: ADPRL, pp. 1–6 (2014)
20. Wilson, N., Razak, A., Marinescu, R.: Computing possibly optimal solutions for multi-objective constraint optimisation with tradeoffs. In: IJCAI, pp. 815–822 (2015)
21. Wu, H., Liu, X.: Double Thompson sampling for dueling bandits. In: NIPS, pp. 649–657 (2016)
22. Yahyaa, S.Q., Drugan, M.M., Manderick, B.: Thompson sampling in the adaptive linear scalarized multi objective multi armed bandit. In: ICAART, pp. 55–65 (2015)
23. Zoghi, M., Whiteson, S., Munos, R., De Rijke, M.: Relative upper confidence bound for the k-armed dueling bandit problem. In: ICML, pp. 10–18 (2014)

Towards a Protocol for Inferring Preferences Using Majority-rule Sorting Models

Alexandru-Liviu Olteanu[1]([✉]), Patrick Meyer[1], Ann Barcomb[2],
and Nicolas Jullien[3]

[1] IMT Atlantique, Lab-STICC, Univ. Bretagne Loire, 29238 Brest, France
{al.olteanu,patrick.meyer}@imt-atlantique.fr
[2] Lero, University of Limerick, Limerick, Ireland
ann@barcomb.org
[3] IMT Atlantique, LEGO-M@rsouin, Univ. Bretagne Loire, 29238 Brest, France
nicolas.jullien@imt-atlantique.fr

Abstract. In Multi-Criteria Decision Aiding, one of the current challenges involves the proper integration and tuning of the preference models in real-life contexts. In this article, we consider the multi-criteria sorting problem where the decision maker's preferences fall within the outranking paradigm. Following recent advances on extensions of classical majority-rule sorting models, we propose a methodology for adapting them to the perspective of the decision maker. We illustrate the application of the methodology on a real-world problem linked to the evaluation of contributors within Free/Libre Open Source Software communities. The experiments that we have carried out show that the various considered model extensions appear to be useful from the perspective of decision makers in a real-life preference elicitation process, and that the proposed methodology gives useful indications that can serve as guidelines for analysts involved in other elicitation processes.

1 Introduction

Multi-Criteria Decision Aiding (MCDA) provides operational tools which simplify the decision making task for people facing complex decisions involving multiple and conflicting perspectives. MCDA therefore proposes a methodology and algorithms that cover the early stages of situating and understanding the decision question, all the way to the construction of a mathematical representation of the problem and the proposal of an operational decision recommendation.

We consider in this article a particular decision problem, called multi-criteria sorting or classification, in which a finite set of decision alternatives is evaluated on a finite set of criteria, and where the goal of a decision maker (DM) is to assign each of these alternatives into predefined preference-ordered categories or classes. In MCDA, to achieve this assignment task, one uses a preference model, which aggregates the criteria, and which can be based on either one of the following paradigms: Multi-attribute Value Theory (MAVT) [7], the outranking approach [12], or decision rules using "if-then" statements, which can, for example be inferred using a dominance-based rough set approach [6].

© Springer International Publishing AG 2017
J. Rothe (Ed.): ADT 2017, LNAI 10576, pp. 35–49, 2017.
DOI: 10.1007/978-3-319-67504-6_3

We suppose here that, when comparing two alternatives, the decision maker uses the *outranking paradigm*. In other words, (s)he considers that an alternative a outranks an alternative b when a weighted majority of criteria validates the fact that a is performing at least as good as b and there is no criterion where b seriously outperforms a. The majority-related condition is usually called concordance, whereas the second condition is called discordance or veto principle. From a computational point of view, various implementations of these conditions, and their conjunction, have been proposed in the literature (see, e.g., [13]).

In order to solve the sorting problem according to this outranking paradigm, each alternative is compared to predefined *category profiles* according to the outranking relation, and is assigned to one of the categories through an *assignment rule*. Again, different implementations of the category profiles (limit, central) and the assignment rules (pessimistic, optimistic) can be found (see, e.g., [11]).

In the previous quite general description of the outranking relation, very negative aspects in the comparison of two alternatives can lead to the invalidation of an outranking statement, even if one alternative stands out as at least as good as the other one on a significant set of criteria. The influence of very positive aspects in the comparison of two alternatives, leading to the validation of the outranking statement, even if a majority of negative aspects are not in accordance with this outperformance, have also been recently studied. The classical Electre outranking relations are augmented by Roy and Słowiński [14] in order to take into account "reinforced preference" and "counter-veto" effects. They allow to strengthen the coalition of criteria in favor of the outranking statement, which in certain cases may lead to countering a veto situation. Bisdorff [2] has proposed to increase or decrease the confidence degree of a bipolar-valued outranking relation when large performance differences are present. Finally, the concept of dictator, and its interaction with the concept of veto, have been studied by Meyer and Olteanu [8], and multiple, increasingly complex, preference models have been proposed.

In this article, we follow this latter work, by focusing more particularly on the algorithms which learn the parameters of the preference models from *assignment examples* provided by the DM. It is clear that these algorithms represent only a small step in a real-life, complex and iterative preference elicitation process. The natural questions which arise from the work of Meyer and Olteanu [8] are:

- Is there any added value in using more complex sorting models (containing veto and dictator effects) in practice?
- How can the analyst be guided to select the right preference model?
- How should the analyst, who leads the decision aiding process, apply the parameter tuning algorithms when facing a real DM?
- Is it possible to help the analyst to select appropriate assignment examples for the parameter tuning algorithms?

These questions involve real analysts and DMs, and therefore it is important that their answers are based on observations made in real decision-making contexts. We therefore propose in this article a *methodology* for eliciting majority-rule sorting models with large performance differences in *practice*, deduced from multiple real-world preference elicitation processes.

The rest of the article is structured as follows. In Sect. 2 we start by providing a state-of-the-art on majority-rule sorting models, followed by the methodology for learning their parameters in practice in Sect. 3. In Sect. 4 we illustrate the application of the methodology on a real-world problem linked to the evaluation of contributors within Free/Libre Open Source Software communities, before drawing some conclusions and highlighting the perspectives for the future in Sect. 5.

2 Majority-Rule Sorting Models: State-of-the-Art

2.1 Majority-Rule Sorting

The model that this work is based on is a simplified version of the Electre Tri [12] method which is close to the version axiomatized by Bouyssou and Marchant in [3,4]. We define the elements of the problem and those of the model below:

$$
\begin{cases}
A = \{a_1, \ldots, a_n\} & \text{, a set of alternatives} \\
J = \{1, \ldots, m\} & \text{, a set of criteria indexes} \\
C = \{c_1, \ldots, c_k\} & \text{, a set of ordered categories (from worst (1) to best } (k)) \\
W = \{w_j \colon \forall j \in J\} & \text{, a set of criteria weights with } \sum_{j \in J} w_j = 1 \\
\lambda & \text{, a majority threshold with } \lambda \in [0.5, 1] \\
B = \{b_i \colon \forall i \in 0..k\} & \text{, a set of category limits} \\
V = \{v_i \colon \forall i \in 0..k\} & \text{, a set of category vetoes}
\end{cases}
$$

We consider that the alternatives, the category limits and the category vetoes are defined through their evaluations using the function $g_j, \forall j \in J$. We assume, without loss of generality, that the performances are supposed to be such that a higher value denotes a better performance.

Each category $c_h \in C$ is defined by the performances of its lower frontier (the category limit b_{h-1} and the category veto v_{h-1}) and its upper frontier (the category limit b_h and the category veto v_h). Furthermore, the performances on the frontiers are non-decreasing, i.e. $\forall j \in J, h \in 1..k : g_j(b_{h-1}) \leqslant g_j(b_h)$ and $g_j(v_{h-1}) \leqslant g_j(v_h)$, while additionally $\forall j \in J, h \in 0..k : g_j(b_h) > g_j(v_h)$.

Two rules to assign an alternative to a class may be found in the literature, the pessimistic and the optimistic assignment rules, out of which the first is the most commonly used. In this case, an alternative $a \in A$ is assigned to the highest possible category $c_h \in C$ such that a outranks the category's lower frontier. This means that a should hold performances at least as good as b_{h-1} on a sufficient coalition of criteria (based on W and λ) while at the same time not holding performances below or equal to v_{h-1}. In order to guarantee that an alternative will be assigned to at least the bottom category and at most the top one, the lower frontier of the bottom category will be set to the worst possible evaluations on all criteria, while the upper frontier of the top category will be set to the best possible evaluations on all criteria.

Roy and Słowiński [14] have augmented the outranking relations used by Electre methods to take into account "reinforced preference" and "counter-veto" effects. An additional set of thresholds have been added to reflect the veto thresholds and measure very good performances of an alternative over another. Such performances are used to strengthen the coalition of criteria in favor of the outranking statement, which in certain cases may lead to countering a veto situation.

A similar set of thresholds has also been explored by Bisdorff [2] in the context of bipolar valued outranking relations. In addition to the two credibility levels of classical outranking relations (true and false), these relations contain an additional intermediate level (indetermination). In this case, the effects of large performance differences increase or decrease the confidence degree of an outranking relation when they are in concordance with the coalition of criteria in its favor, and reduce it to a state of indetermination when they are conflicting.

Meyer and Olteanu [8] introduced the complementary and symmetric notion to that of a veto in majority-rule sorting models, which they call *dictator*. While denoting with MR-Sort the model containing only a simple majority rule, and with MRV-Sort the model also containing the veto effect, an MR-Sort model with dictators (denoted with MRD-Sort) involves the construction of a dictator relation between a and b_{h-1} (denoted with D), instead of a veto relation. This relation is built with respect to a dictator profile b_{h-1}^v, which represents the minimum level of performance that an alternative needs to have in order to be allowed into category c_h despite an insufficient weighted coalition of criteria in favor of this assignment. The authors also propose several models which integrate both veto and dictator effects:

- MR-Sort with vetoes weakened by dictators (MRv-Sort):
  This model is identical to the classical MR-Sort model, except that when both a veto and a dictator are triggered, the dictator has an effect of invalidating the veto.
- MR-Sort with dictators weakened by vetoes (MRd-Sort):
  This model is identical to an MR-Sort model with dictators, except that when both a veto and a dictator are triggered, the veto has an effect of invalidating the dictator.
- MR-Sort with dictators and dominating vetoes (MRdV-Sort):
  When only one type of effect is triggered, this model behaves either like a classical MR-Sort model with vetoes or one with dictators. When both effects occur at the same time, only the veto is taken into account.
- MR-Sort with vetoes and dominating dictators (MRDv-Sort):
  This model is identical to the previous one, except that, when both effects are triggered, only the dictators are taken into account.
- MR-Sort with conflicting vetoes and dictators (MRdv-Sort):
  This model is also identical to the previous two, except that when both effects are triggered, they cancel each other out.

2.2   Parameter Elicitation

Several works have been previously proposed in order to infer the parameters of outranking-based multicriteria sorting models as an alternative to directly eliciting them. Most of these results are linked to Electre Tri models, but can be quite easily reused for MR-Sort models.

Mousseau and Słowiński [10] proposed to find the model parameters through the use of *assignment examples*. More specifically, the decision maker is asked in a first step to assign a few *well known* alternatives to the predefined categories. Then, from these assignment examples, the model parameters are extracted using linear, mixed integer linear or non-linear programs.

Other more robust approaches compute for each alternative a range of possible categories to which they may be assigned when the parameters of the model are not completely determined [5]. Approaches that deal with inconsistent sets of assignment examples leading to non-existing preference model solutions have also been explored by Mousseau et al. [9].

When taking into account large performance differences in majority-rule sorting, exact elicitation approaches have been proposed by Meyer and Olteanu [8], which have additionally been integrated in widely used software packages, such as R [1].

While such approaches become more and more accessible to the wide public through initiatives as those previously mentioned, the question of how to apply them in practice is still left open. The large majority of the literature focuses on one-time use of these elicitation approaches, whereas the matter of applying them in reality involves multiple interactions with a decision maker. The question of which model to choose, be it a simple MR-Sort model without vetoes, one with vetoes or one handling large performance differences, has also been rarely handled.

3   Proposed Methodology

For all of the reasons presented before, we propose a methodology for applying the parameter elicitation approaches of outranking-based multicriteria sorting models. The construction of this proposal is based on the case study that we present in Sect. 4 (as well as our expertise from other applications). One should note that the resolution of this real decision problem and the construction of the elicitation methodology have been performed in parallel: the application fed the methodology by generating new research questions, and the updated methodology allowed us to advance in the resolution of the decision problem and the preference modeling.

The methodology that we propose may also be easily adapted to other types of problems and preference models, however, within this paper, we will only focus on the family of MR-Sort models, i.e. MR-Sort with a simple majority rule, MR-Sort with vetoes or with dictators and MR-Sort with both vetoes and dictators, along with the different interactions between them. The steps of the methodology are illustrated in Fig. 1.

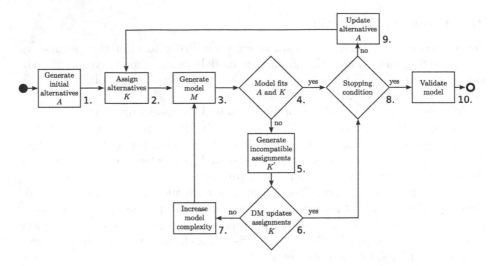

Fig. 1. Methodology for inferring the parameters of an MR-Sort model

The proposed methodology can be summarized as follows:

1. Generate an initial set of alternatives A;
2. The DM assigns the alternatives to classes K;
3. Generate an MR-Sort model M;
4. If the model fits then go to step 8;
5. Generate all minimal sets of incompatible assignments;
6. If the DM accepts to change the assignments as per one incompatible assignment then go to step 8;
7. Select a more complex model and go to step 3;
8. If a stopping condition is met then go to 10;
9. Update the set of alternatives A by generating additional ones;
10. Validate the model and finish.

The first step consists in randomly generating an initial set of alternatives, followed by the DM assigning these alternatives to classes and then fitting an MR-Sort model with a simple majority rule over them. If a model cannot be constructed, we extract all minimal sets of incompatible assignments and ask the DM whether (s)he would agree to change any assignments accordingly. If the answer is negative, we then proceed to considering more complex models.

We consider the MR-Sort model with a simple majority rule to be the simplest, followed by the models which take into account a single large performance difference effect (MR-Sort with vetoes or with dictators) and then by models that take into account both. Among the latter models, we consider that the models that take into account a single effect but which may be weakened by another (MR-Sort with vetoes weakened by dictators and MR-Sort with dictators weakened by vetoes) are the least complex, followed by models that have one dominating effect (MR-Sort with vetoes dominating dictators and MR-Sort with dictators dominating vetoes), and finally the model which equally balances the two effects together (MR-Sort with conflicting vetoes and dictators). When multiple models fit the assignments, we choose based on which model is better able to fit the set of assignment examples, i.e. the one which minimizes the number of incorrectly assigned alternatives. When this criterion is not sufficient in order to differentiate between such models, we select one at random.

After we are able to select a model that fits the assignment examples of the DM, or at least one that minimizes the number of incompatible assignments, we check whether a stopping condition is met. This condition may be based on subjective factors linked to the willingness of the DM to proceed further, or on factors linked to the fit and convergence of the model.

If a stopping condition is not met, we increase the set of alternatives by generating new ones in order to validate the category boundary profiles of the M model. Given, for instance, an MR-Sort model with a simple majority rule and a boundary profile b_{h-1} between categories c_h and c_{h-1}, we generate all alternatives that have either the same evaluations as the profile on some criteria, and slightly lower ones on the others, such that a minimal coalition of criteria supports the statement that a outranks b_h. These alternatives are barely classified as belonging to category c_h (decreasing their evaluation on even one criterion belonging this coalition would alter their assignment), hence having the DM validate them means that the boundary b_h cannot be raised any higher. Similarly, we generate another set of alternatives such that a minimal coalition of criteria supports the statement that a does not outrank b_h. These alternatives are barely classified as belonging to c_{h-1} (increasing their evaluation on even one criterion belonging this coalition would alter their assignment), hence having the DM validate their assignment would fix the b_{h-1} profile from being lowered. When considering MR-Sort models with vetoes and dictators, we additionally need to take into account the corresponding veto and dictator profiles.

4   Experimental Framework

The context of this application is that of online communities of *software developers*, which we call *contributors*. They are usually managed and animated by so-called *community managers* (CM), who are facing the problem of evaluating the contributors to their community. For a community to be effective and dynamical, the evaluation of these contributors can usually not be summarized to the amount of code which they produce. Consequently, this evaluation problem involves multiple perspectives, and also depends on the preferences of the CM.

In our experimental framework, 6 CM have been interviewed and had to undergo (and at the same time contributed to the development of) the methodology that we are proposing in this article. Below, we illustrate one of these experiments, for a CM who identified five criteria as important from his perspective:

- Commitment to the project (c_1);
- Ability to work with others (c_2);

- Quality of produced code (c_3);
- Understanding of the tools, technologies, domain and process behind the project (c_4);
- Documentation skills (c_5).

These criteria, which will be used to generate fictive contributor profiles (the assignment examples of Sect. 3), are defined on ordinal scales with 5 levels: very bad (vb), bad (b), neutral (n), good (g) and very good (vg).

The CM was interested in identifying both good and bad contributors, while additionally allowing for neither good nor bad contributors to be identified. Hence, the categories to which the contributors of his community should be assigned were defined as {Good, Neutral, Bad}.

4.1 Generating and Evaluating the Initial Set of Contributors Profiles

We began by generating an initial set of 25 contributor profiles so that each of the 5 ordinal levels of the criteria scales was uniformly distributed on each criterion. The CM was asked to assign these profiles to one of the three selected categories. We illustrate this dataset and the CM's assignments in Table 1.

Table 1. The initial set of contributor profiles and their assignment by the CM.

Profile number	Criteria c_1	c_2	c_3	c_4	c_5	Category	Profile number	Criteria c_1	c_2	c_3	c_4	c_5	Category
1	vg	g	vb	vg	n	Bad	14	vg	n	vb	b	g	Bad
2	b	vg	n	vb	n	Neutral	15	n	b	b	vb	n	Bad
3	b	b	b	b	g	Bad	16	b	vg	g	vg	vb	Bad
4	b	b	vb	vg	n	Bad	17	n	b	n	g	n	Bad
5	g	vb	vg	b	b	Neutral	18	vg	vb	vg	g	b	Neutral
6	vg	g	vg	n	vg	Good	19	vg	vb	n	n	vb	Bad
7	g	n	b	n	vg	Neutral	20	vb	vg	vg	b	vg	Neutral
8	n	n	g	b	g	Good	21	vb	n	vb	n	vb	Bad
9	n	vg	n	g	b	Bad	22	vb	b	vb	vg	g	Bad
10	vb	g	vg	vb	b	Bad	23	vb	vb	g	g	vg	Neutral
11	g	g	g	vb	vg	Good	24	g	vb	b	g	vb	Bad
12	n	g	g	vb	g	Neutral	25	b	n	b	vg	b	Bad
13	g	g	n	n	vb	Bad							

4.2 Determining the Complexity of the First Model

We continue by testing whether an MR-Sort model with a simple majority rule was able to represent the provided assignments and we found that only at most 23 out of the 25 were captured. We therefore determine sets of incompatible assignments (each containing 2 elements), along with the class that they should

Table 2. Sets of incompatible assignments, with the CM assignment and an MR-Sort-compatible assignment.

	Profile number	Criteria					Category	
		c_1	c_2	c_3	c_4	c_5	CM	MR-Sort
First set	2	b	vg	n	vb	n	Neutral	Bad
	8	n	n	g	b	g	Good	Neutral
Second set	2	b	vg	n	vb	n	Neutral	Bad
	11	g	g	g	vb	vg	Good	Neutral
Third set	2	b	vg	n	vb	n	Neutral	Bad
	12	n	vg	g	vb	g	Neutral	Good

have been assigned to by the CM, if he had done the evaluation according to an MR-Sort model. These sets of profiles are illustrated in Table 2.

Using these sets, we devise a series of questions to ask the CM, in order to simplify his task of accepting or rejecting the MR-Sort-compatible assignments. Since the second contributor profile appears in all of the three sets, we decide to first ask him whether he would agree to change his assignment of this profile from Neutral to Bad. The CM agrees to change his assignment if needed, confirming that he had initially hesitated between these two categories. We continue by asking him whether he would also agree to change the assignments of any of other profiles from each of the three sets, however, in all cases, he disagrees.

At this point, we had two options: either the CM could accept one assignment error and validate an MR-Sort-compatible model, or we could increase the complexity of the preference model. As we were still in the beginning of the elicitation process, the second option was chosen, and we found that a model with vetoes was able to capture all 25 profile assignments, while a model with dictators was not.

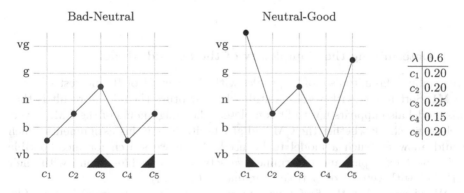

Fig. 2. First preference model (MR-Sort with vetoes), majority threshold (λ) and criteria weights on the right.

The resulting model is illustrated in Fig. 2, where we have divided in two the elements delimiting the first two classes (Bad and Neutral), and those delimiting the last two classes (Neutral and Good). The lines correspond to the delimiting profiles, while the filled in areas represent the ranges of values which would trigger a veto.

Looking at the model, we can deduce multiple rules that characterize good, neutral or bad contributors. For example, focusing on the good contributors (the graph to the right), simply by looking at the veto thresholds, we can say that a good contributor cannot be very bad in c_1, c_3 or c_5, regardless of how they are evaluated on the other criteria. When this is not the case, a good contributor would need to have evaluations at least as good as the delimiting profile on a weighted majority of criteria, such as for instance c_2, c_3 and c_4.

As we have only traversed one iteration of the methodology, and as the CM is willing to continue, we decided to start a new iteration to increase the accuracy of the preference model.

4.3   Generating and Evaluating an Additional Set of Profiles

An additional set of 10 profiles is generated according to the rule described in Sect. 3, based on the previously created model. This set is presented to the community manager who then assigns them as seen in Table 3.

Table 3. The second set of contributor profiles and their assignment.

Profile number	Criteria					Category	Profile number	Criteria					Category
	c_1	c_2	c_3	c_4	c_5			c_1	c_2	c_3	c_4	c_5	
26	vg	g	n	b	b	Bad	31	vb	b	vg	vg	b	Bad
27	vg	b	n	vg	b	Bad	32	vb	b	n	vg	vg	Bad
28	vg	b	n	b	vg	Neutral	33	vg	vg	vb	vg	vg	Neutral
29	vb	vg	n	vg	b	Bad	34	vg	vg	vg	vg	vb	Good
30	vb	vg	n	b	vg	Bad	35	b	vb	g	vb	n	Neutral

4.4   Determining the Complexity of the Second Model

We combine the initial set of 25 profiles with the new set of 10 and test whether an MR-Sort model with vetoes is still able to capture them. The result is that this model also appears not to fit completely the assignments of the CM. Nevertheless, we check whether the CM had any hesitations in his assignments which would allow for such a model to be used. Again we search for incompatible assignment examples and determine five sets of two profiles, along with their MR-Sort with vetoes-compatible assignments (Table 4).

We observe that the first four sets contain the second contributor profile, which the CM already agreed to change if needed. Therefore, we continue by iteratively determining whether he would also agree to accept an alternative

Table 4. Sets of incompatible assignments.

	Profile	Criteria					Category	
	number	c_1	c_2	c_3	c_4	c_5	CM	MRV-Sort
First set	2	b	vg	n	vb	n	Neutral	Bad
	16	b	vg	g	vg	vb	Bad	Neutral
Second set	2	b	vg	n	vb	n	Neutral	Bad
	33	vg	vg	vb	vg	vg	Neutral	Bad
Third set	2	b	vg	n	vb	n	Neutral	Bad
	34	vg	vg	vg	vg	vb	Good	Bad
Fourth set	2	b	vg	n	vb	n	Neutral	Bad
	35	b	vb	g	vb	n	Neutral	Bad
Fifth set	30	vb	vg	n	b	vg	Bad	Neutral
	33	vg	vg	vb	vg	vg	Neutral	Bad

assignment for the remaining profiles in these sets. The CM does not accept changing the assignment of profiles 16, 33 or 34, especially since for the third one the alternative assignment strongly contradicts the initial assignment. Nevertheless, he does agree to switch the assignment of profile 35 to the Bad category, hence we continued to use an MR-Sort model with vetoes (Fig. 3).

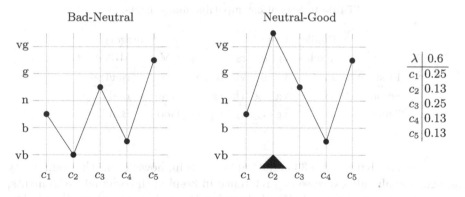

Fig. 3. Second preference model (MR-Sort with vetoes).

We finish another iteration of the preference modeling process and check whether we should finish the process or start a new iteration. Looking at the number of remaining profiles that may be generated around the categories limits, which are only 8 in total, the community manager agrees to continue.

4.5   Generating and Evaluating a Second Complementary Set of Profiles

We generate these 8 new profiles, and ask the community manager to assign them to one of the three categories. The assignments are shown in Table 5.

Table 5. The third set of contributor profiles and their assignment.

Profile number	c_1	c_2	c_3	c_4	c_5	Category	Profile number	c_1	c_2	c_3	c_4	c_5	Category
36	b	vg	n	vg	g	Bad	40	vg	vb	vg	vg	vg	Good
37	n	b	g	vb	vg	Neutral	41	n	vb	g	vb	vb	Bad
38	n	vb	vb	b	vg	Bad	42	n	b	g	b	vb	Bad
39	vb	vb	g	b	vg	Neutral	43	b	vg	n	vb	vg	Good

4.6   Determining the Complexity of the Third Model

After adding the 8 new profiles and their assignments to the existing ones, we find that an MR-Sort model with vetoes is again not able to represent all of these assignments. Still, we find that changing the assignment of only one profile at a time is enough in order to use such a model (Table 6).

Table 6. Sets of incompatible assignments.

	Profile number	c_1	c_2	c_3	c_4	c_5	CM	MRV-Sort
First set	14	vg	n	vb	b	g	Bad	Neutral
Second set	33	vg	vg	vb	vg	vg	Neutral	Bad
Third set	34	vg	vg	vg	vg	vb	Good	Bad

We observe that profiles 33 and 34 appear again, however, as the community manager has already expressed a preference in keeping the original assignments, we only inquire on the possibility of changing the assignment of profile 14. The community manager feels strongly about keeping this profile in the Bad category, therefore motivating us to test a more complex model. We apply an MR-Sort model with vetoes weakened by dictators, as it is the model that is closest to the one previously used. This model, illustrated in Fig. 4 is able to reflect all of the assignments of the community manager.

We finish yet another iteration of the preference modeling process and check whether we should finish the process or start a new iteration. There are still 14 profiles that may be generated around the categories limits, however the community manager wishes to review the model we have generated so far.

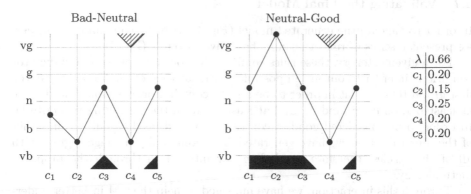

Fig. 4. Third preference model (MR-Sort with vetoes weakened by dictators).

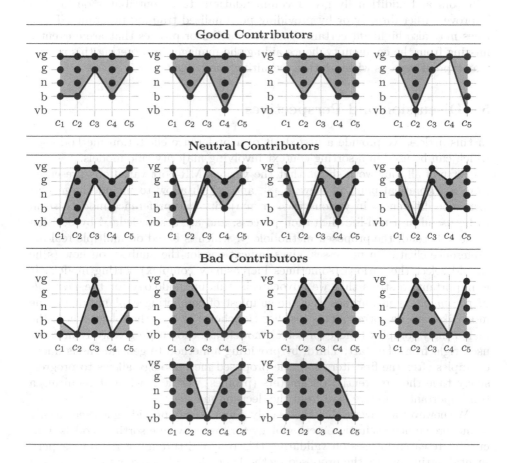

Fig. 5. Assignment rules.

4.7   Validating the Final Model

In order to validate the generated model (Fig. 4) with the community manager, we present a series of rules that may be derived from it (Fig. 5).

For a given category, these rules combine the concordance, veto and dictator conditions in order to output all possible contributors profiles. For example, the first graph tells us that in order to be a good contributor, one would need to be at least bad on c_1, c_2 and c_4, and at least good on the remaining two criteria. In order to be considered good, a contributor would need to fall within either of the four rules that we have generated. The community manager agrees with all of these rules, therefore concluding our interaction through the proposed methodology.

Through this interaction, we have managed to help the CM in better understanding the way in which he perceives the contributors to his community. Through the generated rules, the CM can more consistently evaluate new contributors and additionally give recommendations to a contributor on how to improve, either directly or by providing personalized training programs. These rules may also highlight certain types of contributor profiles that are currently missing from the community, hence aiding the manager in targeting this particular type of persons within future recruiting campaigns.

5   Discussion and Perspectives

In this article, we provide a first draft of a preference elicitation methodology which can be used in a sorting context involving majority-rule models.

First of all, our work shows that the various MR-Sort variations presented from a theoretical point of view in [8], and which allow to include large performance differences in various ways in majority-rule preference models, seem to be useful in a real-life elicitation process, and are acknowledged by decision makers. Second, the proposed methodology, which is based on multiple real-life preference elicitation processes, gives indications for the analyst, on how (s)he should select the various (sometimes increasingly complex) sorting models. It is important to stress that we advocate in this methodology for the *principle of parsimony*, which argues that the simplest of competing explanations is the most likely to be correct. This justifies the choice of a methodology where the increasingly complex (and expressive) models are chosen progressively and not used right from the start. Third, by providing a method to generate assignment examples after the first iteration, the proposed methodology allows to progressively tune the preferential parameters (profiles, weights), without requiring a too important cognitive load from the decision maker.

We however are aware that this is only a first step in providing a scientifically sound preference elicitation methodology for majority-rule sorting models. Our current research focuses on validating this proposal through a rigorous experimental setting, where the proposed methodology is confronted with large sets of artificially generated data and decision maker preferences.

Acknowledgments. This work was supported, in part, by Science Foundation Ireland grants 10/CE/I1855 and 13/RC/2094 to Lero - the Irish Software Research Centre (www.lero.ie).

References

1. Bigaret, S., Hodgett, R.E., Meyer, P., Mironova, T., Olteanu, A.-L.: Supporting the multi-criteria decision aiding process: R and the MCDA package. EURO J. Decis. Processes 1–26 (2017)
2. Bisdorff, R.: On polarizing outranking relations with large performance differences. J. Multi-Criteria Decis. Anal. **20**(1–2), 3–12 (2013)
3. Bouyssou, D., Marchant, T.: An axiomatic approach to noncompensatory sorting methods in MCDM, I: the case of two categories. Eur. J. Oper. Res. **178**(1), 217–245 (2007)
4. Bouyssou, D., Marchant, T.: An axiomatic approach to noncompensatory sorting methods in MCDM, II: more than two categories. Eur. J. Oper. Res. **178**(1), 246–276 (2007)
5. Dias, L.C., Mousseau, V., Figueira, J., Clímaco, J.N.: An aggregation/disaggregation approach to obtain robust conclusions with ELECTRE TRI. Eur. J. Oper. Res. **138**(2), 332–348 (2002)
6. Greco, S., Matarazzo, B., Slowinski, R.: A new rough set approach to multicriteria and multiattribute classification. In: Polkowski, L., Skowron, A. (eds.) RSCTC 1998. LNCS, vol. 1424, pp. 60–67. Springer, Heidelberg (1998). doi:10.1007/3-540-69115-4_9
7. Keeney, R.L., Raiffa, H.: Decisions with Multiple Objectives: Preferences and Value Tradeoffs. Wiley, New York (1976)
8. Meyer, P., Olteanu, A.-L.: Integrating large positive and negative performance differences into multicriteria majority-rule sorting models. Comput. Oper. Res. **81**, 216–230 (2017)
9. Mousseau, V., Dias, L.C., Figueira, J.: Dealing with inconsistent judgments in multiple criteria sorting models. 4OR **4**(3), 145–158 (2006)
10. Mousseau, V., Słowiński, R.: Inferring an ELECTRE TRI model from assignment examples. J. Glob. Optim. **12**(2), 157–174 (1998)
11. Mousseau, V., Słowiński, R., Zielniewicz, P.: A user-oriented implementation of the ELECTRE TRI method integrating preference elicitation support. Comput. Oper. Res. **27**(7–8), 757–777 (2000)
12. Roy, B.: The outranking approach and the foundations of ELECTRE methods. Theor. Decis. **31**, 49–73 (1991)
13. Roy, B.: Multicriteria Methodology for Decision Aiding. Kluwer Academic, Dordrecht (1996)
14. Roy, B., Słowiński, R.: Handling effects of reinforced preference and counter-veto in credibility of outranking. Eur. J. Oper. Res. **188**(1), 185–190 (2008)

Dominance Based Monte Carlo Algorithm for Preference Elicitation in the Multi-criteria Sorting Problem: Some Performance Tests

Tom Denat[✉] and Meltem Öztürk

Université Paris Dauphine, PSL Research University, CNRS, LAMSADE,
75016 Paris, France
{tom.denat,meltem.oztur}@lamsade.dauphine.fr

Abstract. In this article, we study the Dominance Based Monte Carlo algorithm, a model-free Multi-Criteria Decision Aiding (MCDA) method for sorting problems, which was first proposed in Denat and Öztürk (2016). The sorting problem consists in assigning each object to a category, both the set of objects and the set of categories being predefined. This method is based on a sub-set of objects which are assigned to categories by a decision maker and aims at being able to assign the remaining objects to categories according to the decision makers preferences. This method is said model-free, which means that we do not assume that the decision maker's reasoning follows some well-known and explicitly described rules or logic system. It is assumed that monotonicity should be respected as well as the learning set. The specificity of this approach is to be stochastic. A Monte Carlo principle is used where the median operator aggregates the results of independent and randomized experiments. In a previous article some theoretical properties that are met by this method were studied. Here we want to assess its performance through a k-fold validation procedure and compare this performance to those of other preference elicitation algorithms. We also show how the result of this method converges to a deterministic value when the number of trials or the size of the learning set increases.

1 Introduction

This article is a study of some practical performances of the Dominance Based Monte Carlo algorithm. The Dominance Based Monte Carlo algorithm is a preference elicitation algorithm for the multi-criteria sorting problem. The sorting problem consists in assigning individually several objects to predefined and ordered categories. Each of these objects is evaluated on several ordered criteria. In practice, several applications of this problem can be found. An example of such problem while performing an evaluation of the moral behavior of companies on a scale of 5 value levels (from $C1$ to $C5$) based on three criteria, the treatment of workers, the impact on the environment and the financial transparency each of them being expressed on a scale of 5 value levels.

© Springer International Publishing AG 2017
J. Rothe (Ed.): ADT 2017, LNAI 10576, pp. 50–64, 2017.
DOI: 10.1007/978-3-319-67504-6_4

In the literature one can find different methods adapted to this problem, such as ELECTRE TRI (Roy and Bouyssou 1993), rule based methods (Greco et al. 2011) or utility based methods (Keeney and Raiffa 1994). Each of these methods contains preference parameters (such as weights, thresholds etc.) that are supposed to make the results obtained fit as well as possible to the decision maker's judgement. Disaggregation approaches for preference elicitation consists in asking the decision maker to provide her expectations on the results such as "I think that the object a should be assigned to category 2, the object b should be assigned to category 1, etc.". Then theses methods try to find the parameters that suit to this preference information (see for instance (Jacquet-Lagreze and Siskos 1982; Greco et al. 2011)). The attractiveness of such approaches comes from the fact that they only require an instinctive and subjective judgement that suits quite well to the decision makers expectations. Nevertheless, there can be a very large number of parameters that suit to the expressed preferences or, at the opposite; it can be that no parameter would return these preferences. That could mean that the decision maker's preferences are not representable through the chosen model.

The Dominance Based Monte Carlo algorithm combines two properties that are not frequently met by the other methods for preference elicitation: a model free approach and a stochastic approach.

By model-free, we mean that we do not assume that the decision maker's reasoning follows some well known and explicitly described rules or logic system. According to the context this property may be see as desirable or not. It depends on whether or not one of the existing Multi Criteria Aggregation Procedure (MCAP) such as the ones cited on the previous paragraph can be rigorously defended and depending on whether or not there is a high need for justification for the decision to be taken. In some contexts, for instance when we classify instinctively companies into categories representing how globally responsible they are, it is possible that our intuitive reasoning is not based on an additive utility or a majority or logical rule but it is a mix of all of them with some additional noise.

Our method is based on a stochastic approach using a Monte Carlo procedure. It proposes a complete assignment that may be seen as a "center" of all the complete assignments that respect some expected properties, which are the monotonicity (improving an object on one or several criteria cannot lower its overall evaluation) and the learning set (the objects that were assigned to a category by the decision maker in the learning set should be assigned to the same category by the method). Many authors, mainly from psychology and behaviour analysis (Regenwetter et al. 2011; Luce 1995; Carbone and Hey 2000), defend the idea that preferences are subject to randomness. There exist in the literature some MCDA methods using a probabilistic approach such as SMAA methods (Stochastic Multicriteria Acceptability Analysis, Lahdelma et al. (1998)). However, SMAA methods are not model-free, they analyze the stability of the results of a given MCAP method (SMAA-TRI for ELECTRE TRI for instance, see Tervonen et al. (2009)).

As we already mentioned, we want our complete assignment to satisfy the monotonicity and the learning set. Monotonicity means that improving an object on one or several criteria cannot lower its overall evaluation. In a multi-criteria sorting problem, we aim not only at studying the decision makers reasoning but also at creating a method that recommends an overall judgement. Thus a method that does not respect monotonicity cannot be accepted. The respect of the learning set is also generally considered as a good property in MCDA (contrary to the machine learning). We say that a method respects the learning set if the resulting complete assignment respects the partial assignment given by the decision maker. Most of the elicitation methods for multi-criteria sorting respects the learning set as long as this learning set is compatible with their associated MCAP. While in other multi-criteria methods these two concepts are only seen as good properties, in our algorithm, they are imposed and compose the only real frame of this method.

The article is organized as follows: Sect. 2 introduces some basic concepts and the functioning of the DBMC algorithm, Sect. 3 is dedicated to several tests to evaluate the effectiveness of the convergence and the performance of the k-fold validation on real learning sets. Section 4 concludes the article with some perspectives.

2   Notations and Presentation of the Algorithm

The Dominance Based Monte Carlo algorithm (DBMC) works as follows. At first, the decision maker provides a learning set (a subset of objects that are assigned to categories). These objects are directly affected to their categories. Then, iteratively an object is chosen randomly among the remaining ones and it is assigned randomly to a category without violating monotonicity. This step is repeated until every object is fixed in a category. This random complete assignment will be considered as a trial. Obviously, randomness has an important impact in this complete assignment. In order to reduce this impact and to converge we make a large number of trials, probabilistically independent to each other. We collect the information about the consecutive results of the trials and we aggregate it in order to get a single complete assignment. We are now about to give a formal definition of the DBMC algorithm.

We define a sorting context as a $5 - tuple$ $S =< N, V, A, C, L >$:

- $N = \{1, ..., n\}$ is a finite set of criteria.
- Each criterion i is expressed on a discrete and finite scale v_i. V is the union of criterion scales.
- A is the set of objects to be sorted. Here it is considered that any combination of values on the criteria must be sorted i.e. $A = \prod_{i \in N} v_i$ and $|A| = m$.
- $C = \{1, 2, ..., r\} \subset \mathbb{N}$ is a set of r ordered categories in which the objects are to be sorted. We will thereafter assume that the higher a category is, the better it is.

- L represents the learning set: $L= < \Theta, f_l >$ where $\Theta \subseteq A$ is the subset of examples and $f_l : \Theta \to C$ is the assignment of these examples.
- The expected output of this problem is a complete assignment. A complete assignment is a function $f : A \to C$ that assigns every possible object a to a category $f(a)$. Thereafter, we will denote by $DBMC(S,T)$, the result of the Dominance Based Monte Carlo algorithm on the sorting context S with T.

Monotonicity

Given two objects a, b, we say that a weakly dominates b, and denote by aDb, if a is at least as good as b on every criterion. We will thereafter simply use the term "dominates" to mention weak domination.

For any object $a \in A$ we call dominating cone of a, also denoted by $D^+(a) = \{a' \in A : a'Da\}$, the set of all objects that dominate a. For any object $a \in A$ we call dominated cone of a, also denoted by $D^-(a) = \{a' \in A : aDa'\}$, the set of all objects that are dominated by a. We say that a complete assignment f respects monotonicity if, for any $a, b \in A$ such as aDb, we have $f(a) \geq f(b)$.

The Learning Set

Given that we impose the respect of monotonicity and the learning set, the possible remaining complete assignments are then constrained to a smaller space. For instance, if $a'Da$ and aDa'' with a' already assigned to category 4 and a'' already assigned to category 2, then we can deduce that a can only be assigned to category 2, 3 or 4 (we can say that a can be assigned to the interval of categories $[2,4]$). In order to introduce this notion of "interval assignment" not violating the monotonicity and respecting the learning set we define the notion of a "necessary interval assignment", also denoted by $\gamma : A \to \Delta$, (Δ being the set of category intervals). In other words this necessary interval assignment represents the fact that any object a will be assigned to a category at least as good as the best sorted object that is dominated by a and at most as good as the worst sorted object that weakly dominates a. An object a is said fixed with an interval assignment γ if $\gamma_{min}(a) = \gamma_{max}(a)$.

We give an illustration in Fig. 1 of the interval assignment that we may have in the beginning of DBMC algorithm on a problem with two criteria (10 levels for each) and three categories. As we see 5 objects belong to the learning sets (one of them in category 1, two of them in category 2 and two of them in category 3). These five assignments constrain the interval assignments of the remained objects: red ones in category 3, orange-colored ones in interval $[2,3]$, yellow ones in category 2, light green ones in interval $[1,2]$, ...

The space of the possible complete assignments will obviously be empty if the learning set does not it self respect monotonicity. This point will be treated in Subsect. 3.2. Therefore, from now on, we assume that the learning set respects monotonicity.

Process of a Trial

In the DBMC algorithm each trial will be a random completion of the learning set that respects monotonicity. To do so, we iteratively choose uniformly a random

	1	2	3	4	5	6	7	8	9	10
1	1	1	1	1	[1,2]	[1,2]	[1,3]	[1,3]	[1,3]	[1,3]
2	1	1	1	1	[1,2]	[1,2]	[1,3]	[1,3]	[1,3]	[1,3]
3	1	1	1	1	[1,2]	2	[2,3]	[2,3]	[2,3]	[2,3]
4	[1,2]	[1,2]	[1,2]	[1,2]	[1,3]	[2,3]	[2,3]	[2,3]	[2,3]	[2,3]
5	[1,2]	[1,2]	[1,2]	[1,2]	[1,3]	[2,3]	[2,3]	[2,3]	[2,3]	[2,3]
6	[1,2]	[1,2]	[1,2]	2	[2,3]	[2,3]	[2,3]	[2,3]	[2,3]	[2,3]
7	[1,3]	[1,3]	[1,3]	[2,3]	[2,3]	[2,3]	[2,3]	[2,3]	3	3
8	[1,3]	[1,3]	[1,3]	[2,3]	[2,3]	3	3	3	3	3
9	[1,3]	[1,3]	[1,3]	[2,3]	[2,3]	3	3	3	3	3
10	[1,3]	[1,3]	[1,3]	[2,3]	[2,3]	3	3	3	3	3

Fig. 1. Illustration of the necessary interval complete assignment with two criteria, three categories and 5 objects in the learning set (in the blue squares) (Color figure online)

object and assign it uniformly to a random category among the categories in which it could be sorted. This step is repeated until every object is assigned to a category. Algorithm 1 presents this procedure.

Algorithm 1. Random completion - trial

Data: Sorting context $S =< N, V, A, C, L >$

Result: Assignment $f_1 : A \rightarrow C$ monotonic and compatible with L

1 **while** $\exists a \in A$ *such that* $\gamma_{min}(a) \neq \gamma_{max}(a)$ **do**
2   Choose randomly an object χ with a uniform distribution over A;
3   Choose randomly a category Δ with a uniform discrete distribution between $\gamma_{min}(\chi)$ and $\gamma_{max}(\chi)$;
4   Add the information $< \chi, \Delta >$ to the learning set;
5      $\gamma_{max}(\chi) \leftarrow \Delta$;
6      $\gamma_{min}(\chi) \leftarrow \Delta$;
7   **for** $a^- \in D^-(\chi)$ **do**
8          $\quad \gamma_{max}(a^-) \leftarrow min\{\Delta, \gamma_{max}(a^-)\}$
9   **for** $a^+ \in D^+(\chi)$ **do**
10         $\quad \gamma_{min}(a^+) \leftarrow max\{\Delta, \gamma_{min}(a^+)\}$

We say that we add an information $< a, c >$, $a \in A, c \in C$ to the learning set when we impose the fact that the object a should be sorted in category c, i.e. $\Theta \leftarrow \Theta \cup \{a\}$ and $f_l(a) \leftarrow c$.

Collecting and Aggregating the Trial's Information. To apply the DBMC algorithm, we first complete randomly T times the original sorting context. Then, for every object, we note in a DBMC vector in what category it has been assigned at every trial. Finally, we aggregate these vectors to assign each object to the category in which it has "globally" been assigned during the T trials.

Definition 1 (DBMC vector). *We call a DBMC vector of S, $\varphi : A \times \mathbb{N} \to C^T$ the vector in which we store the results of T random completions of S such as $\forall T \in \mathbb{N}, a \in A, k \in \{1, ..., T\}$, $\varphi_k(a)$ represents the category in which a has been assigned at the kth trial among T trials.*

Formally we get a DBMC vector of S as it is explained in Algorithm 2.

Algorithm 2. Building a DBMC vector

    Data: Sorting context $S = < N, V, A, C, L >$
    Result: DBMC vector, $\varphi : A \times \mathbb{N} \to \mathbb{N}^T$
1 **for** j *from* 1 *to* T **do**
2     $S' \leftarrow S$;
3   Complete randomly S' using Algorithm 1;
4   **for** $a \in A$ **do**
5        $\varphi_j(a) \leftarrow \gamma^{S'}_{min}(a)$ or $\gamma^{S'}_{max}(a)$;

The DBMC vector contains all the information about the results of T trials. In order to have a single result about the assignment of objects of A, we need to aggregate this information. The reason why the median operator was chosen to aggregate the DBMC vector is explained in Denat and Öztürk (2016).

Recall of the Theoretical Properties of this Method
In Denat and Öztürk (2016) several theoretical properties were demonstrated. At first, the DBMC algorithm will run and find a complete assignment if and only if the learning set which is given respects monotonicity. This is one of the reason why the median operator was chosen, the mode operator being possibly non-monotonic. Then, the authors showed that computational complexity of the DBMC algorithm is in $O(T \times m^2)$. It was demonstrated that the result of the DBMC algorithm is monotonic and that it respects the learning set. Finally, despite its stochastic nature, the result of the DBMC algorithm converges almost surely when $T \to \infty$.

3 Practical Tests on the DBMC Algorithm

It is always useful to make practical tests on a preference elicitation algorithm, to illustrate how it reacts in practice and evaluate its performances in addition to theoretical proven properties. This is especially true while speaking of an algorithm that may be seen by the user as a black box, which increases the need for a justification. Here the tests that we present aim at answering two questions. At first the Dominance Based Monte Carlo algorithm being a stochastic algorithm, we would like to know to which extent randomness impacts its result. Then, we made tests to evaluate the ability of this algorithm to restore a part of a learning set while looking at the rest of it.

3.1 Stability

It was proved in Denat and Öztürk (2016) that the result of the Dominance Based converges almost surely when the number of trials grows to infinity. But saying that does not necessarily mean that this convergence is observed in practice (the result could converge almost surely but start converging when the number of trials is higher than 10^{100}). Thus, we here make tests to assess the practical stability of the algorithm. To test the stability of the DBMC algorithm on a sorting context $S =< N, V, A, C, L >$ with T trials, Ω stability rounds and a learning set of size τ, we proceed as follows. We iteratively do Ω times the following stability round in which two complete assignments provided by the DBMC algorithm in similar contexts are compared. To do so, a random complete assignment f is chosen that respects monotonicity. We simply perform a random completion of the sorting context S (as described in Algorithm 1) with an empty leaning set. Then, we randomly (uniformly) choose a set of object $A' \subset A$ such that $|A'| = \tau$ which is considered as the learning set. With this learning set, the Dominance Based Monte Carlo algorithm is ran twice and we obtain two complete assignments on A, f_1 and f_2. We count the number of objects in A that are not assigned to the same category with f_1 and f_2. Then we perform the next stability round i.e. the same experience with a new random complete assignment f. At the end of the algorithm, we look at the average percentage of objects sorted differently in f_1 and f_2 across the stability rounds. This number will be called the stability score (a low stability score means that the algorithm is stable). The formal description of this test is given in Algorithm 3.

Algorithm 3. Stability test

Data: Sorting context $S =< N, V, A, C, L = \emptyset >$, number of trial T, number of
  stability rounds Ω
Result: Average number of difference between two complete assignments
  provided by the DBMC algorithm

1 $counter \leftarrow 0$
2 **for** i *from 1 to* Ω **do**
3   Create a random complete assignment f of the sorting context S (algorithm 1).
4   Select randomly $A' \subset A$ (with a uniform distribution) such that $|A'| = \tau$.
5      $L \leftarrow < A', f >$
6      $f_1 \leftarrow DBMC(S, T)$
7      $f_2 \leftarrow DBMC(S, T)$
8   **for** $a \in A$ **do**
9   **if** $f_1(a) \neq f_2(a)$ **then**
10             $counter \leftarrow counter + 1$
11 **return** $\frac{counter}{\Omega}$

In order to illustrate the evolution of the stability with the number of trials, we applied the stability test on a model of 3 criteria, both of them expressed on a scale of 10 value levels with 50 stability rounds, with several fixed values for the number of categories (2, 5 and 7) and the size of the learning set (0 and 50). This test was made several times with different numbers of trials so that we can plot it with the stability and observe the correlation. The results of these tests are shown in Figs. 2 and 3.

nb Trials	10	100	200	500	1000
2 categories	12.9%	4.5%	3.3%	1.3%	0.5%
5 categories	37.5%	14%	9.2%	5.1%	4.2%
7 categories	49.9%	20.3%	13.5%	7.9%	5.5%

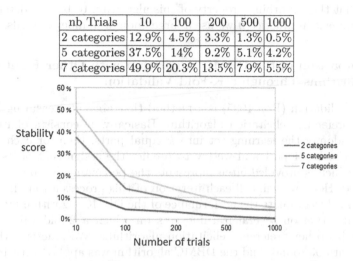

Fig. 2. Result of the stability test with 50 stability rounds in a context with 3 criteria $(10 \times 10 \times 10)$ and 0 elements in the learning set. The number of trials varies from 10 to 1000.

nb Trials	10	100	200	500	1000
2 categories	5%	1.6%	1.3%	0.7%	0.5%
5 categories	16%	5.2%	3.7%	2.3%	1.5%
7 categories	21.3%	7.2%	5.1%	3.3%	2.2%

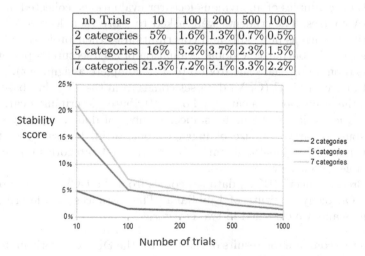

Fig. 3. Result of the stability test with 50 stability rounds in a context with 3 criteria $(10 \times 10 \times 10)$ and 50 elements in the learning set. The number of trials varies from 10 to 1000.

The reader may observe that the convergence is effective in practice when the number of trials increases. Indeed, the convergence starts being rather good with 100 trials. Therefore, in the following we will use 100 trials while applying the DBMC algorithm. Furthermore, we can see that the more categories there are the less stable the DBMC algorithm is. By the way we can also see that the bigger the training set is the more stable the DBMC algorithm is. To conclude, we think that the disturbing property of this algorithm to be non deterministic has a low impact in practice if we use a large enough number of trials.

3.2   Comparison of the DBMC Algorithm with Other Elicitation Algorithms Through a K-Fold Validation

The k-fold validation (Ron 1995) is a method that aims at measuring how efficient is a preference elicitation algorithm. Basically, it consists of iteratively dividing randomly the learning set into k equal parts, then for each part, we learn with the $k - 1$ part and we try to predict the assignment of its objects. The score which is provided often represents the percentage of objects that are misclassified. Here, we will call each iteration of this process a round.

We decided to evaluate the performance of the DBMC algorithm by applying a 2-fold validation on several learning sets on which a 2-fold validation was practised with other preference elicitation algorithms. We practised the 2-fold validation with 50 rounds and the DBMC algorithm was applied with 100 trials. Three learning sets were chosen[1]:

- The "Lecture evaluation" data set (LEV) comes from the Weka data base. It contains examples of anonymous lecturer evaluations, collected at the end of MBA courses. The students were asked to score their lecturers based on four criteria such as oral skills and contribution to their professional/general knowledge. The output is a global evaluation of each lecturer's performance, expressed on an ordinal scale from 0 to 4 (a complete assignment).
- The "Car Evaluation" (CEV) data set comes from the UCI database. It represents the evaluation of a car based on 6 attributes describing a car, namely, buying price, price of the maintenance, number of doors, capacity in terms of persons to carry, the size of luggage boot, estimated safety of the car. The output is the global assignment of the car in 4 categories unacceptable, acceptable, good, very good.
- The "Breast cancer" (BCC) data set comes from the UCI database provided by the Oncology Institute of Ljubljana. The instances are described by 7 attributes and are classified in two categories.

In order to compare the results obtained with the DBMC algorithm to results obtained with other methods we created similar learning sets with only two categories as it is made in Sobrie et al. (2015) i.e. for car evaluation binarized this evaluation into unacceptable versus not unacceptable (acceptable, good or very

---

[1] Available at https://github.com/oso/pymcda/tree/master/datasets.

good) and for lecturers evaluation we binarized the output value by distinguishing between good (score 3 to 4) and bad evaluation (score 0 to 2). The choice of these learning sets was due to the relatively low number of combination of criteria (less than 5 000) which make the DBMC algorithm run in an acceptable time (some few seconds). The reader can observe that the lecture evaluation database (LEV) contains more objects in the learning sets than the number of combinations of criteria (number of theoretically possible objects). When an object is found twice in the learning set with different assignments the relaxed approximation allow the DBMC algorithm to run as explained in Subsect. 3.2.

Table 1. Several properties of the datasets used for the evaluation and the comparison of the elicitation algorithms. "Size LS" represents the number of complete assignments provided in the learning set. "Nb Crit" is the number of criteria, "Nb Cat" is the number of categories, "Nb comb" is the number of combinations on the criteria (number of theoretically possible objects). The column "Monot Viol" gives an idea of how violated is monotonicity in the learning set. It represents the percentage of the pairs of objects a, b with aDb that violate the monotonicity.

Dataset	Size LS	Nb Crit	Nb Cat	Nb Comb	Monot Viol
Car evaluation (CEV)	1728	6	4	1728	<0.1%
Breast cancer (BBC)	278	7	2	4536	7.5%
Lectures evaluation (LEV)	1000	4	5	625	5.2%

The Other Preference Elicitation Algorithms
The result of a k-fold validation presented apart may be difficult to interpret. Indeed, we do not know what is a good performance for this test given that it may depend on various factors (the number of criteria, the size of the learning set, the number of categories, etc.). Thus, it is useful to compare the performances of the DBMC algorithm with those of other preference elicitation algorithms for multi-criteria sorting problem on the same learning sets. Hence, we practised the 2-fold validation with several other algorithms for ordinal preference elicitation: UTADIS (an method based on utility) (Devaud et al. 1980), logistic regression (Fallah Tehrani and Huellermeier 2013), choquistic regression (Fallah Tehrani and Huellermeier 2013) (methods based on statistical regression), Dominance Based Rough Set Approach (Greco et al. 2002) (DRSA) (a method based on rough sets), the heuristic algorithm for Majority Role Sorting (Sobrie et al. 2015) (also know as MR-Sort, an outranking which can be seen as a simplified version of ELECTRE TRI) and an heuristic for 2-additive Non compensatory Sorting (Sobrie et al. 2015) (also know as NCS, an outranking method). The results of the 2-fold validation of the algorithms logistic regression, choquistic regression were found in Fallah Tehrani and Huellermeier (2013) while the results provided for the heuristic algorithm for MR-Sort and an heuristic for 2-additive NCS were found in Sobrie et al. (2015). The 2-fold validation is provided by jMaf software

for DRSA. We programmed it for UTADIS based on the code of the UTADIS program that was provided Patrick Meyer, Sébastien Bigaret, Richard Hodgett and Alexandru-Liviu Olteanu on their MCDA package for the GNU R statistical software on github[2].

Approximations to Deal with Violations of Monotonicity
The DBMC algorithm requires a monotonic learning set while violations of monotonicity are quite common in ordinal learning sets and there are violations of monotonicity in the learning set that we used as shown in Table 1. This property may be seen as a weak point of the DBMC algorithm compared to other methods that allow the use of non-monotonic learning sets (such as DRSA or the MIP for MR-Sort for instance). However, given that the complete assignment obtained by all these methods is expected to be monotonic, it is impossible for a preference elicitation algorithm to obtain a complete assignment which is fully compatible with the learning set, as the DBMC algorithm does, and simultaneously to accept non-monotonic learning sets. In order to deal with learning set involving some violations of monotonicity, we proposed a solution named *relaxed approximation* of the sorting context. The *relaxed approximation* of a sorting context consists in creating a sorting context in which every pair of objects a and b such that a dominates b and b is sorted in a better category than a is relaxed. Here we mean by relaxed that the complete assignment of the objects a and b are intervals and that $\gamma_{max}(a) \leftarrow \gamma_{max}(b)$ and $\gamma_{min}(b) \leftarrow \gamma_{min}(a)$. For instance if an object a that was assigned to category 2 dominates an object b that was assigned to category 4 they will both be assigned to the interval $[2,4]$ i.e. $\gamma_{min}(a) = \gamma_{min}(b) = 2$ and $\gamma_{max}(a) = \gamma_{max}(b) = 4$. In case where an object would be found twice (or more times) in the learning set with different assignments, given that every object weakly dominates itself, then the *relaxed approximation* would assign it between the lowest and the highest category in which it is assigned in the learning set.

During the k-fold validation that we made with the Dominance Based Monte Carlo algorithm, the training data set was modified so that the algorithm could be ran on it. However the testing data set remained unchanged and thus may contain violation of monotonicity. Hence, while comparing the Dominance Based Monte Carlo algorithm to other preference elicitation algorithms, the learning set that was used was similar for all the algorithms.

Results of the k-fold Validation Tests
The results of the 2-fold validation, practised with several preference elicitation algorithms on the previously described data sets, is shown on Table 2. By 2-fold validation we mean that 50% of the learning set is used as a training dataset while the other 50% is used as a test dataset. In each cell the number at the left represent the average percentage of misclassification across the rounds while the number at the right of the cell represents the standard deviation of the percentage of misclassification. Here the category set was binarized as explained

---

[2] Find this package at: https://github.com/paterijk/MCDA.

above. Looking at the results on Table 2, we can observe that all the preference elicitation algorithms perform better on the car evaluation dataset (CEV) than on the lecture evaluation data set (LEV) and have their worst performance on the breast cancer dataset (BCC). Several explanations can be proposed. At first we can observe on the Table 1 that this is consistent with the proportion of violations of monotonicity. Indeed, all these methods providing a monotonic output, they are unable to restore any non monotonic assignment in the learning set. We can also observe that the breast cancer has the smallest learning set with the highest number of combinations of criteria which makes the elicitation algorithms learn with less examples on a model where more possible complete assignments are possible.

Table 2. Results of the 2-fold validation tests. Here the categories were binarized as described in this document. Percentage of misclassification with its standard deviation.

	DRSA	NCS	MR-Sort	UTADIS	DBMC
CEV	$4.91 \pm 0.41\%$	$12.6 \pm 2.63\%$	$13.9 \pm 1.19\%$	$6.9 \pm 0.71\%$	$3.72 \pm 0.28\%$
LEV	$18.76 \pm 0.35\%$	$14.92 \pm 1.88\%$	$15.92 \pm 1.22\%$	$15.01 \pm 1.31\%$	$18.67 \pm 1.12\%$
BCC	$25.95 \pm 1.33\%$	$26.72 \pm 3.45\%$	$27.5 \pm 3.79\%$	$28.70 \pm 1.11\%$	$25.92 \pm 0.63\%$

We can observe that the performances of the DBMC algorithm is relatively good on the car evaluation database (CEV). This algorithm being based on monotonicity it may appear relevant to think that it performs well with model with little violations of monotonicity. The results on the breast cancer database (BCC) are really similar with all the algorithms, DRSA and the DBMC being a bit better than the other three algorithms. The results of the DRSA algorithm and of the DBMC algorithm on the lecture evaluation database (LEV) are less good than the results with the three others.

We also wanted to assess the performance of the DBMC algorithm of data set with more than 2 categories. Thus we ran the k-fold validation test with the car evaluation dataset (CEV) and the lecture evaluation dataset (LEV) without binarizing the category set. The result presented in Table 3 does not include the breast cancer dataset given that it is already binarized in its initial state.

Table 3. Results of the 2-fold validation tests (percentage of misclassification with its standard deviation). Here the categories were not binarized.

	DRSA	UTADIS	DBMC
CEV (5 cat)	$22.11 \pm 0.54\%$	$9.88 \pm 0.43\%$	$6.61 \pm 0.41\%$
LEV (7 cat)	$54.21 \pm 0.78\%$	$41.23 \pm 1.97\%$	$60.07 \pm 2.37\%$

We can see in Table 3 that the binarization of the categories that was made in Table 2 had a real impact on the performances of the elicitation algorithms.

Indeed all the results of k-fold validation are clearly higher than those observed on Table 2. In particular, the rate of misclassification with DRSA on the car evaluation increases dramatically. Given that the categories are being binarized with the category 1 on the one hand and the categories 2, 3 and 4 on the other hand an explanation may be that many misclassifications happen between these three last categories and thus were not counted in Table 2.

We were also interested in knowing, when objects are assigned to the wrong category, at "How far are we from the good category?". In order to answer this question we presented on Table 4 the result of the 2-fold validation with the $L1$ measure for the Dominance Based Monte Carlo algorithm and compared with the results of the ordinal logistic regression (OLR) and ordinal choquistic regression (OCR), UTADIS and DRSA. The $L1$ measure in the k-fold validation is a measure of misclassification in which each misclassification is weighted by the distance between the assignment in the learning set and the assignment found by the preference elicitation algorithm. The reader may observe that when there are only two categories, the $L1$ 2-fold validation returns the same result as the normal 2-fold validation (also called 0–1 measure for the 2-fold validation).

Table 4. Results of the 2-fold validation tests with $L1$ measure. Here the categories were not binarized. Average L1-loss measure with its standard deviation.

	OLR	OCR	DRSA	UTADIS	DBMC
CEV (5 cat)	23.10 ± 0.75	10.97 ± 3.6	26.32 ± 0.92	10.32	7.2 ± 1.19
LEV (7 cat)	42.64 ± 1.48	41.84 ± 1.87	74.69 ± 3.36	45.2	72.27 ± 4.01

While looking at Table 4 we see that, once again, the performance of the DBMC algorithm and the DRSA algorithm on the lecture evaluation database (LEV) are very close. By the way we also observe that on the lecture evaluation database their L1 measure are dramatically higher that the 0–1 measure observed in Table 3 which means that the objects that are assigned by these elicitation algorithms in the wrong category are often assigned to a category which is not adjacent to the category in which they were assigned in the learning set. By comparison we can see that le L1 measure of the UTADIS method is only 4 point higher than the 0–1 measure (percentage of misclassifications) which means that the objects that assigned to the wrong category with UTADIS are mainly assigned in an adjacent category. On this database, the DBMC algorithm and the DRSA algorithm show bad performances compared to the other three elicitation algorithm. Concerning the car evaluation database (CEV) we can remark that the L1 measures of UTADIS and the DBMC algorithm have performance that are relatively close to the one of choquistic regression, the DBMC algorithm being a little better. On this database with the DBMC, UTADIS and DRSA the L1 measure is relatively close to the 0–1 measure (percentage of misclassifications) given in Table 3 which means that the object are mainly assigned to the good category or to an adjacent one.

4   Conclusion

We presented a preference elicitation algorithm for the sorting problem based on a Monte Carlo principle and we showed some practical performance results in term of effective convergence and ability to restore a learning set by looking at one part of it.

On the convergence, we saw that the convergence proved theoretically in Denat and Öztürk (2016) is observed in practice, the result being quite stable from 100 trials onward. We also saw that the bigger the learning set is the more stable the DBMC algorithm is.

Looking at the results of the k-fold validation, several conclusion can be made. At first we can say that the DBMC algorithm is quite efficient on the car evaluation (CEV) while it is less efficient on the lecture evaluation (LEV) data set. On the breast cancer (BCC) data sets all the algorithms that are studied here are rather not efficient and the results of the DBMC algorithm on this database is relatively similar to those of the other algorithms although it seems to be a little better. The reader may observe that the lecture evaluation contains more assignments in the learning set than the number of combinations on the criteria. While applying a 2-fold validation with the DBMC algorithm, the learning set is cut in two and the part used as a training data set (with which the DBMC will learn) contains almost the same number of assignments than the number of combinations. Therefore, the DBMC algorithm has a very little limited room to manoeuvre. One possibility to explain the relatively bad performance of the DBMC algorithm on the Lecture evaluation data set (LEV) may be that this algorithm does not perform well when the number of assignment in the learning set is too high compared to the number of combinations possibly due to its incapacity to reassign object to different categories when the learning set is violated.

Then we can observe that the performance of the DBMC algorithm are very close to the performances of DRSA (excepted on the car evaluation dataset (CEV) with 5 categories). This similarity can be due to the fact that these methods are both model free and based on monotonicity.

As a perspective we are now developing other tests to evaluate the performance of the Dominance Based Monte Carlo algorithm such as a modified version of the idiosyncrasy test described in Sobrie et al. (2013). Our test, which is quite similar to the k-fold validation, consists in iteratively assigning every object of a model with an elicitation method (UTADIS, MR-SORT...) that we call the "assigning method" and random parameters for this method. Then the DBMC would learn with a given proportion of the objects (the training data set) and then would predict the assignments of the other objects. The idea would be to test the performance of the DBMC on this test and compare these results the those that would be obtained with the same test using the assigning method.

We also look at different ways to deal with violations of monotonicity such as canceling all the pairs that are concerned by them.

References

Carbone, E., Hey, J.D.: Which error story is best? J. Risk Uncertain. **20**(2), 161–176 (2000)

Denat, T., Öztürk, M.: Dominance based montecarlo algorithm for preference learning in the multi-criteria sorting problem Theoretical properties. In: DA2PL (2016)

Devaud, J., Groussaud, G., Jacquet-Lagreze, E.: (utadis): Une methode de construction de fonctions d'utilite additives rendant compte de jugements globaux. In: European Working Group on MCDA, Bochum, Germany (1980)

Fallah Tehrani, A., Huellermeier, E.: Ordinal choquistic regression. In: EUSFLAT Conference (2013)

Greco, S., Matarazzo, B., Slowinski, R.: Rough sets theory for multicriteria decision analysis. Eur. J. Oper. Res. **129**(1), 1–47 (2001)

Greco, S., Matarazzo, B., Slowinski, R.: Multicriteria classification by dominance-based rough set approach. In: Kloesgen, W., Zytkow, J. (eds.) Handbook of Data Mining and Knowledge Discovery. Oxford University Press, New York (2002)

Jacquet-Lagreze, E., Siskos, J.: Assessing a set of additive utility functions for multi-criteria decision-making, the uta method. Eur. J. Oper. Res. **10**(2), 151–164 (1982)

Keeney, R., Raiffa, H.: Decisions with multiple objectives preferences and value trade-offs. Behav. Sci. **39**(2), 169–170 (1994)

Lahdelma, R., Hokkanen, J., Salminen, P.: SMAA - stochastic multiobjective acceptability analysis. Eur. J. Oper. Res. **106**(1), 137–143 (1998)

Luce, R.D.: Four tensions concerning mathematical modeling in psychology. Annu. Rev. Psychol. **46**(1), 1–27 (1995)

Regenwetter, M., Dana, J., Davis-Stober, C.P.: Transitivity of preferences. Psychol. Rev. **118**(1), 42 (2011)

Ron, K.: A study of cross validation and bootstrap for accuracy estimation and model selection. In: IJCAI (1995)

Roy, B., Bouyssou, D.: Aide Multicritère à la Décision : Méthodes et Cas. Economica, Paris (1993)

Sobrie, O., Mousseau, V., Pirlot, M.: Learning a majority rule model from large sets of assignment examples. In: Perny, P., Pirlot, M., Tsoukiàs, A. (eds.) ADT 2013. LNCS, vol. 8176, pp. 336–350. Springer, Heidelberg (2013). doi:10.1007/978-3-642-41575-3_26

Sobrie, O., Mousseau, V., Pirlot, M.: Learning the parameters of a non compensatory sorting model. In: Walsh, T. (ed.) ADT 2015. LNCS (LNAI), vol. 9346, pp. 153–170. Springer, Cham (2015). doi:10.1007/978-3-319-23114-3_10

Tervonen, T., Figueira, J.R., Lahdelma, R., Dias, J.A., Salminen, P.: A stochastic method for robustness analysis in sorting problems. Eur. J. Oper. Res. **192**(1), 236–242 (2009)

A Heuristic Approach to Test the Compatibility of a Preference Information with a Choquet Integral Model

Lucie Galand and Brice Mayag[(✉)]

University Paris-Dauphine, PSL Research University LAMSADE, CNRS, UMR 7243, Place du Maréchal de Lattre de Tassigny, 75775 Paris Cedex 16, France

Abstract. This work deals with the problem of the existence of a Multicriteria Decision Aiding model, based on the Choquet integral, that represents the preferences of a decision maker. Given some preferences on a set of actions, our aim is to determine if those preferences are compatible with a Choquet integral model, where the utility function associated to each criterion and the capacity on the subsets of criteria are to be defined. Computing simultaneously the utility functions and the capacity leads to solving a mixed integer program with some quadratic constraints, which can not be performed efficiently. We propose here to solve this problem by using a linear approximation of the quadratic terms given by the Taylor's formula, and then apply a standard mixed integer programming solver. We illustrate and analyze our approach with some numerical experiments.

Keywords: MCDA · Preference modeling · Choquet integral · interaction

1 Introduction

MultiCriteria Decision Aiding (MCDA) aims at representing the preferences of a Decision-Maker (DM) on a set of alternatives (or actions, options) X evaluated over a finite set of criteria $N = \{1, \ldots, n\}$ $(n > 1)$ often conflicting. An alternative can be identified as an element $x = (x_1, \ldots, x_n)$ of the Cartesian product $X = X_1 \times \cdots \times X_n$, where $X_1, \ldots, X_n$ represent a set of points of view or attributes. The Multi-Attribute Utility Theory (MAUT) [6] is one of the decision models usually used to represent the preferences of the DM. In practice, MAUT elaborates a preference relation $\succsim_X$ over X by asking to the DM some pairwise comparisons of alternatives on a finite subset of X, called a reference set or a set of learning data. The question is then: is this preference information $\succsim_X$ compatible with an appropriate aggregation function modeling the DM's preferences?

The arithmetic mean is the most classical and widely used aggregation function modeling the preferences of a DM. It has the major drawback of assuming

© Springer International Publishing AG 2017
J. Rothe (Ed.): ADT 2017, LNAI 10576, pp. 65–80, 2017.
DOI: 10.1007/978-3-319-67504-6_5

independence among criteria [10]. Among multiple models proposed in the literature to overcome this limitation, the Choquet integral is one of the most expressive model. It is able to model the interaction phenomena ranging from veto, favor to complementarity or redundancy between criteria [10–12]. The representation of a preference information by a Choquet integral needs the identification of two types of parameters: the capacity (also called a fuzzy measure) and the utility functions associated to the set of attributes. Since the number of capacity's parameters grows fast according to the number of criteria, the 2-additive Choquet integral model has been introduced in order to take into account only interaction between two criteria, the interaction between more than two of them are ignored. This model appears as a good compromise between complexity and usability [11].

The usual approach to elicit a MCDA model based on the Choquet integral consists in two separate steps: first, construct the utility function for each attribute, and then construct the capacity once the utility functions are known [9,12,13]. Most of the works in the literature have focused on the second phase, transforming it into an optimization problem, once the DM has provided a preference information. As proved in [4] it is more convenient to elicit simultaneously both capacity and utility function. To our knowledge, there exist, in the case of a 2-additive Choquet integral, two heuristic approaches trying to tackle this problem. The authors of the first one [1] proposed a method based on stochastic approaches (Monte Carlo or genetic algorithm). In the second approach [7], the authors propose an approximate algorithm based on a fixed-point approach consisting in transforming the original optimization problem into two iterative linear problems.

We propose in this paper a new heuristic approach based on a linear approximation of the quadratic terms (product of the utilities functions with Shapley values and interaction indexes) by the Taylor's formula [2]. It consists to test the compatibility of DM's preferences with a 2-additive Choquet integral model by computing an approximate value of its parameters from a neighborhood of a given reference point. The aim of this algorithm is not to find the most appropriate values of utility functions and capacity. It only tries to check whether there exist values of the parameters which fulfill the preferences expressed as a partial weak order on X. A negative result returned by this algorithm does not mean that the preferences are not representable by a 2-additive Choquet integral. That is why it is a heuristic. Compared to the other approaches proposed in [1,7], we test our heuristic on some randomly generated hard instances.

The next section presents basic concepts we need in the representation of a preference information by a 2-additive Choquet integral, while Sects. 3 and 4 are respectively dedicated to the presentation of our approach and the experiments we made. We end the paper by a conclusion.

2 Settings

2.1 The 2-additive Choquet Integral

The 2-additive Choquet integral is a particular case of the well known Choquet integral [11,13]. Its main property is to model interactions between two criteria. These interactions are simple and more meaningful than those produced by using the Choquet integral. This aggregation function is based on the notion of *capacity* μ defined as a set function from the powerset of criteria 2^N to $[0, 1]$ such that:

- $\mu(\emptyset) = 0$
- $\mu(N) = 1$
- $\forall A, B \in 2^N$, $[A \subseteq B \Rightarrow \mu(A) \leq \mu(B)]$ (monotonicity).

We adopt the notations $\mu_i := \mu(\{i\})$, $\mu_{ij} := \mu(\{i, j\})$ for all $i, j \in N$, $i \neq j$. Whenever we use i and j together, it always means that they are different.

A capacity μ on N is said to be *2-additive* if its *Möbius transform* $m : 2^N \to \mathbb{R}$ defined by

$$m^\mu(T) := \sum_{K \subseteq T} (-1)^{|T \backslash K|} \mu(K), \forall T \in 2^N.$$

satisfies these two conditions:

1. For all subset T of N such that $|T| > 2$, $m(T) = 0$;
2. There exists a subset B of N such that $|B| = 2$ and $m^\mu(B) \neq 0$.

A 2-additive capacity is fully characterized with $n + \dfrac{n(n+1)}{2}$ Möbius transform coefficients (instead of $2^n - 2$). Given an alternative $x := (x_1, ..., x_n) \in X$, the expression of the 2-additive Choquet integral is the following [11]:

$$C_\mu(u(x)) = \sum_{i=1}^{n} \phi_i^\mu u_i(x_i) - \frac{1}{2} \sum_{\{i,j\} \subseteq N} I_{ij}^\mu |u_i(x_i) - u_j(x_j)| \tag{1}$$

where

- $u_i : X_i \to \mathbb{R}_+$ is an utility function, nondecreasing in its arguments, associated to the attribute X_i, for all $i \in N$;
- $u(x) = (u_1(x_1), \ldots, u_n(x_n))$ for $x = (x_1, ..., x_n) \in X$;
- $I_{ij}^\mu = \mu_{ij} - \mu_i - \mu_j$ is the interaction index, w.r.t. capacity μ, between the two criteria i and j [8,14].
- $\forall i \in N$, the importance of criterion i corresponding to the Shapley value of μ [15] is given by the formula:

$$\phi_i^\mu = \sum_{K \subseteq N \backslash i} \frac{(n - |K| - 1)!|K|!}{n!} (\mu(K \cup i) - \mu(K))$$

$$= \mu_i + \frac{1}{2} \sum_{\{i,j\} \subseteq N, i \neq j} (\mu_{ij} - \mu_i - \mu_j) = \mu_i + \frac{1}{2} \sum_{j \in N \backslash \{i\}} I_{ij}^\mu.$$

In this expression, the 2-additive Choquet integral appears as a good compromise between the arithmetic mean and the general expression of the Choquet integral.

2.2   Preference Information

We assume that the DM can express his preferences by giving a partial order $\succsim_X$ on X allowing to compute two preference relations $\succ$ and $\sim$ defined by:

$$\succ = \{(x, y) \in X \times X : \text{ DM strictly prefers } x \text{ to } y\} \qquad (2)$$

$$\sim = \{(x, y) \in X \times X : \text{ DM is indifferent between } x \text{ and } y\} \qquad (3)$$

We denote by $X_R = \{x, y \in X : x \succsim_X y\}$ the reference subset of X.

Definition 1. *A preference information $\{\succ, \sim\}$ on X is said to be representable by a 2-additive Choquet integral (C_2-compatible for short) if there exists a 2-additive capacity μ and utilities functions $u_i : X_i \to \mathbb{R}_+$ such that: $\forall x = (x_1, \ldots, x_n), y = (y_1, \ldots, y_n) \in X$,*

$$x \succ y \Rightarrow C_\mu(u_1(x_1), \ldots, u_n(x_n)) > C_\mu(u_1(y_1), \ldots, u_n(y_n)) \qquad (4)$$

$$x \sim y \Rightarrow C_\mu(u_1(x_1), \ldots, u_n(x_n)) = C_\mu(u_1(y_1), \ldots, u_n(y_n)). \qquad (5)$$

2.3   Problem Formulation

Given a finite set of criteria N, a finite set of m n-dimensional alternatives $X = \{x^1, x^2, \ldots, x^m\}$, where x_i^k is the value of alternative x^k for criterion i, and a reference subset X_R of X, we want to test if a preference information $\{\succ, \sim\}$ is representable by a 2-additive Choquet integral model. We denote by $\mathcal{P}(X_R)$ this problem. Solving $\mathcal{P}(X_R)$ amounts to determining if the following domain of variables (defined by a system of equalities/inequalities) is feasible:

$$\sum_{i=1}^n \phi_i^\mu u_i(x_i) - \frac{1}{2} \sum_{\{i,j\} \subset N} I_{ij}^\mu |u_i(x_i) - u_j(x_j)| > \sum_{i=1}^n \phi_i^\mu u_i(y_i) - \frac{1}{2} \sum_{\{i,j\} \subset N} I_{ij}^\mu |u_i(y_i) - u_j(y_j)|$$
$$\forall x, y \in X, \text{ s.t. } x \succ y \in X_R \qquad (6)$$

$$\sum_{i=1}^n \phi_i^\mu u_i(x_i) - \frac{1}{2} \sum_{\{i,j\} \subset N} I_{ij}^\mu |u_i(x_i) - u_j(x_j)| = \sum_{i=1}^n \phi_i^\mu u_i(y_i) - \frac{1}{2} \sum_{\{i,j\} \subset N} I_{ij}^\mu |u_i(y_i) - u_j(y_j)|$$
$$\forall x, y \in X, \text{ s.t. } x \sim y \in X_R \qquad (7)$$

$$u_i(x_i) \geq u_i(x_i') \qquad \text{if } x_i \text{ is at least as good as } x_i', \quad \forall i \in N, \quad \forall x_i, x_i' \in X_i \qquad (8)$$

$$\phi_i^\mu = \mu_i + \frac{1}{2} \sum_{\{i,j\} \subseteq N} I_{ij} \qquad \forall i \in N \qquad (9)$$

$$\sum_{i=1}^n \phi_i^\mu = 1 \qquad (10)$$

$$I_{ij}^\mu = \mu_{ij} - \mu_i - \mu_j \qquad \forall i, j \in N \qquad (11)$$

$$\sum_{i \in A \setminus \{k\}} (\mu_{ik} - \mu_i) \geq (|A| - 2)\mu_k \qquad \forall A \subseteq N, |A| \geq 2, \forall k \in A \qquad (12)$$

$$1 \geq \mu_i \geq 0, \qquad \phi_i^\mu \geq 0 \qquad \forall i \in N \qquad (13)$$

$$1 \geq \mu_{ij} \geq 0, \qquad 1 \geq I_{ij}^\mu \geq 0 \qquad \forall i, j \in N \qquad (14)$$

$$u_i(x_i) \geq 0 \qquad \forall i \in N, \forall x \in X \qquad (15)$$

Monotonicity constraints of the 2-additive capacity are given by Eq. (12). The aim of this program is not to find the most appropriate values of utility functions and capacity. We only check here if this system is feasible.

Furthermore this program is nonlinear since utilities and capacities are unknown. Our objective is to propose in the sequel, see Sect. 3, a heuristic approach to tackle this kind of problem. We illustrate through the following classic example the nonlinear program given above.

Example 1 *(A classic example* [10]*). Four students of a faculty are evaluated on three subjects Mathematics (M), Statistics (S) and Language skills (L). All marks are taken from the same scale from 0 to 20. The evaluations of these students are given by the following table:*

	1 : Mathematics (M)	2 : Statistics (S)	3 : Language (L)
a	16	13	7
b	16	11	9
c	6	13	7
d	6	11	9

To select the best students, the dean of the faculty expresses his preferences. For a student good in Mathematics, Language is more important than Statistics: $a \prec b$. For a student bad in Mathematics, Statistics is more important than Language: $d \prec c$. These two preferences lead to a contradiction with the arithmetic mean model. Indeed the following system

$$\begin{cases} a \prec b \Rightarrow u_M(16) \, w_M + u_S(13) \, w_S + u_L(7) \, w_L < u_M(16) \, w_M + u_S(11) \, w_S + u_L(9) \, w_L \\ d \prec c \Rightarrow u_M(6) \, w_M + u_S(11) \, w_S + u_L(9) \, w_L < u_M(6) \, w_M + u_S(13) \, w_S + u_L(7) \, w_L. \end{cases}$$

implies this contradiction: $u_S(13) \, w_S + u_L(7) \, w_L < u_S(11) \, w_S + u_L(9) \, w_L$ and $u_S(11) \, w_S + u_L(9) \, w_L < u_S(13) \, w_S + u_L(7) \, w_L$. To test if this preference informa-tion, $a \prec b$ and $d \prec c$, is C_2-compatible, we compute the following constraints:

$$C_\mu(u(a)) < C_\mu(u(b)), \ C_\mu(u(d)) < C_\mu(u(c))$$

$$C_\mu(u(a)) = \phi_1^\mu u_1(16) + \phi_2^\mu u_2(13) + \phi_3^\mu u_3(7) - \frac{1}{2} I_{12}^\mu |u_1(16) - u_2(13)|$$
$$- \frac{1}{2} I_{13}^\mu |u_1(16) - u_3(7)| - \frac{1}{2} I_{23}^\mu |u_2(13) - u_3(7)|$$

$$C_\mu(u(b)) = \phi_1^\mu u_1(16) + \phi_2^\mu u_2(11) + \phi_3^\mu u_3(9) - \frac{1}{2} I_{12}^\mu |u_1(16) - u_2(11)|$$
$$- \frac{1}{2} I_{13}^\mu |u_1(16) - u_3(9)| - \frac{1}{2} I_{23}^\mu |u_2(11) - u_3(9)|$$

$$C_\mu(u(c)) = \phi_1^\mu u_1(6) + \phi_2^\mu u_2(13) + \phi_3^\mu u_3(7) - \frac{1}{2} I_{12}^\mu |u_1(6) - u_2(13)|$$
$$- \frac{1}{2} I_{13}^\mu |u_1(6) - u_3(7)| - \frac{1}{2} I_{23}^\mu |u_2(13) - u_3(7)|$$

$$C_\mu(u(d)) = \phi_1^\mu u_1(6) + \phi_2^\mu u_2(11) + \phi_3^\mu u_3(9) - \frac{1}{2}I_{12}^\mu |u_1(6) - u_2(11)|$$
$$- \frac{1}{2}I_{13}^\mu |u_1(6) - u_3(9)| - \frac{1}{2}I_{23}^\mu |u_2(11) - u_3(9)|$$

$$0 \le u_1(6) \le u_1(16) \le 1, \ 0 \le u_2(11) \le u_2(13) \le 1, \ 0 \le u_3(7) \le u_3(9) \le 1$$

$$\phi_1^\mu = \mu_1 + \frac{1}{2}I_{12} + \frac{1}{2}I_{13}, \ \phi_2^\mu = \mu_2 + \frac{1}{2}I_{12} + \frac{1}{2}I_{23}, \ \phi_3^\mu = \mu_3 + \frac{1}{2}I_{13} + \frac{1}{2}I_{23}$$

$$I_{12}^\mu = \mu_{12} - \mu_1 - \mu_2, \ I_{13}^\mu = \mu_{13} - \mu_1 - \mu_3, \ I_{23}^\mu = \mu_{23} - \mu_2 - \mu_3$$

$$\mu_{12} + \mu_{13} \ge \mu_1 + \mu_2 + \mu_3, \ \mu_{12} + \mu_{23} \ge \mu_1 + \mu_2 + \mu_3, \ \mu_{13} + \mu_{23} \ge \mu_1 + \mu_2 + \mu_3$$

$$\mu_1 \ge 0, \mu_2 \ge 0, \mu_3 \ge 0, \mu_{12} \ge 0, \mu_{13} \ge 0, \mu_{23} \ge 0$$

3   Our Proposition

In the sequel, we assume that $\forall i \in N$, X_i is a numerical attribute such that $\forall x_i \in X_i$, $0 \le x_i \le 1$ and $0 \le u_i(x_i) \le 1$. We also suppose $I_{ij} \ge 0$ for all $i, j \in N$. This latter assumption could be justified since it is proved in [13] that, a preferential information without indifference on binary alternatives[1] is always representable by nonnegative interaction index. In order to efficiently solve problem $\mathcal{P}(X_R)$, we linearize its mathematical program, namely the absolute values, and the quadratic terms.

3.1   Linearization of the Absolute Values

The absolute value terms in Constraints (6) and (7) can be linearized by introducing integer variables. Let $z = |x - y|$ be the absolute value of the difference of two nonnegative numbers $x, y \in [0, 1]$. The real z can be computed by the following mixed integer linear constraints:

$$
\begin{aligned}
&z \ge x - y \\
&z \ge y - x \\
&z \le x - y + 1 - \alpha \\
&z \le y - x + 1 - \beta \\
&\alpha + \beta = 1 \\
&z, x, y \in [0, 1] \\
&\alpha, \beta \in \{0, 1\}
\end{aligned}
\tag{16}
$$

---

[1] A binary action is a fictitious alternative which takes either the neutral value **0** for all criteria, or the neutral value **0** for all criteria except for one or two criteria for which it takes the satisfactory value **1**. The binary actions are used in many applications through the MACBETH methodology [3,5].

Therefore the computation of the absolute value function $u_{ij}^{x_i x_j} = |u_i(x_i) - u_j(x_j)|$ in (1) is linearized like this:

$$
\begin{aligned}
u_{ij}^{x_i x_j} &\geq u_i(x_i) - u_j(x_j) \\
u_{ij}^{x_i x_j} &\geq u_j(x_j) - u_i(x_i) \\
u_{ij}^{x_i x_j} &\leq u_i(x_i) - u_j(x_j) + 1 - \alpha_{ij} \\
u_{ij}^{x_i x_j} &\leq u_j(x_j) - u_i(x_i) + 1 - \beta_{ij} \\
\alpha_{ij} + \beta_{ij} &= 1 \\
u_{ij}^{x_i x_j}, u_i(x_i), u_j(x_j) &\in [0, 1] \\
\alpha_{ij}, \beta_{ij} &\in \{0, 1\}
\end{aligned}
\tag{17}
$$

3.2   Linearization of the Quadratic Terms

The heuristic we propose in this paper aims at computing an approximate value of the 2-additive Choquet integral function, given by Eq. (1), when utility functions and capacity are unknown. It only tries to check whether there exist values of the parameters which fulfill the learning examples. This computation is based on a linear approximation of the product of two variables, by using the Taylor's formula [2]. We propose indeed to approximate the terms $\phi_i^\mu u_i(x_i)$ and $I_{ij}^\mu |u_i(x_i) - u_j(x_j)|$ in (1) by the Taylor's formula of a function of two variables. The general expression of this formula, for a given function f of two variables x and y defined in a neighborhood of a vector (a, b), is the following:

$$
L(x, y) = f(a, b) + f_x(a, b)(x - a) + f_y(a, b)(y - b)
\tag{18}
$$

where f_x and f_y represent the partial derivatives of f.

$L(x, y)$ is the linear function called the linearization of f at (a, b) and the approximation

$$
f(x, y) \approx f(a, b) + f_x(a, b)(x - a) + f_y(a, b)(y - b)
\tag{19}
$$

is called the linear approximation of f at (a, b).

The function f is said to be differentiable if

$$
\lim_{x \to a, y \to b} \frac{|f(x, y) - L(x, y)|}{\sqrt{(x - a)^2 + (y - b)^2}} = 0
\tag{20}
$$

Theorem 1 (Taylor [2]). *If the partial derivatives f_x and f_y exist in a neighborhood of (a, b) and are continuous at (a, b), then f is differentiable at (a, b).*

Example 2. *Is it not difficult to see that the function $f(x, y) = xy$ is differentiable at any point (a, b) such that $a, b \in [0, 1]$ since $f_x(x, y) = y$ and $f_y(x, y) = x$. For instance, it is differentiable at $(\frac{1}{3}, \frac{1}{5})$ and the linearization is*

$$
L(x, y) = f(\frac{1}{3}, \frac{1}{5}) + f_x(\frac{1}{3}, \frac{1}{5})(x - \frac{1}{3}) + f_y(\frac{1}{3}, \frac{1}{5})(y - \frac{1}{5})
$$

i.e.

$$L(x,y) = \frac{1}{15} + \frac{1}{5}(x - \frac{1}{3}) + \frac{1}{3}(y - \frac{1}{5}) = \frac{1}{5}x + \frac{1}{3}y - \frac{1}{15}.$$

We can use this function to approximate the points $(\frac{1}{3} + \frac{1}{20}, \frac{1}{5} - \frac{1}{20})$ *and*
$(\frac{1}{3} + \frac{1}{10}, \frac{1}{5} + \frac{1}{10})$ *in the neighborhood of* $(\frac{1}{3}, \frac{1}{5})$:

- $f(\frac{1}{3} + \frac{1}{20}, \frac{1}{5} - \frac{1}{20})$: *It follows that* $f(\frac{1}{3} + \frac{1}{20}, \frac{1}{5} - \frac{1}{20}) \approx \frac{1}{5}(\frac{1}{3} + \frac{1}{20}) + \frac{1}{3}(\frac{1}{5} -$
  $\frac{1}{20}) - \frac{1}{15} \simeq 0.06$ *and its real value is* 0.0575.

- $f(\frac{1}{3} + \frac{1}{10}, \frac{1}{5} + \frac{1}{10})$: *It follows that* $f(\frac{1}{3} + \frac{1}{10}, \frac{1}{5} + \frac{1}{10}) \approx \frac{1}{5}(\frac{1}{3} + \frac{1}{10}) + \frac{1}{3}(\frac{1}{5} +$
  $\frac{1}{10}) - \frac{1}{15} \simeq 0.10667$ *and its real value is* 0.13.

To adapt this result to our context, we consider these linear approximations:

1. The product $z_i^{x_i} = \phi_i^\mu u_i(x_i)$ is differentiable at a point $(\bar{\phi}_i, \bar{u}_i)$ and approximated by

$$z_i^{x_i} = (\bar{\phi}_i \times \bar{u}_i) + \bar{u}_i(\phi_i^\mu - \bar{\phi}_i) + \bar{\phi}_i(u_i(x_i) - \bar{u}_i) = \bar{u}_i \times \phi_i^\mu + \bar{\phi}_i \times u_i(x_i) - \bar{\phi}_i \bar{u}_i \quad (21)$$

2. The product $z_{ij}^{x_i x_j} = I_{ij}^\mu u_{ij}^{x_i x_j}$, where $u_{ij}^{x_i x_j}$ is the variable used in (17), is differentiable at $(\bar{I}_{ij}, |\bar{u}_i - \bar{u}_j|)$ and approximated by

$$z_{ij}^{x_i x_j} = (\bar{I}_{ij}|\bar{u}_i - \bar{u}_j|) + |\bar{u}_i - \bar{u}_j|(I_{ij}^\mu - \bar{I}_{ij}) + \bar{I}_{ij}(u_{ij}^{x_i x_j} - |\bar{u}_i - \bar{u}_j|)$$
$$= I_{ij}^\mu |\bar{u}_i - \bar{u}_j| + \bar{I}_{ij} \times u_{ij}^{x_i x_j} - \bar{I}_{ij}|\bar{u}_i - \bar{u}_j| \quad (22)$$

Example 3. *In Example 1, the evaluation of the student a is given by:*
$C_\mu(u(a)) = \phi_1^\mu u_1(16) + \phi_2^\mu u_2(13) + \phi_3^\mu u_3(7) - \frac{1}{2}I_{12}^\mu |u_1(16) - u_2(13)| - \frac{1}{2}I_{13}^\mu |u_1(16) -$
$u_3(7)| - \frac{1}{2}I_{23}^\mu |u_2(13) - u_3(7)|$. *Using our linear approximations and the points*
$(\bar{\phi}, \bar{u}) = (\frac{1}{n}, x_i)$ *and* $(\bar{I}, |\bar{u}_i - \bar{u}_j|) = (\frac{1}{10}, |x_i - x_j|)$, *we replace the product*
$\phi_1^\mu u_1(16)$ *with*

$$z_1^{\frac{16}{20}} = \frac{16}{20} \times \phi_1^\mu + \frac{1}{3} \times u_1(\frac{16}{20}) - \frac{16}{60}$$

and replace the product $I_{12}^\mu |u_1(16) - u_2(13)|$ *with*

$$z_{12}^{\frac{16}{20}\frac{13}{20}} = \left|\frac{16}{20} - \frac{13}{20}\right| \times I_{12}^\mu + \frac{1}{10} \times u_{12}^{\frac{16}{20}\frac{13}{20}} - \frac{\left|\frac{16}{20} - \frac{13}{20}\right|}{10} = \frac{3}{20} \times I_{12} + \frac{1}{10} \times u_{12}^{\frac{16}{20}\frac{13}{20}} - \frac{3}{200}.$$

Hence the linear approximation of $C_\mu(u(a))$ *is given by*

$$\tilde{C}_\mu(u(a)) \approx z_1^{\frac{16}{20}} + z_2^{\frac{13}{20}} + z_3^{\frac{7}{20}} - \frac{1}{2}z_{12}^{\frac{16}{20}\frac{13}{20}} - \frac{1}{2}z_{13}^{\frac{16}{20}\frac{7}{20}} - \frac{1}{2}z_{23}^{\frac{13}{20}\frac{7}{20}}$$

The points $(\bar{\phi}_i, \bar{u}_i)$ in (21) and $(\bar{I}_{ij}, |\bar{u}_i - \bar{u}_j|)$ in (22) are chosen arbitrarily. The idea of these approximations is to try to find a utility value $u_i(x_i)$ not far from the numerical value $\bar{u}_i$, the Shapley value ϕ_i^μ in the neighborhood of $\bar{\phi}_i$ and the interaction index I_{ij}^μ in the neighborhood of $\bar{I}_{ij}$. To translate these notions of neighborhood, we add the following constraints:

$$\bar{u}_i - \varepsilon^{x_i} \leq u_i(x_i) \leq \bar{u}_i + \varepsilon^{x_i}$$

$$\bar{\phi}_i - \varepsilon^{\phi_i^\mu} \leq \phi_i^\mu \leq \bar{\phi}_i + \varepsilon^{\phi_i^\mu}$$

$$\bar{I}_{ij} - \varepsilon^{I_{ij}^\mu} \leq I_{ij}^\mu \leq \bar{I}_{ij} + \varepsilon^{I_{ij}^\mu}$$

$$\varepsilon^{x_i} \geq 0, \varepsilon^{\phi_i^\mu} \geq 0, \varepsilon^{I_{ij}^\mu} \geq 0$$

We denote by $\mathcal{T}(X_R, \bar{s})$ the problem of determining a feasible solution to the mathematical program obtained with the linearization of the quadratic constraints of $\mathcal{P}(X_R)$ in the neighborhood of a given reference point $\bar{s}$ thanks to the Taylor's formula, and with the linearization of the absolute values.

3.3   Global Procedure

Our proposition consists in solving problem $\mathcal{P}(X_R)$ using problem $\mathcal{T}(X_R, \bar{s})$. A solution to problem $\mathcal{P}(X_R)$ is an instantiation of the capacity and the utility functions of a Choquet integral. It can be equivalently described by the instantiation of the Shapley values, the interaction indexes and the utility function. In the sequel, a solution is therefore a point $s = (\phi, I, u)$ where the vector $\phi = (\phi_1, \ldots, \phi_n)$ represents the Shapley values of the criteria, the vector $I = (I_{11}, I_{12}, \ldots, I_{1n}, I_{21}, \ldots, I_{2n}, \ldots, I_{nn})$ represents the interaction indexes of the pairs of criteria, where $I_{ii} = 0$, and the vector $u = (u_1(x^1), \ldots, u_n(x^1), u_1(x^2), \ldots, u_n(x^2), \ldots, u_1(x^m), \ldots, u_n(x^m))$ represents the utility values of the actions $x^j \in X$ on the n criteria. Like a solution to $\mathcal{T}(X_R, \bar{s})$ (or $\mathcal{P}(X_R)$), the reference point $\bar{s} = (\bar{\phi}, \bar{I}, \bar{u})$ is defined by the Shapley values $\bar{\phi}$, the interaction indexes $\bar{I}$ and the utility function $\bar{u}$.

Remark 1. *An optimal solution s to Problem $\mathcal{T}(X_R, \bar{s})$ can be infeasible to Problem $\mathcal{P}(X_R)$.*

Example 4. *Consider Example 1 (classic example). Let $\bar{s}$ be defined by:*

- *$\bar{\phi}_1 = \bar{\phi}_2 = 0.2$, $\bar{\phi}_3 = 0.6$ and $\bar{I}_{12} = 0$, $\bar{I}_{13} = 0.2$, $\bar{I}_{23} = 0.6$*
- *$\bar{u}(a) = (0.8, 0.65, 0.35)$, $\bar{u}(b) = (0.8, 0.55, 0.45)$, $\bar{u}(c) = (0.3, 0.65, 0.35)$, $\bar{u}(d) = (0.3, 0.55, 0.45)$ (where here $u_i(x_i) = x_i/20$ for all i)*

Let us define the objective function of $\mathcal{T}(X_R, \bar{s})$ as the minimization of the distance to $\bar{s}$ with respect to the L^1-norm. An optimal solution $s = (\phi, I, u)$ to $\mathcal{T}(X_R, \bar{s})$ is:

- *$\phi = \bar{\phi}$ and $I_{12} = 0$, $I_{13} = 0.2$, $I_{23} = 0.4$*
- *$u(a) = (0.8, 0.65, 0.4375)$, $u(b) = \bar{u}(b)$, $u(c) = (0.3, 0.65, 0.4375)$, $u(d) = \bar{u}(d)$*

This solution gives $C(a) = 0.473751$ and $C^T(a) = 0.482501$, $C(b) = 0.485$ and $C^T(b) = 0.485$, $C(c) = 0.396251$ and $C^T(c) = 0.405001$, $C(d) = C^T(d) = 0.405$, where $C^T(x)$ is the value of the Choquet integral of solution $x \in X$ with the linearization. The C^T values are close to the real Choquet values, however the error of linearization for action c makes this solution s infeasible to $\mathcal{P}(X_R)$ since $C(c) < C(d)$, which contradicts the preference of the DM, whereas $C^T(c) > C^T(d)$, which makes solution s feasible to $\mathcal{T}(X_R, \bar{s})$.

Remark 2. *Given a point $\bar{s}$, a feasible solution s to Problem $\mathcal{P}(X_R)$ can be infeasible to Problem $\mathcal{T}(X_R, \bar{s})$.*

Example 5. *Consider here again Example 1. Let $\bar{s}$ be defined by:*

- $\bar{\phi}_1 = 0$, $\bar{\phi}_3 = 0.8$, $\bar{\phi}_2 = 0.2$ and $\bar{I}_{12} = 0$, $\bar{I}_{13} = 0$, $\bar{I}_{23} = 0.2$
- $\bar{u}(a) = (0.8, 0.65, 0.35)$, $\bar{u}(b) = (0.8, 0.55, 0.45)$, $\bar{u}(c) = (0.3, 0.65, 0.35)$, $\bar{u}(d) = (0.3, 0.55, 0.45)$ *(where here $u_i(x_i) = x_i/20$ for all i)*

A feasible solution $s = (\phi, I, u)$ to $\mathcal{P}(X_R)$ is defined by:

- $\phi_1 = 0.2$, $\phi_2 = 0.2$, $\phi_3 = 0.6$ and $I_{12} = 0$, $I_{13} = 0.00002$, $I_{23} = 0.2$
- $u(a) = (0.8, 0.65, 0.35)$, $u(b) = (0.8, 0.55, 0.3642857)$, $u(c) = (0.3, 0.65, 0.35)$, $u(d) = (0.3, 0.55, 0.3642857)$

It gives $C(a) = 0.469995$, $C(b) = 0.469996$, $C(c) = 0.37$ and $C(d) = 0.369999$, which makes s feasible to $\mathcal{P}(X_R)$. However we get: $C^T(a) = 0.54$, $C^T(b) = 0.47$, $C^T(c) = 0.44$, $C^T(d) = 0.49$, where $C^T(a) > C^T(b)$ and $C^T(d) > C^T(c)$ which contradicts the two preferences of the DM.

Nevertheless if Problem $\mathcal{P}(X_R)$ admits a feasible solution s then there exists a point $\bar{s}$ such that s is a feasible solution to Problem $\mathcal{T}(X_R, \bar{s})$. It obviously suffices to take $\bar{s} = s$. Moreover, in this case, if the objective function of Problem $\mathcal{T}(X_R, \bar{s})$ is defined as minimizing a distance to $\bar{s}$, s is an optimal solution to $\mathcal{T}(X_R, \bar{s})$. Those observations show the importance of the choice of point $\bar{s}$ when solving Problem $\mathcal{T}(X_R, \bar{s})$. In order to find a C_2-compatible model with the preference information, we solve Problem $\mathcal{T}(X_R, \bar{s})$ for some points $\bar{s}$ until the optimal solution to $\mathcal{T}(X_R, \bar{s})$ is feasible to $\mathcal{P}(X_R)$.

4   Illustration and Experiments

In this section, we test if the approximation with the Taylor's formula enables us to assess the C_2-compatibility of a DM. When a 2-additive Choquet integral model compatible with the reference set is computed with a given reference point, one can claim that the preferences of the DM are C_2-compatible. To perform those tests, one generates random instances with m actions and n criteria. The value of each criterion of each action is randomly generated between 0 and 50. Then each pair of actions x^i, x^j is considered: if x^i (resp. x^j) Pareto-dominates x^j (resp. x^i) then $x^i \succ x^j$ (resp. $x^j \succ x^i$) is added into the reference set R, otherwise $x^i \succ x^j$ (or $x^j \succ x^i$ or $x^i \sim x^j$) is added into R with a probability

$\frac{1}{4*m}$. If this newly generated preference or indifference relation creates a cycle in the preference relation X_R induced by R, it is not added into R. The preference relation X_R obtained from the randomly generated reference set R is therefore compatible with Pareto-dominance (when an action x^i Pareto-dominates another action x^j, x^i is preferred to x^j) and does not contain any cycle. Note that the preference relation X_R may be compatible with an additive model. In this case, the instance associated to X_R is not considered. Therefore, in this section, all the tests are performed on non-additive randomly generated instances.

4.1   Definition of the Reference Points

To solve problem $\mathcal{P}(X_R)$ with the global procedure, we can define iteratively different points $\bar{s}$ and solve problem $\mathcal{T}(X_R, \bar{s})$ for each $\bar{s}$. We can set the values of $\bar{s}$ as follows: at each iteration $\bar{s}_j$ is one value in $\{0, \delta_j, 2\delta_j, \ldots, 1\}$, with $\delta_j \in]0, 1]$, for any dimension j of point $\bar{s}$. Let d be the dimension of $\bar{s}$, the number of distinct reference points $\bar{s}$ is therefore in $O(\frac{1}{\delta^d})$ where $\delta = \max_j \delta_j$. Thus the number of $\mathcal{T}(X_R, \bar{s})$ to be solved in this approach is up to $O(\frac{1}{\delta^d})$.

In this section, we study three approaches for defining some reference points $\bar{s}$ in the global procedure. We have tested the approaches on several instances of different sizes and the conclusions about the definition of the different reference points were the same for all the tested instances. Therefore, in all this section, we illustrate our results on the following small non-additive instance with 5 actions and 3 criteria: the 5 actions are $x^1 = (25, 24, 26)$, $x^2 = (40, 2, 3)$, $x^3 = (4, 27, 37)$, $x^4 = (47, 47, 23)$ and $x^5 = (35, 6, 43)$, and the preference information is $x^1 \sim x^4$, $x^2 \succ x^5$ and $x^4 \succ x^2$. The resulting mathematical program involves 128 variables, 15 of which are binary, and 216 constraints.

The first approach tested here, named *App1* hereafter, consists in solving $\mathcal{T}(X_R, \bar{s})$ for several reference points $\bar{s}$, and looking for a feasible solution the closest to $\bar{s}$ as possible. The idea behind looking for a feasible point the closest to $\bar{s}$ as possible is to "minimize" the error when computing the approximation of a product with the Taylor's formula by staying in the neighborhood of the reference point $\bar{s}$. More precisely we define the objective function in problem $\mathcal{T}(X_R, \bar{s})$ by:

$$\min L^p(\varepsilon^{x_1^1}, \ldots, \varepsilon^{x_n^m}, \varepsilon^{\phi_1^\mu}, \ldots, \varepsilon^{\phi_n^\mu}, \varepsilon^{I_{11}^\mu}, \ldots, \varepsilon^{I_{nn}^\mu})$$

where the ε values represent the gap between the reference point and the solution (see the end of Sect. 3.2). In this first pool of tests, we compare the results obtained with App1 for different values of δ_j that define the different reference points $\bar{s}$, and for two distances defined with L^1-norm or L^∞-norm. We set $\delta_j = \delta_{j'}$ for all j, j'. The results are summarized in Table 1. For any δ value and L^p-norm it is indicated the execution time in seconds (col. CPU(s)) and the number of reference points $\bar{s}$ tested before obtaining an optimal solution to $\mathcal{T}(X_R, \bar{s})$ that is feasible to $\mathcal{P}(X_R)$ (col. #it.). Note that the maximal number of problems $\mathcal{T}(X_R, \bar{s})$ to be solved is here $(\frac{1}{\delta} + 1)^{3+3+15}$ (we omit the terms I_{ii} in the computation of the dimension of the reference point since their values

Table 1. Results obtained with App1

	$\delta = 0.1$		$\delta = 0.3$		$\delta = 0.5$		$\delta = 0.7$	
	#it	CPU (s)	#it	CPU (s)	#it	CPU (s)	#it	CPU (s)
L^1-norm	175702	17141.1	1794	143.21	245	14.06	2053	187.728
L^∞-norm	80440	2876.98	33568	1342.11	1424	45.22	31481	1820.42

are 0). For instance, when $\delta = 0.5$, the number of $T(X_R, \bar{s})$ to be solved in App1 is up to 3^{21}.

These results show that App1 is the most efficient with $\delta = 0.5$. A solution to $\mathcal{P}(X_R)$ is indeed determined in 14 s with the L^1-norm. Decreasing the value of δ makes the number of iterations (and consequently the execution time) increase. This is easily explained by the increasingness of the number of possible reference points in this case. We can note however that the number of iterations with $\delta = 0.7$ is also greater than with $\delta = 0.5$. It means that δ should be neither too large nor too small. Selecting an appropriate value of δ appears therefore like a challenge. Besides, these results show that using L^1-norm in the objective function seems to be more efficient than using L^∞-norm.

As it could be expected, App1 involves a huge number of solutions of problem $T(X_R, \bar{s})$. At each iteration (*i.e.* solution of a problem $T(X_R, \bar{s})$), one looks for a solution the closest to the point $\bar{s}$ as possible. Consequently when the reference point is "far" from being feasible to $\mathcal{P}(X_R)$, one could expect the optimal solution to $T(X_R, \bar{s})$ with App1 to be infeasible to $\mathcal{P}(X_R)$. As we only need to find an optimal solution to $T(X_R, \bar{s})$ that is feasible to $\mathcal{P}(X_R)$, being close to an arbitrary given reference point could be a guide in the search but not a goal to reach. We propose thus to relax the closeness to the reference point by minimizing the distance between the solution and the reference point only on a part of their components. More precisely, in this approach, called hereafter *App2*, one requires the distance between the optimal solution and the reference point to be minimized only on the Shapley values and/or on the interactions indexes. Table 2 shows the results obtained on the 5 actions-3 criteria instance. The symbol $-$ means that the execution of the global procedure has been stopped after 9 hours of running without finding any solution to problem $\mathcal{P}(X_R)$.

Table 2. Results obtained with App2

Min. dist. to	$\bar{\phi} + \bar{I}$		$\bar{\phi}$		$\bar{I}$	
norm/δ	#it	CPU (s)	#it	CPU (s)	#it	CPU (s)
$L^1/0.3$	1185614	43495.7	-	-	17434	636.03
$L^\infty/0.3$	33568	1347.15	33568	1386.18	33568	1382.11
$L^1/0.5$	414907	10693.1	-	-	179	3.76
$L^\infty/0.5$	1424	44.95	1424	45.29	1424	46.21

Those results show that omitting some terms in the objective function makes the global procedure much less efficient. For example, it takes 17434 iterations with L^1-norm and $\delta = 0.3$ when the objective function is only defined with ε^{I^μ} terms, instead of 1794 (see Table 1). With $\delta = 0.5$, no solution was found in a reasonable time with L^1-norm when the objective function only contains ε^{ϕ^μ} terms, whereas it only takes 143 s and 14 s when all terms are in the objective function. These results can be explained by the fact that allowing the values of some components of the optimal solution to be far from the reference values can increase the value of the error in the linearization of the quadratic terms. Besides, we can note that modifying the terms that define the objective function does not change the number of iterations with L^∞-norm.

The main issue of the two previous approach is the huge number of possible reference points. In order to drastically decrease the number of reference points to be used in the global procedure, we propose in this approach to fix the values of some of its components, and to select different values for the other components only. This approach is named *App3* hereafter. Since we assume that the values of the actions on each criteria x_i^j are real numbers, we set $\bar u_i(x_i^j) = x_i^j/x_i^{max}$ for all $i \in N$ and $j \in \{1,\ldots,m\}$, where $x_i^{max} = max_j x_i^j$, when the utility functions values $\bar u$ are fixed in App3. When the interaction indexes are fixed in App3, we set $I_{ij} = 0.2$ for any $i,j \in N$. When the Shapley values are fixed in App3, we set $\phi_i = \frac{1}{n}$ for any $i \in N$. In all the cases, the objective function is to minimize the distance to the all the components of $\bar s$. Table 3 shows the results obtained with App3. In this table, NF (x) means that no feasible solution to problem $\mathcal{P}(X_R)$ has been found in x iterations.

Table 3. Results obtained with App3

Variations of values	$\bar\phi$ and $\bar I$		$\bar\phi$		$\bar I$		$\bar u$	
norm/ε	#it CPU(s)	#it	CPU (s)	#it	CPU (s)	#it	CPU (s)
$L^1/0.1$	1	0.04	14	1.74	1	0.02	133	5.64
$L^\infty/0.1$	37	1.99	43	3.09	3	0.13	593	17.49
$L^1/0.3$	1	0.12	5	0.55	1	0.02	274	22.25
$L^\infty/0.3$	NF (630)	20.68	NF (10)	0.75	1	0.03	1377	41.37
$L^1/0.5$	1	0.06	4	0.53	1	0.03	13	0.59
$L^\infty/0.5$	53	1.4	NF (6)	0.41	1	0.03	712	23.93
$L^1/0.7$	1	0.03	NF (3)	0.34	2	0.08	5	0.18
$L^\infty/0.7$	NF (21)	0.61	NF (3)	0.15	NF (7)	0.54	1742	99.87

These results show that fixing some components of the reference points enables the procedure to be much more efficient. The maximal number of iterations is significantly decreased, but it does not prevent to find a solution, in general. Note however that when varying the values of only ϕ, a solution cannot be found with highest values of δ. Moreover, we can observe that using L^1-norm appears, here again, more efficient than using L^∞-norm.

The tests of this section show that the performance of the global procedure depends on the definition of the objective function of the linear program and on the used reference points. Those tests, illustrated here with a simple instance, were performed on several instances of different size. We come to the same conclusions for all tested instances: varying the utility values of the reference point and using L^∞-norm is not efficient. We propose therefore to minimize the distance between the reference point and the solution with respect to the L^1-norm. However, the components of the reference points that should be taken into account in the objective function depends on the size of the problem. In our experiments, it turns out that it is more efficient to minimize the distance between the reference point and the solution over their Shapley values and interaction indexes for bigger size instances. Furthermore, in the different iterations of the global procedure, we fix the reference values of the utility functions, and we vary the reference values of the Shapley values and of the interaction indexes.

4.2   Illustration on Random Non-additive Instances

In this section, we illustrate the use of the Taylor's formula in the linearization of the Choquet integral. We therefore apply our global procedure to randomly generated instances (generated as previously explained). The reference points are generated either by a sampling on a regular multidimensional grid (see Sect. 4.1) or randomly. Note that the aim of this pool of tests is not to provide the most efficient procedure for testing the C_2-compatibility of a preference information, but only to illustrate the application of the linearization of a Choquet integral model with Taylor's formula. In the following experiments, one minimizes the distance between the reference point and the solution over their Shapley values and interaction indexes with respect to the L^1-norm. Besides one sets $\delta = 0.05$ for the regular multidimensional grid used in the sampling of the reference points. Table 4 indicates the results obtained. The results are given as average over 5 randomly generated non-additive instances of each size. In the random generation of the reference points, one randomly generates each Shapley values and interaction indexes. We set the maximal number of reference points to be generated to 1000, that is at most 1000 problems $\mathcal{T}(X_R, \bar{s})$ have to be solved in this approach. In both approaches we stop the running of the procedure at the first iteration where the feasible solution to $\mathcal{T}(X_R, \bar{s})$ is also feasible to $\mathcal{P}(X_R)$. The results show that the two selection strategies of the reference point (regular grid and random generation) in the global procedure enables us to solve problem $\mathcal{P}(X_R)$ with a linearization based on the Taylor's formula on some instances in a few iterations. There are some other instances for which one cannot solve the problem without increasing the maximal number of reference points to use in the global procedure. The definition of the reference point is therefore a crucial issue in this approach. In further works, one could design an efficient global procedure, where a crucial issue would be to sample the reference points in a relevant way.

Table 4. Illustration of the approach

Generation of ref. points	Regular grid		Random	
$m \times n$	#iterations	CPU(s)	#iterations	CPU(s)
10×3	10.4	23.33	75	4.64
20×3	45.4	15.1	24.6	18.4
30×3	22	29.9	207	169

5   Conclusion

In this work, we proposed an approach which aims at finding values of the utility functions and of the capacity that represent a preference information given by a partial order on a set of alternatives. Our heuristic approximates the Choquet integral value by using a linear approximation of the product of utilities functions with capacity, based on the Taylor's formula. We have tested our approach on some randomly non-additive instances. The tests are encouraging since our heuristic can solve some hard instances. However, they also highlight a limit of this method, which is the choice of the reference point in the linear approximation. This limitation could be overcome by elaborating an appropriate strategy to choose this point.

References

1. Angilella, S., Greco, S., Lamantia, F., Matarazzo, B.: Assessing non-additive utility for multicriteria decision aid. Eur. J. Oper. Res. **158**(3), 734–744 (2004)
2. Apostol, T.M.: Calculus: Multi-Variable Calculus and Linear Algebra with Applications to Differential Equations and Probability, vol. 2, 2nd edn. Wiley, New York (1969)
3. Bana e Costa, C.A., Correa, E.C., De Corte, J.-M., Vansnick, J.-C.: Facilitating bid evaluation in public call for tenders: a socio-technical approach. Omega **30**, 227–242 (2002)
4. Bouyssou, D., Couceiro, M., Labreuche, C., Marichal, J.-L., Mayag, B.: Using Choquet integral in machine learning: What can mcda bring? In: DA2PL 2012 Workshop: From Multiple Criteria Decision Aid to Preference Learning, Mons, Belgique (2012)
5. Clivillé, V., Berrah, L., Mauris, G.: Quantitative expression and aggregation of performance measurements based on the MACBETH multi-criteria method. Int. J. Prod. Econ. **105**, 171–189 (2007)
6. Dyer, J.S.: MAUT - multiattribute utility theory. In: Figueira, J., Greco, S., Ehrgott, M. (eds.) Multiple Criteria Decision Analysis: State of the Art Surveys, pp. 265–285. Springer, Boston, Dordrecht, London (2005)
7. Goujon, B., Labreuche, C.: Holistic preference learning with the Choquet integral. In: Montero, J., Pasi, G., Ciucci, D. (eds.) Proceedings of the 8th conference of the European Society for Fuzzy Logic and Technology, EUSFLAT-13, Milano, Italy, September 11–13, 2013. Atlantis Press (2013)

8. Grabisch, M.: k-order additive discrete fuzzy measures and their representation. Fuzzy Sets Syst. **92**, 167–189 (1997)

9. Grabisch, M., Kojadinovic, I., Meyer, P.: A review of methods for capacity identification in Choquet integral based multi-attribute utility theory: applications of the Kappalab Rpackage. Eur. J. Oper. Res. **186**(2), 766–785 (2008)

10. Grabisch, M., Labreuche, C.: Fuzzy measures and integrals in MCDA. In: Figueira, J., Greco, S., Ehrgott, M. (eds.) Multiple Criteria Decision Analysis: State of the Art Surveys, pp. 565–608. Springer, New York (2005)

11. Grabisch, M., Labreuche, C.: A decade of application of the Choquet, Sugeno integrals in multi-criteria decision aid. 4OR **6**, 1–44 (2008)

12. Mayag, B., Grabisch, M., Labreuche, C.: A characterization of the 2-additive Choquet integral through cardinal information. Fuzzy Sets Syst. **184**(1), 84–105 (2011)

13. Mayag, B., Grabisch, M., Labreuche, C.: A representation of preferences by the Choquet integral with respect to a 2-additive capacity. Theory Decis. **71**(3), 297–324 (2011)

14. Murofushi, T., Soneda, S.: Techniques for reading fuzzy measures (III): interaction index. In: 9th Fuzzy System Symposium, Sapporo, Japan, pp. 693–696, May 1993. (In Japanese)

15. Shapley, L.S.: A value for n-person games. In: Kuhn, H.W., Tucker, A.W. (eds.) Contributions to the Theory of Games, Vol. II. Annals of Mathematics Studies, vol. 28, pp. 307–317. Princeton University Press, Princeton (1953)

An Alternative View of Importance Indices for Multichoice Games

Mustapha Ridaoui[1]($\boxtimes$), Michel Grabisch[1], and Christophe Labreuche[2]

[1] Paris School of Economics, Université Paris I - Panthéon-Sorbonne, Paris, France
{mustapha.ridaoui,michel.grabisch}@univ-paris1.fr
[2] Thales Research & Technology, Palaiseau, France
christophe.labreuche@thalesgroup.com

Abstract. We consider MultiCriteria Decision Analysis (MCDA) models where the underlying attributes are discrete. Without any additional feature, such general models are equivalent to multichoice games in cooperative game theory. Our aim is to define an importance index for attributes. In specific models based on capacities, fuzzy measures, the Shapley value is often taken as an importance index. We show that in our general framework, taking the Shapley value extended to multichoice games is not meaningful, due to the efficiency axiom which has no natural interpretation in MCDA. We propose instead an importance index based on variational calculus and give an axiomatization of it.

Keywords: Multicriteria decision analysis · Multichoice game · Shapley value

1 Introduction

In MultiCriteria Decision Analysis (MCDA), one of the major issues is to be able to give an interpretation or explanation of the model which has been obtained through learning data and/or elicitation of preference. A basic way of explaining the model is to be able to quantify the overall importance of the attributes describing the alternatives under consideration. While this is easy when simple additive models are used (like the additive utility model, or any model based on the weighted arithmetic mean), the task is less obvious with more complex models.

A large class of complex models are based on nonadditive measures (capacities, fuzzy measures, etc.), where nonadditive integrals like the Choquet integral [1], the Sugeno integral [15], etc., are used in place of the weighted arithmetic mean (see [4] for a survey of these models). For those models, the question of defining an importance index has been solved by borrowing concepts from game theory, namely the Shapley value [14], as suggested probably the first time by Murofushi [11]. In game theory, for a (cooperative) game v defined on a set of players N which expresses the benefit achieved by any coalition of players, the Shapley value is a payoff vector in $\mathbb{R}^N$, representing a sharing among all players of the total benefit $v(N)$ achieved by cooperation. It obeys some rational principles like: a player i whose marginal contribution $v(S \cup i) - v(S)$ to coalition S

© Springer International Publishing AG 2017
J. Rothe (Ed.): ADT 2017, LNAI 10576, pp. 81–92, 2017.
DOI: 10.1007/978-3-319-67504-6_6

is null for every coalition S should receive a null payoff (this is called the *null axiom*); the sum of payoffs should be equal to $v(N)$ (this is called *efficiency*). Translated in the domain of MCDA, N is the set of attributes, and for any group of attributes S, $v(S)$ expresses the evaluation of an alternative being satisfactory on all attributes in S, and unsatisfactory otherwise. Therefore, $v(N)$ (taken by convention to be equal to 1) is the evaluation of an alternative being satisfactory for all attributes ("ideal" alternative). Then the Shapley value gives for each attribute its percentage of importance in realizing the ideal alternative. Later, it was proved by Grabisch et al. [7] that the Shapley value is also the average partial derivative w.r.t. attribute i of the Choquet integral, linking the Shapley value to variational calculus. Other complex models exist, like the GAI (Generalized Additive Independence) models proposed by Fishburn [2,3].

In this paper, we depart from these studies and make no assumption on the type of the model. Simply, we consider that attributes take a finite number of values (discrete attributes), which can be ordered. In particular, we do not assume that the evaluation of an alternative is monotonically increasing with the value of the attributes. This is relevant in many situations, typically when the best value of an attribute is not located at the boundaries of the range, but somewhere in the middle. We will show in the paper that such general models on discrete attributes can be assimilated to what is called *multichoice game* in game theory [8]. Then, the definition of an importance index for such general models amounts to defining a kind of Shapley value for multichoice games. The literature has already brought many definitions of values for multichoice games, which all of them generalize the classical Shapley value. However, none of these definitions seem to be adequate for our purpose. Indeed, in this general situation, as monotonicity is not assumed, the value of the model v for an alternative which has on each attribute the highest level has no reason to be the ideal alternative. Therefore, defining an importance index as a sharing of this value does not make sense, especially when this value is close to 0. Our approach to the problem is then to take the way of variational calculus: we define the importance index of attribute i as the average of the variation of v along the i axis. The present paper is a follow-up of Ridaoui et al. [13], addressing the same problem with a similar philosophy. In the latter, the importance index of an attribute was defined as its average variation along the axis of this attribute, considering all possible values of the other attributes. This index, which has a natural interpretation as a collection of "local" Shapley values defined for each cell of the grid of attributes values, solves only partially the problem because single-peaked preferences on attributes may cause the importance index to be very small or even zero (see Example 2 below). The distinguishing point of the present paper is to consider a *cumulative absolute* variation on an attribute, with which the previous problem does not exist.

The paper is organized as follows. Section 2 gives the necessary background and establishes the notation. Section 3 explains the motivation, the main idea and previous approaches. Section 4 gives the axiomatization of the family of importance indices we propose. Section 5 starts a more general approach where the importance index is the norm of the derivative of the multichoice game.

2   Preliminaries

We consider a Multi-Criteria Decision Analysis (MCDA) problem characterized by a set $N = \{1, \ldots, n\}$ of attributes. We suppose that each attribute $i \in N$ takes values in a finite set L_i, as it is often the case in MCDA. For convenience, the elements of L_i are represented by integer values: $L_i = \{0, 1, \ldots, k_i\}$. We can think of the integer values as label indices. The alternatives are represented as elements of the Cartesian product $L = L_1 \times \cdots \times L_n$. An alternative is thus written as a vector $x = (x_1, x_2, \ldots, x_n)$ where $x_i \in L_i$ for all $i \in N$.

The set L is equipped with the dominance relation $\leq$: for $x, y \in L$, $x \leq y$ means that $x_i \leq y_i$ for every $i \in N$. Moreover $<$ is the asymmetric part of $\leq$ ($x < y$ if $x \leq y$ and $x_i < y_i$ for some $i \in N$). For $S \subseteq N$ and $x \in L$, x_S is the restriction of x to S. We denote by L_{-i} the set $\times_{j \neq i} L_j$. For each $y_{-i} \in L_{-i}$, and any $\ell \in L_i$, (y_{-i}, ℓ_i) denotes the compound alternative x such that $x_i = \ell_i$ and $x_j = y_j, \forall j \neq i$. The vector $0_N = (0, \ldots, 0)$ is the null alternative of L, and $k_N = (k_1, \ldots, k_n)$ is the top element of L. For each $x \in L$, we denote by $S(x) = \{i \in N \mid x_i > 0\}$ the support of x, and by $K(x) = \{i \in N \mid x_i = k_i\}$ the kernel of x (locus of maximal values). Their cardinalities are respectively denoted by $s(x)$ and $k(x)$.

The preferences of a Decision Maker (DM) over the alternatives are supposed to be represented by a function $v : L \to \mathbb{R}$. In practice, some assumptions are often made on the form of v in order to ease the elicitation phase. Indeed, eliciting a general function $v : L \to \mathbb{R}$ containing $k_1 \times \cdots \times k_n$ parameters would require far too many preference information from the DM. In order to reduce the elicitation burden, special forms of v are used in practice: an additive model $\sum_{i \in N} u_i(x_i)$ [9], a combination of a Choquet integral and marginal utility functions $C_\mu(u_1(x_1), \ldots, u_n(x_n))$ [4] or a Generalized Additive Independence (GAI) model [2,3], among many other models. We make here no such restriction, as we are interested in the interpretation of v (and not its elicitation), and the index that will be defined shall be agnostic to the precise expression of v.

We make a normalization condition:

$$v(0_N) = 0. \tag{1}$$

This condition is not a restriction, as most of numerical representations are unique up to a positive affine transformation.

The concept of criterion is paramount in MCDA and is related to some notion of monotonicity. It can be described by the relation: the larger the value of an attribute the better. Formally this gives

$$v(x) \geq v(x') \text{ whenever } x \geq x', x, x' \in L. \tag{2}$$

We do not assume such conditions in our framework as we are particularly interested in the cases where preferences are not monotonic. The following example illustrates a situation where the preferences of the DM are not simply increasing or decreasing with respect to each attribute.

Example 1 (From [13]*).* The sensation of ambient comfort can be described by three attributes: temperature of the air (X_1), humidity of the air (X_2) and velocity of the air (X_3). Let $v(x_1, x_2, x_3)$ measure the comfort level. Clearly, the value of v is not monotonically increasing nor decreasing with the level of the attributes: for given values of two attributes, say humidity and velocity, there is an "ideal" value of the remaining attribute (temperature; say, around 23°C). When departing from this value, the comfort level decreases as the distance to this ideal point increases. The same conclusion holds for the two other attributes.

In the previous example, preferences are single-peaked, with an optimal value over each attribute. The simplest single-peaked function is the δ_x function defined by, for any $x \in L$,

$$\delta_x(y) = \begin{cases} 1, \text{ if } y = x \\ 0, \text{ otherwise} \end{cases}$$

Evidently, the set of delta functions δ_x, $x \in L \setminus \{0_N\}$, forms a basis of the set of functions v:

$$v = \sum_{\substack{x \in L \\ x \neq 0_N}} v(x)\delta_x. \tag{3}$$

For convenience, we assume from now on that all attributes have the same number of elements, i.e., $k_i = k$ for every $i \in N$ ($k \in \mathbb{N}$). Note that if this is not the case, we set $k = \max_{i \in N} k_i$, and we extend $v : L \to \mathbb{R}$ to $v' : \{0, \ldots, k\}^N \to \mathbb{R}$ by

$$v'(x) = v(y) \text{ where } y_i = \min(x_i, k_i) \ \forall i \in N.$$

This amounts to duplicating the last element k_i of L_i when $k_i < k$. Under this assumption, we recover well-known concepts. When $k = 1$, v is a *pseudo-Boolean function* $v : \{0, 1\}^N \to \mathbb{R}$ vanishing at 0_N. It can be put in the form of a function $\mu : 2^N \to \mathbb{R}$, which is a game in cooperative game theory. When monotonicity is enforced, we obtain capacities. When $k \geq 1$, $v : L \to \mathbb{R}$ fulfilling (1) corresponds exactly to the concept of *multi-choice game* [8]. They are also called k-*ary games*. A k-*ary capacity* is a k-ary game that is monotone (see (2)) [5]. We denote by $\mathcal{G}(L)$ the set of k-ary games defined on L, and by $\mathcal{G}_M(L)$ the set of monotone k-ary games. Finally, the *derivative* of v w.r.t $i \in N$ at $x \in L$ such that $x_i < k$ is defined by

$$\Delta_i v(x) = v(x + 1_i) - v(x).$$

We recall from the above that $\{\delta_x\}_{x \in L \setminus \{0_N\}}$ is a basis of the set of multichoice games. Another basis of interest is obtained from *unanimity games* u_x, which are defined for any $x \in L \setminus \{0_N\}$ by

$$u_x(y) = \begin{cases} 1, & \text{if } y \geq x \\ 0, & \text{otherwise.} \end{cases} \tag{4}$$

Then, for any game $v \in \mathcal{G}(L)$,

$$v = \sum_{x \in L \setminus \{0_N\}} m^v(x)u_x. \tag{5}$$

The coordinates of v in this basis define a function $m^v : L \to \mathbb{R}$ with $m^v(0_N) := 0$, which is the *Möbius transform* of v. Let us consider $\mathcal{G}_+(L)$ the set of games whose Möbius transform is nonnegative, and note that $\mathcal{G}_+(L) \subseteq \mathcal{G}_M(L)$ is a cone. It is easy to see, as for classical games, that any multichoice game $v \in \mathcal{G}(L)$ can be uniquely decomposed into

$$v = v^+ - v^- \tag{6}$$

with $v^+, v^- \in \mathcal{G}_+(L)$ (v^+, v^- simply correspond respectively to the nonnegative coefficients and negative coefficients of the Möbius transform of v).

3 Aim and Related Work

As indicated in Sect. 2, we place ourselves in a context of multicriteria decision aid, considering that our numerical representation of preferences can be reduced to a multichoice (k-ary) game on L. Our aim is to define an importance index for criteria.

When a classical capacity ($k = 1$) is used for representing the preference, or with Choquet integral-based models, the Shapley value [14], a concept borrowed from game theory, is used. Generally speaking, a *value* is a mapping $\phi : \mathcal{G}(2^N) \to \mathbb{R}^N$ assigning a payoff vector to any game v. When the game is simple, that is, valued in $\{0, 1\}$, the Shapley value coincides with the Shapley-Shubik index, which is seen as a power index. This name comes from the fact that in this context N is the set of voters, and $\phi_i(v)$ is the *power* of voter i, i.e., to what extent the fact that i votes 'yes' makes the final decision to be 'yes'. Obviously, power index in voting theory is close to *importance index* in MCDA, which explains the use of the Shapley value in this context. Its expression is given by, for any $v \in \mathcal{G}(2^N)$,

$$\phi_i^{Sh}(v) = \sum_{S \subseteq N \setminus i} \frac{(n - s - 1)! s!}{n!} \big(v(S \cup i) - v(S)\big), \forall i \in N. \tag{7}$$

The Shapley value has been extended to multichoice games, and several generalizations exist, e.g., Hsiao and Raghavan [8] (historically the first one), van den Nouweland et al. [16], Klijn et al. [10], Peters and Zank [12], Grabisch and Lange [6], etc. Interestingly, the value proposed by Peters and Zank is defined through the Möbius transform:

$$\phi_i^{PZ}(v) = \sum_{\substack{x \in L \\ x_i > 0}} \frac{m^v(x)}{s(x)}, \tag{8}$$

where $s(x)$ is the size of the support of x.

All these values, for classical games or multichoice games, are characterized by axioms, among which we always find linearity (ϕ is a linear operator on $\mathcal{G}(L)$), a null axiom (a player with zero marginal contribution receives zero), a symmetry axiom (the numbering of the players has no influence on the payoff) and efficiency (the sum of the payoffs for all players should be equal to $v(k_N)$, where k_N is the top element of L). While the first three axioms are easily transposable in the MCDA framework, the efficiency axiom is rooted in game theory and there is no convincing interpretation in MCDA. Indeed, efficiency means that the total benefit of the cooperation, represented by $v(k_N)$, should be shared among the players with no waste. In MCDA, the concept of sharing is not natural, and there is no reason why the sum of importance indices should be equal to $v(k_N)$. When v is monotone, then $v(k_N)$ is the evaluation of the best possible alternative, and in this case it may be meaningful to consider the importance index of a criterion as a percentage of the best possible evaluation. However, this interpretation fails when v is not monotone. Especially, in Example 1, the value of $v(k_N)$ should be close to 0, as the highest values of the attributes are not considered as comfortable.

This is why in [13] the authors have introduced another index of importance, which does not satisfy efficiency. It is defined as follows:

$$\phi_i^{\mathrm{RGL}}(v) = \sum_{x_{-i} \in L_{-i}} \frac{(n - s(x_{-i}) - 1)! k(x_{-i})!}{(n + k(x_{-i}) - s(x_{-i}))!} \big(v(x_{-i}, k_i) - v(x_{-i}, 0_i)\big), \forall i \in N.$$

(9)

This importance index measures the impact of attribute i on v as the weighted average of the difference between the highest and lowest value of attribute i, when x_{-i} is varying over the domain L_{-i}. It can be also seen as the average variation along the axis of the attribute.

We present the axioms characterizing the importance index ϕ^{RGL}, as some of them will be used in the next section.

The first one says that ϕ is a linear operator on the set of games.

Linearity axiom (L): ϕ is linear on $\mathcal{G}(L)$, i.e., $\forall v, w \in \mathcal{G}(L), \forall \alpha \in \mathbb{R}$,

$$\phi_i(v + \alpha w) = \phi(v) + \alpha \phi(w).$$

The second one says that an attribute for which an increment of 1 does not improve the evaluation is not important. Formally, an attribute $i \in N$ is said to be null for $v \in \mathcal{G}(L)$ if

$$v(x + 1_i) = v(x), \forall x \in L, x_i < k.$$

Null axiom (N): If an attribute i is null for $v \in \mathcal{G}(L)$, then $\phi_i(v) = 0$.

The third axiom says that the numbering of the attributes should have no influence on their importance. Let σ be a permutation on N. For all $x \in L$, we denote $\sigma(x)_{\sigma(i)} = x_i$. For all $v \in \mathcal{G}(L)$, the game $\sigma \circ v$ is defined by $\sigma \circ v(\sigma(x)) = v(x)$.

Symmetry axiom (S): For any permutation σ of N, $\phi_{\sigma(i)}(\sigma \circ v) = \phi_i(v)$, $\forall i \in N$.

The next axiom is an invariance property. It says that the calculus of the importance index does not depend on the position on the "grid" L. It is another kind of symmetry axiom, relative to the levels $0, 1, \ldots, k$, not to the attributes.

Invariance axiom (I): Let us consider two games $v, w \in \mathcal{G}(L)$ such that, for some $i \in N$,

$$v(x + 1_i) - v(x) = w(x) - w(x - 1_i), \forall x \in L, x_i \notin \{0, k\}$$

$$v(x_{-i}, 1_i) - v(x_{-i}, 0_i) = w(x_{-i}, k_i) - w(x_{-i}, k_i - 1), \forall x_{-i} \in L_{-i}.$$

Then $\phi_i(v) = \phi_i(w)$.

The axiom says that two games v, w should have the same importance index for i if v, w have the same derivative w.r.t. i, up to a shift of one unit.

An important intermediary result, which will be used in the sequel, is the following:

Proposition 1 *Under axioms (L), (N), (I) and (S), $\forall v \in \mathcal{G}(L), \forall i \in N$,*

$$\phi_i(v) = \sum_{x_{-i} \in L_{-i}} p_{n(x_{-i})}\big(v(x_{-i}, k_i) - v(x_{-i}, 0_i)\big),$$

where $n(x_{-i}) = (n_0, n_1, \ldots, n_k)$ with n_j the number of components of x_{-i} being equal to j.

Finally, the "efficiency" axiom used here is the following one:

Efficiency axiom (E): For all $v \in \mathcal{G}(L)$,

$$\sum_{i \in N} \phi_i(v) = \sum_{\substack{x \in L \\ x_j < k}} \big(v(x + 1_N) - v(x)\big).$$

It can be explained as follows: taking an alternative $x \in L$ and increasing the value of each attribute by one unit, i.e., going to $x + 1_N$, the amount of variation is due to the contribution of all attributes, and the sum of all importance indices should be equal to the sum of this variation for all alternatives x. The axiom shows that the approach taken here is more in the spirit of the calculus of variation than game theory. Interestingly, the axiom is nevertheless not so far from the original efficiency axiom because when taking $k = 1$, it reduces to the classical efficiency axiom $\sum_i \phi_i(v) = v(N)$.

Although all axioms seem to fit well our MCDA context, the obtained importance index fails to satisfactorily represent importance in every case. Indeed, it does not solve well Example 1, because in this case, supposing that x has no coordinate equal to 0 or k, it follows that $\delta_x(y_{-i}, k_i) = \delta_x(y_{-i}, 0_i) = 0$, so that $\phi_i(\delta_x) = 0$ for all $i \in N$. Evidently, this is counterintuitive because each of the attributes (temperature, humidity, wind) has a nonnull importance to determine comfort.

We try to overcome this drawback and propose a new kind of importance index. The basic idea is also rooted in the calculus of variation: the importance index of an attribute i is the average variation of v when the value of attribute i is increased by a unit. However, contrarily to (9), the variation of v is cumulated by taking the absolute value of the variations. Doing so, there is no cancellation effect, i.e., a positive variation at x cannot be cancelled by a negative variation of the same amount at x'. Therefore the counterintuitive result obtained on Example 1 with (9) cannot happen any more.

4   Axiomatization

Based on the previous considerations, we propose an axiomatic approach to define an importance index which has the following general form:

$$\phi_i(v) = \sum_{\substack{x \in L \\ x_i < k}} p_x^i |v(x + 1_i) - v(x)|, \tag{10}$$

where p_x^i are real coefficients for every $i \in N$ and $x \in L, x_i < k$. We follow the approach of Weber [17] for the axiomatization, that is, we introduce the axioms one by one and determine the family of importance indices satisfying the introduced axioms at each step.

The major difficulty in axiomatizing (10) is that ϕ does not satisfy linearity. However, linearity is the very first requirement in order to start an axiomatization à la Weber. We circumvent the difficulty by remarking that, if v is monotone, then $|v(x + 1_i) - v(x)| = v(x + 1_i) - v(x)$ for every $x \in L, x_i < k$. However, as $\mathcal{G}_M(L)$ is *not* a linear subspace of $\mathcal{G}(L)$ but a convex cone, we cannot directly apply the linearity axiom on $\mathcal{G}_M(L)$. The idea is the following: using the expression of v in the basis of unanimity games (5), this expression turns to be a conic combination iff v is in $\mathcal{G}_+(L)$. As any game can be written as the difference of two games in $\mathcal{G}_+(L)$ (see (6)), it is then possible to extend this expression to monotone games. Hence, ϕ should commute with conic combination and differences of games in $\mathcal{G}_+(L)$.

Conic Combination axiom (CC): For every $v, w \in \mathcal{G}_+(L)$, for every $\alpha \in \mathbb{R}_+$,

$$\phi(v + \alpha w) = \phi(v) + \alpha \phi(w).$$

Decomposition axiom (D): If $v, v' \in \mathcal{G}_+(L)$ and $v - v'$ is monotone, then $\phi(v - v') = \phi(v) - \phi(v')$.

These two axioms play the role of (L) in the axiomatization of ϕ^{RGL} and permit to obtain the following result.

Proposition 2 *Under axioms (CC) and (D), for all $i \in N$, there exists constants $a_x^i \in \mathbb{R}$, for all $x \in L$, such that $\forall v \in \mathcal{G}_M(L)$,*

$$\phi_i(v) = \sum_{x \in L} a_x^i v(x). \tag{11}$$

We introduce now the nullity axiom **(N)**, the symmetry axiom **(S)**, and the invariance axiom **(I)** as given in Sect. 3, up to the notable difference that now these axioms apply to games in $\mathcal{G}_M(L)$, not $\mathcal{G}(L)$. To make the distinction apparent, we label the new axioms by **(N')**, **(S')** and **(I')**. As the proof of Proposition 1 is not changed if monotone games are used, we can immediately deduce the following result.

Proposition 3 *Under axioms* **(CC)**, **(D)**, **(N')**, **(S')**, **(I')**, $\forall v \in \mathcal{G}_M(L)$, $\forall i \in N$,

$$\phi_i(v) = \sum_{x_{-i} \in L_{-i}} p_{n(x_{-i})} \big(v(x_{-i}, k_i) - v(x_{-i}, 0_i)\big), \tag{12}$$

where $n(x_{-i}) = (n_0, n_1, \ldots, n_k)$ *with* n_j *the number of components of* x_{-i} *being equal to* j.

Taking two k-ary games v and w for which the marginal contribution of a player i to a game v is the same or the opposite of that to a game w, the average importance of attribute i shall be the same for v and w. We propose the following axiom.

Marginal contribution axiom (MC): Let $i \in N$ and $v, w \in \mathcal{G}(L)$ such that

$$|\Delta_i(v)(x)| = |\Delta_i(w)(x)|, \forall x \in L, x_i < k.$$

Then

$$\phi_i(v) = \phi_i(w).$$

Remark 1 Axioms **(MC)** and **(CC)** (or **(MC)** and **(D)**) imply the null axiom **(N')**. Indeed, from **(CC)** or **(D)** we deduce $\phi(0) = 0$, while $\Delta_i(v) = 0 = \Delta_i(0)$ if i is null for v. We may then discard **(N')** from the list of axioms.

Theorem 1 *Under axioms* **(CC)**, **(D)**, **(S')**, **(I')** *and* **(MC)**, *for all* $v \in \mathcal{G}(L)$

$$\phi_i(v) = \sum_{\substack{x \in L \\ x_i < k}} p_{n(x_{-i})} |v(x + 1_i) - v(x)|, \forall i \in N,$$

where $n(x_{-i}) = (n_0, n_1, \ldots, n_k)$ *with* n_j *the number of components of* x_i *being equal to* j.

The previous axioms **(CC)**, **(D)**, **(S')**, **(I')** and **(MC)** have determined the desired form (10): the importance index of attribute i is a weighted sum of the absolute variations v when the value of attribute i is increased by one unit. The determination of the weights is a matter of convention or normalization, and depends on how the numerical value of the importance index is fixed for some typical games. Here, we propose two different ways of achieving this.

The most elementary one seems to take as remarkable game δ_x. Observe that if $x_i \neq 0, k$, the sum of absolute variations along the i axis is 2, otherwise it is 1. Normalizing by the total number of points in the grid L_{-i}, which is $(k+1)^{n-1}$ so that the result is not dependent of the size of the grid, we obtain the following first version of the "calibration" axiom:

Calibration axiom 1st version (C1): For every $x \in L \setminus \{0_N\}$

$$\phi_i(\delta_x) = \begin{cases} 2/(k+1)^{n-1} & \text{if } i \in S(x) \setminus K(x) \\ 1/(k+1)^{n-1} & \text{otherwise.} \end{cases}$$

Proposition 4 *Under axioms* **(CC)**, **(D)**, **(S')**, **(I')**, **(MC)** *and* **(C1)**, *for all* $v \in \mathcal{G}(L)$

$$\phi_i(v) = \frac{1}{(k+1)^{n-1}} \sum_{\substack{x \in L \\ x_i < k}} |v(x + 1_i) - v(x)|, \forall i \in N,$$

Example 2 We illustrate Proposition 4 and we compare our importance index with that of Ridaoui et al. [13]. Let $N = \{1, 2, 3\}, k = 2$, and consider $x = (2, 1, 1)$.

$$\phi_1(\delta_x) = \frac{1}{9}, \phi_2(\delta_x) = \phi_3(\delta_x) = \frac{2}{9}$$

We note that $\forall i \in N, \phi_i(\delta_x)$ is different from zero. However,

$$\phi_1^{RGL}(\delta_x) = 1, \phi_2^{RGL}(\delta_x) = \phi_3^{RGL}(\delta_x) = 0$$

This result is counterintuitive because each attributes must have a non null importance.

We may now use the unanimity games instead of the δ_x games, in a spirit similar to the value proposed by Peters and Zank (see (8)).

Calibration axiom 2nd version (C2): $\forall x \in L \setminus \{0_N\}, \forall i \in S(x)$,

$$\phi_i(u_x) = \frac{1}{s(x)}.$$

Proposition 5 *Under axioms* **(CC)**, **(D)**, **(S')**, **(I')**, **(MC)** *and* **(C2)**, *for all* $v \in \mathcal{G}(L)$

$$\phi_i(v) = \sum_{\substack{x_{-i} \in \{0,k\}^{N \setminus \{i\}} \\ x_i \in L_i, x_i < k}} \frac{(n - s(x_{-i}) - 1)! s(x_{-i})!}{n!} |v(x + 1_i) - v(x)|, \forall i \in N.$$

We remark that the coefficients which are obtained are close to the coefficients of the classical Shapley value. This is no surprise because (8) in the axiomatization of Peters and Zank implies the classical efficiency axiom.

5   Towards a More General Approach

The fundamental idea behind the general expression of the importance index for an attribute i given in Theorem 1 is to cumulate the magnitude of the variations of v when the value of attribute i varies from 0 to k. We have taken the absolute value of the variation as magnitude, but it is clear that other definitions can be taken as well. For example the square of the variations can be taken, so that the importance index could be defined as the square root of the sum of the square of the variations. Remarking that this is the L_2 norm, while the importance index obtained through (C1) (Proposition 4) is, up to a normalization factor, the L_1 norm of the vector of variations, a general approach would be to define the importance index as the norm of the vector of variations, for some given norm:

$$\phi_i(v) = \|\Delta_i(v)\| \qquad (i \in N). \tag{13}$$

Note that the use of a norm different from L_1 forbids to take an axiomatic approach similar to the one we used in Sect. 4, because there would exist no class of games where a property similar to linearity would hold. Nevertheless, it is possible to obtain a general form through a number of axioms which are presented below. In the rest of this section $i \in N$ is fixed.

Nonnegativity (NN): The importance index takes nonnegative values, i.e., $\phi_i : \mathcal{G}(L) \to \mathbb{R}_+$.

Absolute Homogeneity (AH): For every $\alpha \in \mathbb{R}$ and every game $v \in \mathcal{G}(L)$,

$$\phi_i(\alpha v) = |\alpha| \phi_i(v)$$

Subadditivity (SA): For any games $v, w \in \mathcal{G}(L)$,

$$\phi(v + w) \le \phi(v) + \phi(w)$$

Strong Null axiom (SN): $\phi_i(v) = 0$ if and only if i is null for v.

The nonnegativity axiom says that importance indices are nonnegative quantities. Absolute homogeneity says that multiplying a game by a constant just multiplies the importance index by the magnitude of this constant. The subadditivity axiom expresses the fact that summing two games v, w may hinder the importance of an attribute by some hedging effect: the positive variation of i at some point x for v can be cancelled by a negative variation at the same point for w. Lastly, the strong null axiom is a strong version of the usal null axiom, in the sense that *only* games whose attribute i is null can lead to a null importance index for i.

We obtain the following.

Theorem 2 *Under axioms (NN), (AH), (SA) and (SN), there exists a norm* $\| \cdot \|$ *on* $\mathbb{R}^{k(k+1)^{n-1}}$ *and a linear one-to-one mapping h on* $\mathbb{R}^{k(k+1)^{n-1}}$ *such that*

$$\phi_i(v) = \|h \circ \Delta_i(v)\|.$$

Additional axioms may be used to determine a particular norm or class of norms. Note however that the precise determination of h through calibration or efficiency axioms seems to be difficult as h lies inside the norm.

6 Conclusion and Future Works

We have proposed an importance index for MCDA models with discrete attributes, without any additional assumption on the model, which works even in the case of nonmonotonic models. Its original feature is that it cumulates the absolute value of variations along the values of the attributes. We have given an axiomatization of it, proposed several ways of calibration, and proposed the fundamentals for a more general approach based on the norm of the derivative. Future work will consist in pursuing this general approach using norms, and extending our model and importance index to the case of continuous attributes.

References

1. Choquet, G.: Theory of capacities. Annales de l'institut Fourier **5**, 131–295 (1953)
2. Fishburn, P.: Interdependence and additivity in multivariate, unidimensional expected utility theory. Int. Econ. Rev. **8**, 335–342 (1967)
3. Fishburn, P.: Utility Theory for Decision Making. Wiley, New York (1970)
4. Grabisch, M., Labreuche, C.: A decade of application of the Choquet and Sugeno integrals in multi-criteria decision aid. Ann. Oper. Res. **175**, 247–286 (2010)
5. Grabisch, M., Labreuche, C.: Capacities on lattices and k-ary capacities. In: 3rd International Conference of the European Society for Fuzzy Logic and Technology (EUSFLAT 2003), Zittau, Germany, pp. 304–307, September 2003
6. Grabisch, M., Lange, F.: Games on lattices, multichoice games and the Shapley value: a new approach. Math. Methods Oper. Res. **65**(1), 153–167 (2007)
7. Grabisch, M., Marichal, J.-L., Roubens, M.: Equivalent representations of set functions. Mathe. Oper. Res. **25**(2), 157–178 (2000)
8. Hsiao, C.R., Raghavan, T.E.S.: Shapley value for multi-choice cooperative games, I. Games Econ. Behav. **5**, 240256 (1993)
9. Keeney, R.L., Raiffa, H.: Decision with Multiple Objectives. Wiley, New York (1976)
10. Klijn, F., Slikker, M., Zarzuelo, J.: Characterizations of a multi-choice value. Int. J. Game Theor. **28**(4), 521–532 (1999)
11. Murofushi, T.: A technique for reading fuzzy measures (I): the Shapley value with respect to a fuzzy measure. In: 2nd Fuzzy Workshop, Nagaoka, Japan, pp. 39–48, October 1992. In Japanese
12. Peters, H., Zank, H.: The egalitarian solution for multichoice games. Ann. Oper. Res. **137**(1), 399–409 (2005)
13. Ridaoui, M., Grabisch, M., Labreuche, C.: Axiomatization of an importance index for generalized additive independence models. In: Proceedings of ECSQARU 2017, Lugano, Switzerland, July 2017
14. Shapley, L.S.: A value for n-person games. In: Kuhn, H.W., Tucker, A.W. (eds.) Contributions to the Theory of Games. Annals of Mathematics Studies, vol. 2(28), pp. 307–317. Princeton University Press (1953)
15. Sugeno, M.: Theory of fuzzy integrals and its applications. Ph.D thesis, Tokyo Institute of Technology (1974)
16. van den Nouweland, A., Tijs, S., Potters, J., Zarzuelo, J.: Cores and related solution concepts for multi-choice games. Zeitschrift für Oper. Res. **41**(3), 289–311 (1995)
17. Weber, R.J.: Probabilistic values for games. In: Roth, A.E. (ed.) The Shapley Value: Essays in Honor of Lloyd S. Shapley, pp. 101–120. Cambridge University Press (1988)

Anytime Algorithms for Adaptive Robust Optimization with OWA and WOWA

Nadjet Bourdache and Patrice Perny[✉]

Sorbonne Universités, UPMC Univ Paris 06, UMR 7606, LIP6 CNRS, UMR 7606,
LIP6, 4 Place Jussieu, 75005 Paris, France
nadjet.bourdache@gmail.com, patrice.perny@lip6.fr

Abstract. We consider optimization problems in graphs where the utilities of solutions depend on different scenarios. In this context, we study incremental approaches for the determination of robust solutions, i.e. solutions yielding good outcomes in all scenarios. Our approach consists in interleaving adaptive preference elicitation methods aiming to assess the attitude of the Decision Maker towards robustness or risk with combinatorial optimization algorithms aiming to determine a robust solution. Our work focuses on the use of ordered weighted average (OWA) and weighted ordered weighted average (WOWA) to respectively model preferences under uncertainty and risk while accounting for the idea of robustness. These models are parameterized by weighting coefficients or weighting functions that must be fitted to the value system of the Decision Maker. We introduce and justify anytime algorithms for the adaptive elicitation of these parameters until a robust solution can be determined. We also test these algorithms on the robust assignment problem.

Keywords: Robust optimization · Preference elicitation · OWA · WOWA · Ranking algorithms · Assignment problem

1 Introduction

The practice of decision support in complex environments has shown the importance of developing new models and algorithms for optimization under partial information. Uncertainty is pervasive in discrete optimization and various contributions concern robust optimization problems [11] in which multiple instances of the same problem must be considered simultaneously. These instances correspond to possible scenarios and the goal is to determine a feasible solution that remains as good as possible in all scenarios.

In graph optimization, several problems have been revisited under this perspective. For example, assuming that uncertainty only impacts the valuation of the graph (and not on its structure), several contributions address the discrete scenario case, and propose various reformulations of shortest-path problems, assignment problems, minimum spanning tree problems, as a min-max or min-max regret optimization problems, see [1,5,11,15,28,29]. For the same problems,

© Springer International Publishing AG 2017
J. Rothe (Ed.): ADT 2017, LNAI 10576, pp. 93–107, 2017.
DOI: 10.1007/978-3-319-67504-6_7

a similar work has been carried out in the case of graphs valued with interval data, which corresponds to an infinity of possible scenarios [1,2,11,14,24,27]. In this paper we consider the discrete scenario case and propose new (interactive) approaches for robust optimization, with an implementation on the assignment problem. Let us introduce an instance of the robust assignment problem, for illustrative purpose:

Example 1. *We consider a problem with 4 items that must be assigned to 4 agents (one item per agent and one agent per item). The utility of every item for every agent depends on the context which remains unknown. Three scenarios $\{s_1, s_2, s_3\}$ are considered leading to the following three utility matrices:*

$$U_1 = \begin{pmatrix} 10 & 3 & 6 & 4 \\ 10 & 1 & 0 & 1 \\ 9 & 9 & 4 & 0 \\ 4 & 2 & 3 & 4 \end{pmatrix} \quad U_2 = \begin{pmatrix} 10 & 5 & 0 & 2 \\ 6 & 7 & 1 & 4 \\ 7 & 0 & 7 & 3 \\ 1 & 9 & 10 & 2 \end{pmatrix} \quad U_3 = \begin{pmatrix} 0 & 2 & 8 & 0 \\ 2 & 0 & 6 & 5 \\ 4 & 9 & 4 & 2 \\ 7 & 8 & 3 & 9 \end{pmatrix}$$

where U_s is the utility matrix giving in row i and column j the utility u_{ij}^s of item j for agent i in scenario s, for $i = 1, \ldots, 4$, $j = 1, \ldots, 4$, $s = 1, \ldots, 3$. We want to find an assignment that remains as good as possible in all possible scenarios.

Any solution of an $n \times n$ assignment problem is characterized by a one-to-one mapping α defined from the set of agents to the set of items and associating item $\alpha(i)$ to agent i for $i = 1, \ldots n$. Equivalently assignment α will be represented by the set of arcs $\{(i, \alpha(i)), i = 1, \ldots, n\}$ in the bi-partite assignment graph. Given an assignment α, its total utility in scenario s is denoted as $u_s(\alpha)$ and defined by $u_s(\alpha) = \sum_{i=1}^{n} u_{i\alpha(i)}^s$, $s = 1, \ldots, n$. If q distinct scenarios are considered, then any assignment α is characterized by the utility vector $u(\alpha) = (u_1(\alpha), \ldots, u_q(\alpha))$ representing the possible utilities in the q scenarios. For example, in the problem described in Example 1, the assignment $\alpha = \{(1,3),(2,1),(3,2),(4,4)\}$ leads to the utility vector $u(\alpha) = (29, 8, 28)$ whereas $\alpha' = \{(1,3),(2,4),(3,1),(4,2)\}$ leads to the utility vector $u(\alpha') = (18, 20, 25)$. Hence the comparison of α and α' amounts to comparing utility vectors $u(\alpha)$ and $u(\alpha')$.

In order to be able to compare the utility vectors attached to feasible solutions, we need to assess the attitude of the Decision Maker (DM) toward uncertainty or risk (the latter situation occurring when the probabilities of the scenarios are known), and to decide how the outcomes attached to the q scenarios must be aggregated to define the overall value of a solution. One standard criterion for decision making under uncertainty is the *Laplace criterion* that consists in maximizing the average utility taken over all scenarios. In Example 1, the optimal assignment w.r.t. the Laplace criterion is α with an average of $65/3$.

A more cautious approach consists in choosing a solution maximizing the utility measured in its worst scenario. This corresponds to max-min optimization, a standard approach in robust optimization (equivalent to min-max optimization when costs are considered). In Example 1, the max-min optimal assignment is α' introduced above. It can be seen as more robust than α since it guarantees a utility greater or equal to 18 whereas we could obtain 8 with α in scenario s_2.

However the max-min criterion is often considered as overpessimistic and poorly discriminating. In particular, we may obtain solutions having a much lower average utility than with the Laplace criterion. On the other hand, by compensating high and low utilities, the Laplace criterion does not provide any guarantee on the robustness of solutions. In order to have a better flexibility in modelling the DM's attitude toward uncertainty, and to find compromise attitudes between overpessimism and full compensation, we will use an *ordered weighted average* (OWA) to aggregate the utilities obtained in the different scenarios. OWA operators have been introduced by Yager [26] and axiomatically justified in the context of robust discrete optimization in [18]. They are also widely used in fair multiagent optimization for their ability to generate Pareto-optimal solutions with well-balanced profiles [8,12,17].

As we shall see later more formally, the OWA of a given vector is a kind of weighted sum where the weights are not attached to positions of components in the vector but to their ranks. This allows the importance attached to good or bad outcomes to be controlled and provides a continuum of attitudes in the aggregation, ranging from the Laplace criterion (modelled by the average) to pure pessimism (modelled by the min). These various attitudes are defined by the weighting vector parameterizing the OWA model.

When the probabilities of the scenarios are known, the scenarios do not play symmetric roles. In this case we will use the WOWA model [22] also known as Yaari's model [25] which is a non-symmetric extension of OWA allowing weights to be attached to components (here scenarios). In the latter model, the DM's attitude toward risk is controlled by a probability weighting function.

Whether OWA or WOWA is used, assessing the weighting parameter of the model is a critical issue. The aim of this paper is to propose an incremental elicitation method for facilitating the parametrization of these models in order to determine solutions that are well fitted to the value system of the DM. Our aim is not to determine precise values of these parameters for the DM, prior to the optimization stage. Instead, we propose to interleave preference queries with the exploration of solutions in order to progressively reduce the uncertainty attached to these weighting parameters until a robust solution can be found. These preference queries consist in asking the DM to compare between pairs of solutions, his preferences can then be easily translated into linear constraints that we can add to our model. By integrating the elicitation to the exploration, we aim to save a large part of the elicitation burden.

The paper is organized as follows: we introduce in Sect. 2 some background on OWA and WOWA. Then we present in Sect. 3 a ranking algorithm for the determination of possibly optimal solutions when the weighting parameters of OWA or WOWA are imprecise. In Sect. 4 we present an incremental elicitation algorithm to determine or approximate an OWA-optimal or WOWA-optimal solution. In Sect. 5 we implement the proposed approach on the robust assignment problem and present numerical tests, as well as some preliminary results on the robust shortest path problem.

2   Models for Robust Optimization

In this section, we consider a general robust optimization problem involving q distinct scenarios and we discuss the evaluation of solutions represented by vectors of type $(x_1, \ldots, x_q)$ where x_i represents the utility of solution x in scenario i. In particular we recall some background on OWA and WOWA operators.

OWA Optimization. In robust optimization problems with discrete scenarios, one basically looks for Pareto-optimal solutions having a well-balanced utility profile. This vision of robustness in the context of uncertainty can be related to the notion of fairness in social choice (the scenarios acting as different agents providing different views on solutions). In particular, the preference $\succ$ of the DM is expected to satisfy *monotonicity w.r.t. ϵ-transfers*, a standard axiom used in inequality measurement that reads as follows: for all j, k $j \neq k$, for all ϵ such that $0 < x_k - x_j < \epsilon$, $(x_1, \ldots, x_j + \epsilon, \ldots, x_k - \epsilon, \ldots, x_q) \succ (x_1, \ldots, x_j, \ldots, x_k, \ldots, x_q)$.

Moving from a solution to another one using such ϵ-transfers contributes to reducing the utility gap between pairs of scenarios and thus makes the DM better off. It is well known that the minimal preference relation satisfying this condition is the Lorenz dominance relation (L-dominance) [13] defined by $x \succ_L y$ if and only if $L(x) \succ_P L(y)$ where $\succ_P$ is the Pareto dominance and $L(x)$ is the Lorenz vector the ith component of which is defined by $L_i(x) = \sum_{k=1}^{i} x_{\sigma(k)}$, where σ is the permutation of $(1, \ldots, q)$ that reorders the components of x by increasing order $(x_{\sigma(i)} \leq x_{\sigma(i+1)}, i = 1, \ldots, q-1)$. However, L-dominance is a partial order and many solutions remain incomparable. A natural way to extend this partial order, axiomatically justified in the context of robust optimization [18], is to resort to Ordered Weighted Averages (OWA for short).

OWA is an aggregation function that weights the components of a vector in function of their rank. Let $w \in \mathbb{R}_+^q$ be a weighting vector. The OWA defined by w reads: $f(x, w) = \sum_{i=1}^{q} w_i x_{\sigma(i)}$. This function is symmetric because the weights are not attached to the components of x but to the components of the reordered vector $(x_{\sigma(1)}, \ldots, x_{\sigma(q)})$. The OWA family of aggregation functions $f(x, w), w \in \mathbb{R}_+^q$ includes the minimum, the maximum, the median and all order statistics as particular cases.

OWA is widely used in fair optimization because it enables a linear extension of the Lorenz Dominance order. Remark indeed that $x_{\sigma(i)} = L_i(x) - L_{i-1}(x)$ for all $i > 1$. Hence we have: $f(x, w) = \sum_{i=1}^{q-1} (w_i - w_{i+1}) L_i(x) + w_q L_q(x)$. Thus, function f is nothing else but a linear combination of the components of the Lorenz vector. Then, if the weights are decreasing (i.e. $w_i > w_{i+1}$ for all $i = 1, \ldots, q-1$), then the coefficients $w_i - w_{i+1}$ are strictly positive. Therefore, an f-optimal solution is necessary L-non-dominated. Thus, OWA used with strictly decreasing weights w_i leads to Pareto-optimal solutions that cannot be improved in terms of ϵ-transfers reducing inequalities. This favours solutions having a well-balanced utility vector. Considering the properties of OWA recalled above, robust optimization under uncertainty can soundly be reformulated as the problem of maximizing $f(x, w)$ for some w with decreasing components.

Example 1 (continued). *Let us compare solutions α and α' such that $u(\alpha) = (29, 8, 28)$ and $u(\alpha') = (18, 20, 25)$ using $f(., w)$ with $w = (1/2, 1/3, 1/6)$ we obtain $f(u(\alpha), w) = 8/2 + 28/3 + 29/6 = 18.17$ and similarly $f(u(\alpha'), w) = 119/6 = 19.83$. Here, we observe that α' is preferred to α because the average loss in utility incurred when passing form $u(\alpha')$ to $u(\alpha)$ is compensated by the improvement of the minimum (18 against 8). Remark that, although α' is the max-min optimal solution in Example 1, it is not OWA-optimal. The OWA-optimal assignment for the chosen weights is $\alpha'' = \{(1, 1), (2, 4), (3, 2), (4, 3)\}$ with $u(\alpha'') = (23, 24, 17)$ and $f(u(\alpha''), w) = 121/6 = 20.17$. This is a compromise between α and α' that improves the average utility of α' but slightly downgrades the worst case utility.*

WOWA Optimization. As explained before, one typical property of OWA is to be a symmetric aggregator. This property seems natural when the same attention or importance is attached to every scenario. This is no longer the case when the probabilities $(p_1, \ldots, p_q)$ of the scenarios are known. In such cases we may naturally consider a weighted extension of OWA defined as follows:

$$g(x, \varphi) = \sum_{i=1}^{q} [x_{\sigma(i)} - x_{\sigma(i-1)}] \varphi(\sum_{k=i}^{q} p_{\sigma(k)}) \tag{1}$$

$$= \sum_{i=1}^{q} [\varphi(\sum_{k=i}^{q} p_{\sigma(k)}) - \varphi(\sum_{k=i+1}^{q} p_{\sigma(k)})] x_{\sigma(i)} \tag{2}$$

where $x_{\sigma(0)} = 0$; function φ is strictly increasing on the unit interval. A solution x is as least as good as a solution y when $g(x, \varphi) \geq g(y, \varphi)$. This formulation is known as the *Yaari's model* in the literature on decision under risk because it has been introduced and axiomatically justified by Yaari [25] in this context. Since φ is increasing on the unit interval, the preferences induced by the Yaari's model are monotonic with respect to first-order stochastic dominance (FSD). This means that if two solutions x and y are such that the probability $G_x(t) = P[u(x) > t]$ is greater or equal than the probability $G_y(t) = P[u(y) > t]$ for all t then $g(x, \varphi) \geq g(y, \varphi)$. Moreover, a necessary and sufficient condition for these preferences to be monotonic with respect to second order stochastic dominance (SSD) is the convexity of φ. This means that x *SSD* y implies $g(x, \varphi) \geq g(y, \varphi)$ whenever φ is convex, where x *SSD* y means that $\int_{-\infty}^{u} G_x(t) dt \geq \int_{-\infty}^{u} G_y(t) dt$ for all u. This property established in [9] has a major importance in the context of robust optimization since monotonicity with respect to SSD is the standard model of strong risk aversion. This is due to the fact that SSD is strongly related to the existence of mean-preserving spread of the utility distribution increasing the risk attached to a solution (just as Lorenz dominance relates to the existence of ϵ-transfers reducing inequalities); for more details see [21]. For this reason, we shall use convex functions φ to account for risk aversion in robust optimization problem with probabilities over scenarios.

Example 2. *We come back to Example 1 and assume now that the probabilities of the 3 scenarios are given by $p = (1/2, 1/6, 1/3)$. Let us compare solutions α*

and α' such that $u(\alpha) = (29, 8, 28)$ and $u(\alpha') = (18, 20, 25)$ using Yaari's model with $\varphi(x) = x^2$. We have: $g(\alpha, \varphi) = 8 + (28 - 8)\varphi(5/6) + (29 - 28)\varphi(1/2) = 22.1$ and $g(\alpha', \varphi) = 18 + (20 - 18)\varphi(1/2) + (25 - 20)\varphi(1/3) = 19.1$. Therefore α is preferred to α'. The fact that scenario s_1 has a high probability compared to the others gives the advantage to α even if it is more risky than α'. Note that the conclusion would be different if the requirement of risk-aversion were strengthened, using a function φ of higher convexity. For example, if $\varphi(x) = x^4$ we have $g(\alpha, \varphi) = 17.7$ whereas $g(\alpha', \varphi) = 18.2$. In this case α' is preferred to α.

Note that when scenarios are equiprobable then $p_i = 1/q$ and the coefficient of $x_{\sigma(i)}$ in Eq. (2) becomes: $w_i = [\varphi(\frac{q-i+1}{q}) - \varphi(\frac{q-i}{q})]$ which is now constant (independent of σ). In this case the Yaari's model reduces to a standard OWA. Hence Yaari's model can be seen as a weighted generalization of OWA which explains the name WOWA due to Torra [22] and used in this paper.

Considering the properties of WOWA recalled above, robust optimization under multiple scenarios of known probabilities can soundly be reformulated as the problem of maximizing $g(x, \varphi)$ for some proper convex weighting function φ. The determination of an f-optimal or g-optimal solution in graph optimization is generally a challenging problem. Standard constructive algorithms based on dynamic programming or greedy search do not apply directly because neither f-optimality nor g-optimality satisfy the Bellman principle. An f-optimal solution for a given vector w can include sub-optimal subsolutions due to the non-linearity of f with respect to outcomes. For example, when $w = \{(1, 0, \ldots, 0)\}$, f-optimization is nothing else but $\sum$-min maximization ($\sum$-max minimization for costs) which is known to be NP-hard for assignment problems but also for shortest-path problems [1,11]. The same remarks holds for g-optimality. Yet some solution methods are available both for OWA and WOWA optimization, e.g., [12,17].

3   Optimization with Imprecise Parameters

Possibly OWA-optimal Solutions. It is often difficult to precisely determine the weighting vector w to be used in the OWA model. Indeed, the only prior information we have is that the weighting vector is positive and strictly decreasing, providing an exponential number of possible weights, and then an exponential number of preference queries to be assessed. Yet, in most cases, some preference information is available, putting some constraints on the set of admissible weighting vectors. Note that, any judgment of type "i prefer x to y" (where x and y are two feasible utility vectors) translates into the following inequality: $f(x, w) \geq f(y, w)$, which is a linear constraint in w bounding the set of admissible weighting vectors W. Thus, we define the *uncertainty set* W as a convex polyhedron including all weighting vectors compatible with the preference information collected so far. Given the uncertainty set W and a set X of utility vectors attached to feasible solutions, we define $PO_W(X)$ as the set of possibly f-optimal solutions in X, i.e., the elements of X which are f-optimal for some weighting vector w in W. More formally: $\forall X \subseteq \mathbb{R}^n, PO_W(X) = \bigcup_{w \in W} \arg \max_{x \in X} f(x, w)$.

The uncertainty set W allows a W-dominance relation over utility vectors to be defined as follows: $x \succ_W y \iff [\forall w \in W, f(x, w) > f(y, w)]$. This relation can be extended to set-wise dominance: a solution $x \in X$ is said to be dominated by a set $Y \subseteq X$ (denoted $Y \succ_W x$) if there exists $y \in Y$ such that $y \succ_W x$. If Y is explicitly defined, we can decide whether a solution x is dominated by Y in polynomial time by testing whether $\min_{w \in W} \max_{y \in Y} f(y, w) - f(x, w) > 0$ which is done by solving a linear program. Hence, when X is given explicitly, $PO_W(X)$ can be computed in polynomial time. It is sufficient to iteratively remove from X any solution x_i dominated by X, for $i = 1, \ldots, n$. This was shown in [3] for decision models based on the minimization of a weighted sum. We kept the same idea while adapting the algorithm for the OWA maximization problem.

In robust optimization problems, the set X is only implicitly defined and computing $PO_W(X)$ is more difficult. To overcome the problem we now introduce a ranking approach to compute $PO_W(X)$. This approach consists of three steps: (1) *linear scalarization*: a scalar valued instance of the problem is constructed by replacing utility vectors attached to the edges of the graph by their average over all scenarios. (2) *ranking*: we perform an enumeration of solutions by decreasing order of utilities; several algorithms are available in the literature to rank the solutions of an optimization problem by decreasing order of preferences, e.g., Murty algorithm [16] for assignment problems or Eppstein algorithm [6] for shortest path problems. (3) *stopping condition*: we stop the enumeration when we can prove that all possibly optimal solutions have been enumerated. The stopping condition used in step 3 is justified by the following propositions:

Proposition 1. *For any weighting vector $w \in \mathbb{R}_+^q$ with decreasing weights such that $\sum_{i=1}^q w_i = 1$, we have, for all $x \in \mathbb{R}^q$, $f(x, w) \leq \frac{1}{q} \sum_{i=1}^q x_i$.*

Proposition 1, that directly derives from the result presented in proposition 3 of [7], allows us to establish the following result:

Proposition 2. *Let X be the set of all feasible vectors, and let $X^k = \{x^1, \ldots, x^k\}$ be the list of the k best elements of X ordered by decreasing average. We have: $\max_{x \in PO_W(X^k)} \min_{w \in W} f(x, w) > \frac{1}{q} \sum_{i=1}^q x_i^k \Rightarrow PO_W(X) \subseteq PO_W(X^k)$.*

Proof. We show that when the "if" condition holds at step k, then any element that does not belong to X^k cannot be optimal. Let us consider $x \in X \setminus X^k$. Then if $\max_{x \in PO_W(X^k)} \min_{w \in W} f(x, w) > \frac{1}{q} \sum_{i=1}^q x_i^k$ then there exists $y \in X^k$ such that $\min_{w \in W} f(y, w) > \frac{1}{q} \sum_{i=1}^q x_i^k$. Moreover since x comes after x^k in the ordered enumeration, we have $\frac{1}{q} \sum_{i=1}^q x_i^k \geq \frac{1}{q} \sum_{i=1}^q x_i \geq f(x, w)$ for any $w \in W$. Hence, for any $w \in W$ we have $f(y, w) > f(x, w)$ and therefore $x \notin PO_W(X)$. This shows that $PO_W(X) \subseteq X^k$. Moreover an element in $PO_W(X)$ cannot be $\succ_W$-dominated in X^k since $X^k \subseteq X$. Hence $PO_W(X) \subseteq PO_W(X^k)$. $\square$

Proposition 2 provides a stopping condition for a ranking algorithm based on the mean of utilities to determine the set $PO_W(X)$. If at step k of the ranking algorithm the condition is fulfilled, then all solutions that would be enumerated after step k are W-dominated by a solution in $PO_W(X^k)$. The enumeration can be stopped since the elements we are looking for are all included in $PO_W(X^k)$ which can easily be computed (as previously explained) since X^k is explicitly known at step k. To summarize, when X is implicitly defined, the set $PO_W(X)$ can still be determined by ranking its elements by increasing average utilities. This ranking is stopped as soon as the condition described above is activated. Remark that, for the robust version of assignment problem [1,11], we cannot expect the stopping condition to be activated within a polynomial number of steps since the problem of determining $PO_W(X)$ is already NP-hard when W is reduced to the single vector $\{(1,\ldots,0)\}$. Nevertheless, we shall see in the section dedicated to numerical tests that the stopping condition is activated after a reasonable number of iterations on average. Moreover, the stopping condition can be relaxed into $\frac{1}{q}\sum_{i=1}^{q} x_i^k - \max_{x \in PO_W(X^k)} \min_{w \in W} f(z,w) \leq \delta$ where δ is a positive threshold representing the maximum admissible error. This possibly saves multiple iterations while providing good approximations of $PO_W(X)$.

Possibly WOWA-optimal Solutions. The ranking approach described above can also be adapted to compute $PO_\Phi(X)$ for g-optimization under strong risk aversion. In this case φ is assumed to be convex. Consistently with the approach proposed for OWA, we consider now an uncertainty set Φ of admissible convex functions (functions compatible with the preferences observed so far). We will show later that, under mild hypothesis, Φ can also be represented by a convex polyhedron. Given the uncertainty set Φ, we define $PO_\Phi(X)$ as the set of possibly g-optimal solutions in X, i.e., the elements of X which are g-optimal for at least one function φ in Φ. More formally: $\forall X \subseteq \mathbb{R}^n$, $PO_\Phi(X) = \bigcup_{\varphi \in \Phi} \arg\max_{x \in X} g(x,\varphi)$. The associated dominance relation is defined by: $x \succ_\Phi y \Leftrightarrow [\forall \varphi \in \Phi, g(x,\varphi) > g(y,\varphi)]$. This relation obviously extends to set-wise dominance. Deciding whether a solution y is dominated by a set X amounts to testing whether $\min_{\varphi \in \Phi} \max_{x \in X} g(x,\varphi) - g(y,\varphi) > 0$ which may be done by linear programming provided that Φ is represented by a convex polyhedron. Moreover, when X is explicitly defined, the set $PO_\Phi(X)$ can easily be computed by iteratively eliminating dominated elements, as done for $PO_W(X)$. Now, in order to install a ranking procedure to determine $PO_\Phi(X)$ when X is implicitly defined, we need to establish a counterpart of Propositions 1 and 2 for g-optimization. This is exactly the role of the two following propositions:

Proposition 3. *For any convex function* $\varphi : [0,1] \to [0,1]$ *such that* $\varphi(0) = 0$ *and* $\varphi(1) = 1$ *we have, for all* $x \in \mathbb{R}^q, g(x,\varphi) \leq \sum_{i=1}^{q} p_i x_i$.

Proof. Since φ is convex we have $\varphi(ta + (1-t)b) \leq t\varphi(a) + (1-t)\varphi(b)$ for all $a,b,t \in [0,1]$. Setting $a = 1$ and $b = 0$ we obtain: $\varphi(t) \leq t$ for all $t \in [0,1]$. Hence $\varphi(\sum_{k=i}^{q} p_{\sigma(k)}) \leq \sum_{k=i}^{q} p_{\sigma(k)}$ for $i = 1,\ldots,q$. Therefore we have: $g(x,\varphi) \leq \sum_{i=1}^{q} [x_{\sigma(i)} - x_{\sigma(i-1)}] \sum_{k=i}^{q} p_{\sigma(k)}$ by Eq. (1) since $x_{\sigma(i)} - x_{\sigma(i-1)} \geq 0$ for $i =$

$1, \ldots, q$. Hence we obtain: $g(x, \varphi) \leq \sum_{i=1}^{q} \left[\sum_{k=i}^{q} p_{\sigma(k)} - \sum_{k=i+1}^{q} p_{\sigma(k)} \right] x_{\sigma(i)} = \sum_{i=1}^{q} p_{\sigma(i)} x_{\sigma(i)} = \sum_{i=1}^{q} p_i x_i$. $\qquad \square$

Proposition 4. *Let X be the set of all feasible vectors and $X^k = \{x^1, \ldots, x^k\}$ be the list of the k best elements of X ordered by decreasing expected utility. Let $p_1, \ldots, p_q$ be the probabilities of the q scenarios. The following property holds:*
$$\max_{x \in PO_\Phi(X^k)} \min_{\varphi \in \Phi} g(z, \varphi) > \sum_{i=1}^{q} p_i x_i^k \Rightarrow PO_\Phi(X) \subseteq PO_\Phi(X^k).$$

The proof is very similar to the one of Proposition 2 and is deliberately omitted. To complete the parallel with the approach proposed for computing $PO_W(X)$, when X implicitly defined, the set $PO_\Phi(X)$ can be determined by ranking the solutions of the instance obtained by replacing the utility vectors attached to the edges of the graph by their expected utility. This approach applies to problems for which a ranking algorithm is available, in particular, assignment problems, shortest path and minimum spanning tree problems.

4   Interleaving Elicitation and Ranking

Incremental Elicitation. As shown in Example 1, preferences induced by $f(\cdot, w)$ or $g(\cdot, \varphi)$ models may be sensitive to variations of their respective parameters, w and φ. It is therefore necessary to design elicitation procedures aiming to reduce the uncertainty set W (resp. Φ) introduced in the previous section. The elicitation of these parameters may require numerous preference queries if it is performed independently on the problem instance to be solved. For this reason, it is preferable to interleave elicitation and search. We suggest inserting preference queries in the ranking algorithm presented above in order to progressively enrich the set of preference statements and the list of constraints defining W (resp. Φ). This will iteratively reduce the W (resp. Φ), and therefore the set $PO_W(X)$ (resp. $PO_\Phi(X)$) until the obtention of a necessarily optimal solution, i.e., a solution that is f-optimal (resp. g-optimal) for all remaining parameter values in the uncertainty set. This incremental elicitation process should save a large part of the elicitation burden since an optimal solution can be identified although the parameters of the models remain largely imprecise.

Regret Minimization. We want to design an *anytime* algorithm that can return a valid solution to the problem even if it is interrupted before it ends. To make such a recommendation upon request at any step of the algorithm we use a standard regret based elicitation approach [23] based on the following definitions:

$$\mathrm{PMR}(x, y, W) = \max_{w \in W} \{ f(y, w) - f(x, w) \} \tag{3}$$

$$\mathrm{MR}(x, X, W) = \max_{y \in X} \mathrm{PMR}(x, y) \tag{4}$$

$$\mathrm{MMR}(X, W) = \min_{x \in X} \mathrm{MR}(x, X, W) \tag{5}$$

The pairwise regret $\mathrm{PMR}(x, y, W)$ is the maximum regret of choosing x instead of y, defined as the maximum gap of OWA values. When X is explicitly known, this regret can easily be computed by linear programming since f is linear in w. The maximum regret (MR) attached to a solution x is the maximum regret of choosing x instead of any other solution. Finally, the minimax regret (MMR) is the minimal MR regret over X. If the algorithm is stopped at a given step with a set X of possibly optimal solutions, then we shall recommend an element x in X that achieves the MMR. This solution will be named the MMR solution hereafter. As suggested in [23], these regrets can also be used to select informative preference queries during an incremental elicitation process to iteratively reduce the MMR to zero. An efficient strategy introduced in [4] under the name the *Current Solution Strategy* (CSS for short) consists in asking the DM to compare the current MMR solution x^* with its strongest challenger defined by $y^* = \arg\max_{y \in X} \mathrm{PMR}(x^*, y, W)$. Whatever the answer to this query, a new constraint will be derived, further restricting the set W.

A similar approach could be implemented for WOWA optimization, using regrets $\mathrm{PMR}(x, y, \Phi)$, $\mathrm{MR}(x, X, \Phi)$ and $\mathrm{MMR}(X, \Phi)$ that simply derive from (3–5) by substituting $f(\cdot, w)$ by $g(\cdot, \varphi)$. However the optimization of such regrets might be challenging because we have to optimize over a continuous set of weighting functions. To overcome this problem, we use a spline representation of function φ. Spline functions are piecewise polynomials whose elements connect with a high degree of smoothness. They are widely used in data interpolation due to their ability to approximate complex shapes [20]. Interestingly enough, spline functions can be generated by linear combinations of basis spline functions. This allows to reduce the elicitation of a spline function to the determination of its weights in the spline basis. The use of spline representations for function φ in WOWA model has been recently introduced in [19]. It enables an efficient incremental elicitation of the model to describe DM's preferences over probabilistic distributions. The proposed construction relies on the definition of φ as a convex combination of m basis spline functions of degree 3, increasing from 0 to 1 on the unit interval, and known as I-spline functions [20]. More precisely we have: $\varphi(x) = \sum_{j=1}^{m} b_j I_j(x)$ where $I_j(x), j = 1, \ldots, m$ are the basic spline functions (see [19,20] for a formal definition of I_j).

Note however that this construction does not completely fit to our context because we have the additional constraint that φ must be convex, in order to enforce strong risk aversion, as explained in the previous section. To overcome this problem, we use another spline basis to generate spline functions that are both increasing and convex on the unit interval. To this end, $\varphi(x)$ is defined by $\varphi(x) = \sum_{j=1}^{m} \phi_j C_j(x), j = 1, \ldots, m$ where C_j are C-spline functions defined as the normalized integrals of the I-spline functions. More precisely: $C_j(x) = \int_0^x I_j(t)dt / \int_0^1 I_j(t)dt$. As the integrals of positive and increasing functions, C_j functions are increasing and convex. Moreover we have $C_j(0) = 0$ and $C_j(1) = 1$ for all j. Therefore, $\varphi(x)$ will also be increasing and convex since coefficients ϕ_j will be constrained to be non-negative. This is the model we use hereafter because g defined in this way is a linear function of coefficients $\phi_j, j = 1, \ldots, m$, a key property for regret optimization. Hence, any preference constraint of type $g(x, \varphi) \geq g(y, \varphi)$ translates into a linear equation in coefficients ϕ_j; thus Φ can be soundly defined as the convex polyhedron of vectors $(\phi_1, \ldots, \phi_m)$ compatible with the preference statements collected so far. We give below on Fig. 1 the I-spline basis and the associated C-spline basis used to generate φ.

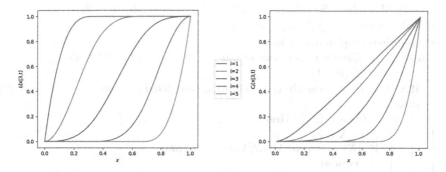

Fig. 1. I-spline and C-spline cubic functions $I_j(x)$ and $C_j(x)$ for $m = 5$

Adaptative Elicitation in a Ranking Algorithm. Our adaptive elicitation procedure of OWA's weights uses progressive reductions of MMR values to discriminate between the solutions generated by the ranking algorithm. The baseline of our algorithm is a ranking procedure enumerating the solutions by decreasing average utilities (or increasing average costs). During the enumeration, if the current solution is not dominated by the previous ones, it is inserted in a bag of solution named X. When $|X| = sb$, preferences queries are generated to discriminate between the element of X and W is reduced accordingly until $\mathrm{MMR}(X, W) = 0$. Then all solutions with a strictly positive MR value are removed from X. The process is iterated until the stopping condition introduced in Proposition 2 holds. Due to this proposition, we know that, at this point, X includes all possibly optimal solutions. Moreover, since there exists $x^* \in X$ such that $\mathrm{MR}(x^*, X, W) = 0$ (since $\mathrm{MMR}(X, W) = 0$), x^* dominates all other solutions in X for all $w \in W$ and is therefore necessarily optimal. This is the solution returned by the algorithm. The pseudo-code is given in Algorithm 1.

Function $next(G, k)$ generates the kth best solution x^k in the ranking process and returns its outcome vector (one component per scenario) or the empty vector if there are no solutions left. Function $\mathrm{MMR}(X, W)$ returns the best MR value in X for the uncertainty set W. We keep asking queries, applying CSS, until the MMR becomes zero. The stopping condition of the ranking process holds when $current\text{-}distance = 0$ where $current\text{-}distance$ is defined in line 20. So δ should be set to 0. However, in practice, the stopping condition can be relaxed by using a positive tolerance threshold δ in line 4 in order to save many iterations. More generally the algorithm can be stopped at any step k of the ranking (anytime property). The current MMR solution x^* will be returned as the current best solution. It is necessarily optimal within the set $\{x^1, \ldots, x^k\}$ of solutions enumerated so far. Moreover, all solutions coming after x^k in the ranking have an f-value lower than the average utility of x^k. Hence $current\text{-}distance$ provides an

---

Algorithm 1. MMR based elicitation in ranking for OWA optimization

---

Input: $G = (V, E)$ undirected graph; $u_i(e) \geqslant 0$ edge utility in scenario i;
q: number of scenarios, sb: size of bags; δ: tolerance threshold
Output: x^*: a near-optimal solution

1 $X \leftarrow \{\}$, $W \leftarrow \{w \in \mathbb{R}_+^q : \sum_{i=1}^q w_i = 1\}$, $x^1 \leftarrow next(G, 0)$, $k \leftarrow 2$,

2 current-distance $\leftarrow \frac{1}{q} \sum_{i=1}^q x_i^k - \min_{w \in W} f(x^1, w)$

3 enumeration-completed $\leftarrow$ false

4 **while** *(current-distance $> \delta$ **and** enumeration-completed = false)* **do**

5     $x^k \leftarrow next(G, k - 1)$

6   **if** $x^k = emptyVector$ **then** enumeration-completed $\leftarrow$ true

7   **else**

8   **if** *not($X \succ_W \{x^k\}$)* **then**

9           $X \leftarrow X \cup \{x^k\}$

10   **if** $|X| = sb$ **and** $MMR(X, W) > 0$ **then**

11   **repeat**

12   ask a preference query to the DM (selected by CSS)

13   restrict W according to the answer

14   update regrets PMR, MR and MMR

15   **until** $MMR(X, W) = 0$

16            $X \leftarrow \{x \in X : MR(x, W) = 0\}$

17   **end**

18   **end**

19   current-distance $\leftarrow \frac{1}{q} \sum_{i=1}^q x_i^k - \max_{x \in X} \min_{w \in W} f(x, w)$

20        $k \leftarrow k + 1$

21   **end**

22 **end**

23 **return** $x^* = \arg\min_{x \in X} MR(x, X, W)$, current-distance

---

upper bound on the gap to optimality in the case of an early interruption of the ranking process. Consequently, when *current-distance* is less than δ, the gap to optimality for x^* is at most δ.

Example 1 (continued). *Let us briefly illustrate the behavior of Algorithm 1 on Example 1. We use bags of size 5 ($sb = 5$) and simulate the answers of the DM using an OWA with $w = (1/2, 1/3, 1/6)$. The best solution according to the mean value is $x^1 = \{(1,3), (2,1), (3,2), (4,4)\}$ with utility vector $u(x^1) = (29, 8, 28)$ and a mean at 21.67. Then, a lower bound of the OWA value of x^1 is obtained by minimizing $f(u(x^1), w) = 8w_1 + 28w_2 + 29w_3$ over all possible weighting vectors. We obtain 8, hence current-distance $= 21.67 - 8 > 0$ and another iteration is necessary. The ordered enumeration continues until step 5 where the algorithm computes the MMR value which is strictly positive. So a preference query is asked to the DM: the MMR solution $x^2 = \{(1,1), (2,4), (3,2), (4,3)\}$ such that $u(x^2) = (23, 24, 17)$ must be compared to its best challenger x^4 chosen as explained before and such that $u(x^4) = (18, 20, 25)$. The DM prefers x^2 therefore x^4 is removed from X and the constraint $f(x^2, w) \geq f(y^4, w)$ is added to the definition of* *polyhedron W i.e.* $17w_1 + 23w_2 + 24w_3 \geq 18w_1 + 20w_2 + 25w_3$ *or equivalently* $-w_1 + 3w_2 - w_3 \geq 0$. *At this time, the MMR is still positive so the algorithm asks another query, this time between* x^2 *and* x^1 *(chosen in the same way as* x^4*), the DM still prefers* x^2*, the algorithm proceeds similarly, removing* x^1 *and adding a constraint according to this preference. After this second question, the MMR is equal to 0, therefore all solutions but* x^2 *are removed from X. Then the ranking algorithm continues until step 15 without inserting new solutions in X because all of them are dominated by* x^2*. At step 15, the stopping condition is activated and the algorithm returns the optimal solution* x^2*.*

A variant of Algorithm 1 can be used for WOWA optimization. It is sufficient to replace f by g and to modify the definition of regrets accordingly. The definition of the current distance must also be adapted not only by substituting f by g but also by replacing the weighted average by $\sum_{i=1}^{q} p_i x_i^k$. The correctness of this variant derives from Proposition 4 instead of Proposition 2.

5 Numerical Tests

We have implemented Algorithm 1 on the robust assignment problem using both OWA and WOWA models. For these tests, function $next(G, k)$ (line 5) was implemented with Murty's algorithm [16]. The complexity of Murthy's algorithm to rank assignments by increasing utility is $\mathcal{O}(Kn^4)$ for n agents and K enumerations. We used the Gurobi library of Python to solve the linear programs required for dominance tests and regret minimization. During the elicitation steps, the DM's answers to preference queries are simulated using a hidden OWA or WOWA model. We evaluate the performance of the algorithm in terms of computation time, number of preference queries and number of ranking steps, we performed tests on multiple instances of different size and number of scenarios. For every case, performances are averaged over 20 runs. The tests are performed on a Intel Core i7-4770 CPU with 11 GB of RAM. Table 1 shows the results obtained for OWA and WOWA elicitation and optimization. The performance are obtained for an error threshold δ set to 10% of the initial error (obtained for solution x^1), with a bag size of 10 and a timeout of 20 min (1200 s). Time is given in seconds in the tables and the gap is the maximal error attached to the returned solution, expressed as a percentage of the range of the valuation scale.

To test the generality of the approach, we also made some preliminary tests on the robust shortest path problem. In this problem, the arcs of the graph are valued by cost vectors with q components corresponding to the cost of the arc in the different scenarios. The definition of OWA and WOWA aggregators as well as Algorithm 1 have been modified to fit to minimization problems. For ranking the paths by increasing average costs, we used a lazy version of Eppstein's algorithm introduced in [10] with a complexity of $\mathcal{O}(m + n \log n)$ in the worst case for a graph with n nodes and m arcs. To give an idea of the performance of Algorithm 1 on robust shortest path problems, we solve instances including 500 nodes and 5 scenarios in 548 s and 5 preference queries with a gap of 0.1 on average (after

Table 1. Tests for OWA and WOWA optimization on the robust assignment problem

Number of agents		q = 3				q = 5				q = 10			
		Time	Steps	Queries	Gap	Time	Steps	Queries	Gap	Time	Steps	Queries	Gap
OWA	20	274	13607	1.3	0	719	34009	2.3	0.5	940	37739	2	1.4
	60	1200	7236	4	1.1	1200	6945	6.7	5	1200	6382	11.6	4.7
	100	1200	2075	5.1	2	1200	2550	11.3	6	1200	2216	17.5	7.5
WOWA	20	270	5338	10.3	1	475	4657	9.45	1.1	865	6743	9.2	1.7
	60	1200	5446	8.95	8.1	1200	3192	5.3	7.6	1200	4381	5.4	6.5
	100	1200	2978	6.15	10.3	1200	2302	5	15.9	1200	1901	4.9	11.3

28460 steps on average). In the case of 10 scenarios, the algorithm needs 681 s on average on similar graphs, 6.3 queries with a gap of 0.1 on average (after 41796 steps on average).

6   Conclusion

We have introduced a new adaptive elicitation approach for OWA and WOWA optimization and tested its practical efficiency on robust assignment problems and on robust shortest path problems in the discrete scenarios case. Our approach is quite general and applies to any other optimization problem for which an efficient ranking algorithm is known. An interesting extension of this work would be to design a similar approach for the incremental elicitation of Choquet integrals under the constraint of convex capacity (this is a more general model to account for robustness in optimization under uncertainty, including OWA and WOWA as special cases). The implementation of the ranking approach for Choquet integrals and the definition of a valid stopping condition are challenging questions because the capacity is imprecisely known.

References

1. Aissi, H., Bazgan, C., Vanderpooten, D.: Complexity of the minmax and minmax regret assignment problems. OR Lett. **33**(6), 634–640 (2005)
2. Aron, I.D., Van Hentenryck, P.: On the complexity of the robust spanning tree problem with interval data. OR Lett. **32**(1), 36–40 (2004)
3. Benabbou, N., Perny, P.: Incremental weight elicitation for multiobjective state space search. In: Proceedings of the Twenty-Ninth AAAI Conference on Artificial Intelligence, pp. 1093–1099 (2015)
4. Boutilier, C., Patrascu, R., Poupart, P., Schuurmans, D.: Constraint-based optimization and utility elicitation using the minimax decision criterion. Artif. Intell. **170**(8–9), 686–713 (2006)
5. Deı, V.G., Woeginger, G.J.: On the robust assignment problem under a fixed number of cost scenarios. OR Lett. **34**(2), 175–179 (2006)
6. Eppstein, D.: Finding the k shortest paths. SIAM J. Comput. **28**(2), 652–673 (1998)

7. Galand, L., Spanjaard, O.: Exact algorithms for OWA-optimization in multiobjective spanning tree problems. Comput. Oper. Res. **39**(7), 1540–1554 (2012)
8. Heinen, T., Nguyen, N.-T., Rothe, J.: Fairness and rank-weighted utilitarianism in resource allocation. In: Walsh, T. (ed.) ADT 2015. LNCS (LNAI), vol. 9346, pp. 521–536. Springer, Cham (2015). doi:10.1007/978-3-319-23114-3_31
9. Hong, C.S., Karni, E., Safra, Z.: Risk aversion in the theory of expected utility with rank dependent probabilities. J. Econ. Theory **42**(2), 370–381 (1987)
10. Jiménez, V.M., Marzal, A.: A lazy version of Eppstein's *K* shortest paths algorithm. In: Jansen, K., Margraf, M., Mastrolilli, M., Rolim, J.D.P. (eds.) WEA 2003. LNCS, vol. 2647, pp. 179–191. Springer, Heidelberg (2003). doi:10.1007/3-540-44867-5_14
11. Kouvelis, P., Yu, G.: Robust Discrete Optimization and Its Applications, vol. 14. Springer Science & Business Media, New York (2013)
12. Lesca, J., Perny, P.: LP solvable models for multiagent fair allocation problems. In: Proceedings of the 2010 Conference on ECAI 2010: 19th European Conference on Artificial Intelligence, pp. 393–398. IOS Press (2010)
13. Marshall, A.W., Olkin, I., Arnold, B.C.: Inequalities: Theory of Majorization and Its Applications, vol. 143. Springer, New York (1979)
14. Montemanni, R., Gambardella, L.M.: An exact algorithm for the robust shortest path problem with interval data. Comput. Oper. Res. **31**(10), 1667–1680 (2004)
15. Murthy, I., Her, S.-S.: Solving min-max shortest-path problems on a network. Nav. Res. Logist. (NRL) **39**(5), 669–683 (1992)
16. Murthy, K.G.: An algorithm for ranking all the assignments in order of increasing costs. Oper. Res. **16**(3), 682–687 (1968)
17. Ogryczak, W.: Multicriteria models for fair resource allocation. Control Cybern. **36**(2), 303–332 (2007)
18. Perny, P., Spanjaard, O., Storme, L.-X.: A decision-theoretic approach to robust optimization in multivalued graphs. Ann. Oper. Res. **147**(1), 317–341 (2006)
19. Perny, P., Viappiani, P., Boukhatem, A.: Incremental preference elicitation for decision making under risk with the rank-dependent utility model. In: Proceedings of UAI 2016 (2016)
20. Ramsay, J.O.: Monotone regression spline in action. Stat. Sci. **3**, 425–441 (1988)
21. Stiglitz, J.E., Rothschild, M.: Increasing risk. I. A definition. J. Econ. Theory **2**(3), 225–243 (1970)
22. Torra, V.: The weighted OWA operator. Int. J. Intell. Syst. **12**, 153–166 (1997)
23. Wang, T., Boutilier, C.: Incremental utility elicitation with the minimax regret decision criterion. In: Proceedings of IJCAI, pp. 309–318 (2003)
24. Wu, W., Iori, M., Martello, S., Yagiura, M.: Algorithms for the min-max regret generalized assignment problem with interval data. In: Proceedings of IEEM, pp. 734–738. IEEE (2014)
25. Yaari, M.E.: The dual theory of choice under risk. Econometrica **55**, 95–115 (1987)
26. Yager, R.R.: On ordered weighted averaging aggregation operators in multicriteria decision making. IEEE Trans. Syst. Man Cybern. **18**, 183–190 (1998)
27. Yaman, H., Karaşan, O.E., Pınar, M.Ç.: The robust spanning tree problem with interval data. Oper. Res. Lett. **29**(1), 31–40 (2001)
28. Gang, Y.: Min-max optimization of several classical discrete optimization problems. J. Optim. Theory Appl. **98**(1), 221–242 (1998)
29. Gang, Y., Yang, J.: On the robust shortest path problem. Comput. Oper. Res. **25**(6), 457–468 (1998)

Fair Proportional Representation Problems with Mixture Operators

Hugo Gilbert[(✉)]

Sorbonne Universités, UPMC Univ Paris 06, CNRS, LIP6 UMR 7606,
4 place Jussieu, 75005 Paris, France
hugo.gilbert@lip6.fr

Abstract. This paper deals with proportional representation problems in which a set of winning candidates must be selected according to the ballots of the voters. We investigate the use of a new class of optimization criteria to determine the set of winning candidates, namely mixture operators. In a nutshell, mixture operators are similar to weighted means where the numerical weights are replaced by weighting functions. In this paper: (1) we give the mathematical condition for which a mixture operator is fair and provide several instances of this operator satisfying this condition; (2) we show that when using a mixture operator as optimization criterion, one recovers the same complexity results as in the utilitarian case (i.e., maximizing the sum of agent's utilities) under a light condition; (3) we present solution methods to find an optimal set of winners w.r.t. a mixture operator under both Monroe and Chamberlin-Courant multi-winner voting rules and test their computational efficiency.

Keywords: Computational social choice · Inequality measurement · Mixture operators · Proportional representation

1 Introduction

This paper deals with multi-winner voting rules where one aims at electing a subset of candidates rather than a single one. In multi-winner election rules, a set of voters express preferences over a set of candidates. The objective is then to determine k winning candidates such that each voter is satisfied by the winning candidate that represents her. Multi-winner election rules are important for both political elections (i.e., electing committees of representatives) and multi-agent recommendation systems (e.g., choosing a set of dinning menus for a conference) [10,27]. A key property for multi-winner voting rules is Proportional Representation (PR), i.e., the proportional support enjoyed by the different candidates should be accurately reflected by the results of the elections.

Two multi-winner voting rules have been designed to account for PR, namely *Monroe*'s Voting Rule [19] (abbreviated by MVR) and *Chamberlin-Courant*'s Voting Rule [5] (abbreviated by CCVR). In these two frameworks, a feasible solution is characterized by a set of k winning candidates and an assignment

© Springer International Publishing AG 2017
J. Rothe (Ed.): ADT 2017, LNAI 10576, pp. 108–123, 2017.
DOI: 10.1007/978-3-319-67504-6_8

from winning candidates to voters. Each voter is then represented by the elected candidate assigned to her. While CCVR does not constraint the possible assignments from winning candidates to voters, MVR imposes that the k sets consisting of the voters represented by the same candidate should be equally sized. The choice of the solution is then based on the ballots, where each voter ranks the candidates from best to worst. Indeed, the solution chosen should maximize the utilities of the voters, where the utility (i.e., satisfaction level) of a voter depends on the rank she gave to the candidate assigned to her. In the utilitarian version of MVR and CCVR, the goal is to find a solution maximizing the sum of voter's utilities. However, maximizing the sum of utilities can yield an unfair solution as it compensates between the utilities of the different voters. Thus, several alternative optimization criteria have been investigated for multi-winner voting rules to address this problem. In CCVR, Betzler et al. proposed to maximize the utility of the least happy voter [4]. This is known as the egalitarian version of CCVR. However, maximizing the minimum utility value of the voters can be considered extreme as it does not take into account the satisfaction of all but one voter. To address this issue, Elkind and Ismaili extended this approach to Ordered Weighted Averages (OWAs) of utilities, which provides a smooth interpolation between the egalitarian version and the utilitarian version of CCVR [11].

In this paper, we investigate the use of another aggregation operator, namely Mixture Operators (MOs), to find an efficient and fair solution with CCVR and MVR. In a nutshell, MOs are similar to weighted means where the numerical weights are replaced by weighting functions. The solution sought with the MO should be efficient in the sense that the vector of voter's satisfactions should be Pareto optimal (i.e., the utility of a voter cannot be improved without decreasing the utility of another voter) and fair in the sense that the vector of voter's satisfactions should be well-balanced (which will be formalized later). MOs have recently been investigated in multi-criteria decision making. Indeed, while those operators (which are instances of Bajraktarević means) are not new, they have received a renewed interest due to successful applications in data fusion [1,24].

Regarding the complexity of PR problems, Procaccia et al. proved that winner determination under the utilitarian versions of MVR and CCVR are both NP-hard problems even if the utility values are based on approval ballots [23]. Similarly, for the egalitarian version of CCVR, Betzler et al. proved that winner determination is NP-hard [4]. More positive results where obtained by resorting to approximation algorithms or special structures of preferences. Approximation algorithms for PR problems were given by Lu and Boutilier [16] and Skowron et al. [28]. Betzler et al. showed that, for single-peaked preferences, winner determination under CCVR is a polynomial time problem by designing a dynamic programming solution method [4]. Their results where extended by Cornaz et al. to the case of clustered single-peaked preferences [7]. Lastly, Skowron et al. showed that, for single-crossing preferences, winner determination under CCVR is also a polynomial time problem [29]. In this paper, we investigate the complexity of solving PR problems under CCVR and MVR with an MO as optimization criterion and show that we obtain the same complexity results as in the utilitarian case under a light condition.

The paper is organized as follows. Section 2 presents PR problems under MVR and CCVR, introduces our notations and discusses the use of several criteria to solve these problems. In Sect. 3, we present MOs and their properties. In Sect. 4, we design several solution procedures to solve PR problems with an MO criterion. Lastly, our numerical tests are presented in Sect. 5.

2   Proportional Representation

Let $V = \{1, \ldots, n\}$ be a set of n voters and $C = \{1, \ldots, m\}$ be a set of m candidates. A solution of the PR problem is a set of k winning candidates $\{c_1, \ldots, c_k\} \subseteq C$, and k sets S_j ($j \in \{c_1, \ldots, c_k\}$), where S_j is the subset of voters represented by candidate j. We recall that, while in MVR the sets S_j should be of the same size, it is not the case in CCVR. Consequently, in CCVR, a voter is always represented by the candidate she likes most in the set of k winning candidates. We will denote by $\mathbf{e} = \{c_1, S_{c_1}; \ldots; c_k, S_{c_k}\}$ any feasible solution and by $\mathcal{E}$ the set of all feasible solutions. Each voter has preferences over candidates. These preferences are expressed by a *preference profile* P of size mn, where the ith column of P is the preference order of voter i. From profile P, we derive nm utility values v_{ij} that represent the level of satisfaction of voter i if she is represented by candidate j. Given a feasible solution $\mathbf{e}$, the utility $v_i(\mathbf{e})$ of voter i is then given by v_{ij} if i belongs to S_j in $\mathbf{e}$. The PR problem aims at determining a solution $\mathbf{e} \in \mathcal{E}$ such that the utilities of the voters are maximized. We assume that all utility values v_i belong to some open interval[1] $\mathsf{D} \subset \mathbb{R}^+$ and we denote by $\mathbf{v}(\mathbf{e}) = (v_1(\mathbf{e}), \ldots, v_n(\mathbf{e})) \in \mathsf{D}^n$ the vector[2] giving the utilities of each voter for solution $\mathbf{e}$. A solution $\mathbf{e}_1$ will then be preferred to another solution $\mathbf{e}_2$ if, globally, $\mathbf{e}_1$ satisfies more the voters than $\mathbf{e}_2$. This is formalized by an aggregation criterion to maximize. More formally, given an operator $F : \mathsf{D}^n \to \mathbb{R}$, a solution $\mathbf{e}_1$ is preferred to a solution $\mathbf{e}_2$ if $F(\mathbf{v}(\mathbf{e}_1)) \geq F(\mathbf{v}(\mathbf{e}_2))$. The problem of finding an optimal solution is then written as follows:

$$\max_{\mathbf{e} \in \mathcal{E}} \quad F(\mathbf{v}(\mathbf{e})) \tag{1}$$

The choice of the operator F is a key but difficult point. A "good" operator should both enable to solve efficiently the maximization problem defined by Eq. 1 and ensure that the optimal solution found satisfies some desirable theoretical properties, as for instance *Pareto optimality*.

Definition 1. *Vector* $\mathbf{y} \in \mathsf{D}^n$ *Pareto-dominates vector* $\mathbf{y}' \in \mathsf{D}^n$ *if:*

$$(i) \ \forall i \in \{1, \ldots, n\}, y_i \geq y_i' \qquad (ii) \ \exists i \in \{1, \ldots, n\}, y_i > y_i'$$

A solution $\mathbf{e}$ *is said to be* Pareto-optimal *iff there is no* $\mathbf{e}' \in \mathcal{E}$ *such that* $\mathbf{v}(\mathbf{e}')$ *Pareto-dominates* $\mathbf{v}(\mathbf{e})$.

---

[1] Considering an open interval simplifies the writing of Proposition 1 in Sect. 3.

[2] Note that we follow the convention to use bold letters to represent vectors.

Pareto optimality ensures that the solution is efficient in the sense that the utility of a voter cannot be increased without decreasing the utility of another voter. We will say that an operator F is Pareto-compatible if any optimal solution of maximization problem 1 w.r.t. operator F is Pareto-optimal. To ensure Pareto-compatibility, operator F should be strictly increasing in each of its variables. In this case, we will say that the operator is *increasing*. A simple class of increasing operators is the class of Weighted Averages (WA) (with strictly positive weights):

Definition 2. *Let* $\mathbf{w} = (w_1, \ldots, w_n)$ *be a vector of weights. The* $\mathrm{WA}_{\mathbf{w}}(\cdot)$ *operator induced by* $\mathbf{w}$ *is defined by:*

$$\forall \mathbf{x} \in \mathrm{D}^n, \mathrm{WA}_{\mathbf{w}}(\mathbf{x}) = \sum_{i=1}^{n} w_i x_i$$

Given a WA operator, a solution $\mathbf{e}$ *is then evaluated by* $\mathrm{WA}_{\mathbf{w}}(\mathbf{v}(\mathbf{e}))$.

Optimizing a WA criterion is attractive due to the simplicity of this aggregation operator. In fact, the average operator used in the utilitarian version of PR problems is a WA operator, with $w_1 = \ldots = w_n$ in order to treat each voter identically. However, such criterion may lead to an unfair solution as it compensates between the utility values of the different voters. For instance, for two voters and $\mathbf{w} = (1, 1)$, the utility vector $(3, 10)$ is preferred to utility vector $(6, 6)$ for the $\mathrm{WA}_{\mathbf{w}}$ criterion. This is not satisfying as fairness should be an important property for multi-winner voting problems. A natural condition that can be satisfied by an operator to favor fairness is the Pigou-Dalton transfer principle [20]:

Definition 3. *Pigou-Dalton Transfer Principle. Let* $\mathbf{x} \in \mathrm{D}^n$ *such that* $x_i > x_j$ *for some* i, j. *Then, for all* ϵ *such that* $0 < \epsilon < x_i - x_j$, $\mathbf{x} - \epsilon \mathbf{b}_i + \epsilon \mathbf{b}_j$ *should be strictly preferred to* $\mathbf{x}$ *where* $\mathbf{b}_i$ *and* $\mathbf{b}_j$ *are the vector whose ith (resp. jth) component equals 1, all others being null.*

The transfer principle states that a transfer from a "more satisfied" voter to a "less satisfied" voter should improve a solution. Indeed, such a transfer reduces inequality while keeping the arithmetic mean of the vector constant. We will say that an operator is *fair* if it satisfies the Pigou-Dalton transfer principle.

A well known class of fair operators whose optimization yields a Pareto-optimal solution is the class of *Ordered Weighted Average* (OWA) operators with strictly decreasing weights [32].

Definition 4. *Let* $\mathbf{w} = (w_1, \ldots, w_n)$ *be a vector of weights. The* $\mathrm{OWA}_{\mathbf{w}}(\cdot)$ *operator induced by* $\mathbf{w}$ *is defined by:*

$$\forall \mathbf{x} \in \mathrm{D}^n, \mathrm{OWA}_{\mathbf{w}}(\mathbf{x}) = \sum_{i=1}^{n} w_i x_{\sigma(i)}$$

where σ *is a permutation of* $\{1, \ldots, n\}$ *such that* $x_{\sigma(1)} \leq x_{\sigma(2)} \leq \ldots \leq x_{\sigma(n)}$. *Given an OWA operator, a solution* $\mathbf{e}$ *is then evaluated by* $\mathrm{OWA}_{\mathbf{w}}(\mathbf{v}(\mathbf{e}))$.

Note that both the average operator used in the utilitarian version of PR problems and the min operator used in the egalitarian version of PR problems are OWA operators obtained by setting $\mathbf{w} = (1, \ldots, 1)$ and $\mathbf{w} = (1, 0, \ldots, 0)$ respectively. PR problems with OWA operators as optimization criteria have been investigated by Elkind and Ismaili [11]. The authors investigated several classes of weights $\mathbf{w}$ defining families of OWA operators that allows for a compromise between the utilitarian and the egalitarian versions of PR problems under CCVR. Interestingly, they showed that if the preferences of the voters present particular structures (e.g., single-crossing, single-peaked), then it becomes possible to design polynomial or pseudo-polynomial solution methods for some of these OWA families. However, the validity conditions of these methods and the algorithms themselves depend on the family of weights used.

In the next section, we present the class of operators that we will use as optimization criteria, namely *mixture operators*. Contrary to OWA operators, the way mixture operators weight the different components of a vector do not require any reordering. After recalling the main properties of this class of operators, we will give the condition under which they are fair. Interestingly, we will see that mixture operators have some descriptive advantages over OWA operators.

3   Mixture Operators

In this section, we present Mixture Operators (MOs) and their relations to several other operators.

Definition 5. *Let $w : \mathsf{D} \to (0, \infty]$ be a positive weighting function. The mixture operator $M_w(\cdot)$ induced by function w is defined as follows:*

$$\forall \mathbf{x} \in \mathsf{D}^n, M_w(\mathbf{x}) = \sum\nolimits_{i=1}^{n} \frac{w(x_i)}{\sum_{j=1}^{n} w(x_j)} x_i$$

Given a mixture operator, a solution $\mathbf{e} \in \mathcal{E}$ is then evaluated by $M_w(\mathbf{v}(\mathbf{e}))$.

MOs are special instances of Losonczi means [15], Bajraktarević means and generalized mixture functions [25]. On the other hand, MOs extend several averaging operators as the Gini mean or the Lehmer mean. MOs resemble OWA and WA operators but their weights depend on the *values* that are at stake in the evaluated vector and not on the *ranks* of the components. Note that if function w is constant then the MO boils down to the average operator used in the utilitarian version of PR problems. Furthermore, the special case of the Lehmer mean is defined by $w(x) = x^{p-1}$ where p is a parameter. If p tends towards $-\infty$, then the Lehmer mean tends towards the min operator used in the egalitarian version of PR problems. The following example illustrates the way an MO distorts the weights of the voters.

Example 1. *Let* $D = (0, U)$ *and consider the weight function* w *defined by* $w(x) = 2U - x$. *Then, given a solution* $\mathbf{e}$ *of the PR problem, the weight* w_i *associated to voter* i *is given by:*

$$w_i = \frac{w(v_i(\mathbf{e}))}{\sum_{j=1}^{n} w(v_j(\mathbf{e}))} = \frac{2U - v_i(\mathbf{e})}{\sum_{j=1}^{n} 2U - v_j(\mathbf{e})} = \frac{2U - v_i(\mathbf{e})}{n(2U - E)}$$

where $E = \sum_{j=1}^{n} v_j(\mathbf{e})/n$ *is the arithmetic mean of vector* $\mathbf{v}(\mathbf{e})$. *Thus,* $w_i \geq$ *(resp* $\leq$*)* $1/n$ *if* $v_i(\mathbf{e}) \leq$ *(resp.* $\geq$*)* E. *Stated differently, if a voter's utility is less than the average utility of the voters, then she is given a greater weight in operator* M_w. *More generally, the more* w *is decreasing, the more operator* M_w *will focus on the least satisfied voters. This property enables us to find compromises between the utilitarian and the egalitarian variants of PR problems.*

Moreover, by using a weighting function w depending on voters' utilities, MOs are able to account for preferences that cannot be represented by OWA or WA operators, as illustrated by the following example.

Example 2. *Let* $n = 2$ *and consider four solutions* $\mathbf{e}_1, \mathbf{e}_2, \mathbf{e}'_1$ *and* $\mathbf{e}'_2$ *such that:*

$$\mathbf{v}(\mathbf{e}_1) = (4, 8) \qquad \mathbf{v}(\mathbf{e}_2) = (2, 12)$$
$$\mathbf{v}(\mathbf{e}'_1) = (8, 8) \qquad \mathbf{v}(\mathbf{e}'_2) = (6, 12).$$

Note that $\mathbf{v}(\mathbf{e}'_1)$ *and* $\mathbf{v}(\mathbf{e}'_2)$ *are obtained from* $\mathbf{v}(\mathbf{e}_1)$ *and* $\mathbf{v}(\mathbf{e}_2)$ *by adding 4 to the utility value of the first voter. In solution* $\mathbf{e}_2$, *voter 1 is strongly unsatisfied, so one could prefer the more balanced solution* $\mathbf{e}_1$ *even if the average utility value of the voters is lower with* $\mathbf{e}_1$ *than with* $\mathbf{e}_2$. *Conversely, in both solutions* $\mathbf{e}'_1$ *and* $\mathbf{e}'_2$, *voters 1 and 2 are quite satisfied. Hence, one could prefer* $\mathbf{e}'_2$ *to* $\mathbf{e}'_1$ *as it yields the highest average utility value. Accounting for both of these preferences is not possible with a WA operator nor an OWA operator. Indeed, whatever the values of the weights* $\mathbf{w} = (w_1, w_2)$:

$$\mathrm{WA}_{\mathbf{w}}(\mathbf{v}(\mathbf{e}'_1)) = \mathrm{WA}_{\mathbf{w}}(\mathbf{v}(\mathbf{e}_1)) + 4w_1 \quad \mathrm{WA}_{\mathbf{w}}(\mathbf{v}(\mathbf{e}'_2)) = \mathrm{WA}_{\mathbf{w}}(\mathbf{v}(\mathbf{e}_2)) + 4w_1$$
$$\mathrm{OWA}_{\mathbf{w}}(\mathbf{v}(\mathbf{e}'_1)) = \mathrm{OWA}_{\mathbf{w}}(\mathbf{v}(\mathbf{e}_1)) + 4w_1 \quad \mathrm{OWA}_{\mathbf{w}}(\mathbf{v}(\mathbf{e}'_2)) = \mathrm{OWA}_{\mathbf{w}}(\mathbf{v}(\mathbf{e}_2)) + 4w_1$$

Hence, if preferences are represented by a WA or an OWA, the preference holding between $\mathbf{e}_1$ *and* $\mathbf{e}_2$ *should be the same as the one holding between* $\mathbf{e}'_1$ *and* $\mathbf{e}'_2$. *However, by considering* $w(x) = 24 - x$, *one obtains:*

$$M_w(\mathbf{v}(\mathbf{e}_1)) \approx 5.78 \qquad M_w(\mathbf{v}(\mathbf{e}_2)) \approx 5.53$$
$$M_w(\mathbf{v}(\mathbf{e}'_1)) = 8 \qquad M_w(\mathbf{v}(\mathbf{e}'_2)) \approx 8.4$$

which is consistent with the desired preferences.

MOs have recently received some attention in multi-criteria decision making where their mathematical properties have been studied (e.g. monotonicity, orness, ...) [14,25].

Increasingness. Many works have been focused on the monotonicity of MOs [2, 3]. Indeed, MOs are not increasing (and therefore not Pareto-compatible) in general. For this reason, sufficient conditions to ensure the monotonicity of this operator have been found. For instance, if $D = (0, U)$ and if $w(x) \geq \frac{dw}{dx}(x)(U-x)$ for all $x \in (0, U)$ with w increasing and piecewise differentiable, then M_w is increasing [18]. A simpler condition to impose monotonicity, if $D \subset \mathbb{R}^+$, is to have function $x \rightarrow w(x)$ decreasing and function $x \rightarrow w(x)x$ increasing.

Interestingly, MOs have also been studied in decision making under risk. Indeed, they are instances of both the Weighted Expected Utility (WEU) model [6] and the decomposable Skew-Symmetric Bilinear (SSB) functions investigated by Nakamura [21]. The properties of WEU functions and SSB functions w.r.t. risk-sensitivity and stochastic dominance have been thoroughly investigated [6, 12, 21] and these results can directly be used to entail results on MOs.

Fairness. We now give the condition under which MOs are fair (i.e., they satisfy the Pigou-Dalton transfer principle). This condition is a consequence of a result by Chew (Corollary 6 in [6]), who studied the consistency of WEU functions with stochastic dominance. Indeed, the Pigou-Dalton transfer principle in inequality measurement coincides with consistency with second order stochastic dominance in decision making under risk.

Proposition 1. *([6]) Assume that functions $x \rightarrow w(x)$ and $x \rightarrow w(x)x$ as well as their first derivatives are continuous and bounded on D, then operator $M_w(\cdot)$ is fair iff function $x \rightarrow w(x)(x - y)$ is strictly concave on D for all y in D.*

Proof. For completeness, we give the sketch of the proof.
Sufficiency: Note that given vectors $\mathbf{x}, \mathbf{y} \in D^n$:

$$M_w(\mathbf{x}) \geq M_w(\mathbf{y}) \Leftrightarrow \sum_{i=1}^n \frac{w(x_i)}{\sum_{j=1}^n w(x_j)} x_i \geq \sum_{i=1}^n \frac{w(y_i)}{\sum_{j=1}^n w(y_j)} y_i$$

$$\Leftrightarrow \sum_{i=1}^n \sum_{j=1}^n w(y_j)w(x_i)x_i \geq \sum_{i=1}^n \sum_{j=1}^n w(y_i)w(x_j)y_i$$

$$\Leftrightarrow \sum_{i=1}^n \sum_{j=1}^n w(y_j)w(x_i)(x_i - y_j) \geq 0$$

In particular, for any vector $\mathbf{x} \in D^n$,

$$\sum_{i=1}^n \sum_{j=1}^n w(x_j)w(x_i)(x_i - x_j) = 0.$$

Assume that function $x \rightarrow w(x)(x - y)$ is strictly concave on D for all y in D, then function $\phi_{\mathbf{y}} : x \rightarrow \sum_{j=1}^n w(y_j)w(x)(x - y_j)$ is also strictly concave on D for all $\mathbf{y}$ in D^n as $w(y) > 0$ for all y in D. Hence, by Lemma 2 in [8], if $\mathbf{x}^\epsilon$ is obtained from $\mathbf{x} \in D^n$ by an ϵ-transfer (i.e., $\mathbf{x}^\epsilon = \mathbf{x} + \epsilon(\mathbf{b}_j - \mathbf{b}_i)$ with $0 < \epsilon < x_i - x_j$ and where $\mathbf{b}_i$ is the ith canonical vector) then:

$$\sum_{i=1}^n \sum_{j=1}^n w(x_j)w(x_i^\epsilon)(x_i^\epsilon - x_j) > \sum_{i=1}^n \sum_{j=1}^n w(x_j)w(x_i)(x_i - x_j) = 0$$

Thus, $M_w(\mathbf{x}^\epsilon) > M_w(\mathbf{x})$ and the MO satisfies the Pigou-Dalton transfer principle.

Necessity: We recall that D is an open interval. By contradiction, assume there exists $s \in$ D such that $x \to w(x)(x - s)$ is not strictly concave. Thus, there exists $\hat{x}$ and $\epsilon > 0$ such that $[\hat{x} - \epsilon, \hat{x} + \epsilon] \subset$ D and $\phi_s : x \to w(s)w(x)(x - s)$ is convex on $[\hat{x} - \epsilon, \hat{x} + \epsilon]$. Assume w.l.o.g. that $s > \hat{x}$ and consider $t \in$ D such that $t > s$. As $\mathbb{Q}$ is dense in $\mathbb{R}$ and as w is bounded and continuous on D, t can be chosen such that there exists $k, l \in \mathbb{N}^*$ with $\phi_s(\hat{x}) + (k/l)\phi_s(t) = 0$. Set $n = 2(l + k)$ and consider vectors $\mathbf{x} = (\underbrace{\hat{x}, \hat{x}, \ldots, \hat{x}}_{2l}, \underbrace{t, \ldots, t}_{2k})$ and $\mathbf{s} = (\underbrace{s, \ldots, s}_{2(l+k)})$.

Then, by construction, $M_w(\mathbf{x}) = M_w(\mathbf{s})$. Consider $\mathbf{x}^\epsilon$ obtained by transferring ϵ from one of the $\hat{x}$ terms to another (increasing inequality). As in the sufficiency part, as ϕ_s is convex on $[\hat{x} - \epsilon, \hat{x} + \epsilon]$, we have $M_w(\mathbf{x}^\epsilon) \geq M_w(\mathbf{s}) = M_w(\mathbf{x})$ which violates the Pigou-Dalton transfer principle and concludes the proof.   □

The conditions of Proposition 1 can easily be met. For instance, a sufficient condition, if D $\subset \mathbb{R}^+$, is to have function $x \to w(x)x$ concave and function $x \to w(x)$ convex (with at least one property being strict). Under this sufficient condition, it is easy to see that the MO will be fair, as a Pigou-Dalton transfer will increase (resp. decrease) $\sum_{i=1}^n w(x_i)x_i$ (resp. $\sum_{i=1}^n w(x_i)$).

If D $= (0, U)$, examples of MOs that are increasing and fair on D can be defined by using $w(x) = 1/(1 + x)$ or $w(x) = (\alpha + 1)U^\alpha - x^\alpha$ with $0 < \alpha \leq 1$. Indeed, in these cases, function $x \to w(x)$ is convex and decreasing on D and function $x \to w(x)x$ is strictly concave and increasing on D.

In the next section, we investigate the complexity of winner determination in multi-winner voting rules with mixture operators.

4   Complexity and Solution Methods

Let u denote the function defined on D by $u(x) = w(x)x$. By abuse of notation, given a vector $\mathbf{x} \in$ D^n, we denote by $u(\mathbf{x})$ the sum $\sum_{i=1}^n u(x_i)$ and by $w(\mathbf{x})$ the sum $\sum_{i=1}^n w(x_i)$. By using these notations, the definition of an MO $M_w(\cdot)$ can be rewritten as follows:

$$\forall \mathbf{x} \in \text{D}^n, M_w(\mathbf{x}) = \frac{\sum_{i=1}^n u(x_i)}{\sum_{j=1}^n w(x_j)} = \frac{u(\mathbf{x})}{w(\mathbf{x})} \qquad (2)$$

Thus optimizing an MO entails the maximization of a ratio. Fractional programming is a subfield of operational research dedicated to this type of objective functions [9, 26, 30]. Several solution methods and techniques have been developed in this domain to optimize objective functions taking the form of a ratio of two linear objective functions. In this section, we will adapt and present two of these methods. Note that the two algorithms we present could also be used with the more general class of Losonczi means. The first one relies on a linearization trick that we will use to design a mixed integer linear program to solve PR problems w.r.t. an MO. The second one is a parametric approach that we will use to design polynomial time algorithms to solve PR problems with an MO criterion when a special structure of preferences makes it possible to solve the utilitarian version of the PR problem in polynomial time.

4.1 A Mixed Integer Linear Program

Note that if function w is constant, then one recovers the utilitarian version of PR problems which are NP-hard. Thus, it follows that PR problems under CCVR or MVR with an MO are also NP-hard. Yet, the NP-hardness of winner determination under CCVR and MVR has not prevented researchers from designing solution procedures for these problems. Indeed, Brams and Potthoff investigated the use of integer linear programs to solve PR problems [22]. We denote by $\mathcal{IP}$ the integer program they proposed. Program $\mathcal{IP}$ is given below on the left side of the page. It includes nm binary variables x_{ij} where x_{ij} takes value 1 if voter i is represented by candidate j and m binary variables z_j where z_j takes value 1 if candidate j represents at least one voter. Constraint (4) ensures that the election has k winning candidates. The n constraints in (5) make sure that each voter is only represented by one candidate. Lastly, constraints (6) and (7) specify a lower bound L and an upper bound U on the number of voters that can be represented by the same candidate. The pair (L, U) in MVR (resp. CCVR) is equal to $(\lfloor n/k \rfloor, \lceil n/k \rceil)$ (resp. $(0, n)$).

Program $\mathcal{IP}$ can be adapted to obtain a Mixed Integer Linear Program (MILP) $\mathcal{MIP}^{MO}$ (given below on the right side of the page) to solve a PR problem w.r.t. an MO. We now describe how it has been obtained. With the MO M_w, the objective function becomes:

$$\frac{\sum_{i,j\in V\cdot C} u(v_{ij})x_{ij}}{\sum_{i,j\in V\cdot C} w(v_{ij})x_{ij}} \tag{3}$$

$\mathcal{IP}$
$$\begin{cases} \max \sum_{i,j\in V\cdot C} v_{ij}x_{ij} \\[2mm] \sum_{j\in C} z_j = k \qquad\qquad (4) \\[2mm] \sum_{j\in C} x_{ij} = 1, \quad \forall i \in V \quad (5) \\[2mm] \sum_{i\in V} x_{ij} \geq Lz_j, \quad \forall j \in C \quad (6) \\[2mm] \sum_{i\in V} x_{ij} \leq Uz_j, \quad \forall j \in C \quad (7) \\[2mm] z_j \in \{0,1\}, \quad \forall j \in C \\[2mm] x_{ij} \in \{0,1\}, \quad \forall i,j \in V\cdot C \end{cases}$$

$\mathcal{MIP}^{MO}$
$$\begin{cases} \max \lambda \\[2mm] \sum_{j\in C} z_j = k \\[2mm] \sum_{j\in C} x_{ij} = 1, \quad \forall i \in V \\[2mm] \sum_{i\in V} x_{ij} \geq Lz_j, \quad \forall j \in C \\[2mm] \sum_{i\in V} x_{ij} \leq Uz_j, \quad \forall j \in C \\[2mm] \sum_{i,j\in V\cdot C} (w(v_{ij})y_{ij} - u(v_{ij})x_{ij}) = 0 \\[2mm] \sum_{j\in C} y_{ij} = \lambda, \quad \forall i \in V \\[2mm] y_{ij} \leq x_{ij}\lambda^u, \quad \forall i,j \in V\cdot C \\[2mm] z_j \in \{0,1\}, \quad \forall j \in C \\[2mm] x_{ij} \in \{0,1\}, \quad \forall i,j \in V\cdot C \\[2mm] y_{ij} \in \mathbb{R}^+, \quad \forall i,j \in V\cdot C \\[2mm] \lambda \in \mathbb{R} \end{cases}$$

To linearize this objective function, we use the following general method proposed by Williams [31]. Introduce a continuous variable λ into the problem to represent the expression given in Eq. 3. The objective is then to maximize this variable. By definition of λ, the following condition should hold:

$$\sum_{i,j\in V\cdot C} w(v_{ij})\lambda x_{ij} - \sum_{i,j\in V\cdot C} u(v_{ij})x_{ij} = 0$$

However, this equation is not linear in the variables of the problem because of the quadratic terms λx_{ij}. Thus, to enforce this equation in program $\mathcal{MIP}^{MO}$, we introduce nm continuous variables y_{ij} taking values in $\mathbb{R}^+$ to replace expressions λx_{ij}. We then impose that $y_{ij} = \lambda x_{ij}$ with the following constraints:

$$y_{ij} \leq x_{ij}\lambda^u, \qquad \forall i,j \in V \cdot C \tag{8}$$

$$\sum_{j\in C} y_{ij} = \lambda, \qquad \forall i \in V \tag{9}$$

where λ^u denotes an upper bound on λ. Such an upper bound can easily be obtained by computing $(\max u(v_{ij}))/(\min w(v_{ij}))$. While Eq. 8 ensures that $y_{ij} = 0$ if $x_{ij} = 0$, Eq. 9 ensures that $y_{ij} = \lambda$ if $x_{ij} = 1$. Indeed, in Eq. 9, only one of the y_{ij} is non null due to constraints 5 in program $\mathcal{IP}$ and Eq. 9 imposes that this variable equals λ. The final program $\mathcal{MIP}^{MO}$ involves $nm + 1$ additional continuous variables and $nm + n + 1$ additional constraints.

4.2 A Parametric Approach

We now assume that the preferences of the voters have a particular structure which makes it possible to solve the utilitarian version of the PR problem with a polynomial time algorithm denoted by $\mathcal{A}$. This is for instance the case for single-peaked or single-crossing preferences with CCVR. Using algorithm $\mathcal{A}$, we provide two polynomial time methods to solve the PR problem w.r.t. an MO. The first one follows from a method proposed by Megiddo [17] and the second one is a cutting plane method. The only light condition required by these two methods on the MO is that M_w should be an increasing MO with a decreasing function w and an increasing function u. Indeed, both of these methods require to solve utilitarian versions of the PR problem with utilities defined by $\tilde{v}_{ij}^\lambda = u(v_{ij}) - \lambda w(v_{ij})$ ($\lambda \in \mathbb{R}^+$). The previous condition on the monotony of w and u ensures that utilities $\tilde{v}_{ij}^\lambda$ are consistent with the preferences of the voters (i.e., if voter i prefers candidate j_1 to candidate j_2 then $\tilde{v}_{ij_1}^\lambda \geq \tilde{v}_{ij_2}^\lambda$).

Megiddo's method. The first method we present follows from a general method designed by Megiddo [17]. This method is based on the following observation (recasted in our PR setting):

Observation 1. *Let $\lambda \in \mathbb{R}$ and consider utility values $\tilde{v}_{ij}^\lambda = u(v_{ij}) - \lambda w(v_{ij})$. Let $\mathbf{e}^\lambda$ and $v^\lambda = u(\mathbf{v}(\mathbf{e}^\lambda)) - \lambda w(\mathbf{v}(\mathbf{e}^\lambda))$ be the optimal solution and the optimal* *value for the utilitarian version of the PR problem with utility values $\tilde{v}_{ij}^{\lambda}$. Then, the sign of v^{λ} is determined by the position of λ w.r.t. λ^{*}, where λ^{*} is defined as the optimal value w.r.t. M_{w} (i.e. $\lambda^{*} = \max_{e \in \mathcal{E}} M_{w}(\mathbf{v}(\mathbf{e}))$).*

1. *If $\lambda = \lambda^{*}$, then $v^{\lambda} = 0$ and $\mathbf{e}^{\lambda}$ is an optimal solution according to MO M_{w}.*
2. *If $\lambda > \lambda^{*}$, then v^{λ} will be strictly negative. Indeed, there exists no feasible solution $\mathbf{e} \in \mathcal{E}$ such that $u(\mathbf{v}(\mathbf{e}))/w(\mathbf{v}(\mathbf{e})) > \lambda$.*
3. *On the contrary, if $\lambda < \lambda^{*}$, then v^{λ} will be strictly positive.*

Thus, the problem of solving a PR problem w.r.t. an MO reduces to the one of solving the utilitarian version of the PR problem with utility values $\tilde{v}_{ij}^{\lambda^{*}} = u(v_{ij}) - \lambda^{*} w(v_{ij})$ (case 1 of Observation 1). However, while values $u(v_{ij})$ and $w(v_{ij})$ are known, the value of λ^{*} is not. Thus, the values $\tilde{v}_{ij}^{\lambda^{*}}$ are incompletely specified. Nevertheless, if the utilitarian version of the PR problem can be solved by an algorithm $\mathcal{A}$ relying on a polynomial number of additions/subtractions and comparisons (which is the case for PR problems under CCVR with single-peaked or single-crossing preferences), then this problem can be solved via Megiddo's method. In short, Megiddo's method applied to the PR problem mimics algorithm $\mathcal{A}$ to solve the utilitarian version of the PR problem with utility values $\tilde{v}_{ij}^{\lambda^{*}}$. However, the method redefines the addition and the comparison operations to handle the fact that λ^{*} is unknown.

Management of the imprecisely known value of λ^{*}. Instead of having precise values for $\tilde{v}_{ij}^{\lambda^{*}}$, the algorithm works with pairs of values $(u(v_{ij}), w(v_{ij}))$ for all i in V and j in C and maintains a lower bound λ^{l} and an upper bound λ^{u} over λ^{*} (originally 0 and ∞). Thus $\tilde{v}_{ij}^{\lambda^{*}}$ is only known to be in the interval $[u(v_{ij}) - \lambda^{u} w(v_{ij}), u(v_{ij}) - \lambda^{l} w(v_{ij})]$.

Redefinition of the addition operation. The additions and subtractions required by algorithm $\mathcal{A}$ are simply replaced by componentwise additions and subtractions of pairs. Stated differently, the sum of $(u(v_{ij}), w(v_{ij}))$ and $(u(v_{kl}), w(v_{kl}))$ is $(u(v_{ij}) + u(v_{kl}), w(v_{ij}) + w(v_{kl}))$. Indeed, whatever the value of λ^{*}:

$$u(v_{ij}) - \lambda^{*} w(v_{ij}) + u(v_{kl}) - \lambda^{*} w(v_{kl}) = u(v_{ij}) + u(v_{kl}) - \lambda^{*}(w(v_{ij}) + w(v_{kl})).$$

Redefinition of the comparison operation. To compare two pairs $(u(v_{ij}), w(v_{ij}))$ and $(u(v_{kl}), w(v_{kl}))$, Megiddo's method uses the following routine.

A fist step consists in checking if $u(v_{ij}) - \lambda w(v_{ij})$ is less (resp. greater) than $u(v_{kl}) - \lambda w(v_{kl})$, regardless of the value of $\lambda \in [\lambda^{l}, \lambda^{u}]$. Indeed, in that case (illustrated on the left side of Fig. 1 below), the algorithm can conclude that $u(v_{ij}) - \lambda^{*} w(v_{ij}) \leq$ (resp. $\geq$) $u(v_{kl}) - \lambda^{*} w(v_{kl})$. Note that $u(v_{ij}) - \lambda w(v_{ij})$ and $u(v_{kl}) - \lambda w(v_{kl})$ are linear functions of λ. Therefore, the method needs only to check the inequality for values λ^{l} and λ^{u}.

If this first step does not conclude which pair is the minimum (case illustrated on the right side of Fig. 1), then the algorithm considers the value $\hat{\lambda}$ such that $u(v_{ij}) - \hat{\lambda} w(v_{ij}) = u(v_{kl}) - \hat{\lambda} w(v_{kl})$. Then, if we assume w.l.o.g.[3] that $u(v_{ij}) -$

---

[3] if, $u(v_{ij}) - \lambda^{l} w(v_{ij}) > u(v_{kl}) - \lambda^{l} w(v_{kl})$, just reverse inequalities 10 and 11.

$\lambda^l w(v_{ij}) < u(v_{kl}) - \lambda^l w(v_{kl})$, we have:

$$\forall \lambda \in [\lambda^l, \hat{\lambda}], u(v_{ij}) - \lambda w(v_{ij}) \leq u(v_{kl}) - \lambda w(v_{kl}) \tag{10}$$

$$\forall \lambda \in [\hat{\lambda}, \lambda^u], u(v_{ij}) - \lambda w(v_{ij}) \geq u(v_{kl}) - \lambda w(v_{kl}). \tag{11}$$

Now, for the algorithm to conclude, it needs only to check if $\lambda^* \in [\lambda^l, \hat{\lambda}]$ (in which case, λ^u is set to $\hat{\lambda}$) or $\lambda^* \in [\hat{\lambda}, \lambda^u]$ (in which case, λ^l is set to $\hat{\lambda}$). This is done by testing the sign of the optimal value of the utilitarian version of the PR problem with utility values $\tilde{v}_{ij}^{\lambda}$ (see Observation 1), which is computed by using algorithm $\mathcal{A}$. Note that this operation updates either λ^l or λ^u, which refines the value of λ^* and enables us to perform more comparisons without rerunning algorithm $\mathcal{A}$.

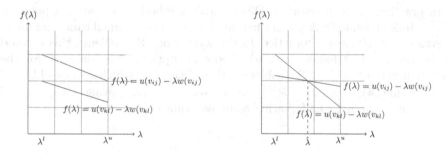

Fig. 1. Illustration of the two possible cases that can occur in the comparison routine.

Polynomial time complexity of the method. As Megiddo's method mimics algorithm $\mathcal{A}$, it relies on a polynomial number of (redefined) additions and comparisons. As algorithm $\mathcal{A}$ (which is used in the comparison operator) is of polynomial complexity, these two operations can be performed in polynomial time. This proves the polynomial complexity of the method.

Cutting plane method. As shown by Eq. 2, for a solution $\mathbf{e}^*$ maximizing $M_w(\mathbf{v}(\mathbf{e}))$, we have $\frac{u(\mathbf{v}(\mathbf{e}^*))}{w(\mathbf{v}(\mathbf{e}^*))} \geq \frac{u(\mathbf{v}(\mathbf{e}))}{w(\mathbf{v}(\mathbf{e}))}$ for all $\mathbf{e}$ in $\mathcal{E}$. Replacing $u(\mathbf{v}(\mathbf{e}^*))/w(\mathbf{v}(\mathbf{e}^*))$ by λ^*, these inequalities rewrite as follows:

$$u(\mathbf{v}(\mathbf{e})) - \lambda^* w(\mathbf{v}(\mathbf{e})) \leq 0, \qquad \forall \mathbf{e} \in \mathcal{E} \tag{12}$$

and the constraint is tight for $\mathbf{e} = \mathbf{e}^*$. Therefore, λ^* is the minimal value such that all inequalities in 12 are satisfied. This analysis yields the following Linear Program (LP) $\mathcal{LP}^{MO}$:

$$\mathcal{LP}^{MO} \begin{cases} \min_{\lambda} \lambda \\ u(\mathbf{v}(\mathbf{e})) - \lambda w(\mathbf{v}(\mathbf{e})) \leq 0 \qquad \forall \mathbf{e} \in \mathcal{E} \\ \lambda \in \mathbb{R} \end{cases} \tag{13}$$

Even if the number of constraints in 13 may be exponential in m (as there is one constraint per feasible solution), these constraints can be handled efficiently by resorting to a *cutting plane algorithm*. A cutting plane algorithm makes it possible to solve an LP involving a set S of constraints, the size of which is exponential, provided there exists a *separation oracle*. A separation oracle dynamically generates a violated constraint in S given the current optimal solution according to the previously generated constraints, or states that there is no violated constraint (in which case the current solution is optimal). In program $\mathcal{LP}^{MO}$, a separation oracle to determine a most violated constraint amounts to solve the utilitarian version of the PR problem with utilities defined by $\tilde{v}_{ij}^\lambda = u(v_{ij}) - \lambda w(v_{ij})$. This problem can be solved by algorithm $\mathcal{A}$. As the complexity of $\mathcal{A}$ is polynomial, the complexity of solving $\mathcal{LP}^{MO}$ is polynomial by the polynomial time equivalence of optimization and separation by resorting to the ellipsoid method [13].

In practice, one can resort to Dinkelbach's method [9] for solving program $\mathcal{LP}^{MO}$. Indeed, while Dinkelbach's method has not a polynomial time guarantee, it reveals more efficient in practice. In the setting of PR problems, this method can be described as follows. Let $\mathbf{e}^\lambda$ denote an optimal solution in $\mathcal{E}$ for the utilitarian variant of the PR problem with utilities $\tilde{v}_{ij}^\lambda$ (which can be obtained via algorithm $\mathcal{A}$). Program $\mathcal{LP}^{MO}$ can be solved by computing a sequence of solutions in $\mathcal{E}$ through the following recursive equation:

$$\mathbf{e}_{t+1} = \mathbf{e}^{\lambda_t}$$

where $\lambda_t = u(\mathbf{v}(\mathbf{e}_t))/w(\mathbf{v}(\mathbf{e}_t)) = M_w(\mathbf{v}(\mathbf{e}_t))$. The key point of this approach is that the solutions generated in this way are of increasing values w.r.t the MO operator. Indeed, a direct corollary from Observation 1 is that while $M_w(\mathbf{v}(\mathbf{e}_t)) < \lambda^*$, $M_w(\mathbf{v}(\mathbf{e}_{t+1})) > M_w(\mathbf{v}(\mathbf{e}_t))$. Note that, by definition of λ^*, we cannot have $M_w(\mathbf{v}(\mathbf{e}_t)) > \lambda^*$. Therefore, the sequence $(M_w(\mathbf{v}(\mathbf{e}_t)))_{t \in \mathbb{N}}$ is strictly increasing until reaching λ^*. Value λ^* is always reached after a finite number of iterations as there is a finite number of values in $\{M_w(\mathbf{v}(\mathbf{e})) : \mathbf{e} \in \mathcal{E}\}$. After a finite number of iterations, we will thus have $M_w(\mathbf{v}(\mathbf{e}_t)) = M_w(\mathbf{v}(\mathbf{e}_{t+1}))$ which means that an optimal solution w.r.t. the MO has been found.

We now turn to our numerical tests in which the efficiency of the different solution methods are compared.

5   Numerical Tests

In this section, we compare the execution times of the proposed solution methods in two different experiments[4]. In both experiments, values v_{ij} are set to $m -$ $\mathtt{rk_i(j)}$ (where $\mathtt{rk_i(j)}$ denotes the rank of candidate j in the preference order of voter i) and the weighting function w of the MO is defined by $w(x) = 2m - x$,

---

[4] All methods were implemented in C++ using Gurobi version 5.6.3 to solve the LPs. Times are wall-clock times on a 2.4 GHz Intel Core i5 machine with 8GB of RAM.

so that the transfer principle holds. Lastly, for an election with n voters, values m and k were set to $n/5$ and $n/20$.

In the first experiment, we restrict ourselves to randomly generated single-peaked preferences in the CCVR framework, and we compare the execution times of the two methods presented in Sect. 4.2, denoted by CP (for Cutting Plane) and MEG (for Megiddo). The method CP is solved by using the Dinkelbach method. In methods CP and MEG, the algorithm $\mathcal{A}$ used (see Sect. 4.2) is the dynamic programming algorithm proposed by Betzler et al. [4]. The average computation times (in seconds) over 50 instances, as well as the average number of calls to algorithm $\mathcal{A}$, are given in Table 1 for instances with a number of voters ranging from 100 to 1000.

We observe that both methods CP and MEG are very fast and require very few calls to algorithm $\mathcal{A}$. On these instances, method CP seems to perform best and requires less calls to algorithm $\mathcal{A}$.

Table 1. Evolution of the average computation time in seconds and the average number of calls to algorithm $\mathcal{A}$ (nbc in the table) for methods CP and MEG as n increases.

n		100	200	300	400	500	600	700	800	900	1000
CP	time	<0.001	0.005	0.017	0.037	0.074	0.132	0.276	0.416	0.618	1.064
	nbc	2.62	2.86	2.94	3.00	2.98	2.98	3.00	3.00	3.00	3.00
MEG	time	0.001	0.017	0.064	0.171	0.370	0.676	1.461	2.360	3.872	6.981
	nbc	3.42	7.18	8.90	11.32	11.94	12.40	13.34	14.32	16.20	15.04

In a second experiment, we compare the execution times of the two instantiations of $\mathcal{MIP}^{MO}$ in the CCVR and MVR frameworks. We denote these programs by $\mathcal{M}^{MO}_{CCVR}$ and $\mathcal{M}^{MO}_{MVR}$. For each instance, the preferences of the voters were generated uniformly at random. The average computation times (in seconds) over 50 instances are given in Table 2 for instances with a number of voters ranging from 50 to 100.

Obviously, we observe that solving those two programs are computationally demanding. For instance, solving programs $\mathcal{M}^{MO}_{CCVR}$ and $\mathcal{M}^{MO}_{MVR}$ takes more than 25 s for elections with 100 voters.

Table 2. Evolution of the average computation time in seconds to solve programs $\mathcal{M}^{MO}_{CCVR}$ and $\mathcal{M}^{MO}_{MVR}$ as n increases.

n	50	60	70	80	90	100
$\mathcal{M}^{MO}_{CCVR}$	0.79	1.19	2.22	3.35	13.45	28.52
$\mathcal{M}^{MO}_{MVR}$	1.01	1.60	3.30	5.34	14.78	25.66

6 Conclusion

We studied PR problems with MOs. We presented these operators and the conditions under which they make it possible to find a fair solution for a PR problem. We designed a MILP to solve PR problems w.r.t. an MO and presented two other solution methods that are of polynomial complexity if the preferences of the voters abid to a particular structure enabling to solve the utilitarian version of the PR problem in polynomial time.

As future work, it would be worth investigating the extent to which the methods presented here can be adapted to other domains requiring fairness. For instance, it seems that they could also be used for the assignment problem. However, using MO for solving allocation problems where agents can receive several goods seems more challenging as the weighting function would be applied to a sum of utilities, which triggers new technical difficulties. Moreover, the results presented here can be extended to Losonczi means and we plan to investigate if those more general operators can yield new desirable properties for PR problems.

Acknowledgements. This work is supported by the ANR project CoCoRICo-CoDec ANR-14-CE24-0007-01.

References

1. Angelov, P., Yager, R.: Density-based averaging-a new operator for data fusion. Inf. Sci. **222**, 163–174 (2013)
2. Beliakov, G., Špirková, J.: Weak monotonicity of Lehmer and Gini means. Fuzzy Sets Syst. **299**, 26–40 (2016)
3. Beliakov, G., Wilkin, T.: On some properties of weighted averaging with variable weights. Inf. Sci. **281**, 1–7 (2014)
4. Betzler, N., Slinko, A., Uhlmann, J.: On the computation of fully proportional representation. J. Artif. Intell. Res. **47**, 475–519 (2013)
5. Chamberlin, J.R., Courant, P.N.: Representative deliberations and representative decisions: proportional representation and the Borda rule. Am. Polit. Sci. Rev. **77**(3), 718–733 (1983)
6. Chew, S.: A generalization of the quasilinear mean with applications to the measurement of income inequality and decision theory resolving the Allais paradox. Econometrica J. Econ. Soc. **51**(4), 1065–1092 (1983)
7. Cornaz, D., Galand, L., Spanjaard, O.: Bounded single-peaked width and proportional representation. In: Proceedings ECAI 2012, pp. 270–275. IOS Press (2012)
8. Dasgupta, P., Sen, A., Starrett, D.: Notes on the measurement of inequality. J. Econ. Theor. **6**(2), 180–187 (1973)
9. Dinkelbach, W.: On nonlinear fractional programming. Manage. Sci. **13**(7), 492–498 (1967)
10. Elkind, E., Faliszewski, P., Skowron, P., Slinko, A.: Properties of multiwinner voting rules. In: Proceedings AAMAS 2014, pp. 53–60. IFAAMAS (2014)
11. Elkind, E., Ismaili, A.: OWA-based extensions of the chamberlin–courant rule. In: Walsh, T. (ed.) ADT 2015. LNCS, vol. 9346, pp. 486–502. Springer, Cham (2015). doi:10.1007/978-3-319-23114-3_29

12. Fishburn, P.C.: Dominance in SSB utility theory. J. Econ. Theor. **34**(1), 130–148 (1984)
13. Grötschel, M., Lovász, L., Schrijver, A.: The ellipsoid method and its consequences in combinatorial optimization. Combinatorica **1**(2), 169–197 (1981)
14. Liu, X.: The orness measures for two compound quasi-arithmetic mean aggregation operators. Int. J. Approximate Reasoning **51**(3), 305–334 (2010)
15. Losonczi, L.: General inequalities for nonsymmetric means. Aequationes Math. **9**(2–3), 221–235 (1973)
16. Lu, T., Boutilier, C.: Budgeted social choice: from consensus to personalized decision making. In: Proceedings IJCAI 2011, pp. 280–286. AAAI Press/IJCAI (2011)
17. Megiddo, N.: Combinatorial optimization with rational objective functions. Mathe. Oper. Res. **4**(4), 414–424 (1979)
18. Mesiar, R., Špirková, J.: Weighted means and weighting functions. Kybernetika **42**(2), 151–160 (2006)
19. Monroe, B.L.: Fully proportional representation. Am. Polit. Sci. Rev. **89**(04), 925–940 (1995)
20. Moulin, H.: Axioms of Cooperative Decision Making. Cambridge University Press, Cambridge (1991)
21. Nakamura, Y.: Risk attitudes for nonlinear measurable utility. Ann. Oper. Res. **19**(1), 311–333 (1989)
22. Potthoff, R.F., Brams, S.J.: Proportional representation: Broadening the options. J. Theor. Polit. **10**(2), 147–178 (1998)
23. Procaccia, A.D., Rosenschein, J.S., Zohar, A.: On the complexity of achieving proportional representation. Soc. Choice Welfare **30**(3), 353–362 (2008)
24. Ribeiro, R.A., Falcão, A., Mora, A., Fonseca, J.M.: FIF: A fuzzy information fusion algorithm based on multi-criteria decision making. Knowl.-Based Syst. **58**, 23–32 (2014)
25. Ribeiro, R.A., Pereira, R.A.M.: Generalized mixture operators using weighting functions: a comparative study with WA and OWA. Eur. J. Oper. Res. **145**(2), 329–342 (2003)
26. Schaible, S., Ibaraki, T.: Fractional programming. Eur. J. Oper. Res. **12**(4), 325–338 (1983)
27. Skowron, P., Faliszewski, P., Lang, J.: Finding a collective set of items: From proportional multirepresentation to group recommendation. Artif. Intell. **241**, 191–216 (2016)
28. Skowron, P., Faliszewski, P., Slinko, A.: Achieving fully proportional representation: approximability results. Artif. Intell. **222**, 67–103 (2015)
29. Skowron, P., Yu, L., Faliszewski, P., Elkind, E.: The complexity of fully proportional representation for single-crossing electorates. In: Vöcking, B. (ed.) SAGT 2013. LNCS, vol. 8146, pp. 1–12. Springer, Heidelberg (2013). doi:10.1007/978-3-642-41392-6_1
30. Stancu-Minasian, I.: Fractional Programming: Theory, Methods and Applications. Springer, London (2012)
31. Williams, H.P.: Experiments in the formulation of integer programming problems. In: Balinski, M.L. (ed.) Approaches to Integer Programming, vol. 2, pp. 180–197. Springer, Heidelberg (1974)
32. Yager, R.R.: On ordered weighted averaging aggregation operators in multicriteria decisionmaking. IEEE Trans. Syst. Man. Cybern. **18**, 183–190 (1988)

On the Complexity of Chamberlin-Courant on Almost Structured Profiles

Neeldhara Misra[✉], Chinmay Sonar, and P. R. Vaidyanathan

Indian Institute of Technology Gandhinagar, Gandhinagar 382355, India
{neeldhara.m,sonar.chinmay,pr.vaidyanathan}@iitgn.ac.in

Abstract. The Chamberlin-Courant voting rule is an important multi-winner voting rule. Although NP-hard to compute on general profiles, it is known to be polynomially solvable on single-crossing and single-peaked electorates by exploiting the structures of these domains. We consider the problem of generalizing the domain on which the voting rule admits efficient algorithms.

On the one hand, we show efficient algorithms on profiles that are k candidates or k voters away from the single-peaked and single-crossing domains. In particular, for profiles that are k candidates away from being single-peaked or single-crossing, we show algorithms whose running time is FPT in k. For profiles that are k voters away from being single-peaked or single-crossing, our algorithms are XP in k. These algorithms are obtained by a careful extension of known algorithms on structured profiles [2,12]. This provides a natural application for the work by Elkind and Lackner in [9], who study the problem of finding deletion sets to single-peaked and single-crossing profiles.

In contrast to these results, for a different, but equally natural way of generalizing these domain, we show severe intractability results. In particular, we show that the problem is NP-hard on profiles that can be "decomposed" into a constant number of single-peaked profiles. Also, if the number of crossings per pair of candidates in a profile is permitted to be at most three (instead of one), the problem continues be NP-hard. This stands in contrast with other attempts at generalizing these domains (such as single-peaked or single-crossing width), as it rules out the possibility of fixed-parameter (or even XP) algorithms when parameterized by the number of peaks, or the maximum number of crossings per candidate pair.

Keywords: NP-hardness · Chamberlin Courant · Single-crossing profiles · Single-peaked profiles · Fixed-parameter algorithms · Voting rules

1 Introduction

A traditional election setting consists of voters expressing their preferences over alternatives, where preferences can be modelled in several ways (approval ballots, ternary ballots, top-truncated lists, total orders, and so forth). Usually,

The first author acknowledges support from the INSPIRE Faculty Scheme, DST India (project IFA12-ENG-31).

© Springer International Publishing AG 2017
J. Rothe (Ed.): ADT 2017, LNAI 10576, pp. 124–138, 2017.
DOI: 10.1007/978-3-319-67504-6_9

given such a scenario, we would like to identify a winning alternative. In many applications, however, we need to identify not one, but a fixed *set of alternatives* that best represent the interests of the voters. Such a problem arises in a variety of scenarios like committee selection, parliamentary elections, movie recommendation systems, and so forth.

There are several ways of measuring how well a committee fares against a set of votes. When votes are approval ballots, for instance, the maximum or the sum of Hamming distances is often used as a measure of quality. We consider the setting of votes given as complete rankings, and focus on the well-studied Chamberlin-Courant rule [6], which achieves proportional representation. The way this voting rule works is the following. We begin by fixing a notion of a "dissatisfaction function" $\alpha : \mathbb{N} \to \mathbb{N}$, which simply specifies, by $\alpha(i)$, how unhappy a voter is when she is represented by a candidate who is ranked at the ith position on her list. Given a committee with k candidates, a voter is represented by the candidate that she ranks the highest among candidates from X. If $\phi(v)$ denotes the candidate that is representing voter v, the optimal committee under the Chamberlin-Courant voting rule seeks to minimize either the sum or the maximum value of $\alpha(\text{pos}_v(\phi(v)))$, taken over all voters v (where $\text{pos}_v(c)$ denotes the ranking of the candidate c in the vote v).

The Chamberlin-Courant rule (and the closely related Monroe voting rule which we do not consider in the present work) has several desirable properties. It has been argued [12] that rules that achieve proportional representation are particularly well-suited for electing committees that need to make unanimous decisions, and in particular, that takes minority candidates into account. However, it turns out that finding an optimal committee under this rule is NP-hard, and it is therefore unlikely to admit an efficient algorithm.

On the other hand, there have been promising developments showing that the optimal Chamberlin-Courant committees can be computed efficiently on structured profiles which are commonly encountered in practical scenarios. Two such restrictions that have been particularly successful are the single-peaked and single-crossing domains. In a parallel development, [9] showed various efficient algorithms for detecting profiles that are close to being structured (in this case, the notion of closeness is that these profiles exhibit the structure of the domain on all but a small number of candidates or voters). More generally, the notion of closeness to a domain is well-studied and has been defined in various ways [11].

We combine these scenarios to address the following question: how well do the efficient algorithms on the restricted domains extend to profiles that are of the latter type, that is, they exhibit the properties of the domain on all but a small number of candidates or voters? We now turn to our findings in the context of this question and closely related issues.

Our Contributions and Methodology. A natural framework for addressing the problem of how well algorithms on structured domains scale up to nearly-structured ones is parameterized complexity [8]. To begin with, we show efficient algorithms on profiles that k candidates or k voters away from the single-peaked and single-crossing domains. In particular, for profiles that are k candidates away from being single-peaked or single-crossing, we show algorithms whose running time is FPT in k. For profiles that are k voters away from being single-peaked or single-crossing, our algorithms are XP in k. These algorithms are obtained by a careful extension of the the known algorithms [2,12] on the structured profiles. This provides a natural application for the work by Elkind and Lackner in [9], who study the problem of finding deletion sets to single-peaked and single-crossing profiles.

In contrast to these results, for a different, but equally natural way of generalizing these domain, we show severe intractability results. In particular, we show that the problem is NP-hard on profiles that can be "decomposed" into a constant number of single-peaked profiles. Also, if the number of crossings per pair of candidates in a profile is permitted to be at most three (instead of one), the problem continues be NP-hard. This stands in contrast with other attempts at generalizing these domains (such as single-peaked or single-crossing width [7,12]), as it rules out the possibility of fixed-parameter (or even XP) algorithms when parameterized by the number of peaks, or the maximum number of crossings per candidate pair.

Related Work. Our work builds primarily on two lines of work from before. We appeal to the known algorithms that determine the optimal Chamberlin-Courant committees on single-peaked profiles [2] and single-crossing profiles [12]. These results have been be extended to other multiwinner voting rules, which we do not consider in the present work. Also, efficient algorithms have been shown on more general preference restrictions such as single-peakedness on trees, or single-crossing width.

2 Technical Preliminaries

In this section, we introduce some of the notation and definitions that we will use. For a more detailed introduction to notions relating to restricted domains and voting rules, we refer the reader to the appropriate chapters in [4], and for a comprehensive introduction to parameterized algorithms, we refer the reader to [8].

For a positive integer ℓ, we denote the set $\{1, \ldots, \ell\}$ by $[\ell]$. We first define some general notions relating to voting rules. Let $V = \{v_i : i \in [n]\}$ be a set of n *voters* and $C = \{c_j : j \in [m]\}$ be a set of m *candidates*. If not mentioned otherwise, we denote the set of candidates, the set of voters, the number of candidates, and the number of voters by C, V, m, and n respectively.

Every voter v_i has a *preference* $\succ_i$ which is a complete order over the set C of candidates. We say voter v_i prefers a candidate $x \in C$ over another candidate $y \in C$ if $x \succ_i y$. We denote the set of all preferences over C by $\mathcal{L}(C)$. The n-tuple $(\succ_i)_{i \in [n]} \in \mathcal{L}(C)^n$ of the preferences of all the voters is called a *profile*. Note that a profile, in general, is a multiset of linear orders. For a subset $M \subseteq [n]$, we call $(\succ_i)_{i \in M}$ a sub-profile of $(\succ_i)_{i \in [n]}$. For a subset of candidates $D \subseteq C$, we use $\mathcal{P}|_D$ to denote the projection of the profile on the candidates in D alone. A *domain* is a set of profiles.

The rest of this section is organized as follows. We first define the Chamberlin-Courant voting rule. We then introduce the domain restrictions that are of interest to us, and the notion of closeness to a restricted domain. We finally define the problems that we will study subsequently.

Chamberlin-Courant. The Chamberlin–Courant voting rule is based on the notion of a *dissatisfaction function* or a *misrepresentation function*. This function specifies, for each $i \in [m]$, a voter's dissatisfaction from being represented by candidate she ranks in position i.

Definition 1. *For an m-candidate election, a* dissatisfaction function *is given by a non-decreasing function* $\alpha \colon [m] \to \mathbb{N}$ *with* $\alpha(1) = 0$.

A popular dissatisfaction function is Borda, given by $\alpha_B^m(i) = \alpha_B(i) = i - 1$. We now turn to the notion of an assignment function. Let k be a positive integer. A k-*CC-assignment function* for an election $E = (C, V)$ is a mapping $\Phi \colon V \to C$ such that $\|\Phi(V)\| \leqslant k$. For a given assignment function Φ, we say that voter $v \in V$ is *represented* by candidate $\Phi(v)$ in the chosen committee. There are several ways to measure the quality of an assignment function Φ with respect to a dissatisfaction function α; we use the following two:

1. $\ell_1(\Phi) = \sum_{i=1,\ldots,n} \alpha(\mathrm{pos}_{v_i}(\Phi(v_i)))$, and
2. $\ell_\infty(\Phi) = \max_{i=1,\ldots,n} \alpha(\mathrm{pos}_{v_i}(\Phi(v_i)))$.

We are now ready to define the Chamberlin-Courant voting rule, which is the primary focus of this paper.

Definition 2. *For every family of dissatisfaction functions* $\alpha = (\alpha^m)_{m=1}^{\infty}$, *and every* $\ell \in \{\ell_1, \ell_\infty\}$, *the* α-ℓ- CC *voting rule is a mapping that takes an election* $E = (C, V)$ *and a positive integer* k *with* $k \leqslant \|C\|$ *as its input, and returns a k-CC-assignment function* Φ *for* E *that minimizes* $\ell(\Phi)$ *(if there are several optimal assignments, the rule is free to return any of them).*

Chamberlin and Courant [6] originally proposed the utilitarian variants of their rules with a focus on the Borda dissatisfaction function. The egalitarian variant was considered by, for instance, Betzler et al. [2].

Single-peaked Profiles. A preference profile is said be single-peaked if there exists an ordering σ over the candidates C such that the preference of every voter v has the following structure: v has a favorite candidate c (sometimes called the "peak" for v), and the further away a candidate $d \neq c$ is from c in σ, the less it is preferred by the voter v. The notion of single-peaked preferences was introduced by Black [3] and a formal definition is as follows.

Definition 3 (Single-peaked Domain). *A preference* $\succ \in \mathcal{L}(C)$ *over a set of candidates C is called* single-peaked *with respect to an order* $\succ' \in \mathcal{L}(C)$ *if, for every pair of candidates $x, y \in C$, we have $x \succ y$ whenever we have either $c \succ' x \succ' y$ or $y \succ' x \succ' c$, where $c \in C$ is the candidate at the first position of $\succ$. A profile* $\mathcal{P} = (\succ_i)_{i \in [n]}$ *is called single-peaked with respect to an order* $\succ' \in \mathcal{L}(C)$ *if $\succ_i$ is single-peaked with respect to $\succ'$ for every $i \in [n]$.*

We now turn to the definition of a k-composite single-peaked profile, which is a natural generalization of the single-peaked notion above. We say that a profile is k-composite single-peaked if there is an ordering of the candidates σ and a partition of the candidate set into at most k parts such that each part induces a single-peaked profile on σ restricted to that part. We note, importantly, that this is different from the more well-studied notion of multipeaked profiles, where we have the additional constraint that the k parts have to additionally form intervals on a fixed global ordering. A similar notion called k-additional axis where the votes(rather than the candidates) are divided into k buckets and each bucket is single-peaked, has been studied in [10].

Single-crossing Profiles. A preference profile is said to belong to the single-crossing domain if it admits a permutation of the voters such that for any pair of candidates a and b, there is an index $j\langle a, b\rangle$ such that either all voters v_j with $j < j\langle a, b\rangle$ prefer a over b and all voters v_j with $j > j\langle a, b\rangle$ prefer b over a, or vice versa. The formal definition is as follows.

Definition 4 (Single-crossing Domain). *A profile* $\mathcal{P} = (\succ_i)_{i \in [n]}$ *of* n *preferences over a set* C *of candidates is called a* single-crossing *profile if there exists a permutation* σ *of* [n] *such that, for every pair of distinct candidates* $x, y \in C$, *whenever we have* $x \succ_{\sigma(i)} y$ *and* $x \succ_{\sigma(j)} y$ *for two integers* i *and* j *with* $1 \leqslant \sigma(i) < \sigma(j) \leqslant n$, *we have* $x \succ_{\sigma(k)} y$ *for every* $\sigma(i) \leqslant k \leqslant \sigma(j)$.

As we did with single-peaked profiles, we generalize the notion of single-crossing domains to r-single-crossing domains in the following natural way: for every pair of candidates (a, b), instead of demanding one index where the preferences "switch" from one way to the other, we allow for r such switches. More formally, a profile is r-single-crossing if for every pair of candidates a and b, there exist r indices $j_0\langle a, b\rangle, j_1\langle a, b\rangle, \ldots, j_r\langle a, b\rangle, j_{r+1}\langle a, b\rangle$ with $j_0\langle a, b\rangle = 1$ and $j_{r+1}\langle a, b\rangle = n + 1$, such that for all $1 \leqslant i \leqslant r + 1$, all voters v_j with $j_i\langle a, b\rangle \leqslant j < j_{i+1}\langle a, b\rangle$ are unanimous in their preferences over a and b.

Nearly Structured Domains. Let $\mathcal{D} = \{SP, SC\}$ be a fixed domain, where SP refers to single-peaked domains, and SP denotes single-crossing domains. We say that a profile $\mathcal{P}$ over candidates C has a candidate (voter) modulator of size k to $\mathcal{D}$ if there exists a subset of at most k candidates (voters) such that the restriction of the profile to all but the chosen candidates (voters) belongs to the domain $\mathcal{D}$. Whenever a profile admits a k-sized candidate modulator to $\mathcal{D}$, we say that it is k-*close to* $\mathcal{D}$ *via candidates*. The notion of being k-*close to* $\mathcal{D}$ *via voters* is analogously defined.

The work of [5,10] shows that it is polynomial-time to find the smallest candidate (voter) modulator to the domain of single-peaked (single-crossing) profiles respectively. The work of [9] addressed the NP-hard variants and showed 2-approximation and 6-approximation algorithms for finding the smallest voter and candidate modulator to the domains of single-peaked and single-crossing profiles, respectively. Therefore, in all our problem formulations, we assume that we are given an instance of an election with a modulator to either domain as a part of the input — since it is tractable to find such modulators in all cases.

Parameterized Complexity. A parameterized problem is denoted by a pair $(Q, k) \subseteq \Sigma^* \times \mathbb{N}$. The first component Q is a classical language, and the number k is called the parameter. Such a problem is *fixed–parameter tractable* (FPT) if there exists an algorithm that decides it in time $O(f(k)n^{O(1)})$ on instances of size n. On the other hand, a problem is said to belong to the class XP if there exists an algorithm that decides it in time $n^{O(f(k))}$ on instances of size n. We refer the reader to [8] for a more detailed introduction to the parameterized paradigm.

Problem Definition. We now define the main problem that we address in this work, which we denote by $\ell, \mathcal{D}$-CC Via χ, where ℓ is an aggregation function, $\mathcal{D}$ is a domain and χ is either candidates or voters, referring to the type of the modulator we are given as a part of the input.

$\ell, \mathcal{D}$-CC Via χ **Parameter:** k

Input: An election $E = (C, V)$, a committee size b, a target misrepresentation score R, a misrepresentation function α, and a k-sized χ modulator X to the domain $\mathcal{D}$.

Question: Is there a committee of size b whose ℓ-misrepresentation score under the function α is at most R?

---

3 Tractability on Nearly Structured Preference Profiles

The goal of this section is to establish the following theorem.

Theorem 1. *For all $\ell \in \{\ell_1, \ell_\infty\}$ and for all $\mathcal{D} \in \{\mathsf{SP}, \mathsf{SC}\}$, the $(\ell, \mathcal{D})$-CC Via Candidates problem is in* FPT *and the $(\ell, \mathcal{D})$-CC Via Voters problem is in* XP.

We describe now informally our overall approach for solving the $(\ell, \mathcal{D})$-CC Via χ problem. First, we brute force through all possible "behaviors" of the solution on the modulator. Next, instead of solving the "vanilla" Chamberlin-Courant optimization problem on the part of the profile that is structured (according to the domain $\mathcal{D}$), we adapt our solution to account for the guessed behavior on the modulator. For ease of presentation, we define an intermediate auxiliary problem, which is an extension version of the original problem, described below.

In the extension problem corresponding to $(\ell, \mathcal{D})$, we are given, as usual, an election $E = (C, V)$, a committee size k, a target misrepresentation score R and a misrepresentation function α. In addition, we are also given a subset of candidates X and a partition of X into G and B. The promise is that the election induced by the votes V when restricted to the candidates $C \setminus X$ is structured according to the domain $\mathcal{D}$. The goal is to find an optimal Chamberlin-Courant committee among the ones that contain all candidates in G and contain none of the candidates in B. The formal definition is as follows. In the following, we say that a committee respects a partition $(D \uplus G \uplus B)$ of the candidate set C if it contains all of G and none of B.

$(\ell, \mathcal{D})$–CC Extension

Input: An election $E = (C, V)$, a partition of the candidates into $(D \uplus G \uplus B)$, a committee size b, a target misrepresentation score R, a misrepresentation function α; such that the election induced by (D, V) belongs to the domain $\mathcal{D}$.

Question: Is there a committee of size b that respects $(D \uplus G \uplus B)$ and whose ℓ-misrepresentation score under the function α is at most R?

Before describing how to solve the $(\ell, \mathcal{D})$–CC Extension problem, we first establish that it is indeed useful for solving the $(\ell, \mathcal{D})$–CC Via χ problem. Let $\mathcal{D}$ be a fixed domain from {Single-Peaked, Single-Crossing}. First, consider the $(\ell, \mathcal{D})$–CC Via χ problem where we are given a k-sized candidate modulator as input, or that χ is fixed to be candidates. Let $(E = (C, V), b, R, \alpha, X)$, denoted by $\mathcal{I}$, be an instance of $(\ell, \mathcal{D})$–CC Via χ. Recall that X is a candidate modulator to the domain $\mathcal{D}$, in other words, the election induced by $(C \setminus X, V)$ has the structure of $\mathcal{D}$. Our algorithm proceeds as follows. For a subset of candidates $Y \subseteq X$, let:

$$\mathcal{I}_Y := (E = (C, V); (C \setminus X, Y, X \setminus Y), b, R, \alpha).$$

If $\mathcal{I}_Y$ is a YES-instance of $(\ell, \mathcal{D})$–CC Extension for some $Y \subseteq X$, then our algorithm returns YES and aborts. If, on the other hand, for every subset $Y \subseteq X$ of candidates it turns out that $\mathcal{I}_Y$ is a NO-instance of $(\ell, \mathcal{D})$–CC Extension, then we return NO. It is easy to see that whenever the algorithm returns YES, assuming the correctness of the $(\ell, \mathcal{D})$–CC Extension procedure used, there exists a committee that has the desired misrepresentation score.

To argue the correctness of the algorithm, we show that if $\mathcal{I}$ is a YES-instance then the algorithm does indeed produce a committee that can achieve the desired misrepresentation score. To this end, let C^* be a committee whose ℓ-misrepresentation score under the function α is at most R. Let Y^* denote $C^* \cap X$. Then note that C^* is a committee that respects the partition $D := C \setminus X$, $G := Y^*$, and $B := X \setminus Y^*$. Further, note that since X is a candidate modulator to $\mathcal{D}$, the election induced by (D, V) belongs to the domain $\mathcal{D}$. Clearly, the instance $(E = (C, V); (D, G, B), b, R, \alpha)$ is a well-formed input to the $(\ell, \mathcal{D})$–CC Extension problem, and C^* is a valid solution to it. Assuming again the correctness of the $(\ell, \mathcal{D})$–CC Extension procedure used, we are done. Observe that the running time of our algorithm here is $2^k q(n, m)$, where $q(n, m)$ is the time required by the $(\ell, \mathcal{D})$–CC Extension procedure on an instance of size $n + m$.

We now turn to the $(\ell, \mathcal{D})$–CC Via χ problem where we are given a k-sized voter modulator as input, or that χ is fixed to be voters. Here a direct brute-force approach as in the previous case does not suggest itself, because of which we suffer a greater overhead in our running time. For simplicity, we first describe our algorithm for the egalitarian variant, that is, we fix $\ell = \ell_\infty$. We later describe the changes we need to make when we deal with the utilitarian variant.

Let $(E = (C, V), b, R, \alpha, X)$, denoted by $\mathcal{I}$, be an instance of $(\ell, \mathcal{D})$–CC Via χ. Recall that X is a voter modulator to the domain $\mathcal{D}$, in other words, the election induced by $(C, V \setminus X)$ has the structure of $\mathcal{D}$. For every voter, we guess the candidate who represents that voter in an arbitrary but fixed, and valid, Chamberlin-Courant committee. For such a guess μ, let Y_μ denote the set of at most k candidates who have been chosen to represent the voters in the modulator. More specifically, a voter $v \in X$, let $\mu(v)$ denote the candidate that we have guessed as the representative for the voter v, and let $d(v)$ denote the set of candidates ranked higher than $\mu(v)$ by the voter v. Note that Y_μ is simply $\cup_{v \in X} \mu(v)$.

We first run the following easy sanity check: if, for $u, v \in X$, $u \neq v$, we have that $\mu(v) \in d(u)$, then we reject the guess Y. Otherwise, define $B_\mu := \cup_{v \in X} d(v)$ and $G_\mu := Y_\mu$, and let $D_\mu := C \setminus (G \cup B)$. Observe that B_μ and G_μ are disjoint because of the sanity check. Further, let:

$$\mathcal{I}_\mu := (E = (C, V \setminus X); (D_\mu, G_\mu, B_\mu), b, R, \alpha).$$

It is easily checked that $\mathcal{I}_\mu$ is a well-formed instance for $(\ell, \mathcal{D})$–CC Extension. As before, we return YES if and only if there exists a guess μ for which $\mathcal{I}_\mu$ is a YES instance of $(\ell, \mathcal{D})$–CC Extension. To see the correctness of this approach, let C^* be a committee whose ℓ-misrepresentation score under the function α is at most R. For each voter $v \in X$, let $\mu^*(v)$ denote the top-ranking candidate from C^* in the vote of v. Let Y^* be given by $\cup_{v \in X} \mu^*(v)$, and let B^* be the set of all candidates ranked higher than $\mu^*(v)$ in the votes v from X. Observe that C^* does not contain any candidates from B^* by the definition of μ^*.

Now, as before, define: $G := Y^*$, $B := B^*$, and $D := C \setminus (G \cup B)$. Clearly, the instance $(E = (C, V \setminus X); (D, G, B), k, R, \alpha)$ is a well-formed input to the $(\ell, \mathcal{D})$–CC Extension problem, and C^* is a valid solution to it. Assuming again the correctness of the $(\ell, \mathcal{D})$–CC Extension procedure used, we are done. Observe that the running time of our algorithm here is $n^k q(n, m)$, where $q(n, m)$ is the time required by the $(\ell, \mathcal{D})$–CC Extension procedure on an instance of size $n + m$. For the utilitarian version of the problem (where $\ell = \ell_1$), the procedure is identical, except that we use R' instead of R in the definition $\mathcal{I}_\mu$, where R' is $R - R_{X,\mu}$, and $R_{X,\mu}$ is the sum of the misrepresentation score of the candidate $\mu(v)$ with respect to the voter v, and the sum is over $v \in X$. It is easily verified that the other details work out in the same fashion.

The rest of this section is section is devoted to showing that the $(\ell, \mathcal{D})$–CC Extension problem can be solved in polynomial time by adapting suitably the known algorithms for the Chamberlin-Courant problem on the relevant domain $\mathcal{D}$. These adaptations are sometimes subtle and in particular for the single-peaked case, we have to treat the utilitarian and the egalitarian variants separately (corresponding to $\ell = \ell_1$ and $\ell = \ell_\infty$ respectively).

3.1 $(\ell, \mathcal{D})$-CC Extension for the Single-Crossing Domain

In this section we demonstrate a polynomial time algorithm for the $(\ell, \mathcal{D})$–CC Extension problem for the case when $\mathcal{D} = SC$. This builds closely on the algorithm shown by [12]. First, we show a structural property which is an easy adaptation of Lemma 5 in [12]. The statement corresponding to single-crossing profiles states that there is an optimal committee for which an optimal assignment assigns candidates in contiguous blocks over the single-crossing order. For the $(\ell, \mathcal{D})$–CC Extension problem, this continues to be the case for candidates c from $\mathcal{D}$ except that some candidates in the contiguous block may be assigned to candidates in G instead of being assigned to c. We now state this formally. In the statement below, an optimal b-CC assignment is considered only among committees that respect the semantics of (D, G, B) in the given instance $\mathfrak{J}$ of $(\ell, \mathcal{D})$–CC Extension.

Lemma 1 $(\star)$. *Let* $\mathfrak{J} = (E = (C, V); (D, G, B), b, R, \alpha)$ *be an instance of* (ℓ,SC)-*CC* Extension. *Suppose* $V = (v_1, \ldots, v_n)$ *is the single-crossing order of the votes and* $C = (c_1, \ldots, c_m)$ *is an ordering of the candidates according to* v_i. *Then for every* $b \in [m]$, *every dissatisfaction function* α *for* m *candidates, and for every* $\ell \in \{\ell_1, \ell_\infty\}$, *there is an optimal* b-CC *assignment* Φ *for* E *under* $\alpha - \ell - \text{CC}$ *such that for each candidate* $c_i \in D$, *if* $\phi^{-1}(c_i) \neq \emptyset$, *then there are two integers* e_i *and* f_i, *with* $e_i < f_i$, *such that for every vote* v *in the set of voters* $V' = \{v_{e_i}, v_{e_i+1}, \ldots, v_{f_i}\}$, $\phi(v) \in \{c_i\} \cup G$. *Moreover, for each* $i < j$ *such that* $\Phi^{-1}(c_i) \neq \emptyset$ *and* $\Phi^{-1}(c_j) \neq \emptyset$, *it holds that* $e_i < f_i$.

Due to space considerations we omit the proof of the technical claim above, however, we note that it is along the lines of the proof in [12]. In particular, observe that if there are voters u, v, w appearing in that order in the single-crossing ordering, and for two candidates $c_1, c_2 \in D$, if u and w were to be assigned to c_1 and v were to be assigned to c_2, then this would imply that $c_1 \succ_u c_2$ and $c_1 \succ_w c_2$, while $c_2 \succ_v c_1$, violating the single-crossing structure of the election restricted to D. Since the only other assignments allowed are to candidates in D, the claim follows. We now have the following natural consequence.

Lemma 2. *(ℓ,SC)-CC* Extension *admits a polynomial time algorithm, both for when* $\ell = \ell_1$ *and when* $\ell = \ell_\infty$.

Proof. (Sketch) For Single-Crossing profiles we propose a modified version of the dynamic programming routine which was originally developed in [12]. Here, for $i \in \{0\} \cup [n]$, $j \in [m - |G| - |B|]$ and $t \leq b - |G|$, we define $A[i, j, t]$ as the best possible misrepresentation score that can be achieved by a committee of size $t + |G|$ that respects the semantics of (G, B, D) formed using a subset of first j candidates considering first i votes, where the candidates of D are ordered according to the ranking of the first voter in the single-crossing ordering and the voters are ordered according to the single-crossing ordering. The recurrence for single-crossing orders works by "guessing" the first voter v to be represented by the candidate c_j, and the optimal representation of the preceding voters is found recursively. In our setting, this approach continues to work, except that instead of simply adding up the misrepresentation score of c_j for all voters in the interval starting from v and ending at v_i, we check (for every vote in this interval) if there is a candidate from G who is ranked above c_j, and appropriately adjust the calculation of the misrepresentation score for such voters. The time complexity of above algorithm turns out to be $O(mn^2k)$ (as calculating the misrepresentation score for each voter can take $O(n)$ time). □

3.2 $(\ell, \mathcal{D})$−CC Extension for the Single-Peaked Domain

For the single-peaked domain, as alluded to earlier, we need to consider the utilitarian and egalitarian variants separately. We first consider $\ell = \ell_1$. In the following discussion the terms *first* and *last* are with respect to the societal order, which we denote by $\sqsupset$. A candidate c_i is said to be *smaller* than another candidate c_j if the candidate c_i appears before c_j in the societal order $\sqsupset$, and a candidate is said to be *larger* if it appears after the other candidate. Betzler et al. [2] proposed separate algorithms for the utilitarian and egalitarian variants. To solve $(\ell, \mathcal{D})$−CC Extension in this setting, we extend the dynamic programming algorithm proposed by Betzler et al. for the utilitarian setting.

Lemma 3. *(ℓ_1,SP)-CC* Extension *admits a polynomial time algorithm.*

Proof. Recall that we are given an instance $(E = (C, V); G, B, D, b, r, R, \ell)$ of (ℓ_1,SP)-CC Extension. If $b = |G|$, then there is nothing to do. If $b > |G|$, we assume without loss of generality that there is at least one voter whose top candidate does not belong to G, otherwise we may simply return Yes since the committee G is already good enough for any reasonable R^1. The main semantics of the DP table employed previously is the following. For $i \in [m]$ and $j \in 1, \ldots, \min(i, k)$, we define $z(i, j)$ to be the total misrepresentation for a set of j winners from $\{c_1, \ldots, c_i\}$ including c_i. The final answer is given by $\min_{i \in \{k, \ldots, m\}} z(i, k)$.

We let d denote $|D|$ and let $c_1 \succ c_2 \succ \cdots c_d$ be the single-peaked order. As before, for $i \in [m]$ and $j \in 1, \ldots, \min(i, k)$, we define a modified DP table as follows: let $z(i, j)$ be the total misrepresentation for a set of j winners from $\{c_1, \ldots, c_i\}$ including $\{c_i\} \cup G$. Now, note that the final answer is given by $\min_{i \in \{b', \ldots, m\}} z(i, b')$, where $b' = |G| - b$. Observe that our solution respects the partition (G, B, D), since the semantics of z are such that the candidates G are always incorporated and no candidate from B is ever chosen. Towards describing the recurrence, we establish some notation. First, let $g^*(v)$ denote the highest-ranked candidate from G in the ordering of the voter v. Also, define:

$$g(p, i) := \sum_{v \in V} \max\{0, \min\{r(v, c_p) - r(v, c_i), r(v, g^*(v)) - r(v, c_i)\}\}$$

Intuitively, $g(p, i)$ gives the potential gain of assigning candidate i to the voter v, assuming that the voter v was previously assigned to either the candidate c_p or $g^*(v)$. Both $d(p, i)$ and $g(i)$ can be precomputed in time $O(nm^2)$ by performing

---

[1] If $R < \alpha(1)*n$, for instance, then it is already impossible to achieve for any committee.

one pass over the votes and two passes over the candidates. We are now ready to describe the main recurrence:

$$z[i,j] = \min_{j-1 \leqslant p \leqslant i-1} \left(z[p,j-1] - g(p,i) \right),$$

with the base case:

$$z[i,1] = \min(r(v,c_i), r(v,g^*(v))).$$

Due to space constraints, our argument for correctness only focuses on the part that needs to be adapted appropriately from the proof of [2]. Let C^* be a committee that witnesses the value of $z[i,j]$. Let p be the largest index smaller than i (in the societal ordering) which is such that $c_p \in C^*$ and let $g^*(v)$ be c_q. If for a voter v it holds that $r(v,c_i) < r(v,c_p)$ and $r(v,c_i) < r(v,c_q)$, then note that $r(v,c_i) < r(v,c_t)$ for all $t < p$. Then the contribution of such a voter v to the misrepresentation of $z[p,i-1]$ is $\min(r(v,c_p), r(v,c_q))$. This implies that the improvement in the misrepresntation score of this voter obtained by reassigning the voter to the candidate c_i is precisely given by $g(p,i)$. For all other voters, an assignment to c_i does not improve their misrepresentation, so the algorithm does nothing in these situations. The correctness follows from the fact that the algorithm tries all possible values of p, and the inductively assumed correctness of $z[p,j-1]$. The time complexity of the core algorithm is $O(m^2)$, as both i and j can take at most m values, coupled with the time to precompute $d(p,i)$ and $g(i)$, the total time complexity is $O(nm^2)$. $\qquad\square$

We now turn to the egalitarian version of the rule, that is, $\ell = \ell_\infty$. Here again, the solution involves a straightforward adaptation of the approach of [2] to account for the constraints imposed by the semantics of (G,B,D) in the extension problem.

Lemma 4. *(ℓ_∞,SP)-CC* EXTENSION *admits a polynomial time algorithm.*

Proof (Sketch). Let q be the largest integer for which $\alpha(q) \leqslant R$. We first remove voters who have a candidate from G in their top q positions. Let V' denote the remaining set of voters. For a voter $v \in V'$, let $T_q(v)$ denote the top q candidates in v's ranking. Consider the set $M(v) := T_q(v) \setminus B$. Note that any valid committee must contain a candidate from $M(v)$ for all $v \in V'$. However, observe that the set $M(v) \subseteq D$, and therefore forms a continuous interval on the societal ordering of candidates in D. Therefore our problem reduces to finding a clique cover of size at most $b - |G|$ on the interval graph that is naturally defined by the votes in V', which can be found in time $O(nm)$. $\qquad\square$

4   Hardness for Generalized Restrictions on the Domain

4.1   3-composite Single-Peaked Domains.

To show the hardness of computing an optimal ℓ_∞-CC committee on double-peaked domains, we reduce from the following variant of SAT, which is called

LSAT. In an LSAT instance, each clause has at most three literals, and further the literals of the formula can be sorted such that every clause corresponds to at most three consecutive literals in the sorted list, and each clause shares at most one of its literals with another clause, in which case this literal is extreme in both clauses. The hardness of LSAT was shown in [1]. For ease of description, we will assume in the following reduction that every clause has exactly three literals, although it is easy to see that the reduction can be extended to account for smaller clauses as well.

Theorem 2. *Computing an optimal ℓ_∞-CC committee with respect to the Borda misrepresentation score is* NP-*hard even when the domain is a three-composite single-peaked domain.*

Proof (Sketch). Let ϕ be an instance of LSAT with variables $x_1, \ldots, x_n$ and clauses $C_1, \ldots C_m$. Towards constructing the election instance, we introduce one candidate for every literal in ϕ. Let p_i and q_i denote the candidates corresponding to the variable x_i. We also introduce $(n+1)$ dummy candidates for each variable (which is a total of $n(n+1)$ dummy candidates). Let $d[i,j]$ denote the jth dummy candidate corresponding to the variable x_i. We use C to denote the $2n$ candidates corresponding to the literals, and D to denote the set of dummy candidates. P and Q denote the candidates corresponding to the positive and the negated literals respectively.

Let us fix the ordering σ on the candidates as follows. The first $2n$ candidates are from C arranged according to the LSAT ordering. The last $n(n+1)$ candidates are from D and are arranged in an arbitrary but fixed order. Let σ' be the reverse of σ. For a subset of candidates X, the notation $\overline{X}$ refers to an ordering of X according to σ. For a subset of candidates $X \subset C$, who occupy adjacent positions in the LSAT ordering projected over C, the notation $\overrightarrow{C \setminus X}$ refers to an ordering according to σ of the candidates from $C \setminus X$ who appear after X in the LSAT ordering and similarly $\overleftarrow{C \setminus X}$ refers to an ordering according to σ' of the candidates from $C \setminus X$ who appear before X in the LSAT ordering. This notation easily yields an ordering which is single-peaked — $\overline{X} \succ \overrightarrow{C \setminus X} \succ \overleftarrow{C \setminus X}$. For a subset of candidates $X \subset C$, who occupy adjacent positions in the LSAT ordering projected over C, the notation $\overleftrightarrow{C \setminus X}$ refers to an ordering according to σ of the candidates from $C \setminus X$ who appear after X in the LSAT ordering followed by an ordering according to σ' of the candidates from $C \setminus X$ who appear before X in the LSAT ordering. This notation allows us to easily express an ordering which is single-peaked — $\overline{X} \succ \overrightarrow{C \setminus X}$.

We would now like to setup the votes in such a way that a winning committee corresponds to a valid satisfying assignment. We introduce one vote for every clause as follows. Suppose the clause c consists of the literals (ℓ_1, ℓ_2, ℓ_3), and let the candidates corresponding to these literals be t_1, t_2, t_3 respectively. If $\ell_1 < \ell_2 < \ell_3$ in the LSAT ordering, then we introduce the following vote:

$$v(c) := t_2 \succ t_1 \succ t_3 \succ \overrightarrow{(C \setminus \{t_1, t_2, t_3\})} \succ \overline{D}$$

For every variable x_i, we also introduce the following $(n + 1)$ votes, with $1 \leqslant j \leqslant (n + 1)$:

$$v(x_i, j) := d[i, j] \succ p_i \succ q_i \succ \overrightarrow{(P \setminus \{p_i\})} \succ \overleftarrow{(Q \setminus \{q_i\})} \succ \overleftarrow{D \setminus \{d[i, j]\}}$$

This completes a description of the profile. We fix the Borda misrepresentation target score at two and the committee size is set to n. It is easily checked that this profile is three-composite single-peaked with respect to the partition (P, Q, D). First we look at $v(c)$ – the votes based on a clause. $v(c)$ when projected on D is trivially single-peaked. $v(c)$ when projected on C is single-peaked, and hence when projected on $P, Q \subset C$ will remain single-peaked. Now we look at $v(x_i, j)$ – the votes based on variables, which are clearly single-peaked when projected over P, Q and D individually. We now prove the equivalence of these two instances.

In the forward direction, we simply pick the literals corresponding to a satisfying assignment. If a satisfying assignment does not set a variable, then we pick either p_i or q_i. This clearly satisfies every vote based on a clause $v(c)$, if a vote is not satisfied, then the corresponding clause will also not be satisfied. This trivially satisfies the votes based on variables $v(x_i, j)$, as we pick at least one from p_i and q_i satisfying $v(x_i, j)$ for all $1 \leqslant j \leqslant n + 1$.

In the reverse direction, let W be a committee whose score is at most two. Observe that W must choose at least one of p_i or q_i, for all $1 \leqslant i \leqslant n$. Indeed, if not, then such a committee is forced to pick every $d[i, j]$, $1 \leqslant j \leqslant n+1$, which is a violation of the committee size. Since the committee has at most n candidates, it follows by a standard pigeon-hole argument that $|W \cap \{p_i, q_i\}| \leqslant 1$ for all $1 \leqslant i \leqslant n$, which implies that we pick exactly one of p_i or q_i. Therefore, the committee corresponds naturally to an unambiguous assignment of the variables. It is easily checked that this satisfies every clause, because an unsatisfied clause c would correspond to a voter $v(c)$ whose Borda misrepresentation score would exceed two. This completes the proof.   □

4.2   3-Crossing Profiles

In this section, we show the hardness of computing an optimal ℓ_∞-CC committee with respect to the Borda misrepresentation score with respect to three-crossing domains. The reduction is again from LSAT, and the construction is similar to the one used in the proof of Theorem 2 in that we again have candidates corresponding to literals and votes representing clauses. A committee corresponds to a satisfying assignment precisely when its misrepresentation score is at most two. The main difference from before is in how the candidates are ordered in the preferences of the voters.

Theorem 3. *Computing an optimal ℓ_∞-CC committee with respect to the Borda misrepresentation score is NP-hard even when the domain is three-crossing domain.*

Proof (Sketch). Let ϕ be an instance of LSAT with variables $x_1, \ldots, x_n$ and clauses $C_1, \ldots C_m$. Without loss of generality, let us assume that the ordering of the clauses in the LSAT instance is also given by $C_1, \ldots, C_m$. Towards constructing the election instance, we introduce one candidate for every literal in ϕ. Let p_i and q_i denote the candidates corresponding to the variable x_i. We also introduce $(n+1)$ dummy candidates for each variable (which is a total of $n(n+1)$ dummy candidates). Let $d[i, j]$ denote the jth dummy candidate corresponding to the variable x_i. We use C to denote the $2n$ candidates corresponding to the literals, and D to denote the set of dummy candidates.

Towards describing the votes, let us fix an ordering σ on the candidates as follows. The first $2n$ candidates are from C arranged according to the LSAT ordering. The last $n(n+1)$ candidates are from D and are arranged in an arbitrary but fixed order. For a subset of candidates X, the notation $\overline{X}$ refers to an ordering of X according to σ. We would now like to setup the votes in such a way that a winning committee corresponds to a valid satisfying assignment. For $1 \leqslant i \leqslant m-1$, let G_i denote literals in the set $C_i \setminus C_{i+1}$, while we let G_m denote the literals in C_m. We are now ready to describe the votes. For every $1 \leqslant i \leqslant m$, we introduce the vote v_i, which has the literals of the clause C_i in the top three positions, and the remaining candidates are ranked as follows:

$$v_i := \overline{G_i} \succ \overline{G_{i+1}} \succ \cdots \succ \overline{G_m} \succ \overline{G_{i-1}} \succ \cdots \succ \overline{G_1} \succ \overline{D}$$

It is useful to note that the vote v_{i+1} can be thought of as a ranking obtained from the vote v_i by "pushing back" the tuple $\overline{G_i}$ to just behind $\overline{G_m}$. Therefore, the ordering among the G_i's in v_m is reverse of their ordering in v_1. Observe that if a literal occurs in $C_i \cap C_{i+1}$, then it appears among the top three positions of both v_i and v_{i+1}.

We now turn to the second part of our profile, which consists of votes corresponding to the variables. Here, for a subset of candidates X, we will use $\overline{\overline{X}}$ to refer to an ordering of X according to v_m. Now, for every variable x_i, we introduce the following $(n+1)$ votes, with $1 \leqslant j \leqslant (n+1)$.

$$v_{i,j} := d[i, j] \succ p_i \succ q_i \succ \overline{\overline{(C \setminus \{p_i, q_i\})}} \succ \overline{D \setminus \{d[i, j]\}}$$

This completes a description of the profile. We fix the Borda misrepresentation target score at two and the committee size is set to n. It can be shown, by a careful case analysis, that this profile is three-crossing with respect to the following ordering of the votes:

$$v_1, v_2, \ldots, v_m, v_{1,1}, \ldots, v_{1,n+1}, \ldots, v_{i,1}, \ldots, v_{i,n+1}, \ldots, v_{n,1}, \ldots v_{n,n+1}$$

This completes the description of the construction. Due to lack of space, we defer the case analysis alluded to above and the proof of equivalence. $\square$

5   Concluding Remarks

We have made some progress in demonstrating that the Chamberlin-Courant voting rule can be computed efficiently on nearly-structured domains, and there are some notions of being "almost structured" for which the rule remains hard. Several specific problems remain open. The most pertinent issue is whether the problem admits a FPT algorithm when parameterized by the size of a voter modulator to either single-peaked or single-crossing profiles. The complexity of the utilitarian version of the voting rule on composite profiles or k-crossing profiles is also open.

References

1. Arkin, E.M., Banik, A., Carmi, P., Citovsky, G., Katz, M.J., Mitchell, J.S.B., Simakov, M.: Choice is hard. In: Elbassioni, K., Makino, K. (eds.) ISAAC 2015. LNCS, vol. 9472, pp. 318–328. Springer, Heidelberg (2015). doi:10.1007/978-3-662-48971-0_28
2. Betzler, N., Slinko, A., Uhlmann, J.: On the computation of fully proportional representation. J. Artif. Intell. Res. **47**, 475–519 (2013)
3. Black, D.: On the rationale of group decision-making. J. Polit. Econ. **56**, 23–34 (1948)
4. Brandt, F., Conitzer, V., Endriss, U., Lang, J., Procaccia, A.: Handbook of Computational Social Choice. Cambridge University Press, Cambridge (2016)
5. Bredereck, R., Chen, J., Woeginger, G.J.: Are there any nicely structured preference profiles nearby? Math. Soc. Sci. **79**, 61–73 (2016)
6. Chamberlin, J.R., Courant, P.N.: Representative deliberations and representative decisions: proportional representation and the borda rule. Am. Polit. Sci. Rev. **77**(03), 718–733 (1983)
7. Cornaz, D., Galand, L., Spanjaard, O.: Bounded single-peaked width and proportional representation. In: Proceedings of the 20th European Conference on Artificial Intelligence (ECAI). Frontiers in Artificial Intelligence and Applications, vol. 242, pp. 270–275 (2012)
8. Cygan, M., Fomin, F.V., Kowalik, Ł., Lokshtanov, D., Marx, D., Pilipczuk, M., Pilipczuk, M., Saurabh, S.: Parameterized Algorithms. Springer, Cham (2015). doi:10.1007/978-3-319-21275-3
9. Elkind, E., Lackner, M.: On detecting nearly structured preference profiles. In: Proceedings of the Twenty-Eighth AAAI Conference on Artificial Intelligence, pp. 661–667 (2014)
10. Erdélyi, G., Lackner, M., Pfandler, A.: Computational aspects of nearly single-peaked electorates. In: Proceedings of the Twenty-Seventh AAAI Conference on Artificial Intelligence (2013)
11. Faliszewski, P., Hemaspaandra, E., Hemaspaandra, L.A.: The complexity of manipulative attacks in nearly single-peaked electorates. Artif. Intell. **207**, 69–99 (2014)
12. Skowron, P., Yu, L., Faliszewski, P., Elkind, E.: The complexity of fully proportional representation for single-crossing electorates. Theoret. Comput. Sci **569**, 43–57 (2015)

Learning Agents for Iterative Voting

Stéphane Airiau[1]([⊠]), Umberto Grandi[2], and Filipo Studzinski Perotto[2]

[1] Université Paris-Dauphine, PSL Research University, LAMSADE, Paris, France
stephane.airiau@dauphine.fr
[2] IRIT, University of Toulouse, Toulouse, France
umberto.grandi@irit.fr, filipo.perotto@gmail.com

Abstract. This paper assesses the learning capabilities of agents in a situation of collective choice. Each agent is endowed with a private preference concerning a number of alternative candidates, and participates in an iterated plurality election. Agents get rewards depending on the winner of each election, and adjust their voting strategy using reinforcement learning. By conducting extensive simulations, we show that our agents are capable of learning how to take decisions at the level of well-known voting procedures, and that these decisions maintain good choice-theoretic properties when increasing the number of agents or candidates.

Keywords: Computational social choice · Iterative voting · Bandit algorithms

1 Introduction

In a situation of collective choice, we say that an agent is voting strategically, or that she is manipulating, when the agent does not submit her sincere view to the voting system in order to obtain a collective result that she prefers to the one that would be obtained had she voted sincerely. A classical result in social choice theory showed that all sensible voting rules are susceptible to strategic voting [6,16]. In fact, strategic voting may be exploited to make better decisions in several situations where, for instance, agents are confronted with a sequence of repeated elections, from where an interesting compromising candidate can be elected.

The **plurality rule**, aka. first-past-the-post, selects the alternatives that have been voted for by the highest number of agents. Its computation is quick and its communication costs very low, but in view of its simplicity it suffers from numerous problems. For example, it is possible that the plurality winner would lose in pairwise comparison against all other alternatives[1]. If however the plurality rule is used in an iterative fashion, staging sequential elections in which at each point in time one of the voters is allowed to manipulate, then it constitutes an effective tool for selecting an outcome at equilibrium with good properties [11]. This setting is known as **iterative voting**, and several recent

---

[1] See, e.g., [13] for a proof.

© Springer International Publishing AG 2017
J. Rothe (Ed.): ADT 2017, LNAI 10576, pp. 139–152, 2017.
DOI: 10.1007/978-3-319-67504-6_10

papers explored its convergence using different voting rules, and assessed the quality of the winner [7,9,10,12,14,15].

Most works in this field suffer from two main drawbacks. First, agents are highly myopic in not taking into account the history of their interactions, and in having an horizon for strategic thinking of one single iteration. It creates an artificial asymmetry between the available knowledge and the strategic behavior. Second, to ensure convergence it is required that agents manipulate **one at a time**, a property that is difficult to enforce.

In this paper, we tackle both aspects by studying a **concurrent manipulation process** in which agents have the capability of **learning from their past interaction**. In our setting, iterative voting is seen as a repeated game in which voters use reinforcement learning to cast their ballots. We limit the information available to the learning agents to only the winner of each iteration step (when classic iterative voting methods require more information). Our goal is to show that multiagent learning can be a solution in the context of iterative voting, even when the information available to agents is severely limited: a learning agent bases her decisions on the history of past interactions, and because of the learning rate, the choice of the vote is not purely myopic. In addition, in our model all the learning agents are allowed to change their ballot at the same time. Such possibility cannot be given to classic iterative voting methods because there may not be convergence to a single winner [11]. The question we ask in this paper is whether learning can help making a good collective decision [17]: do we observe convergence, and is the winner good according to choice-theoretic criteria?

We show experimentally that our learning agents are able to learn how to make collective decisions under standard measures of decision quality, such as the Condorcet efficiency and the Borda score. The **contribution** of this paper is twofold: we show that iterative learning (1) outdo all iterative voting methods using less information, and (2) is comparable to a well-known procedure called single transferable vote.

The paper is organized as follows. Section 2 provides the basic definitions and reviews the literature on iterative voting and multiagent reinforcement learning. Section 3 presents the specifics of our simulation setting, and Sect. 4 discusses the obtained results. Section 5 concludes the paper.

2   Basic Definitions and Related Works

We now provide all definitions that are needed for the construction of our setting. We introduce the basics of iterative voting and of multiagent reinforcement learning.

2.1   Voting Rules

Let C be a finite set of m candidates or alternatives and N be a finite set of n agents. Based on their preferences, individuals in N need to make a decision on which alternative in C to choose. Agents are typically assumed to have preferences over candidates in C in the form of a *linear order*, i.e., a transitive, anti-symmetric and complete binary relation over C. We denote with $>_i$ the preference of agent i and with $\boldsymbol{P} = (>_1, \ldots, >_n)$ the profile listing all individual preferences. Hence, we write $b >_i a$ to denote that agent i prefers candidate b to candidate a. A (non-resolute) *voting rule* is a function w that associates with every profile $\boldsymbol{P}$ a non-empty subset of winning candidates $w(\boldsymbol{P}) \in 2^C \setminus \emptyset$. The simplest voting rule, and the one involving as little communication as possible among the agents, is the **plurality** rule: each agent votes for a single candidate, and the candidates with the highest number of votes win. The **Borda** rule is a scoring rule in which a candidate c is given $m - j$ points for each voter that is ranking c in j-th position. The score of a candidate is the sum of her points over all voters. For the **Copeland** rule, the score of a candidate c is the number of pairwise comparisons she wins (i.e., contests between c and another candidate a such that there is a majority of voters preferring c to a) minus the number of pairwise comparisons she loses. For Borda and Copeland, the candidates with the highest score win. Finally, **Single Transferable Vote (STV)** can be viewed as an iterative process: at the first round the candidate that is ranked first by the fewest number of voters gets eliminated (ties are broken following a predetermined order). Votes initially given to the eliminated candidate are then transferred to the candidate that comes immediately after in the individual preferences. This process is iterated until one alternative is ranked first by a majority of voters.

2.2 Iterative Voting

Agents face the choice of submitting their truthful ballot, i.e., a ballot corresponding to their individual preferences, or to vote strategically. In iterative voting, agents start from a voting situation: they fill their ballots and a winner is announced. Then, **one at a time**, an agent changes her ballot, and a new winner is announced, creating a sequence of ballot profiles and consequent winners.

Iterative voting is guaranteed to converge for the plurality rule with linear tie-breaking [11] when agents know the score of each candidate at each round, though for most other voting rules convergence cannot be guaranteed [9]. Restricted dynamics, defined by limiting the possible actions available to players, have therefore been studied to guarantee convergence [7,12,15]. In this paper we focus on iterative voting with the plurality rule and on the following two strategies for individual manipulation:

Best response: at time t, given the plurality score for each candidate c at time $t - 1$, the voter computes her best response(s) and votes for one;

3-Pragmatists [15]: at time t, given the top three plurality candidates at time $t - 1$, the voter manipulates in favour of her preferred candidate amongst them.

Convergence with 3-pragmatists manipulation is guaranteed by the fact that the set of 3 most-voted candidates is not changed by every manipulation step, hence each agent will manipulate the election only once.

Two main critiques have been raised to the setting defined above. First, one agent at a time can change her ballot and the new winner is announced before another agent can change her ballot: this sequential aspect is unrealistic but is key to guarantee the convergence of iterative voting. As was already observed by [11], there is no convergence if individuals are allowed to move at the same time. Second, individuals are highly myopic, since their strategic horizon only considers one-step forward in the iterative process and they do not make use of the history of previous manipulations by other agents when making their next choice.

2.3   Multiagent Learning

Multiagent reinforcement learning has been used both in cooperative domains (where the set of agents share the same goal) and in non-cooperative ones (where each agent is trying to optimize its own personal utility). For cooperative domains, the key issue is that learners obtain a local/personal reward but need to learn a decision that is good for the society of agents. For example, agents that try to optimise air-traffic [1] care about individual preferences as well as the global traffic. In this paper, agents are not concerned about the quality of the outcome for the entire population: each voter would like one of her favourite candidates to win. We are in a non-cooperative setting similar to the one of learning in games: the actions are the different ballots and agents have preferences over the joint actions (i.e. voters have preferences over the candidates). One key difference is that preferences are typically ordinal in voting whereas they are cardinal for games (see Sect. 3.1 describing how we generate cardinal utilities from ordinal ones).

In this paper, we use a basic multiarmed bandit style reinforcement learning algorithm [18] for testing whether agents can learn to make a collective decision, experimenting with different exploration strategies (see Sect. 3 for a detailed description). Many reinforcement learning algorithms have been used for playing normal form games, e.g. joint-action learning [5], gradient-based algorithms such as IGA-WoLF or WoLF-PHC [4], to name a few. Since no algorithm can be claimed to be best, we focus on showing that the most basic learning algorithm is able to perform well. For a similar reason we also choose that agents will only get to observe the current winner, and no other information is available to them, such as the score of all candidates (as done in standard iterative voting).

2.4   Evaluation Criteria

Because there is no consensus on the quality of a collective outcome, we will study the results on multiple criteria. Given a profile of preferences $(>_1, \ldots, >_n)$, a *Condorcet winner (CW)* is a candidate that beats every other candidate in pairwise comparisons. A CW is not guaranteed to exists, but a first parameter in assessing a voting rule is the percentage of profiles in which it elects a CW when there exists one:

Condorcet efficiency: the ratio of profiles where a CW is elected out of all profiles where a CW exists.

Many voting rules are designed to elect a CW whenever it exists, such as the Copeland rule, which hence have a Condorcet efficiency of 1. Other voting rules, such as plurality, Borda and STV may elect a candidate that is not a Condorcet winner. Related work estimates the Condorcet efficiency of plurality and Borda for large electorates using Monte Carlo simulations [8].

A second parameter that can be used to measure the quality of the winner is the Borda score itself:

Borda Score: a candidate c is given $m - j$ points for each agent ranking c in j-th position in her truthful preference

The Borda score provides a good measure of how the rule compromises between top-ranked candidates and candidates ranked lower in the individual preferences. One interpretation of the Borda score is that it estimates the average rank of candidates, and the Borda winner is the candidate with the highest average rank over all candidates. Obviously, the Borda rule is the best rule according to this criterion. When varying the number of voters or candidates, we measure the ratio between the Borda score of the elected winner and the maximal Borda score that can be obtained, i.e., if $B(c)$ is the Borda score of a candidate c then $BR(c) = B(c)/\max_{a \in C}(B(a))$.

3   Learning and Simulation Setting

We now describe the settings of our simulations. Each simulation is defined by the parameters $m = |C|$ (the number of candidates), $n = |N|$ (the number of voters), T as the number of iterations, or repeated elections, the agents dispose to learn. We use iterative voting with plurality rule and lexicographic tie-breaking. Note that the choice of the tie-breaking method has been shown to be an important factor in guaranteeing the convergence of iterative voting rules [9]. We also performed experiments with a randomised tie-breaking rule, obtaining comparable results.

3.1   Preferences and Utilities

While voting is based on ordinal information, reinforcement learning needs cardinal utility. Hence the need to translate a preference order $>_i$ of agent i into a utility function $u_i : C \to \mathbb{R}$. Given an ordering $>$, let $pos(c)$ be candidate c's position, where position 0 is taken by the most preferred candidate, and $|C| - 1$ by the least preferred. We considered three possibilities:

Linear utilities: $u_i^{lin}(c) = 1 - \frac{pos(c)}{|C|-1}$;

Exponential utilities: $u_i^{exp}(c) = \frac{1}{2^{pos(c)}}$;

Logistic utilities: $u_i^{sig}(c) = 1 - \frac{1}{1+e^{-k(pos(c) - \frac{|C|-1}{2})}}$,

The parameter k controls the steepness of the curve in the last definition. These three different methods represent distinct satisfaction contexts. Linear utilities correspond to the Borda values, meaning that the satisfaction with a given candidate decreases linearly following the preference order. Exponential utilities are a more realistic representation, especially in large domains where alternatives at the top bear more importance than those at the bottom. They can also be used to simulate partial orders, since the alternatives below a certain threshold of utility count as non-ranked. In this case, the voters have precise choices, and the satisfaction decreases quickly as soon as the winner is not the preferred candidate. In contrast, logistic utilities decrease slowly in a neighbourhood of the top preferred candidates.

3.2   Profiles Generation

Our experiments are averaged over 10.000 preference profiles generated using the following two distributions:

Impartial culture assumption (IC): linear orders are drawn uniformly at random.

Urn model with correlation $\alpha \in [0, 1)$**:** The preference order of the first voter is drawn with uniform probability among all possible linear orders that are present in an urn. A number of copies of the first drawn preference is then put into the urn depending on the parameter α (more precisely, $\frac{m!}{(\frac{1}{\alpha}-1)}$), and the preference of the second voter is then drawn. The process is repeated until all n preference orders have been selected.

The urn model is also known as the Polya-Eggenberger model [3]. The interest of such scheme is to have some correlation between voters, where some observed preference is more likely to be observed again. The higher the correlation parameter α the more likely it is that a Condorcet winner exists, and the less likely it is that a single voter can change the winner of the plurality rule with a single manipulation. Note that IC is equivalent to the Urn model with $\alpha = 0$.

We developed a generator for the urn model that avoids the manipulation of all permutations over the set of candidates, based on the following intuition. At first the probability of creating a random preference order is equal to 1, and the list of defined voters is empty. At each iteration a new preference is generated, either as a copy of a previous preference, or by generating a random preference. The probability of using a random preference decreases with the number of preference orders generated. The method is described in Algorithm 1, the function *CreateRandomPreference()* returns a sample from the uniform distribution over all possible linear orders, and *CopyPreferenceFrom(N)* picks a voter uniformly at random from N, returning a copy of her preferences.

Algorithm 1. Efficient Urn Model Generator (α, C, n)

/* α is correlation, C is the set of candidates, n is the size of the profile */
$N \leftarrow \emptyset$ // preference profile
$\beta \leftarrow \frac{1}{\alpha} - 1$
for i from 0 to $n - 1$ **do**
   $v \leftarrow \begin{cases} CreateRandomPreference(), & \text{with prob.} \frac{\beta}{\beta + |N|} \\ CopyPreferenceFrom(N), & \text{otherwise} \end{cases}$
   $N \leftarrow N \cup \{v\}$
end for

3.3   Learning Algorithm

Since the voting rule is plurality and only the winner is announced at each iteration, the learner has to decide, at each iteration, the candidate she will vote for. From the point of view of each voter, the proposed setting correspond to the well-known multiarmed bandit problem (MAB), a wide studied case of computational reinforcement learning (RL). A MAB is equivalent to a Markovian Decision Process (MDP) with a single state [2,18].

The learning mechanism must evaluate the utility of each possible action during the sequence of interactions, only based on the feedback suggested by the reward. In our case, the reward is the preference value of the elected candidate (the winner) for the agent. The agent then learns a function $Q : C \rightarrow \mathbb{R}^+$ that estimates the expected utility of voting for each candidate. Once a voter i has voted for candidate c and knows the winner w, it can compute its reward r (the elected winner is w and we have $r = u_i(w)$), and from there, i updates its Q value using the following update rule:

$$Q(c) \leftarrow \beta r + (1 - \beta) \cdot Q(c).$$

where α is the learning rate used to control the impact of new information: when $\beta = 0$, the new information is not used, when $\beta = 1$, only the new information matters. Initially, we fix $\beta = 0.1$.

For controlling the exploration, we have several choices (e.g. using ϵ-greedy, softmax exploration schemes or UCB [2]). We experimented several mechanisms, and the one described below obtained the best results. We use a simple implementation of the "optimism in face of uncertainty" principle for exploration: the voter always picks the candidate with the highest Q-value. To ensure exploration, the Q-values are initialized with a copy of the preference values of the agent, and in case of ties the agent prefers to chose the action that has been tested the least. Thus, a voter is most likely to vote for candidates she prefers at the beginning. If a different candidate wins the iteration, the bad reward decreases the Q-value, and the agent has incentives to try other candidates in the coming iterations.

Both ϵ-greedy and softmax algorithms make the agents very explorative at the beginning, and when an agent explores, sub-optimal actions are chosen (even the least preferred ones), disturbing the learning progress of all the agents.

The same happens with classic version of the *optimistic-greedy method*, where all the utilities are equally initialized to the maximal reward possible. The *UCB* method is also very conservative, the exploration is made sufficient to guarantee near optimal performance in stationary MAB problems. As the environment is not stationary (agents are concurrently learning), exploration is no longer adequate and UCB performs poorly.

4 Simulation Results

We now present the main results, showing that a society of agents provided with very simple learning capabilities can make a "good" collective decision, comparable to that taken by well-known voting rules and often better than what standard iterative voting would recommend. In all results we present next, we use the urn model with $\alpha = 0.1$ for generating the preferences. With the exception of Sect. 4.3, all experiments in this section use exponential utilities, but the results for linear and logistic utilities are similar and not presented here.

4.1 Learning Dynamics

By considering the result of iterated plurality with learning agents as a voting rule *per se* (recall that in our setting iterated plurality is guaranteed to converge), we are able to evaluate its performance in social-choice-theoretic terms, measuring both its Condorcet efficiency and the Borda score of the winner at the end of the iteration. In Figs. 1 and 2 we plot the progress of the first parameter

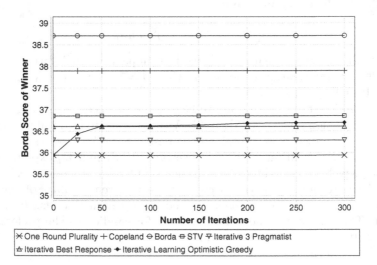

Fig. 1. Performance of learning agents in terms of Borda score of the winner (9 voters, 7 candidates). The number of profiles which actually have a Condorcet winner is 9359 profiles.

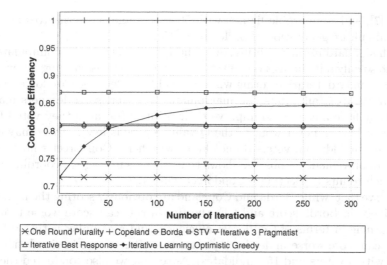

Fig. 2. Performance of learning agents in terms of Condorcet efficiency (9 voters, 7 candidates). The number of profiles which actually have a Condorcet winner is 7331 profiles.

depending on the number of learning iterations that is allowed (a similar figure can be obtained for the second parameter). In order to interpret our findings, we also plot the Condorcet efficiency (CE) and Borda score of one-round plurality, best-response iterative voting, 3-pragmatists iterative voting, STV, Copeland and Borda. We present the results in one scenario with a population of 9 voters and 7 candidates. Utilities of voters are generated using exponential utilities and the results averaged over 10,000 elections.

First, let us consider the Borda score as evaluation criteria in Fig. 1. By design, the Borda voting rule is best. The averaged rank of the winner over all the elections is about 1.7. The next best mechanisms are Copeland and STV. All these voting rules require the knowledge of the complete ordinal preference of the agents. We observe that iterative voting comes next, either in the standard mechanism or our new mechanism with learning agent. Note that our learning agents are using less information as they only know the current winner whereas standard iterative voting uses the plurality score of all candidates. The level of performance is still quite acceptable (the winners' averaged rank is about 1.94). Finally, we observe that our mechanism outperforms one round plurality and 3-pragmatists.

Now, let us turn to Fig. 2 in which we evaluate the performances using Condorcet efficiency. Among the voting rules we consider, only Copeland always elects a Condorcet winner when it exists. For the other rules, STV performs well, followed by iterative voting with learning agents. The best results are obtained for a high number of candidates and low number of voters. Previous work showed that a 10% increase could be obtained with restricted iterative voting in a similar setting (25 candidates and 10 voters, with preferences generated using the urn model) [7], and comparable figures with 50 voters and 5 alternatives using the impartial culture generation of preferences [15].

With standard iterative voting, only one voter at a time can manipulate, and it will do so only if it changes the outcome. If the winner c is winning by two votes or more, standard iterative voting will not be able to elect the Condorcet winner. In our framework, all voters can manipulate. However, some learners may not notice they have a chance to improve their utilities (because they voted for cw in the past but cw never won, so the Q-value for cw is low), others may decide to explore. In addition, voters do not know whether a Condorcet winner exists. But the improvement we observe shows that learners manage to coordinate their vote which results in electing a Condorcet winner.

Observe that, while Borda and Copeland are obviously scoring the maximum, respectively, in Borda score and CE, the Borda rule can score worse than iterative learning in terms of Condorcet efficiency, and a complex voting rule such as STV can score worse under both parameters (this occurs in simulations performed with 3 voters and 15 candidates). Note that we also conducted the same experiments under the impartial culture assumption, obtaining similar results. As a last remark, observe that in view of our initialization the first election is always truthful, i.e. its result coincide with one-round plurality. This is not the case anymore when using other exploration strategies, for which we obtained slower but similar learning dynamics.

4.2 Scalability

One drawback of using learning agents is the number of iterations for convergence. Obviously, it is not reasonable for a human agent to participate in such an iterated process. In the results presented in this section simulations are run with 500 iterations, and we show that learners can still perform well. In addition, the number of voters and the number of candidates are two parameters that, when increased, could significantly deteriorate the performance of learning agents in iterative voting. Figures 3 and 4 shows instead that the deterioration is comparable to those voting rules we considered.

When we keep the number of voters fixed and we add more candidates, the Condorcet efficiency decreases at a similar rate as the other voting rules (results are equivalent to the ones of STV, and learning agents perform better than iterative best response and 3-pragmatist with a similar margin). Typically, in learning in games adding additional actions requires more iterations for learning well (and we usually observe a slight drop in performance). When the number of candidates is high, candidates low in the ranking will have utilities that are negligible compared to the top candidates. Therefore, under those circumstances, the loss in Condorcet efficiency is acceptable.

What is perhaps the most interesting result is that the number of voters does not affect significantly the performance of the learning agents. This is surprising since the environment is less stationary and the noise level is higher with more agents trying to learn concurrently. Typically, it is much more difficult to reach convergence with a high number of voters. On the other hand, with the increasing

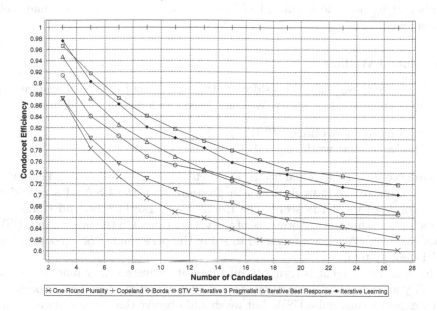

Fig. 3. Scalability of the performance of iterative voting with learning agents at 500 iterations, increasing candidates (Condorcet efficiency, 9 voters).

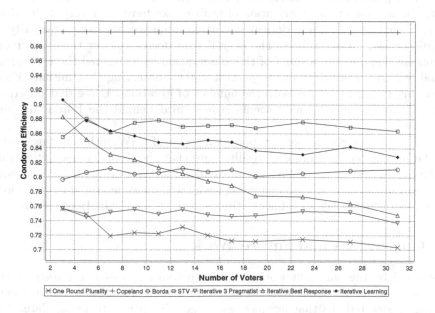

Fig. 4. Scalability of the performance of iterative voting with learning agents at 500 iterations, increasing voters (Condorcet efficiency, 7 candidates)

number of agents, the likelihood of being pivotal decreases, which may help convergence.

4.3 Social Welfare

The last criteria we want to consider are the measures of the social welfare that aggregate the individual utility. We considered the following two definitions:

Utilitarian social welfare: $USW(c) = \sum_{i \in N} u_i(c)$
Egalitarian social welfare: $ESW(c) = \min_{i \in N} u_i(c)$

Observe that if individual utilities are defined as the Borda score, i.e., giving $m - j$ points to the individual in j-th position, then the USW of a candidate corresponds to its Borda score. The Borda score, Borda ratio and Condorcet Efficiency measure the performance of the voting rules. On the other hand, USW is a measure of efficiency, often used to study the performance of a (cooperative) multiagent system.

Each learning agent is trying to maximise its private utility function. Using plurality at each round, a majority of agents will be satisfied, so we do not necessarily expect to maximise USW, but we should observe that a majority of agents improves their utilities. Indeed, this is what we observed in our simulations, in which initially the performance of the learning agents is comparable with those of other voting rules but the USW quickly deteriorates. Remember that the utilities of each candidate (from the most preferred to the least preferred) are 1, $\frac{1}{2}$, $\frac{1}{4}$, etc. If we look at the distribution of utilities, we have more very high values initially (but less than a majority) and at convergence, we have a majority of middle or high utility values. The overall sum decreases over time, but more agents are happier. As a proof of this observation we measured the egalitarian social welfare - the utility of the poorest voter - observing that initially ESW starts pretty low and raises with the number of iterations. Using that criterion, iterative voting with learning agents scores third behind the Borda rule and Copeland.

Our method seems to perform poorly in terms of USW, although this should not be interpreted as a negative result. Note that the second worse voting rule is the Borda rule for the same reason. We also ran experiments with linear and logistic utilities, observing an increasing social welfare as measured by the Borda score (recall that USW with linear utilities is the same as the Borda score).

5 Conclusions and Future Work

Motivated by the emergent works on iterative voting, this paper assesses the learning capabilities of autonomous agents in making good collective decisions. Voting theory tells us that agents always have incentives to manipulate. The idea of iterative voting is then to use a simple voting rule such as plurality, and ask each voter one at a time whether she wants to change her ballot to obtain a preferred winner in the next round.

In this paper we address two weaknesses of existing models of iterative voting: it is not realistic that voters change their ballot one at a time, nor is the myopic-agent assumption that does not allow voters to profit from past interactions. In our model we allow all agents to change their ballots at the same time if they wish to do so. In order to avoid a completely chaotic process, we use learning agents as a mean to learn a good compromise. We show that by using a simple learning algorithm with scarce information (only the winner of the current election is shown to learning agents), the performance of winning candidates are quite good. We evaluate the winner of the iterative process using extensive simulations both in terms of Borda score and of Condorcet efficiency against various voting rules, and we show that we obtain reasonable performances from around 300 iterations. While this number is of course too large for any human to use this method, it is manageable for artificial agents.

We leave it for future work to use more sophisticated learning mechanisms for decreasing the number of iterations to obtain a reasonable performance, and to explore settings in which more information is available to the agents. For instance, using the score of each candidate appears quite reasonable, but has consequences on the learning space.

References

1. Agogino, A.K., Tumer, K.: A multiagent approach to managing air traffic flow. Auton. Agent. Multi-Agent Syst. **24**(1), 1–25 (2012)
2. Auer, P., Cesa-Bianchi, N., Fischer, P.: Finite-time analysis of the multiarmed bandit problem. Mach. Learn. **47**(2–3), 235–256 (2002)
3. Berg, S.: Paradox of voting under an urn model: the effect of homogeneity. Public Choice **47**(2), 377–387 (1985)
4. Bowling, M., Veloso, M.: Multiagent learning using a variable learning rate. Artif. Intell. **136**(2), 215–250 (2002)
5. Claus, C., Boutilier, C.: The dynamics of reinforcement learning in cooperative multiagent systems. In: Proceedings of the 15th National Conference on Artificial Intelligence (AAAI-1998) (1998)
6. Gibbard, A.: Manipulation of voting schemes: a general result. Econometrica **41**(4), 587–601 (1973)
7. Grandi, U., Loreggia, A., Rossi, F., Venable, K.B., Walsh, T.: Restricted manipulation in iterative voting: Condorcet efficiency and Borda score. In: Perny, P., Pirlot, M., Tsoukiàs, A. (eds.) ADT 2013. LNCS, vol. 8176, pp. 181–192. Springer, Heidelberg (2013). doi:10.1007/978-3-642-41575-3_14
8. Lepelley, D., Louichi, A., Valognes, F.: Computer simulations of voting systems. Adv. Complex Syst. **3**(1–4), 181–194 (2000)
9. Lev, O., Rosenschein, J.S.: Convergence of iterative voting. In: Proceedings of the 11th International Conference on Autonomous Agents and Multiagent Systems (AAMAS-2012) (2012)
10. Meir, R., Lev, O., Rosenschein, J.S.: A local-dominance theory of voting equilibria. In: Proceedings of the 15th ACM Conference on Economics and Computation (EC-2014) (2014)

11. Meir, R., Polukarov, M., Rosenschein, J.S., Jennings, N.R.: Convergence to equilibria in plurality voting. In: Proceedings of the Twenty-Fourth Conference on Artificial Intelligence (AAAI-2010) (2010)
12. Obraztsova, S., Markakis, E., Polukarov, M., Rabinovich, Z., Jennings, N.R.: On the convergence of iterative voting: how restrictive should restricted dynamics be? In: Proceedings of the Twenty-Ninth AAAI Conference on Artificial Intelligence (AAAI-2015) (2015)
13. Pacuit, E.: Voting methods. In: Zalta, E.N. (ed.) The Stanford Encyclopedia of Philosophy. Stanford University (2012)
14. Rabinovich, Z., Obraztsova, S., Lev, O., Markakis, E., Rosenschein, J.S.: Analysis of equilibria in iterative voting schemes. In: Proceedings of the Twenty-Ninth AAAI Conference on Artificial Intelligence (AAAI-2015) (2015)
15. Reijngoud, A., Endriss, U.: Voter response to iterated poll information. In: Proceedings of the 11th International Joint Conference on Autonomous Agents and Multiagent Systems (AAMAS-2012), June 2012
16. Satterthwaite, M.A.: Strategy-proofness and arrow's conditions: existence and correspondence theorems for voting procedures and social welfare functions. J. Econ. Theory **10**(2), 187–217 (1975)
17. Shoham, Y., Powers, R., Grenager, T.: If multi-agent learning is the answer, what is the question? Artif. Intell. **171**(7), 365–377 (2007)
18. Sutton, R.S., Barto, A.G.: Introduction to Reinforcement Learning. MIT Press, Cambridge (1998)

The Complexity of Campaigning

Cory Siler, Luke Harold Miles, and Judy Goldsmith[(⊠)]

University of Kentucky, Lexington, KY, USA
jcsi225@g.uky.edu, luke.lambda@uky.edu, goldsmit@cs.uky.edu

Abstract. In "The Logic of Campaigning", Dean and Parikh consider a candidate making campaign statements to appeal to the voters. They model these statements as Boolean formulas over variables that represent stances on the issues, and study optimal candidate strategies under three proposed models of voter preferences based on the assignments that satisfy these formulas. We prove that voter utility evaluation is computationally hard under these preference models (in one case, #P-hard), along with certain problems related to candidate strategic reasoning. Our results raise questions about the desirable characteristics of a voter preference model and to what extent a polynomial-time-evaluable function can capture them.

1 Introduction

In light of some fairly surprising election outcomes around the world, we are very interested in understanding how politicians construct their platforms. For instance, what motivates many candidates to speak in platitudes that reveal little information about their views? On the other hand, what motivates candidates to commit to specific and sometimes audacious policies? The focus of this paper is a logical formalism introduced by Dean and Parikh [6] (and extended by Parikh and Taşdemir [13]) that aims to explain candidates' choices of campaign statements to make; these statements are modeled as propositional formulas over variables representing stances on issues, and Dean and Parikh consider different definitions of voters' utility for a candidate as functions of the possible sets of policies that the candidate might implement based on these statements.

Political scientists have also taken interest in candidates' decisions about what to say when campaigning; petrocik [15] found empirical evidence that a candidate will try to focus on issues where the candidate has a good record and their opponents have bad records. Game theorists have a shared interest with Dean and Parikh in what might motivate a candidate to be ambiguous [1,2,11]. The game-theoretic models take into account the interaction between multiple candidates (and in the case of Manzoni's model [11], voters' uncertainty about their most-preferred policies), but often with simplified representations of a platform (e.g., points on a one-dimensional spectrum or probability distributions over a small set of alternatives). In contrast, Dean and Parikh's framework abstracts away details of the electoral system like multi-agent interactions and

© Springer International Publishing AG 2017
J. Rothe (Ed.): ADT 2017, LNAI 10576, pp. 153–165, 2017.
DOI: 10.1007/978-3-319-67504-6_11

voter strategy to focus on the implications for an individual candidate of communicating using more expressive logic-based statements.

This expressivity, however, brings computational costs. In this paper, we consider the computational complexity of problems related to voter and candidate reasoning. Although Dean and Parikh's formulations capture many desirable characteristics of possible voters, and go a long way toward explaining the assumptions about voters that both vague and overspecific candidates might be making, our results raise some questions about whether these are the right models of how voters evaluate candidates' platforms.

In Sect. 2 we introduce Dean and Parikh's model and some computational complexity classes we will use later. We then find computational complexity results for problems related to the model — in Sect. 3, evaluating voters' utility for a candidate; in Sect. 4, choosing campaign statements that optimize total voter utility for the candidate; and in Sect. 5, choosing campaign statements that motivate enough individual voters to vote for the candidate. In Sect. 6, we conclude with directions for future work in modeling campaigns, including questions about desirable characteristics of voter evaluation.

2   Preliminaries

2.1   Candidates, Voters, and Statements

In Dean and Parikh's model, political views are expressed in terms of Boolean variables (atomic propositions) $\mathcal{X} = \{x_1, \ldots, x_n\}$ (e.g., x_1 = "Every citizen is entitled to a free pony.", x_2 = "Tooth-brushing should be mandatory.", x_3 = "We must invest in zombie-based renewable energy sources."). A candidate makes statements about their platform in the form of propositional formulas over these variables (e.g., $\neg x_1$, or $x_1 \rightarrow x_2$). The candidate's current *theory*[1] T consists of the statements the candidate has issued so far and their logical closure. In our discussion of complexity, we will assume that T is given as a set of statements and that anything in the logical closure besides the statements themselves must be computed. In general we assume that T is self-consistent.

A voter v has a preference function $p_v : \mathcal{X} \rightarrow [-1, 1]$ indicating which direction and how strongly v stands on each issue x_i:[2] A negative $p_v(x_i)$ indicates that v prefers x_i to be false and a positive $p_v(x_i)$ indicates that v prefers x_i to be true, with the magnitude reflecting the strength of preference (and 0 being indifference). If, for example, v was against mandatory tooth-brushing and cared greatly about this issue, then we might have $p_v(x_2) = -0.9$. We assume that candidates have complete knowledge of the voters' preference functions.

We let $\mathcal{W}$ be the set of all possible assignments (*worlds*) to the variables $\mathcal{X}$. Hence, $|\mathcal{W}| = 2^{|\mathcal{X}|} = 2^n$. We denote a specific world as $\omega \in \mathcal{W}$. We say that ω

---

[1] Some sources, particularly in the belief revision literature, use the term *belief base*.
[2] Note that we modify our notation from Dean and Parikh's. In particular, they represent this preference function using two quantities — a weight in $[0, 1]$ and a truth-value preference in $\{-1, 0, 1\}$ — which we combine into the single function p_v.

models a theory T, $\omega \models T$, if ω is consistent with the logical closure of T. For the purpose of defining voter utilities, we treat ω as a function $\omega : \mathcal{X} \to \{-1, 1\}$ where $\omega(x_i) = 1$ if x_i is true in that world and $\omega(x_i) = -1$ if x_i is false in that world. The voter's utility for some $\omega \in \mathcal{W}$ is

$$u_v(\omega) = \sum_{x_i \in \mathcal{X}} p_v(x_i) \cdot \omega(x_i).$$

The voter's utility for a candidate is a function of the possible worlds modeled by the current theory T of what the candidate has said so far. Dean and Parikh consider three classes of voters:

- *Optimistic* voters evaluate the candidate on the best world modeled by the theory, $ut_v(T) = \max\{u_v(\omega) : \omega \models T\}$.
- *Pessimistic* voters evaluate the candidate on the worst world modeled by the theory, $ut_v(T) = \min\{u_v(\omega) : \omega \models T\}$.
- *Expected-value* voters take the average [3] utility over modeled worlds,[4]

$$ut_v(T) = \frac{\sum_{\omega \models T} u_v(\omega)}{|\{\omega : \omega \models T\}|}.$$

Dean and Parikh consider the case of a single candidate who wants to choose statements that maximize the total utility of a population of voters. They prove that with expected-value voters, an optimal strategy involves announcing a stance on every issue, creating a theory that models only one world. (We will refer to such a theory as a *complete* theory; when a theory T' is complete and $T \subseteq T'$, we call T' a *completion* of T.) Furthermore, they observe that with pessimistic voters, it is also advantageous for the candidate to announce a stance on every issue, since eliminating possible worlds can never result in a loss of utility. Only with optimistic voters is it advantageous to remain silent, since eliminating possible worlds can never result in a gain of utility.

2.2   Computational Complexity Classes

We assume familiarity with P and NP. In addition, we will invoke some well-studied but less common complexity notions, which we describe here.

Let C and D be computational complexity classes defined via resource bounds on Turing machines. We denote by C^D those languages or functions computable by a C Turing machine with an *oracle* for D. In other words, we modify a C Turing machine to have an additional tape and state q. If d is a language or function computable by a D Turing machine, then we allow computations of the modified Turing machine to write a string x on the new tape, enter state q, and

---

[3] Dean and Parikh assume that for the purpose of determining "expected value" of the worlds, all worlds are considered equally likely.

[4] One consequence of this definition is that the utility of an empty or otherwise tautological theory is 0 for any expected-value voter.

in the next step, the new tape contains $d(x)$. We write $A \leq_T^P B$ if $A \in P^B$, meaning "A is polynomial-time Turing reducible to B".

The class NPO is an analogue to NP for optimization problems, i.e., problems that are specified in terms of a definition of valid *instances*, a definition of valid *solutions* with respect to an instance, and a *value function* over the solutions, and ask for a solution with maximum or minimum value. Such a problem is in NPO if and only if it meets the following criteria defined by Ausiello et al. [4]: Instances can be verified as valid in polynomial time, solutions can be verified as valid in time polynomial in the instance size, and the value of a solution can be computed in polynomial time. Note that $NPO \subseteq P^{NP}$.

The function class $\#P$, introduced by Valiant [17], contains those problems that are equivalent to determining the number of accepting paths in an NP Turing machine. If a problem is in NP, then the problem of counting how many witness strings satisfy the NP machine for a given instance is in $\#P$. Since a nonzero answer for a $\#P$ problem instance entails a positive answer for the corresponding NP problem instance and a zero answer for the $\#P$ problem entails a negative answer for the NP problem, $NP \leq_T^P \#P$.

We have $P \subseteq NP \subseteq P^{NP} \subseteq P^{\#P} \subseteq PSPACE \subseteq EXPTIME$.

3   Complexity of Finding Voter Utility

One important factor in the epistemology of campaigns that Dean and Parikh's framework (with its implicit assumption of "logical omniscience") does not explicitly model is cognitive complexity (for which we use *computational* complexity as a proxy) as it pertains to the voters. We argue that even a superficially simple series of campaign statements may induce a complex underlying theory: Though we, and Dean and Parikh, have introduced the set of variables $\mathcal{X}$ in terms of high-level issues ($x_1 =$"Every citizen is entitled to a free pony") for explanatory purposes, in reality such issues might more accurately be viewed as complex interplays of finer-granularity subissues ("Every citizen is entitled to a free pony" $= (x_1' \vee x_2' \vee x_3') \wedge (x_1' \rightarrow x_4' \vee \neg x_5') \wedge \ldots$, where x_1', x_2', x_3' are potential taxes to fund the pony giveaway, x_4', x_5' are about the logistics of pony distribution, and so on).

Furthermore, as Dean and Parikh note, additional information can arise from a statement through *implicature* — that which is suggested by a speaker without directly being part of or entailed by "what is said". For instance, when Vermin Supreme says, "When I'm president everyone gets a free pony", we discount the possibility that he plans to give everyone *two* free ponies; if he did, that would not *contradict* his promise, but his omission of information would be *infelicitous*.[5]

---

[5] In Grice's account of implicature [7], participants in a conversation assume each other to be obeying certain maxims of cooperativity (for instance, illustrated here is the maxim of Quantity — roughly, *give as much information as necessary, and do not give more information than necessary*); they interpret each other's statements in light of this mutual assumption.

A potential source of discrepancies between the framework's predictions and the reality of campaigns is that, given an elaborate body of information about a candidate's policies, voters have trouble evaluating the candidate due to the intractability of their utility functions. We will consider the computational complexity of the function problems of determining exact voter utility, but also of decision problems of determining whether the utility meets a given threshold, which are particularly relevant for the "stay-at-home voter" scenario we will discuss in Sect. 5.

3.1   Optimistic Voter Evaluation

Theorem 1. *Given a theory T, an optimistic voter v, and a value k, the problem of deciding whether $ut_v(T) \geq k$ is NP-complete.*

Proof. For *NP* membership, observe that given a world modeled by T for which v's utility is at least k, we can verify the consistency and utility in polynomial time.

We will show *NP*-hardness with a polynomial-time reduction from Boolean satisfiability (SAT). Let ϕ be a propositional formula over a set of variables $\mathcal{X}$. We construct the theory as $T = \{x_* \to \phi\}$, where x_* is a new variable. We construct an optimistic voter v with preferences set as $p_v(x_*) = 1$ and $p_v(x_i) = 0$ for all $x_i \in \mathcal{X}$. And we let $k = 1$. Let $A = \{\omega : \omega \models T\}$. If ϕ is unsatisfiable, then $A = \{\omega : \omega(x_*) = -1\}$, hence $ut_v(T) = -1$. However, if ϕ is satisfiable then there are some $\omega \in A$ where $w(x_*) = 1$ and hence $ut_v(T) = 1$. Finally, we have $ut_v(T) \geq 1 = k$ if and only if ϕ is satisfiable.

Theorem 2. *Given a theory T and an optimistic voter v, the problem of computing $ut_v(T)$ and a corresponding best world modeled by T is NPO-complete.*

Proof. The problem satisfies the criteria for *NPO*-membership [4]: Instances (i.e., the theory and voter specification) and solutions (i.e., worlds modeled by the theory) are recognizable as such in time polynomial in the instance size, and the value function (i.e., voter utility) is computable in polynomial time.

We will show *NPO*-hardness with a polynomial-time reduction from the maximum weighted satisfiability problem (MAX-WSAT),[6] for which Ausiello et al. [4] prove *NPO*-completeness. A MAX-WSAT instance consists of a propositional formula ϕ and a positive weight r_i for each variable x_i; the problem is to find a satisfying assignment that maximizes the total weight of the variables assigned

---

[6] The name "weighted satisfiability" (WSAT) has been used by different sources to refer to two different groups of problems — one where an instance consists only of a propositional formula and the value of a solution is the number of true variables (the Hamming weight), and the generalization we use here where the instance includes weights for the variables. The maximization/minimization versions of the former are sometimes called "maximum number of ones" (MAX-ONES) / "minimum number of ones" (MIN-ONES), and are complete for *NPO-PB* [12], a subclass of *NPO* where the magnitude of a solution's value is polynomially bounded by the size of the input.

to be true. Let $R = \max\{r_i : 1 \leq i \leq n\}$. We construct the theory as $T = \{\phi\}$ and the voter preferences as $p_v(x_i) = r_i/R$ for each x_i.[7] Then the voter's best world ω is the optimal assignment for the MAX-WSAT instance, and given $u_v(\omega) \in [-\sum_i r_i/R, \sum_i r_i/R]$ we can retrieve the corresponding total weight for the MAX-WSAT assignment by mapping this range onto $[0, \sum_i r_i]$.

3.2   Pessimistic Voter Evaluation

Theorem 3. *Given a theory T, a pessimistic voter v, and a value k, the problem of deciding whether $ut_v(T) \geq k$ is coNP-complete.*

Proof. For *coNP* membership, observe that given a world modeled by T for which v's utility is less than k, we can verify the consistency and utility in polynomial time.

We can show *coNP*-hardness by polynomial-time reduction to this problem from Boolean unsatisfiability (UNSAT). Given a formula ϕ, we construct the theory T and the voter v's preferences in the same way as in the proof of Theorem 1, except that v prefers the new variable x_* to be false, $p_v(x_*) = -1$. Then $ut_v(T) \geq 1 = k$ if and only if ϕ is unsatisfiable.

Theorem 4. *Given a theory T and a pessimistic voter v, the problem of computing $ut_v(T)^*$ and a corresponding worst world modeled by T is NPO-complete.*

Proof. *NPO* membership applies by the same argument as in the proof of Theorem 2. We can show *NPO*-hardness by polynomial-time reduction from the minimum weighted satisfiability problem (MIN-WSAT), the minimization counterpart to MAX-WSAT; the mapping is constructed in the same manner as in the proof of Theorem 2.

3.3   Expected-Value Voter Evaluation

Lemma 1. *Given a theory T and an expected-value voter v, the problem of computing $ut_v(T)$ is $\leq_T^P$-hard for #P.*

Proof. Let ϕ be a Boolean formula over $\{x_1, \ldots, x_n\}$ and let $S = \#SAT(\phi)$ be the number of satisfying assignments of ϕ. We define a formula ψ' over $\{x_1, \ldots, x_n\} \cup \{y, z\}$ as follows. First we define $\psi = (\phi \wedge y \wedge z)$, and set

$$\psi' = \psi \vee (y \wedge \neg z \wedge \bigwedge_{i=1}^{n} x_i).$$

Define a voter D with preferences $p_D(y) = p_D(z) = 1$ and $p_D(x_i) = 0$ for each $x_i \in \{x_1, \ldots, x_n\}$. (So this voter is defined over $n + 2$ variables.)

Let $A = \{-1, 1\}^n$ and $B = \{-1, 1\}^{n+2}$ be the possible worlds for ϕ and ψ, respectively. Then we have

$$S = |\{a \in A : a \models \phi\}| = |\{b \in B : b \models \psi\}| = |\{b \in B : b \models \psi'\}| - 1.$$

---

[7] We divide through by the maximum weight so that the $p_v(x_i)$'s are in $[0, 1]$.

That is, ϕ has as many satisfying assignments over n variables as ψ has over $n + 2$ variables, and ψ' has one more satisfying assignment than ψ. Then we get the critical equalities:

$$\frac{ut_D(\{\psi\})}{ut_D(\{\psi'\})} = \frac{(\sum_{b\models\psi} u_D(b))/|\{b \in B : b \models \psi\}|}{(\sum_{b\models\psi'} u_D(b))/|\{b \in B : b \models \psi'\}|}$$

$$= \frac{(\sum_{b\models\psi} u_D(b))/S}{(\sum_{b\models\psi'} u_D(b))/(S+1)}$$

$$= \frac{(\sum_{b\models\psi} u_D(b))/S}{(0 + \sum_{b\models\psi} u_D(b))/(S+1)} = \frac{S+1}{S}.$$

(The third equality is valid because the only world in $\{b \in B : b \models \psi'\} \setminus \{b \in B : b \models \psi\}$ has utility 0 for voter D.) This allows us to derive an equation to get #SAT from ut_D:

$$\frac{1}{ut_D(\psi)/ut_D(\psi') - 1} = \frac{1}{(S+1)/S - S/S} = \frac{1}{1/S} = S = \#SAT(\phi).$$

Thus, we can use two calls to an oracle for expected-value utility to compute $\#SAT(\phi)$ in polynomial time.

Lemma 2. *Given a theory T and an expected-value voter v, the problem of computing $ut_v(T)$ is in $P^{\#P}$.*

Proof. Let T be some theory and let v be some expected-value voter. We assume p_v is represented as a vector of rational binary numbers and we set b as the number of bits in the 'longest' number. We describe an NP Turing machine M whose number of witness strings is proportional to $u_v(T)$.

The machine M takes in a Boolean formula ϕ and a voter's preference function p_v. Then M guesses an assignment ω and a binary integer k where $1 \le k \le 2^b$. If both $\omega \models \phi$ and $k \le u_v(\omega) \cdot 2^b$, then the machine accepts. Otherwise, the machine rejects. Note that both checks take polynomial time.

Given ϕ and b and p_v, how many ways can M accept? If ω does not satisfy, then M cannot accept. If ω does satisfy, then M accepts in exactly $u_v(\omega) \cdot 2^b$ different ways. Hence, $\#M(\phi, p_v, b) = 2^b \sum_{w\models\phi} u_v(\omega)$. Finally, we can compute the utility:

$$ut_v(T) = \frac{\#M(\bigwedge T, p_v, b)}{\#SAT(\bigwedge T) \cdot 2^b}.$$

Theorem 5. *Given a theory T and an expected-value voter v, the problem of computing $ut_v(T)$ is $\le_T^P$-complete for $\#P$.*

Proof. This follows from Lemmas 1 and 2.

4   Complexity of Making an Optimal Theory

Dean and Parikh observe that when all voters are optimistic, a candidate looking to increase total voter utility is best off simply saying nothing; as such,
this situation does not raise any nontrivial computational issues from the candidate's perspective. On the other hand, when appealing to an expected-value or
pessimistic voter population, the candidate is best off taking an explicit stance
on every issue. The candidate's ability to do so, of course, depends on their
knowing *which* stance to take. This has two aspects: Firstly, the candidate must
know the voters' stances on each issue; given the increasing availability of tools
like mass surveys and data analytics that let politicians gauge the attitudes of
their constituents, this is a reasonable assumption (though models of candidate
uncertainty about voter stances are of interest for future study). Secondly, the
candidate must be able to compute the best announcements for appealing to
the overall voter population, given the individual voter preferences; this is the
family of problems we examine here. When we say "optimal theory" in the following results, we mean a theory T that maximizes $\sum_{v \in V} ut_v(T)$ for the voter
population V.

In general, we assume that the candidate starts with an empty theory, and
that the candidate is willing to craft whatever platform is most advantageous
(being what Dean and Parikh call a "Machiavellian" candidate) rather than
being committed to personal beliefs. However, in Sect. 4.3, we show that having to remain consistent with an existing theory raises the complexity of some
relevant problems.

4.1   Appealing to Expected-Value Voters

Theorem 6. *Given n variables and a set $V = \{v_1, \cdots, v_m\}$ of expected-value
voters, a candidate can construct an optimal theory in time $O(n \cdot m)$.*

Proof. In particular, we will describe a procedure for finding a *complete* optimal
theory (since Dean and Parikh have established that with expected-value voters
there always exists a complete theory that is optimal).

We can reformulate the set V of m voters with possibly many different
preference functions into a new set V' of m voters with all the same preference function such that the candidate receives the same total utility, i.e.,
$\sum_{v \in V} ut_v(T) = \sum_{v' \in V'} ut_{v'}(T)$ for any theory T. We define

$$p_{v'}(x_i) = \sum_{v \in V} \frac{p_v(x_i)}{|V|}$$

for each variable x_i, for each $v' \in V'$. This reformulation takes time $O(m)$
to compute for each of the n variables. The candidate can then construct a
theory that models only the world with the preferred assignment to each variable
according to the new preference function.

4.2   Appealing to Pessimistic Voters

Theorem 7. *Given n variables and a set $V = \{v_1, \cdots, v_m\}$ of pessimistic voters, a candidate can construct an optimal theory in time $O(n \cdot m)$.*

Proof. The procedure from the proof of Theorem 6 can also be used to construct a complete optimal theory for pessimistic voters; while the equality between the total utilities of the original and reformulated sets of voters no longer holds in general, it still holds for theories like the constructed one that model only a single world (since the utility for this world is both the pessimistic and expected value for the theory).

4.3   Extending an Existing Theory

Until now, we have assumed that a candidate starts with a "blank slate", able and willing to shape the voters' beliefs with no restrictions. However, there are many reasons why the candidate may instead need to stay consistent with particular set of formulas — the candidate may be an experienced politician who has revealed platform information in prior elections and incumbencies, may be a member of a political party with established doctrine, or may be "tactically honest" [13] — willing to make strategic statements only insofar as they do not contradict certain deeply-held opinions. The strategy of choosing the most informative theory possible to appeal to expected-value or pessimistic voters becomes harder when the candidate must also remain consistent with an existing theory:

Theorem 8. *Given an expected-value or pessimistic voter v and current theory T, the problem of computing an optimal completion of T is NPO-complete.*

Proof. Observe that this problem is equivalent to the optimistic voter utility problem from Theorem 2 (except that instead of yielding a best world ω modeled by a theory, we are yielding a *theory* that models only ω, which can be accomplished by inserting into T a conjunction of literals with their assignments in ω); thus, the proof of Theorem 2 applies here as well.

5   Complexity of Motivating Enough Voters to Vote

While having an enthusiastic constituency is no doubt correlated with a candidate's success, the more direct measure is whether enough supporters actually turn up to vote. A 2006 Pew Research Center study addressed the question of when people vote — since so many people do not, at least in the US. Their findings included the following, which addresses the questions of showing up, rather than the decision about how to vote:

> The Pew analysis identifies basic attitudes and lifestyles that keep these intermittent voters less engaged in politics and the political process. Political knowledge is key: Six-in-ten intermittent voters say they sometimes

don't know enough about candidates to vote compared with 44% of regular voters–the single most important attitudinal difference between intermittent and regular voters identified in the survey. [...] One other key difference: Regular voters are more likely than intermittent voters to say they have been contacted by a candidate or political group encouraging them to vote, underscoring the value of get-out-to-vote campaigns and other forms of party outreach for encouraging political participation.

Pew Research Center [16]

Refusal to vote does not necessarily indicate irrationality on the voter's part; under decision-theoretic models of *expressive voting* [3,8], where a voter's foremost goal is expressing their views rather than bringing about an outcome, abstinence from voting is a rational choice under certain circumstances. Parikh and Taşdemir [13] suggest the presence of "stay-at-home voters", whose utility for a candidate must meet a certain threshold before they will vote, to explain why a candidate might remain silent in situations where Dean and Parikh's model would otherwise suggest a strategy of explicitness.

5.1   Appealing to Optimistic Voters

Theorem 9. *Given an integer h and a set of optimistic voters $V = \{v_1, \cdots, v_m\}$ with thresholds $\{k_1, \cdots, k_m\}$, the problem of deciding the existence of a theory T such that $ut_{v_i}(T) \geq k_i$ for at least h voters is in P.*

Proof. Since the empty theory $T = \emptyset$ has the maximum utility for any optimistic voter, it suffices to compute the utility of each voter v_i's best world, $\sum_{x_j \in \mathcal{X}} |p_{v_i}(x_j)|$, and check whether at least h of these utilities meet their respective voters' thresholds.

5.2   Appealing to Pessimistic Voters

Theorem 10. *Given an integer h and a set of pessimistic voters $V = \{v_1, \cdots, v_m\}$ with thresholds $\{k_1, \cdots, k_m\}$, the problem deciding the existence of a theory T such that $ut_{v_i}(T) \geq k_i$ for at least h voters is NP-complete.*

Proof. For *NP* membership: If there exists a theory for which at least h pessimistic voters meet their thresholds, then these voters also meet their thresholds in any completion of this theory (since eliminating worlds never decreases pessimistic voter utility); given the world modeled by one of these completions, we can verify in polynomial time that the thresholds are met.

We will show *NP*-hardness with a reduction from conjunctive normal form Boolean satisfiability (CNF-SAT). Let ϕ be a Boolean formula in conjunctive normal form. For each clause containing r literals, we construct a pessimistic voter v_i with preferences as follows: For each variable x_j, $p_{v_i}(x_j) = \frac{1}{r}$ if x_j appears in the clause with positive polarity, $p_{v_i}(x_j) = -\frac{1}{r}$ if x_j appears in the clause with negative polarity $(\neg x_j)$, and $p_{v_i}(x_j) = 0$ if x_j does not appear in the clause. We set v_i's threshold as $k_i = -\frac{r-1}{r}$ so that $ut_{v_i}(T) \geq k_i$ if and only if all worlds modeled by T have at least one variable assigned to match its polarity in the clause. Finally, we set h equal to the number of clauses to require that all worlds modeled by T have at least one variable assigned to match its polarity for *every* clause; such a T exists if and only if ϕ is satisfiable.

5.3   Appealing to Mixed Voters

Theorem 11. *Given an integer h, a set of voters $V = \{v_1, \cdots, v_m\} = V^o \cup V^p$ where V^o consists of optimistic voters and V^p consists of pessimistic voters, and thresholds $\{k_1, \cdots, k_m\}$, the problem of deciding the existence of a theory T such that $ut_{v_i}(T) \geq k_i$ for at least h voters is NP-complete.*

Proof. NP-hardness follows from the fact that this is a generalization of the problem with only pessimistic voters from Theorem 10.

For *NP*-membership: Let $V' \subseteq V$ be a set of h or more voters. We could guess an acceptable world (not necessarily distinct) ω_i for each optimistic voter $v_i \in V' \cap V^o$ (i.e., $u_{v_i}(\omega_i) > k_i$), such that ω_i is also acceptable for each pessimistic voter $v_j \in V' \cap V^p$ (i.e., $u_{v_j}(\omega_i) > k_j$). Then the disjunction $T = \{\vee_i \omega_i\}$ would satisfy all the voters in V', and each voter can verify this in polynomial time, since there will only be as many modeled worlds as there are optimistic voters. Furthermore, if there exist theories that satisfy h or more voters, then there exists at least one in the aforementioned form.

6   Conclusions

There are many bodies of research that are, or might be, relevant to whether voters show up to vote, and once there, how they vote. We have explored the computational side of one such theory, and showed that it proposes computationally intensive methods for voter evaluation of platforms and for multi-voter satisficing. Given the basic premise that voter satisfaction or satisficing is a combinatorial problem, the intractability is not surprising.

There are many ways this investigation of the computational complexity of modeling voters' behavior can be extended. They include:

- adding value to informativeness of the candidate(s)' platform when predicting whether a voter will show up to vote, as per the Pew study [16];
- decreasing the complexity of candidates' platforms (conjunctions or disjunctions of atomic propositions; Horn formulas; ...);
- modeling change over time in voter priorities [9] or opinions;
- adding affective variables to the voter models [14];
- investigating social-network models of voter interaction and influence [5];
- using game-theoretic models of candidate-candidate interactions and voter choices [10];
- including group-based identity in the decision to show up as well as the choice of candidate.

In addition, we could start from axiomatic characterizations of models of candidate platforms and voter choice. What are good properties to model? (For instance, if a candidate adds to the specificity of their platform in ways that agree with a voter's preferences, should that increase the likelihood that a voter chooses that candidate, or the likelihood that the voter shows up to vote?) Will we run into Arrow-style impossibility results for achieving all the desiderata we propose?

> I will promise your electorate heart anything you desire, because you are my constituents, you are the informed voting public, and I have no intention of keeping any promise that I make.
>
> Vermin Supreme[8].

Acknowledgments. The authors thank three anonymous reviewers for their helpful feedback and thank Alec Gilbert for catching errors in a late draft. All remaining errors are the responsibility of the authors. This material is based upon work partially supported by the National Science Foundation under Grants No. IIS-1646887 and No. IIS-1649152. Any opinions, findings, and conclusions or recommendations expressed in this material are those of the authors and do not necessarily reflect the views of the National Science Foundation.

References

1. Aragones, E., Neeman, Z.: Strategic ambiguity in electoral competition. J. Theoret. Polit. **12**(2), 183–204 (2000). doi:10.1177/0951692800012002003
2. Aragones, E., Postlewaite, A.: Ambiguity in election games. Rev. Econ. Des. **7**(3), 233–255 (2002). doi:10.1007/s100580200081
3. Aragones, E., Gilboa, I., Weiss, A.: Making statements and approval voting. Theory Decis. **71**(4), 461–472 (2011). doi:10.1007/s11238-010-9208-5
4. Ausiello, G., Crescenzi, P., Gambosi, G., Kann, V., Marchetti-Spaccamela, A., Protasi, M.: Complexity and Approximation: Combinatorial Optimization Problems and Their Approximability Properties. Springer, Heidelberg (1999). doi:10.1007/978-3-642-58412-1
5. Bond, R.M., Fariss, C.J., Jones, J.J., Kramer, A.D.I., Marlow, C., Settle, J.E., Fowler, J.H.: A 61-million-person experiment in social influence and political mobilization. Nature **489**(7415), 295–298 (2012). doi:10.1038/nature11421
6. Dean, W., Parikh, R.: The logic of campaigning. In: Banerjee, M., Seth, A. (eds.) ICLA 2011. LNCS, vol. 6521, pp. 38–49. Springer, Heidelberg (2011). doi:10.1007/978-3-642-18026-2_5
7. Grice, H.P.: Logic and conversation. In: Cole, P., Morgan, J.L. (eds.) Syntax and Semantics 3: Speech Acts, pp. 41–58. Academic Press, New York (1975)
8. Klein, D.: Expressive voting: Modeling a voter's decision to vote. In: ESSLLI Workshop on Logical Models of Group Decision Making (2013)
9. Klein, D., Pacuit, E.: Focusing on campaigns. In: Başkent, C., Moss, L.S., Ramanujam, R. (eds.) Rohit Parikh on Logic, Language and Society. OCL, vol. 11, pp. 77–89. Springer, Cham (2017). doi:10.1007/978-3-319-47843-2_5

---

[8] From Revolution PAC's 2012 interview (https://www.youtube.com/watch?v=9jKszduiK8E)

10. Krueger, J.I., Acevedo, M.: A game-theoretic view of voting. J. Soc. Issues **64**(3), 467–485 (2008). doi:10.1111/j.1540-4560.2008.00573.x
11. Manzoni, E.: Electoral campaigns with strategic candidates: A theoretical and empirical analysis. Ph.D. thesis, London School of Economics and Political Science (United Kingdom) (2010)
12. Panconesi, A., Ranjan, D.: Quantifiers and approximation. Theoret. Comput. Sci. **107**(1), 145–163 (1993). doi:10.1016/0304-3975(93)90259-V
13. Parikh, R., Taşdemir, Ç.: The strategy of campaigning. In: Beklemishev, L.D., Blass, A., Dershowitz, N., Finkbeiner, B., Schulte, W. (eds.) Fields of Logic and Computation II. LNCS, vol. 9300, pp. 253–260. Springer, Cham (2015). doi:10.1007/978-3-319-23534-9_15
14. Parker, M.T., Isbell, L.M.: How I vote depends on how I feel: The differential impact of anger and fear on political information processing. Psychol. Sci. **21**(4), 548–550 (2010). doi:10.1177/0956797610364006
15. Petrocik, J.R.: Issue ownership in presidential elections, with a 1980 case study. Am. J. Polit. Sci. **40**(3), 825–850 (1996). doi:10.2307/2111797
16. Pew Research Center: Who votes, who doesn't, and why: Regular voters, intermittent voters, and those who don't (2006)
17. Valiant, L.G.: The complexity of computing the permanent. Theoret. Comput. Sci. **8**(2), 189–201 (1979). doi:10.1016/0304-3975(79)90044-6

Some Axiomatic and Algorithmic Perspectives on the Social Ranking Problem

Stefano Moretti$^{(\boxtimes)}$ and Meltem Öztürk

Univ. Paris Dauphine, PSL Research University, CNRS, LAMSADE, Paris, France
{stefano.moretti,meltem.ozturk}@dauphine.fr

Abstract. Several real-life complex systems, like human societies or economic networks, are formed by interacting units characterized by patterns of relationships that may generate a group-based social hierarchy. In this paper, we address the problem of how to rank the individuals with respect to their ability to "influence" the relative strength of groups in a society. We also analyse the effect of basic properties in the computation of a social ranking within specific classes of (ordinal) coalitional situations. We show that the pairwise combination of these natural properties yields either to impossibility (i.e., no social ranking exists), or to flattening (i.e., all the individuals are equally ranked), or to dictatorship (i.e., the social ranking is imposed by the relative comparison of coalitions of a given size). Then, we turn our attention to an algorithmic approach aimed at evaluating the frequency of "essential" individuals, which is a notion related to the (ordinal) marginal contribution of individuals over all possible groups.

Keywords: Social ranking · Coalitional power · Ordinal power · Axioms

1 Introduction

Ranking is a fundamental ingredient of many real-life situations, like the ranking of candidates applying to a job, the rating of universities around the world, the distribution of power in political institutions, the centrality of different actors in social networks, the accessibility of information on the web, etc. Often, the criterion used to rank the items (e.g., agents, institutions, products, services, etc.) of a set N also depends on the interaction among the items within the subsets of N (for instance, with respect to the users' preferences over bundles of products or services). In this paper we address the following question: given a finite set N of items and a ranking over its subsets, can we derive a "social" ranking over N according to the "overall importance" of its single elements?

We are grateful to Hossein Khani for pointing out a mistake in the proof of Proposition 2. We also thank four anonymous referees for their valuable suggestions and comments on a former version of this paper. This work benefited from the support of the projects AMANDE ANR-13-BS02-0004 and CoCoRICo-CoDec ANR-14-CE24-0007 of the French National Research Agency (ANR).

© Springer International Publishing AG 2017
J. Rothe (Ed.): ADT 2017, LNAI 10576, pp. 166–181, 2017.
DOI: 10.1007/978-3-319-67504-6_12

For instance, consider a company with three employees 1, 2 and 3 working in the same department. According to the opinion of the manager of the company, the job performance of the different teams $S \subseteq N = \{1,2,3\}$ is as follows: $\{1,2,3\} \succcurlyeq \{3\} \succcurlyeq \{1,3\} \succcurlyeq \{2,3\} \succcurlyeq \{2\} \succcurlyeq \{1,2\} \succcurlyeq \{1\} \succcurlyeq \emptyset$ ($S \succcurlyeq T$, for each $S, T \subseteq N$, means that the performance of S is at least as good as the performance of T). Based on this information, the manager asks us to make a ranking over his three employees showing their attitude to work with others as a team or autonomously. Intuitively, 3 seems to be more influential than 1 and 2, as employee 3 belongs to the most successful teams in the above ranking. Can we state more precisely the reasons driving us to this conclusion? And what can we say if we have to decide who between 1 and 2 is more productive and deserves a promotion? In this paper we analyse different properties of ordinal social rankings in order to get some answers to such questions.

The problem studied in this paper can be seen as an ordinal counterpart of the one about how to measure the power of players in *simple games*, which are coalitional games where coalitions may be winning or not [2,5]. However, our framework is different for at least two reasons: first, we face coalitional situations where only a qualitative (ordinal) comparison of the strength of coalitions is given; second, we look for a ranking over the single objects in N, and we do not require a quantitative assessment of the "power" of the players. As far as we know, the only attempt in the literature to generalize the notions of coalitional game and power index within an ordinal framework has been provided in Moretti [11], where, given a total preorder representing the relative strength of coalitions, a social ranking over the player set is provided according to a notion of *ordinal influence* and using the Banzhaf index [2] of a "canonical" coalitional game.

In the literature of simple games, related questions deal with the ordinal equivalence of power indices (see, for instance, [4,7,10]) and the analysis of the differences between rankings generated by alternative power indices on special classes of simple games (e.g., the papers [9,14]). Similarly to our work, in Taylor and Zwicker [15] the authors investigated alternative notions of ordinal power on different classes of simple games. All the aforementioned papers focus on the notion of simple game, that is a numerical representation of a dichotomous power relation (i.e., winning or losing coalitions), a much more restricted domain than the one considered in this work, where a power relation can be any total preorder over the coalitions. In a still different context, a model of coalition formation has been introduced in Piccione and Razin [12], where the relative strength of disjoint coalitions is represented by an exogenous binary relation and the players try to maximize their position in a social ranking. We also notice a connection with some kind of "inverse problems", precisely, how to derive a ranking over the set of all subsets of N in a way that is "compatible" with a primitive ranking over the single elements of N (see, for instance, [3]; see [10] for an approach using coalitional games).

In this paper, a *social ranking* is defined as a map associating to each *power relation* (i.e., a total preorder over the set of all subsets of N) a total preorder over the elements of N. The properties for social rankings that we analyse in this paper have classical interpretations, such as *symmetry*, basically saying that the relative social ranking of "symmetric"[1] pairs of elements i, j and p, q should coincide (i.e., i is in the social ranking relation with j if and only if p is in the social ranking relation with q); or the *dominance*, saying that an element $i \in N$ should be ranked higher than an element $j \in N$ whenever i dominates j, i.e. a coalition $S \cup \{i\}$ is stronger than $S \cup \{j\}$ for each $S \subset N$ containing neither i nor j. Another property we study in this paper is the *independence of irrelevant coalitions*, saying that the social ranking between two elements i and j should only depend on their respective contributions when added to coalitions containing neither i nor j (in other words, the information needed to rank i and j is provided by the relative comparison of coalitions $U, W \subset N$ such that $U \setminus \{i\} = W \setminus \{j\}$).

We use these properties to axiomatically analyse social rankings on particular classes of power relations. We first notice that two natural properties, precisely, dominance and symmetry, are not compatible over the class of all power relations (see Theorem 1 in Sect. 4), despite the fact that, in some related axiomatic frameworks (see, for instance, [3]), similar axioms have been successfully used in combination. On the other hand, the properties of independence of irrelevant coalitions and symmetry, when applied in combination to a a large class of power relations, determine a flattening of the social ranking, where all the items are equivalent (see Proposition 2 in Sect. 4). Moreover, we prove that the property of independence of irrelevant coalitions and dominance property determine a kind of 'dictatorship of the cardinality' when a relation of strong dominance among coalitions of the same size holds: in this case, the only social ranking satisfying those two properties is the one imposed by the relation of dominance of a given cardinality $s \in \{1, \ldots, |N|\}$ (see Theorem 2 in Sect. 5). Finally, we focus on an alternative algorithmic approach aimed at representing the influence of an item i as the number of coalitions S for which item i results to be essential [13], i.e., $S \cup \{i\}$ is strictly stronger than S.

The structure of the paper is the following. In the next section, we present some related approaches from the literature and our main contributions. Basic notions and definitions are presented in Sect. 2. In Sect. 3 we introduce and discuss some properties for social rankings. In Sect. 4 we study the compatibility of certain axioms and their effect on some elementary notions of social ranking. In Sect. 5 we focus on the analysis of social rankings that satisfy both the dominance property and the property of independence of irrelevant coalitions, and that, on particular power relations, are specified by the ordering of coalitions of the same size. In Sect. 6 we introduce a procedure to define a social ranking based on the cardinality of particular *essential set* and we finally provide some future research directions.

---

[1] Roughly speaking, two pairs of single elements i, j and p, q are said to be symmetric if, for coalitions S with the same cardinality, the number of times that $S \cup \{i\}$ is stronger than $S \cup \{j\}$ equals the number of times that $S \cup \{p\}$ is stronger than $S \cup \{q\}$, and the number of times that $S \cup \{j\}$ is stronger than $S \cup \{i\}$ equals the number of times that $S \cup \{q\}$ is stronger than $S \cup \{p\}$ (for more details, see Definition 3).

2 Preliminaries and Notations

A *binary relation* R on a finite set $N = \{1, \ldots, n\}$ is a collection of ordered pairs of elements of N, i.e. $R \subseteq N \times N$. for all $x, y \in N$, the more familiar notation xRy will be often used instead of the more formal one $(x, y) \in R$. We provide some standard properties for R. *Reflexivity*: for each $x \in N$, xRx; *transitivity*: for each $x, y, z \in N$, xRy and $yRz \Rightarrow xRz$; *totality*: for each $x, y \in N$, $x \neq y$ $\Rightarrow xRy$ or yRx; *antisymmetry*: for each $x, y \in N$, xRy and $yRx \Rightarrow x = y$. A reflexive and transitive binary relation is called *preorder*. A preorder that is also total is called *total preorder*. A total preorder that also satisfies antisymmetry is called *linear order*. The notation $\neg(xRy)$ means that xRy is not true. We denote by 2^N the power set of N and we use the notations T^N and T^{2^N} to denote the set of all total preorders on N and on 2^N, respectively. Moreover, the cardinality of a set $S \in 2^N$ is denoted by $|S|$. In the remaining of the paper, we will also refer to an element $S \in 2^N$ as a *coalition* S. Consider a total preorder $\succeq \subseteq 2^N \times 2^N$ over the subsets of N. Often we will use the notation $S \succ T$ to denote the fact that $S \succeq T$ and $\neg(T \succeq S)$ (in this case, we also say that the relation between S and T is 'strict'), and the notation $S \sim T$ to denote the fact that $S \succeq T$ and $T \succeq S$ (in this case, we say that S and T are indifferent in $\succeq$). For each $i, j \in N$, $i \neq j$, and all $k = 1, \ldots, n - 2$, we denote by $\Sigma_{ij}^k = \{S \in 2^{N \setminus \{i,j\}} : |S| = k\}$ the set of all subsets of N containing neither i nor j with k elements. Moreover, for each $i, j \in N$, we define the set $D_{ij}^k(\succeq) = \{S \in \Sigma_{ij}^k : S \cup \{i\} \succeq S \cup \{j\}\}$ as the set of coalitions $S \in 2^{N \setminus \{i,j\}}$ of cardinality k such that $S \cup \{i\}$ is in relation with $S \cup \{j\}$ (and, changing the ordering of i and j, the set $D_{ji}^k(\succeq) = \{S \in \Sigma_{ij}^k : S \cup \{j\} \succeq S \cup \{i\}\}$).

3 Axioms for Social Rankings

In the remaining of this paper, we interpret a total preorder $\succeq$ on 2^N as a *power relation*, that is, for each $S, T \in 2^N$, $S \succeq T$ stands for 'S is considered at least as strong as T according to the power relation $\succeq$'.

Given a class $\mathcal{C}^{2^N} \subseteq T^{2^N}$ of power relations, we call a map $\rho : \mathcal{C}^{2^N} \longrightarrow T^N$, assigning to each power relation in $\mathcal{C}^{2^N}$ a total preorder on N, a *social ranking solution* or, simply, a *social ranking*. Then, given a power relation $\succeq$, we will interpret the total binary relation $\rho(\succeq)$ associated to $\succeq$ by the social ranking ρ, as the relative power of items (e.g., agents) in a society under relation $\succeq$. Precisely, for each $i, j \in N$, $i\rho(\succeq)j$ stands for 'i is considered at least as influential as j according to the social ranking $\rho(\succeq)$', where the influence of an item is intended as its ability to join coalitions in the strongest positions of a power relation. Note that we require that $\rho(\succeq)$ is a total preorder over the elements of N, that is we always want to express the relative comparison of two items, and such a relation must be transitive. Two elements $i, j \in N$ such that $i\rho(\succeq)j$ and $j\rho(\succeq)i$ are said to be indifferent in $\rho(\succeq)$.

Let $\succeq \in \mathcal{C}^{2^N} \subseteq T^{2^N}$. A social ranking $\rho : \mathcal{C}^{2^N} \longrightarrow T^N$ such that $i\rho(\succeq)j \Leftrightarrow \{i\} \succeq \{j\}$ for each $i, j \in N$ is said to be *primitive* on $\succeq$ (i.e., it neglects any information contained in $\succcurlyeq$ about the comparison of coalitions of cardinality different from 1). A social ranking $\rho : \mathcal{C}^{2^N} \longrightarrow T^N$ such that $i\rho(\succcurlyeq)j$ and $j\rho(\succcurlyeq)i$ for all $i, j \in N$ is said to be *unanimous* on $\succcurlyeq$ (N is an indifference class with respect to $\rho(\succcurlyeq)$).

Now we introduce some properties for social rankings. The first axiom is the *dominance* one: if each coalition S containing item i but not j is stronger than coalition S with j in the place of i, then item i should be ranked higher than item j in the society, for any $i, j \in N$. Precisely, given a power relation $\succcurlyeq \in T^{2^N}$ and $i, j \in N$ we say that i *dominates* j in $\succcurlyeq$ if $S \cup \{i\} \succcurlyeq S \cup \{j\}$ for each $S \in 2^{N \setminus \{i,j\}}$ (we also say that i *strictly dominates* j in $\succcurlyeq$ if i dominates j and in addition there exists $S \in 2^{N \setminus \{i,j\}}$ such that $S \cup \{i\} \succ S \cup \{j\}$).

Definition 1 (DOM). *A social ranking $\rho : \mathcal{C}^{2^N} \longrightarrow T^N$ satisfies the dominance (DOM) property on $\mathcal{C}^{2^N} \subseteq T^{2^N}$ if and only if for all $\succcurlyeq \in \mathcal{C}^{2^N}$ and $i, j \in N$, if i dominates j in $\succcurlyeq$ then $i\rho(\succcurlyeq)j$ (and $\neg(j\rho(\succcurlyeq)i)$ if i strictly dominates j in $\succcurlyeq$).*

The following axiom states that the relative strength of two items $i, j \in N$ in the social ranking should only depend on their effect when they are added to each possible coalition S containing neither i nor j, and the relative ranking of the other coalitions is irrelevant. Formally:

Definition 2 (IIC). *A social ranking $\rho : \mathcal{C}^{2^N} \longrightarrow T^N$ satisfies the Independence of Irrelevant Coalitions (IIC) property on $\mathcal{C}^{2^N} \subseteq T^{2^N}$ iff*

$$i\rho(\succcurlyeq)j \Leftrightarrow i\rho(\sqsupseteq)j$$

for all $i, j \in N$ and all power relations $\succcurlyeq, \sqsupseteq \in \mathcal{C}^{2^N}$ such that for each $S \in 2^{N \setminus \{i,j\}}$

$$S \cup \{i\} \succcurlyeq S \cup \{j\} \Leftrightarrow S \cup \{i\} \sqsupseteq S \cup \{j\}.$$

Let $\succcurlyeq \in T^{2^N}$, and let $i, j, p, q \in N$ be such that $|D_{ij}^k(\succcurlyeq)| = |D_{pq}^k(\succcurlyeq)|$ and $|D_{ji}^k(\succcurlyeq)| = |D_{qp}^k(\succcurlyeq)|$ for each $k = 0, \dots, n-2$. Differently stated, for coalitions S of fixed cardinality, we have that the number of times that $S \cup \{i\}$ is stronger than $S \cup \{j\}$ equals the number of times that $S \cup \{p\}$ is stronger than $S \cup \{q\}$ (and the number of times that $S \cup \{j\}$ is stronger than $S \cup \{i\}$ equals the number of times that $S \cup \{q\}$ is stronger than $S \cup \{p\}$). In this symmetric situation, the following axiom states a principle of equivalence between the pairs $\{i, j\}$ and $\{p, q\}$.

Definition 3 (SYM). *A social ranking $\rho : \mathcal{C}^{2^N} \longrightarrow T^N$ satisfies the symmetry (SYM) property on $\mathcal{C}^{2^N} \subseteq T^{2^N}$ iff*

$$i\rho(\succcurlyeq)j \Leftrightarrow p\rho(\succcurlyeq)q$$

for all $i, j, p, q \in N$ and $\succcurlyeq \in \mathcal{C}^{2^N}$ such that $|D_{ij}^k(\succcurlyeq)| = |D_{pq}^k(\succcurlyeq)|$ and $|D_{ji}^k(\succcurlyeq)| = |D_{qp}^k(\succcurlyeq)|$ for each $k = 0, \dots, n-2$.

Remark 1. Note that if a social ranking ρ satisfies the SYM axiom on $\mathcal{C}^{2^N} \subseteq$ $\mathcal{T}^{2^N}$, then for every $\succcurlyeq \in \mathcal{C}^{2^N}$ and $i, j \in N$, if $|D_{ij}^k(\succcurlyeq)| = |D_{ji}^k(\succcurlyeq)|$ for each $k = 0, \ldots, n-2$, then $i\rho(\succcurlyeq)j$ and $j\rho(\succcurlyeq)i$, that is i and j are indifferent in $\rho(\succcurlyeq)$ (to see this, simply take $p = i$ and $q = j$ in Definition 3).

Remark 2. If we want to check if a given social ranking solution satisfies DOM, IIC, or SYM only partial information on $\succcurlyeq$ is needed. In fact, conditions on the ranking $\rho(\succcurlyeq)$ between two elements i and j only depend on the comparisons of subsets $S \cup \{i\}$ and $S \cup \{j\}$, for all $S \in 2^{N \setminus \{i,j\}}$.

We conclude this section with an example showing that an apparently natural procedure (namely, the majority rule) to rank the items of N may fail to provide a transitive social ranking. We first formally introduce such a procedure.

Definition 4 (Majority rule). *The majority rule (denoted by M) is the map assigning to each power relation $\succcurlyeq \in \mathcal{T}^{2^N}$ the total binary relation $M(\succcurlyeq)$ on N such that*

$$iM(\succcurlyeq)j \Leftrightarrow d_{ij}(\succcurlyeq) \geq d_{ji}(\succcurlyeq).$$

where $d_{ij}(\succcurlyeq) = \sum_{k=0}^{n-2} |D_{ij}^k(\succcurlyeq)|$ for each $i, j \in N$.

Example 1. One can easily check that the majority rule M satisfies the property of DOM, IIC and SYM on the class $\mathcal{T}^{2^N}$. On the other hand, it is also easy to find an example of power relation $\succcurlyeq$ such that $M(\succcurlyeq)$ is not transitive. Consider for instance the power relation $\succcurlyeq \in \mathcal{T}^{2^N}$ with $N = \{1, 2, 3, 4\}$ such that: $2 \succ 1 \succ 3 \succ 23 \succ 13 \succ 12 \succ 14 \succ 34 \succ 24 \succ 134 \sim 124 \sim 234$.

We rewrite the relevant information about $\succcurlyeq$ by means of Table 1 (From now, we will sometimes omit braces and commas to separate elements, for instance, ij denotes the set $\{i, j\}$). Note that $d_{12}(\succcurlyeq) = 2$, $d_{21}(\succcurlyeq) = 3$, $d_{23}(\succcurlyeq) = 2$, $d_{32}(\succcurlyeq) = 3$, $d_{13}(\succcurlyeq) = 3$ and $d_{31}(\succcurlyeq) = 2$. So, we have that $2M(\succcurlyeq)1$, $3M(\succcurlyeq)2$ and $1M(\succcurlyeq)3$, but $\neg(3M(\succcurlyeq)1))$: $M(\succcurlyeq)$ is not a transitive relation. The fact that the majority rule violates transitivity suggests a close affinity of the social ranking set-up with the classical social choice framework: some further similarities with the famous Arrow's impossibility theorem [1] will be further clarified in Sect. 5.

Table 1. The relevant information about $\succcurlyeq$ of Example 1.

1 vs. 2	2 vs. 3	1 vs. 3
$1 \prec 2$	$2 \succ 3$	$1 \succ 3$
$13 \prec 23$	$12 \prec 13$	$12 \prec 23$
$14 \succ 24$	$24 \prec 34$	$14 \succ 34$
$134 \sim 234$	$124 \sim 134$	$124 \sim 234$

4   Primitive and Unanimous Social Rankings

In this section we study the relations between the axioms introduced in the previous section and the social ranking solutions. In the following, we show that DOM and SYM are not compatible in a general case, for $N > 3$ (see Theorem 1), whereas SYM and IIC determine a unanimous social ranking on particular power relations.

We start with showing some consequences of using the axioms introduced in the previous section when the cardinality of the set N is 3 or 4. The analysis for cardinality $|N| = 3$ is easy since we can enumerate all the cases. As we will present in the following, the notion of complementarity plays an important role in this case. We denote by S^* the complement of the subset S ($S^* = N \setminus S$), and we say that a social ranking ρ such that $i\rho(\succcurlyeq)j \Leftrightarrow \{j\}^* \succcurlyeq \{i\}^*$ for each $i, j \in N$ is *complement primitive* on $\succcurlyeq$ (i.e., it neglects any information contained in $\succcurlyeq$ about the comparison of coalitions of cardinality different from $n - 1$).

Proposition 1. *If $|N| = 3$, the only social ranking solution satisfying the DOM and SYM axioms can be either primitive or complement primitive on $\succcurlyeq \in \mathcal{T}^{2^N}$.*

Proof. Let $N = \{1, 2, 3\}$ with $1 \succ 2 \succ 3$. Then six cases may occur in $\succcurlyeq$: case (1) $13 \succcurlyeq 23 \succcurlyeq 12$, case (2) $13 \succcurlyeq 12 \succcurlyeq 23$, case (3) $23 \succcurlyeq 13 \succcurlyeq 12$, case (4) $12 \succcurlyeq 13 \succcurlyeq 23$, case (5) $23 \succcurlyeq 12 \succcurlyeq 13$ and case (6) $12 \succcurlyeq 23 \succcurlyeq 13$.

DOM and SYM impose that:

case (1) by DOM: $1\rho(\succcurlyeq)2$, by SYM ($1\rho(\succcurlyeq)3$ and $2\rho(\succcurlyeq)3$) or ($3\rho(\succcurlyeq)1$ and $3\rho(\succcurlyeq)2$). Hence we have $1\rho(\succcurlyeq)2\rho(\succcurlyeq)3$ (primitive) or $3\rho(\succcurlyeq)1\rho(\succcurlyeq)2$ (complement primitive)

case (2) by DOM: $1\rho(\succcurlyeq)2$ and $1\rho(\succcurlyeq)3$. We can have $2\rho(\succcurlyeq)3$ or $3\rho(\succcurlyeq)2$. Hence we have $1\rho(\succcurlyeq)2\rho(\succcurlyeq)3$ (primitive) or $1\rho(\succcurlyeq)3\rho(\succcurlyeq)2$ (complement primitive)

case (3) by SYM: ($1\rho(\succcurlyeq)2$, $1\rho(\succcurlyeq)3$ and $2\rho(\succcurlyeq)3$) or ($2\rho(\succcurlyeq)1$, $3\rho(\succcurlyeq)1$ and $3\rho(\succcurlyeq)2$).

case (4) by DOM $1\rho(\succcurlyeq)2\rho(\succcurlyeq)3$

case (5) by DOM: $2\rho(\succcurlyeq)3$, by SYM ($1\rho(\succcurlyeq)2$ and $1\rho(\succcurlyeq)3$) or ($2\rho(\succcurlyeq)1$ and $3\rho(\succcurlyeq)1$). Hence we have $1\rho(\succcurlyeq)2\rho(\succcurlyeq)3$ (primitive) or $2\rho(\succcurlyeq)3\rho(\succcurlyeq)1$ (complement primitive)

case (6) by DOM : $1\rho(\succcurlyeq)3$ and $2\rho(\succcurlyeq)3$. We can have $1\rho(\succcurlyeq)2$ or $2\rho(\succcurlyeq)1$. Hence we have $1\rho(\succcurlyeq)2\rho(\succcurlyeq)3$ (primitive) or $2\rho(\succcurlyeq)1\rho(\succcurlyeq)3$ (complement primitive)

Corollary 1. *If $|N| = 3$ and $\succcurlyeq \in \mathcal{T}^{2^N}$ such that for all $S, Q \subseteq N$, $S \succ Q$ implies $Q^* \succ S^*$ (i.e., according to [6], $\succcurlyeq$ is said to be "self-reflecting"), then a social ranking satisfying the DOM property is primitive on $\succcurlyeq$.*

Proof. Let $N = \{i, j, k\}$. Self-reflecting implies that for all $i, j \in N$ $i \succ j \Leftrightarrow j^* \succ i^* \Leftrightarrow ik \succ jk$. By DOM we get for all $i, j, k \in N$ $i\rho(\succcurlyeq)j \Leftrightarrow i \succ j \Leftrightarrow j^* \succ i^* \Leftrightarrow ik \succ jk$.

Next theorem shows that on the class $\mathcal{T}^{2^N}$ (all possible total preorders) the properties of DOM and SYM are not compatible.

Table 2. The relevant information about $\succcurlyeq$ and the elements $1, 2$ and 3.

1 vs. 2	2 vs. 3	1 vs. 3
$1 \sim 2$	$2 \sim 3$	$1 \sim 3$
$13 \succ 23$	$12 \prec 13$	$12 \prec 23$
$14 \sim 24$	$24 \succ 34$	$14 \succ 34$
$134 \sim 234$	$124 \sim 134$	$124 \sim 234$

Theorem 1. *Let $|N| > 3$. There is no social ranking solution $\rho : T^{2^N} \longrightarrow T^N$ which satisfies DOM and SYM on T^{2^N}.*

Proof. We first show a particular situation where DOM and SYM are not compatible. Consider a power relation $\succcurlyeq \in T^{2^N}$ with $N = \{1, 2, 3, 4\}$ and such that

$$1 \sim 2 \sim 3 \succ 13 \succ 23 \succ 12 \succ 24 \sim 14 \succ 34 \succ 1234 \sim 123 \sim 124 \sim 134 \sim 234$$

We rewrite the relevant information about $\succcurlyeq$ and the elements $1, 2$ and 3 by means of the following Table 2. By Remark 1, a social ranking solution $\rho : T^{2^N} \longrightarrow T^N$ which satisfies SYM should be such that $2\rho(\succcurlyeq)3$, $3\rho(\succcurlyeq)2$, $1\rho(\succcurlyeq)3$, $3\rho(\succcurlyeq)1$. By the DOM property, we should have $1\rho(\succcurlyeq)2$, and $\neg(2\rho(\succcurlyeq)1)$, which yields a contradiction with the transitivity of the ranking $\rho(\succcurlyeq)$.

The incompatibility between DOM and SYM also holds for power relations on 2^N with $|N| > 4$. This conclusion directly follows from the fact that one can generate power relations in T^{2^N}, with $N \supseteq \{1, 2, 3, 4\}$, that are obtained from the power relation $\succcurlyeq$ defined above and assigning all the additional subsets of N not contained in $\{1, 2, 3, 4\}$ in the same indifference class. More precisely, the arguments used to show the incompatibility of DOM and SYM on $\succcurlyeq$ also hold for a power relation $\succcurlyeq' \in T^{2^N}$ with $N \supset \{1, 2, 3, 4\}$ and such that

$$U \succcurlyeq' W :\Leftrightarrow U \succcurlyeq W$$

for all the subsets $U, W \subseteq \{1, 2, 3, 4\}$ (i.e., the subsets of $\{1, 2, 3, 4\}$ are ranked in $\succcurlyeq'$ precisely as in $\succcurlyeq$) and

$$U \succcurlyeq' W \text{ and } W \succcurlyeq' U$$

for all the other subsets of N not included in $\{1, 2, 3, 4\}$ (i.e., all the sets not contained in $\{1, 2, 3, 4\}$ are indifferent with respect to the power relation $\succcurlyeq'$).

The following proposition shows that the adoption of properties IIC and SYM yields a unanimous social ranking over all those power relations $\succcurlyeq \in T^N$ such that, for some $i, j \in N$ and $k \in \{0, \ldots, |N| - 2\}$, the relation between $S \cup \{j\}$ and $S \cup \{i\}$ holds strict in the two directions for some $S \in 2^{N \setminus \{i,j\}}$ with $|S| = k$ (precisely, $D_{ij}^k(\succcurlyeq) \setminus D_{ji}^k(\succcurlyeq) \neq \emptyset$ and $D_{ji}^k(\succcurlyeq) \setminus D_{ij}^k(\succcurlyeq) \neq \emptyset$), whereas for all the cardinalities $t \neq k$, we have that $S \cup \{j\}$ and $S \cup \{i\}$ are indifferent for each $S \in 2^{N \setminus \{i,j\}}$ with $|S| = t$ (precisely, $D_{ji}^t(\succcurlyeq) = D_{ij}^t(\succcurlyeq)$).

Proposition 2. *Let* $\rho : T^{2^N} \longrightarrow T^N$ *be a social ranking satisfying IIC and SYM. Let* $\succcurlyeq \in T^{2^N}$, $i, j \in N$ *and* $k \in \{0, \ldots, |N| - 2\}$ *be s.t.* $D_{ij}^k(\succcurlyeq) \setminus D_{ji}^k(\succcurlyeq) \neq \emptyset$ *and* $D_{ji}^k(\succcurlyeq) \setminus D_{ij}^k(\succcurlyeq) \neq \emptyset$, *and s.t.* $D_{ji}^t(\succcurlyeq) = D_{ij}^t(\succcurlyeq)$, *for all* $t \neq k$. *Then* $i\rho(\succcurlyeq)j$ *and* $j\rho(\succcurlyeq)i$.

Proof. Take $i, j \in N$ such that $|D_{ij}^k(\succcurlyeq)| \geq |D_{ji}^k(\succcurlyeq)|$. Define another power relation $\sqsupseteq \in T^{2^N}$ such that

$$S \cup \{i\} \succcurlyeq S \cup \{j\} \Leftrightarrow S \cup \{i\} \sqsupseteq S \cup \{j\}$$

for each $S \in 2^{N \setminus \{i,j\}}$ with $|S| = k$, and $S \sqsupseteq T$ and $T \sqsupseteq S$ for all the other coalitions $S, T \in 2^N$ with $|S| = |T| \neq k + 1$. We still need to define relation $\sqsupseteq$ on the remaining coalitions of size k.

Take $l \in N \setminus \{i, j\}$. Let $\mathcal{D} \subseteq D_{ij}^k(\succcurlyeq)$ be such that $|\mathcal{D}| = |D_{ji}^k(\succcurlyeq)|$. Define the remaining comparisons in $\sqsupseteq$ as follows (an illustrative example of these cases are given in Table 3):

case (1) for each $S \in D_{ji}^k(\succcurlyeq)$ with $l \in S$, let $S \cup \{i,j\} \setminus \{l\} \sqsubseteq S \cup \{j\}$ and $S \cup \{i,j\} \setminus \{l\} \sqsupseteq S \cup \{i\}$;
case (2) for each $S \in D_{ji}^k(\succcurlyeq)$ with $l \notin S$, let $S \cup \{i\} \sqsubseteq S \cup \{l\}$ and $S \cup \{j\} \sqsupseteq S \cup \{l\}$;
case (3) For each $S \in \mathcal{D}$ with $l \in S$, let $S \cup \{i,j\} \setminus \{l\} \sqsubseteq S \cup \{j\}$ and $S \cup \{i,j\} \setminus \{l\} \sqsubseteq S \cup \{i\}$;
case (4) for each $S \in \mathcal{D}$ with $l \notin S$, let $S \cup \{i\} \sqsubseteq S \cup \{l\}$ and $S \cup \{j\} \sqsubseteq S \cup \{l\}$;
case (5) for each $S \in D_{ij}^k \setminus \mathcal{D}$ with $l \in S$, let $S \cup \{i,j\} \setminus \{l\} \sqsupseteq S \cup \{j\}$ and $S \cup \{i,j\} \setminus \{l\} \sqsupseteq S \cup \{i\}$;
case (6) for each $S \in D_{ij}^k \setminus \mathcal{D}$ with $l \notin S$, let $S \cup \{i\} \sqsupseteq S \cup \{l\}$ and $S \cup \{j\} \sqsupseteq S \cup \{l\}$.

Notice that $|D_{ji}^k(\succcurlyeq)| = |D_{li}^k(\sqsupseteq)| = |D_{jl}^k(\sqsupseteq)|$ and $|D_{ij}^k(\succcurlyeq)| = |D_{il}^k(\sqsupseteq)| = |D_{lj}^k(\sqsupseteq)|$. Suppose now that $i\rho(\succcurlyeq)j$. By IIC, we have $i\rho(\sqsupseteq)j$. By SYM, $j\rho(\sqsupseteq)l$ and $l\rho(\sqsupseteq)i$. By transitivity of $\rho(\sqsupseteq)$, $j\rho(\sqsupseteq)i$. By IIC we conclude that $j\rho(\succcurlyeq)i$ too. In a similar way, if we suppose $j\rho(\succcurlyeq)i$, then we end up with the conclusion that $i\rho(\succcurlyeq)j$ too, and the proof follows.

Table 3. An illustrative example of the six possible cases for a power relation $\sqsupseteq$ as the one considered in Proposition 2 with $N = \{1, 2, 3, i, j, l\}$, $k = 2$ and $\mathcal{D} = \{\{1, 2\}, \{2, l\}\}$.

	i vs j	i vs. l	j vs. l
case (1): $S = \{3, l\}$	$\{3, i, l\} \sqsubseteq \{3, j, l\}$	$\{3, i, j\} \sqsubseteq \{3, j, l\}$	$\{3, i, j\} \sqsupseteq \{3, i, l\}$
case (2): $S = \{2, 3\}$	$\{2, 3, i\} \sqsubseteq \{2, 3, j\}$	$\{2, 3, i\} \sqsubseteq \{2, 3, l\}$	$\{2, 3, j\} \sqsupseteq \{2, 3, l\}$
case (3): $S = \{2, l\}$	$\{2, i, l\} \sqsupseteq \{2, j, l\}$	$\{2, i, j\} \sqsubseteq \{2, j, l\}$	$\{2, i, j\} \sqsubseteq \{2, i, l\}$
case (4): $S = \{1, 2\}$	$\{1, 2, i\} \sqsupseteq \{1, 2, j\}$	$\{1, 2, i\} \sqsubseteq \{1, 2, l\}$	$\{1, 2, j\} \sqsubseteq \{1, 2, l\}$
case (5): $S = \{1, l\}$	$\{1, i, l\} \sqsupseteq \{1, j, l\}$	$\{1, i, j\} \sqsupseteq \{1, j, l\}$	$\{1, i, j\} \sqsupseteq \{1, i, l\}$
case (6): $S = \{1, 3\}$	$\{1, 3, i\} \sqsupseteq \{1, 3, j\}$	$\{1, 3, i\} \sqsupseteq \{1, 3, l\}$	$\{1, 3, j\} \sqsupseteq \{1, 3, l\}$
	$|D_{ij}(\sqsupseteq)| = 4$	$|D_{il}(\sqsupseteq)| = 2$	$|D_{jl}(\sqsupseteq)| = 4$
	$|D_{ji}(\sqsupseteq)| = 2$	$D_{li}(\sqsupseteq)| = 4$	$|D_{lj}(\sqsupseteq)| = 2$

5   Dictatorship of the Coalition Size

In this section, we define a class of power relations (namely, the *per size-strong dominant* relations) characterized by the fact that a relation of dominance always exists with respect to coalitions of the same size, but the dominance may change with the cardinality (for instance, an element i could dominate another element j when coalitions of size s are considered, but j could dominate i over coalitions of size $t \neq s$). We first need to introduce the notion of s-strong dominance.

Definition 5. *Let* $\succeq \in T^{2^N}$, $i, j \in N$ *and* $s \in \{0, \ldots, n-2\}$. *We say that* i *s-strong dominates* j *in* $\succeq$, *iff*

$$S \cup \{i\} \succ S \cup \{j\} \text{ for each } S \in 2^{N \setminus \{i,j\}} \text{ with } |S| = s. \tag{1}$$

Definition 6. *We say that* $\succeq \in T^{2^N}$ *is per size-strong dominant (shortly, ps-sdom) iff for each* $s \in \{0, \ldots, n-2\}$ *and all* $i, j \in N$, *we have either*

$$[i \text{ } s\text{-strong dominates } j \text{ in } \succeq] \text{ or } [j \text{ } s\text{-strong dominates } i \text{ in } \succeq].$$

The set of all ps-sdom power relations is denoted by $\mathcal{S}^{2^N} \subseteq T^{2^N}$.

Now, we study the effect of the combination of the properties of DOM and IIC on a specific instance of ps-sdom power relations where there exist elements that are always placed at the top or at the bottom in the rankings of coalitions of equal cardinality.

Example 2. Consider a power relation $\succeq \in \mathcal{S}^{2^N}$ with $N = \{1, 2, 3, 4\}$ and such that

$$1 \succ 2 \succ 3 \succ 4 \succ 34 \succ 24 \succ 14 \succ 23 \succ 13 \succ 12 \succ 123 \succ 134 \succ 124 \succ 234.$$

We rewrite the relevant information about $\succeq$ by means of Table 4.
    Note that for all $S \subseteq N \setminus \{1\}$ and each $l \in N \setminus (S \cup \{1\})$, it holds that $S \cup \{1\} \succ S \cup \{l\}$ if $|S| \in \{0, 2\}$ (i.e., coalition $S \cup \{1\}$ is ranked above coalition $S \cup \{l\}$, for all S containing 0 or 2 elements), whereas $S \cup \{1\} \prec S \cup \{l\}$ if $|S| = 1$ (i.e., coalition $S \cup \{1\}$ is ranked below coalition $S \cup \{l\}$, for all S containing precisely one element). So, elements 1 (or, similar, element 4) is an "extreme" element of N in $\succeq$, where for extreme element we mean an element $i \in N$ such

Table 4. The relevant information about $\succeq$ of Example 2.

1 vs. 2	2 vs. 3	1 vs. 3	1 vs. 4	2 vs. 4	3 vs. 4
$1 \succ 2$	$2 \succ 3$	$1 \succ 3$	$1 \succ 4$	$2 \succ 4$	$3 \succ 4$
$13 \prec 23$	$12 \prec 13$	$12 \prec 23$	$12 \prec 24$	$12 \prec 14$	$13 \prec 14$
$14 \prec 24$	$24 \prec 34$	$14 \prec 34$	$13 \prec 34$	$23 \prec 34$	$23 \prec 24$
$134 \succ 234$	$124 \prec 134$	$124 \succ 234$	$123 \succ 234$	$123 \succ 134$	$123 \succ 124$

that, for all coalitions S of the same size and not containing i, we have either $S \cup \{i\} \succcurlyeq S \cup \{l\}$ for all $l \in N \setminus (S \cup \{i\})$, or, $S \cup \{l\} \succcurlyeq S \cup \{i\}$ for all $l \in N \setminus (S \cup \{i\})$. In Proposition 3 we argue that on this kind of power relations, a social ranking satisfying both DOM and IIC cannot rank "extreme" elements in between two others.

Proposition 3. *Let $\rho : \mathcal{S}^{2^N} \longrightarrow \mathcal{T}^N$ be a social ranking satisfying IIC and DOM on $\mathcal{S}^{2^N}$. Let $\succcurlyeq \in \mathcal{S}^{2^N}$ and $i \in N$ be such that for each $s \in \{0, \ldots, n - 2\}$ either*

$$[S \cup \{i\} \succ S \cup \{j\} \text{ for all } j \in N \setminus \{i\} \text{ and } S \in 2^{N \setminus \{i,j\}} \text{ with } |S| = s] \quad (2)$$

or

$$[S \cup \{j\} \succ S \cup \{i\} \text{ for all } j \in N \setminus \{i\} \text{ and } S \in 2^{N \setminus \{i,j\}} \text{ with } |S| = s]. \quad (3)$$

Then, $[i\rho(\succcurlyeq)j$ for all $j \in N]$ or $[j\rho(\succcurlyeq)i$ for all $j \in N]$.

Proof. Suppose on the contrary that there exist $j, k \in N \setminus \{i\}$, such that

$$j\rho(\succcurlyeq)i \text{ and } i\rho(\succcurlyeq)k. \quad (4)$$

Define $\sqsupseteq \in \mathcal{S}^{2^N}$ such that

$$S \cup \{i\} \sqsupseteq S \cup \{j\} \Leftrightarrow S \cup \{i\} \succ S \cup \{j\} \text{ for all } S \subseteq N \setminus \{i, j\}, \quad (5)$$

$$S \cup \{i\} \sqsupseteq S \cup \{k\} \Leftrightarrow S \cup \{i\} \succ S \cup \{k\} \text{ for all } S \subseteq N \setminus \{i, k\}, \quad (6)$$

and

$$S \cup \{k\} \sqsupseteq S \cup \{j\} \text{ for all } S \subseteq N \setminus \{j, k\}. \quad (7)$$

(note that each coalition $S \cup \{i\}$, with $S \subseteq N \setminus \{i\}$, by condition (2) and (3), is ranked strictly higher or lower than each other coalition $S \cup \{j\}$, $j \neq i$, so condition (7) does not violate the transitivity of $\sqsupseteq$.)

By IIC, we have that $i\rho(\succcurlyeq)j \Leftrightarrow i\rho(\sqsupseteq)j$ and $i\rho(\succcurlyeq)k \Leftrightarrow i\rho(\sqsupseteq)k$. So, by relation (4), $j\rho(\sqsupseteq)i$ and $i\rho(\sqsupseteq)k$. On the other hand, by DOM we have $k\rho(\sqsupseteq)j$ and $\neg(j\rho(\sqsupseteq)k)$, which yields a contradiction with the transitivity of $\rho(\sqsupseteq)$. $\qquad\square$

Proposition 3 shows that if there is an element $i \in N$ having "contradictory" and "radical" behavior depending on the size of coalitions, then the social ranking satisfying IIC and DOM can not give him an intermediate position. In the following, we argue that if a power relation is in $\mathcal{S}^{2^N}$ and a social ranking satisfies both DOM and IIC on the set of ps-sdom power relations $\mathcal{S}^{2^N}$, then it must exist a cardinality $t^* \in \{0, \ldots, n - 2\}$ whose relation of t^*-strong dominance (dictatorially) determines the social ranking. We first introduce the next lemma.

Lemma 1. *Let $i \in N$ and $\rho : \mathcal{S}^{2^N} \longrightarrow \mathcal{T}^N$ be a social ranking satisfying IIC and DOM on $\mathcal{S}^{2^N}$. There exists $t^* \in \{0, \ldots, n - 2\}$ such that*

$$j\rho(\succcurlyeq)k \Leftrightarrow j \ t^*\text{-strong dominates } k \text{ in } \succcurlyeq,$$

for all $j, k \in N \setminus \{i\}$ and $\succcurlyeq \in \mathcal{S}^{2^N}$.

Proof. Given a power relation $\succcurlyeq \in \mathcal{S}^{2^N}$, define another power relation $\succcurlyeq_0 \in \mathcal{S}^{2^N}$ such that for each $S \subseteq N \setminus \{i\}$ we have

$$S \cup \{l\} \succ_0 S \cup \{i\} \text{ for all } l \in N \setminus (S \cup \{i\}), \tag{8}$$

and $U \succcurlyeq_0 W :\Leftrightarrow U \succcurlyeq W$ for all the other possible pairs of coalitions U, W whose comparison is not already considered in (8). Roughly speaking, the only difference between $\succcurlyeq_0$ and $\succcurlyeq$ is that coalitions of size s containing i are placed at the bottom of the ranking induced by $\succcurlyeq$ over the coalitions of the same size. By DOM, it follows that $l\rho(\succcurlyeq_0)i$ for every $l \in N$.

Now, for each $t \in \{0, \ldots, n-2\}$, define a power relation $\succcurlyeq_t \in \mathcal{T}^{2^N}$ such that

$$S \cup \{i\} \succ_t S \cup \{l\} \text{ for each } l \in N \text{ and } S \in 2^{N\setminus\{i,l\}} \text{ with } |S| = s, \tag{9}$$

where $s \in \{0, \ldots, t\}$, and $U \succcurlyeq_t W :\Leftrightarrow U \succcurlyeq_{t-1} W$ for all the other possible pairs of coalitions U, W whose comparison is not already considered in (9). So, the only difference between $\succcurlyeq_t$ and $\succcurlyeq_{t-1}$, for each $t \in \{1, \ldots, n-2\}$, is that in $\succcurlyeq_t$ coalitions of size t containing i are placed at the top of the ranking induced by $\succcurlyeq_{t-1}$ over coalitions of the same size t, and all the remaining comparisons remain the same as in $\succcurlyeq_{t-1}$.

Note that by Proposition 3, we have that either $l\rho(\succcurlyeq_t)i$ for every $l \in N$, or $i\rho(\succcurlyeq_t)l$ for every $l \in N$. Moreover, By DOM, it follows that $i\rho(\succcurlyeq_{n-2})l$ for every $j \in N$. Let t^* be the smallest number in $\{0, \ldots, n-2\}$ such that $l\rho(\succcurlyeq_{t^*-1})i$ for every $l \in N$ and $i\rho(\succcurlyeq_{t^*})l$ for every $l \in N$ (for the considerations above such a t^* must exist, being, at most, $t^* = n - 2$). Next, we argue that for every $j, k \in N \setminus \{i\}$, the social ranking between j and k in $\succcurlyeq$ is imposed by the relation of t^*-strong dominance in $\succcurlyeq$. W.l.o.g., suppose that $S \cup \{j\} \succcurlyeq S \cup \{k\}$ (and, as a consequence, $S \cup \{j\} \succcurlyeq_{t^*} S \cup \{k\}$) for each $S \in 2^{N\setminus\{j,k\}}$, and $|S| = t^*$. Consider another power relation $\sqsupseteq \in \mathcal{T}^{2^N}$ obtained by $\succcurlyeq_{t^*}$ and such that:

$$S \cup \{j\} \sqsupseteq S \cup \{i\} \text{ for each } S \in 2^{N\setminus\{i,j\}} \text{ with } |S| = t^*, \tag{10}$$

$$S \cup \{i\} \sqsupseteq S \cup \{k\} \text{ for each } S \in 2^{N\setminus\{i,k\}} \text{ with } |S| = t^*, \tag{11}$$

$$S \cup \{j\} \sqsupseteq S \cup \{k\} \text{ for each } S \in 2^{N\setminus\{j,k\}} \setminus (2^{N\setminus\{i,j\}} \cup 2^{N\setminus\{i,k\}}), \text{ and } |S| = t^*, \tag{12}$$

and, finally,

$$U \sqsupseteq V :\Leftrightarrow U \succcurlyeq_{t^*} V \tag{13}$$

for all the other relevant pairs of coalitions U, W of size $s \neq t^* + 1$. By IIC $j\rho(\sqsupseteq)i$ (since in $\sqsupseteq$ the comparisons between coalitions containing i and j are precisely as in $\succcurlyeq_{t^*-1}$ and, as previously stated, $j\rho(\succcurlyeq_{t^*-1})i$) and $i\rho(\sqsupseteq)k$ (since in $\sqsupseteq$ the comparisons between coalitions containing i and k are precisely as in $\succcurlyeq_{t^*}$ and, as previously stated, $i\rho(\succcurlyeq_{t^*})k$). Then, by transitivity of $\rho(\sqsupseteq)$ we have $j\rho(\sqsupseteq)k$. Note that by IIC, $j\rho(\sqsupseteq)k \Leftrightarrow j\rho(\succcurlyeq_{t^*})k \Leftrightarrow j\rho(\succcurlyeq)k$. We have then proved that whenever j t^*-dominates k, then $j\rho(\succcurlyeq)k$.

The following theorem states the "dictatorship of the coalition's size".

Theorem 2. *Let* $\rho : \mathcal{S}^{2^N} \longrightarrow \mathcal{T}^N$ *be a social ranking satisfying IIC and DOM on* $\mathcal{S}^{2^N}$. *There exists* $t^* \in \{0, \ldots, n-2\}$ *such that*

$$i\rho(\succcurlyeq)j \Leftrightarrow i \ t^*\text{-strong dominates } j \text{ in } \succcurlyeq,$$

for all $i, j \in N$ *and* $\succcurlyeq \in \mathcal{S}^{2^N}$.

Proof. Given a power relation $\succcurlyeq \in \mathcal{S}^{2^N}$, let $i \in N$ and define $\succcurlyeq_{t^*}$ starting from $\succcurlyeq$ and i precisely as in the proof of Lemma 1.

Now take $k \in N \backslash \{i\}$ and apply Lemma 1 with k in the role of i. Consequently, we have that there exists $\hat{t} \in \{0, \ldots, n-2\}$ such that

$$h\rho(\succcurlyeq)l \Leftrightarrow h \ \hat{t}\text{-strong dominates } l \text{ in } \succcurlyeq,$$

for each $h, l \in N \setminus \{k\}$, and in particular

$$i\rho(\succcurlyeq)l \Leftrightarrow i \ \hat{t}\text{-strong dominates } l \text{ in } \succcurlyeq,$$

for any complete power relation $\succcurlyeq \in \mathcal{S}^{2^N}$. But in the proof of Lemma 1 we have shown that

$$i\rho(\succcurlyeq)l \Leftrightarrow i \ t^*\text{-strong dominates } l \text{ in } \succcurlyeq_{t^*}$$

(remember that t^* in the proof of Lemma 1 is the smallest number in $\{0, \ldots, n-2\}$ such that $l\rho(\succcurlyeq_{t^*-1})i$ for every $l \in N$ and $i\rho(\succcurlyeq_{t^*})l$ for every $l \in N$). Then it must be $\hat{t} = t^*$, and the proof follows.

6   An Algorithmic Approach

In view of the results provided in the previous axiomatic analysis, each combination of two axioms yields either no social ranking or an unsatisfactory one. It is worth noting that all the axioms that we studied in this paper are based on the comparison of subsets having the same number of elements. Therefore, it would be interesting to study properties based on the comparison among subsets with different cardinalities. Following this idea, an interesting property is the notion of *essential alternative* that has been introduced in Puppe [13] as a necessary condition for a power relation representing the preferences of a decision maker over menus (in this context, the preference over menus of a decision maker should reflect her or his freedom to chose a most preferred alternative from any selected menu). Given a power relation $\succcurlyeq \in \mathcal{T}^{2^N}$ and a coalition $S \in 2^N$, an element $i \in N \setminus S$ is said to be *essential* for S if $S \cup \{i\} \succ S$. In our framework, where a power relation represents the relative strength of coalitions, an item i is essential for a coalition S not containing i if coalition $S \cup \{i\}$ is strictly stronger than S. Differently stated, an item i is essential for S (not containing i), if the marginal contribution $v(S \cup \{i\}) - v(S)$ of i to $S \cup \{i\}$ is strictly positive, for every utility function $v : 2^N \to \mathbb{R}$ associated to the power relation $\succcurlyeq$ and such that $v(T) \geq v(U) :\Leftrightarrow T \succcurlyeq U$, for each $T, U \in 2^N$ [11]. Our goal in this section is to assess the influence of items in terms of the number of coalitions in which each item i is essential under a given power relation. More precisely, for each item $i \in N$ we first need to introduce the notion of *essential set* $E_i(\succcurlyeq) := \{S \in 2^{N \setminus \{i\}} : S \cup \{i\} \succ S\}$. Then we define the social ranking solution $\rho^e : \mathcal{T}^{2^N} \longrightarrow \mathcal{T}^N$ such that

$$i\rho^e(\succcurlyeq)j :\Leftrightarrow |E_i(\succcurlyeq)| \geq |E_j(\succcurlyeq)| \tag{14}$$

for each $i, j \in N$ and $\succcurlyeq \in \mathcal{T}^{2^N}$. It is easy to check that ρ^e does not satisfy any of the axioms studied in the previous sections.

Example 3. Consider the power relation $\succcurlyeq \in \mathcal{T}^{2^N}$ with $N = \{1, 2, 3, 4\}$ such that $2 \succ 4 \succ 23 \succ 123 \succ 13 \sim 134 \sim 124 \sim 234 \sim N \sim 12 \succ 14 \succ 1 \succ 3 \succ 34 \succ 24 \succ \emptyset$. Notice that the relevant information presented in Table 1 of Example 1 is still compatible with this power relation. Moreover, the essential sets for players in N are: $E_1(\succcurlyeq) = \{\emptyset, \{3\}, \{2, 4\}, \{3, 4\}\}$, $E_2(\succcurlyeq) = \{\emptyset, \{3\}, \{1\}, \{1, 3\}, \{1, 4\}, \{3, 4\}\}$, $E_3(\succcurlyeq) = \{\emptyset, \{1, 2\}, \{1, 4\}, \{2, 4\}\}$ and $E_4(\succcurlyeq) = \{\emptyset, \{1\}\}$. Consequently, accordingly to the social ranking ρ^e, 2 is the most influential item ($|E_2(\succcurlyeq)| = 6$), followed by 1 and 3 with the same score ($|E_1(\succcurlyeq)| = E_3(\succcurlyeq)| = 4$), and finally by item 4 ($|E_4(\succcurlyeq)| = 2$).

Notice that the definition of an essential set $E_i(\succcurlyeq)$, for all $i \in N$, involves the comparison of 2^{n-1} pairs of coalitions S and $S \cup \{i\}$, with $S \subseteq N \setminus \{i\}$. On the other hand, several coalitions are compared multiple times over different essential sets. So, it is computationally useful to design a procedure aimed at computing the social ranking $\rho^e(\succcurlyeq)$ avoiding those multiple comparisons (see Algorithm 1). To this aim, we first group coalitions over classes of indifferences with respect to $\succcurlyeq$: suppose we have $S_1 \succcurlyeq S_2 \succcurlyeq S_3 \succcurlyeq \cdots \succcurlyeq S_{2^n}$ then we shall write $\Sigma_1 \succ \Sigma_2 \succ \Sigma_3 \succ \cdots \succ \Sigma_l$, to denote the power relation $\succcurlyeq$, but having

---

Algorithm 1. A procedure to find a social ranking based on the essential sets.

> **Input** : A power $\succcurlyeq$ on 2^N in the form of indifference classes
> $\quad\quad \Sigma_1 \succ \Sigma_2 \succ \cdots \succ \Sigma_l$.
> **Output**: A vector $d \in \mathbb{R}^N$ such that $d_i = |E_i(\succcurlyeq)|$ for each $i \in N$.

```
 1  initialisation: d_i := 0 for each i ∈ N ; X := ∅ ;
 2  for k = 1 to l do
 3  │    X := X ∪ Σ_k;
 4  │    for every S ∈ Σ_k do
 5  │    │    for every i ∈ S do
 6  │    │    │    if {S \ {i}} ∉ X then
 7  │    │    │    │    d_i := d_i + 1;
 8  │    │    │    end
 9  │    │    end
10  │    end
11  end
12  return d.
```

grouped in Σ_1 all the coalitions indifferent to S_1 (i.e., all $T \in 2^N$ s.t. $T \succcurlyeq S_1$ and $S_1 \succcurlyeq T$), in Σ_2 all the coalitions indifferent to the first coalition strictly less strong than S_1 in the ranking $\succcurlyeq$, and so on. Then, a coalition S in Σ_k is strictly stronger than any coalition in Σ_{k+1}. Notice that at each iteration k, $k \in \{1, \ldots, l\}$, the test to establish whether i is essential for $S \in \Sigma_k$ is done by means of the **if** condition in line 6 (if $S \setminus \{i\}$ belongs to some Σ_t, $t \le k$, then i is not essential for S).

A possible direction for future research is the open question about which axioms could be used to characterize a social ranking based on the essential sets introduced in this section. It would also be interesting to consider social ranking based on alternative definitions of essential item. For instance, consider a set of items $N = \{1, 2, 3\}$ and a power relation such that $\{2, 3\} \succ \{1, 3\} \succ \{1\} \sim \{2\}$. Clearly items 1 and 2 are essential for $\{1, 3\}$ and $\{2, 3\}$, respectively, but 2 seems "more" essential than 1, in the sense that the contribution of 2 to the power of coalition $\{2, 3\}$ is larger than the contribution of 1 to $\{1, 3\}$ ($\{1\}$ and $\{2\}$ are indifferent, but $\{2, 3\}$ is strictly stronger than $\{1, 3\}$). This kind of considerations about the "intensity" of items' contribution requires a more complex algorithmic analysis of the structure of a power relation aimed at comparing the role of single elements over sets of different cardinality.

References

1. Arrow, K.J.: Social Choice and Individual Values. Yale University Press, New Haven (1963)
2. Banzhaf III, J.F.: Weighted voting doesn't work: A mathematical analysis. Rutgers Law Rev. **19**, 317 (1964)
3. Barberà, S., Bossert, W., Pattanaik, P.K.: Ranking sets of objects. In: Barberà, S., Hammond, P.J., Seidl, C. (eds.) Handbook of Utility Theory, vol. 2, pp. 893–977. Kluwer Academic Publishers (2004)
4. Diffo, L.L., Moulen, J.: Ordinal equivalence of power notions in voting games. Theory Decis. **53**, 313–325 (2002)
5. Dubey, P., Neyman, A., Weber, R.J.: Value theory without efficiency. Math. Oper. Res. **6**, 122–128 (1981)
6. Fishburn, P.C.: Signed orders and power set extensions. J. Econ. Theory **56**, 1–19 (1992)
7. Freixas, J.: On ordinal equivalence of the Shapley and Banzhaf values. Int. J. Game Theory **39**, 513–527 (2010)
8. Kannai, Y., Peleg, B.: A note on the extension of an order on a set to the power set. J. Econ. Theory **32**, 172–175 (1981)
9. Laruelle, A., Merlin, V.: Different least square values, different rankings. Soc. Choice Welfare **19**, 533–550 (2002)
10. Lucchetti, R., Moretti, S., Patrone, F.: Ranking sets of interacting objects via semivalues. TOP **23**, 567–590 (2015)
11. Moretti, S.: An axiomatic approach to social ranking under coalitional power relations. Homo Oeconomicus **32**, 183–208 (2015)
12. Piccione, M., Razin, R.: Coalition formation under power relations. Theoret. Econ. **4**, 1–15 (2009)

13. Puppe, C.: An axiomatic approach to "preference for freedom of choice". J. Econ. Theory **68**, 174–199 (1996)
14. Saari, D.G., Sieberg, K.K.: Some surprising properties of power indices. Games Econ. Behav. **36**, 241–263 (2001)
15. Taylor, A.D., Zwicker, W.S.: Simple Games: Desirability Relations, Trading, Pseudoweightings. Princeton University Press, Princeton (1999)

Complexity of Group Identification with Partial Information

Gábor Erdélyi[✉], Christian Reger, and Yongjie Yang

School of Economic Disciplines, University of Siegen, Siegen, Germany
erdelyi@wiwi.uni-siegen.de

Abstract. In this paper, we study the computational complexity of several problems pertaining to the model of group identification. In group identification, a set of agents try to determine who among themselves are socially qualified for a given task. In particular, we introduce the concepts of possible and necessary winners in the context of group identification and study them for the consent, liberal-start-respecting, and consensus-start-respecting rules.

1 Introduction

Decision making plays an important role in multi-agent systems. For instance, a set of agents (or robots) need to complete a task cooperatively. Due to some reasons (e.g., in order to minimize the cost of the resources), only a few agents can take the job. In this case, all agents need to make a joint decision of which agents are going to take the job. In this paper, we study such a decision making model, in which a set N of individuals desire to select a subset of N. In particular, each individual qualifies or disqualifies every individual in N, and then a social rule is applied to select the socially qualified individuals. This model has been widely studied under the name of *group identification* in economics [5, 6, 15, 19]. In particular, the consent rules and the two procedural rules consensus-start-respecting rule (CSR) and liberal-start-respecting rule (LSR) have been extensively studied in the literature [5, 14, 19]. Consent rules are a class of social rules, where each of them is characterized by two positive integers s and t. Moreover, if an individual qualifies herself, then this individual is socially qualified if and only if there are at least $s - 1$ other individuals who also qualify her. On the other hand, if the individual disqualifies herself, then this individual is not socially qualified if and only if there are at least $t - 1$ other individuals who also disqualify her. The CSR and the LSR social rules recursively determine the socially qualified individuals. In the beginning, the set K^{LSR} of individuals each of whom qualifies herself are considered LSR socially qualified, while the set K^{CSR} of individuals each of whom is qualified by all individuals are considered CSR socially qualified. Then, in each iteration for the social rule LSR (resp. CSR), an individual a is added to K^{LSR} (resp. K^{CSR}) if there is an individual in K^{LSR} (resp. K^{CSR}) qualifying a. The iteration terminates until no new individual can be added to K^{LSR} (resp. K^{CSR}), and the socially qualified individuals are the ones in K^{LSR} (resp. K^{CSR}).

This work is supported in part by the DFG under grant ER 738/2-1.

© Springer International Publishing AG 2017
J. Rothe (Ed.): ADT 2017, LNAI 10576, pp. 182–196, 2017.
DOI: 10.1007/978-3-319-67504-6_13

Given the full information of qualifications or disqualifications of these individuals, the socially qualified individuals with respect to all social rules mentioned above can be calculated in polynomial time. However, in some real-world applications, we are not able to obtain or access full information. For instance, if the number of individuals is extremely large (this happens often on online platforms), then it is not expected that every individual shows her opinion over every individual. Instead, every individual only qualifies or disqualifies a small number of individuals whom she knows well or whom she is particularly interested in. In addition, in some cases, even though the full qualifications and disqualifications exist, they cannot be fully accessed by some specific people, say someone (or a company) who would like to predict the result. In this case, the one who wants to predict the result can only do so based on a part of the information. With the missing of some information, two significant questions arise: who have positive possibility to be socially qualified if the missing information is filled and who are definitely socially qualified regardless of the missing information. Moreover, how much complexity resources we need to achieve an answer to the above questions is of particular importance.

In this paper, we study the complexity of two problems that capture the above two questions. In particular, we study the POSSIBLY QUALIFIED INDIVIDUALS problem (PQI) and the NECESSARILY QUALIFIED INDIVIDUALS problem (NQI). In both problems we are given a set of individuals N each of whom qualifies or disqualifies a subset of N, together with a subset $S \subseteq N$. The former problem asks whether there is an extension of these qualifications and disqualifications with respect to which all individuals in S are socially qualified, and the latter one asks whether all individuals in S are socially qualified with respect to every extension of these qualifications and disqualifications. Here, an extension means that every individual fills out potential gaps regarding individuals' qualifications, i.e., she determines a given qualification for individuals with previously undefined qualifications.

Investigating scenarios with incomplete information are relevant, especially from a practical perspective. The PQI/NQI problems considered here are a natural first step towards more complicated scenarios with incomplete information. The main reason why we study consent rules, the liberal-start-respecting rule, and the consensus-start-respecting rule is that they are the most significant social rules that have been investigated in the literature so far. Importantly, it is shown that they satisfy several fairness properties, respectively, see, e.g., [6, 10, 19].

1.1 Related Work

The model of group identification is related to voting systems. In a voting system, we are given a set of candidates and a set of voters each of whom casts a vote. Then, a voting correspondence is used to select a subset of candidates. From this standpoint, group identification can be considered as a voting system where the individuals are both voters and candidates. Nevertheless, group identification differs from voting systems in many significant aspects. First, the goal of a voting system is to select a subset of candidates, which are often called winners since they are considered as more competitive or outstanding compared with the remaining candidates for some specific purpose. However, in group identification the socially qualified individuals do not imply that they are more competitive or outstanding than the remaining individuals. For instance, in situations where we want to identify left-wing party members among a group of people, the model of group identification is more suitable. In other words, group identification is more close to a classification model. Second, as voting systems aim to select a subset of competitive candidates for some special purpose, more often than not, the number of winners are pre-decided (e.g., in a single-winner voting, exactly one candidate is selected as the winner). As a consequence, many voting systems need to adopt a certain tie-breaking method to break the tie when many candidates are considered equally competitive. However, group identification does not need a tie breaking method, since there is no size bound of the number of socially qualified individuals.

Among many voting systems, group identification is most related to the classic Approval voting system. In an Approval voting, each voter approves or disapproves each candidate. Thus, each voter's vote is represented by a 1–0 vector, where the entries with 1s (resp. 0s) mean that the voter approves (resp. disapproves) the corresponding candidate. The winners are among the candidates which get the most approvals. If the voters and candidates are the same group of individuals, then it seems that Approval voting is a social rule. Nevertheless, as discussed above, Approval voting is more often considered as a single-winner voting system and thus need to utilize a tie breaking method. Recently, several variants of Approval voting have been studied as multi-winner voting systems [12,13,16,18]. However, the number of winners is bounded by (or exactly equals to) an integer k [1]. As such, tie-breaking schemes have to be factored in determining the winning set.

As far as we know, the problem of determining the socially qualified individuals in group identification with partial information has not been investigated so far. Nevertheless, determining winners in voting systems with partial information has been studied in the literature, see, e.g., [1,11,20,22]. In addition, many other voting problems involving partial information have also been studied in the literature [3,4].

Our work is also related to the paper by Yang and Dimitrov [21], where they studied the complexity of constructive group control problems for the social rules studied in this paper. Recently, Erdélyi, Reger and Yang complemented the work of Yang and Dimitrov by investigating the complexity of destructive group control problems and group bribery problems [7]. We refer to [7,21] for further details of the group control problems.

The liberal-start-respecting and the consensus-start-respecting rules were introduced by Dimitrov, Sung and Xu [6].

2   Preliminaries

In this paper, we will need the following notations and definitions. Unless stated otherwise, all numerical data are integers.

Social rule. Let N be a set of individuals and $n = |N|$ throughout this paper. A *profile* φ over N is a mapping from $N \times N$ to $\{0,1\}$. In particular, $\varphi(a,a') = 1$ means that the individual $a \in N$ *qualifies* the individual $a' \in N$, and $\varphi(a,a') = 0$ means that a disqualifies a'. A *social rule* is a function f which associates each profile φ over N with a subset $f(\varphi) \subseteq N$. We call the individuals in $f(\varphi)$ the *socially qualified individuals* with respect to f and φ. In this paper, we mainly study the following social rules.

Consent rules $f^{(s,t)}$. Each consent rule $f^{(s,t)}$ is specified by two positive integers s and t such that for every individual $a \in N$,

1. if $\varphi(a,a) = 1$, then $a \in f^{(s,t)}(\varphi)$ if and only if $|\{a' \in N \mid \varphi(a',a) = 1\}| \geq s$, and
2. if $\varphi(a,a) = 0$, then $a \notin f^{(s,t)}(\varphi)$ if and only if $|\{a' \in N \mid \varphi(a',a) = 0\}| \geq t$.

The two positive integers s and t are referred to as the *consent quotas* of the rule $f^{(s,t)}$. It is worth mentioning that in the original definition of consent rules by Samet [19] there is an additional condition $s+t \leq n+2$ for consent quotas s and t to satisfy. Indeed, the condition $s+t \leq n+2$ is crucial for the consent rules to satisfy the *monotonicity property*[1]. Recall that a social rule is *monotonic* if a socially qualified individual a is still socially qualified when someone who disqualifies a changes her preference to qualify a. Since we mainly study PQI and NQI from the complexity theoretic point of view, we drop this condition from the definition of the consent rules (we indeed achieve results for a more general class of social rules that encapsules the original consent rules defined in [19]).

Consensus-start-respecting rule f^{CSR}. This rule determines the socially qualified individuals iteratively. First, all individuals who are qualified by everyone in the society are considered socially qualified. Then, in each iteration, all individuals who are qualified by at least one of the currently socially qualified individuals are added to the set of socially qualified individuals. The iteration terminates until no new individual is added. Formally, let

$$K_0^{CSR}(\varphi) = \{a \in N \mid \forall a' \in N, \ \varphi(a',a) = 1\}.$$

For each positive integer $\ell = 1, 2, \ldots$, let $K_\ell^{CSR}(\varphi)$ be defined as

$$K_{\ell-1}^{CSR}(\varphi) \cup \{a \in N \mid \exists a' \in K_{\ell-1}^{CSR}(\varphi), \ \varphi(a',a) = 1\}.$$

Then $f^{CSR}(\varphi) = K_\ell^{CSR}(\varphi)$ for some ℓ such that $K_\ell^{CSR}(\varphi) = K_{\ell-1}^{CSR}(\varphi)$.

Liberal-start-respecting rule f^{LSR}. This rule is similar to f^{CSR} with the only difference that the initial socially qualified individuals are those who qualify themselves. In particular, let

$$K_0^{LSR}(\varphi) = \{a \in N \mid \varphi(a,a) = 1\}.$$

For each positive integer $\ell = 1, 2, \ldots$, let $K_\ell^{LSR}(\varphi)$ be defined as

$$K_{\ell-1}^{LSR}(\varphi) \cup \{a \in N \mid \exists a' \in K_{\ell-1}^{LSR}(\varphi), \ \varphi(a',a) = 1\}.$$

Then $f^{LSR}(\varphi) = K_\ell^{LSR}(\varphi)$ for some ℓ such that $K_\ell^{LSR}(\varphi) = K_{\ell-1}^{LSR}(\varphi)$.

Note that $f^{CSR}(\varphi)$ (resp. $f^{LSR}(\varphi)$) is empty if there are no individuals qualified by everyone (themselves) in the society.

Example. Let $N = \{a_1, a_2, a_3, a_4\}$. Consider the profile φ over N as follows (the entry row indexed by a_i and column indexed by a_j is $\varphi(a_i, a_j)$).

---

[1] In fact, the problem occurs only when individual a changes her own assessment from not qualified to qualified. In other words, without the restriction that $s+t \leq n+2$, if a is socially qualified, then she is still socially qualified if an individual other than a disqualifying a in advance changes her assessment to qualify a.

$$\begin{array}{c|cccc} & a_1\ a_2\ a_3\ a_4 \\ \hline a_1 & 1\ \ 1\ \ 1\ \ 1 \\ a_2 & 0\ \ 1\ \ 1\ \ 0 \\ a_3 & 0\ \ 1\ \ 0\ \ 0 \\ a_4 & 0\ \ 1\ \ 1\ \ 0 \end{array}$$

The socially qualified individuals with respect to some of the above social rules f are as follows.

$f^{(1,1)}$	$f^{(1,2)}$	$f^{(2,1)}$	f^{CSR}	f^{LSR}
a_1, a_2	a_1, a_2, a_3	a_2	a_2, a_3	a_1, a_2, a_3, a_4

Partial profile. Intuitively, a partial profile is a profile with several 1 s and 0 s to be replaced with the symbol $*$, where $\varphi(a,b) = *$ means that whether a qualifies b or not is unknown, or a does not hold an opinion on the qualification of b. The formal definition is as follows.

A *partial profile* φ is a mapping $\varphi : N \times N \mapsto \{0,1,*\}$. For each $a \in N$ and $x \in \{0,1,*\}$, $x(a,\varphi)$ is the set of individuals $b \in N$ such that $\varphi(a,b) = x$, i.e., $x(a,\varphi) = \{b \in N \mid \varphi(a,b) = x\}$. A profile ϕ is an *extension* of a partial profile φ if and only if

1. for every $a,b \in N$ such that $\varphi(a,b) \in \{0,1\}$, it holds that $\phi(a,b) = \varphi(a,b)$; and
2. for every $a,b \in N$ such that $\varphi(a,b) = *$, it holds that $\phi(a,b) \in \{0,1\}$.

For a nonnegative integer $r \leq n$, an *r-profile* φ over N is a profile such that for every $a \in N$, it holds that $|1(a,\varphi)| = r$, i.e., each individual qualifies exactly r individuals in N. A partial profile φ' is called an *r-partial profile* if $|1(a,\varphi')| \leq r$ for every $a \in N$. An r-profile ϕ is an *r-extension* of an r-partial profile φ if ϕ is an extension of φ.

Clearly, an r-partial profile φ has an r-extension if and only if $|*(a,\varphi)| \geq r - |1(a,\varphi)|$ for every $a \in N$. Throughout this paper, we consider only r-partial profiles that have r-extensions.

Problem statement. We mainly study the complexity of the following problems.

---

f-Possibly/f-Necessarily Qualified Individuals (f-PQI/f-NQI)

Given: A 3-tuple (N, φ, S) of a set N of individuals, a partial profile φ over N, and a nonempty subset $S \subseteq N$.

f**-PQI:** Is there an extension ϕ of φ such that $S \subseteq f(\phi)$?

f**-NQI:** Does $S \subseteq f(\phi)$ hold for every extension ϕ of φ?

In addition to the above problems, we also study f-r-PQI (resp. f-r-NQI), where the input and question are similar to that of f-PQI (resp. f-NQI) with only the following differences. First, in the input we require φ to be an r-partial profile rather than a partial profile. Second, in the question we replace "extension" with "r-extension". Notice that f-r-PQI (f-r-NQI) is not a special case of f-PQI (f-NQI) as one of the restrictions is on the solution space.

Throughout this paper, we will use $I = (N, \varphi, S \subseteq N)$ to denote the given instance in the problems we study for a social rule f, where $f \in \{f^{(s,t)}, f^{LSR}, f^{CSR}\}$, and will not state this explicitly in the proofs of the theorems. Furthermore, for better readability we will use the diction "PQI/NQI for social rule f" instead of "f-PQI/f-NQI".

Graph. We will also need some basic knowledge on graph theory.

A *digraph* (or *directed graph*) G is a tuple (V, A) where V is the *vertex set* and A is the *arc set*. An arc from a vertex a to a vertex b is denoted by (a, b). We also use $V(G)$ and $A(G)$ for the vertex set and arc set of G, respectively. A *directed path* is a vertex sequence $(v_1, v_2, \ldots, v_t)$ such that $(v_i, v_{i+1}) \in A$ for every $i = 1, 2, \ldots, t-1$. A *Hamiltonian path* is a directed path such that every vertex in the digraph appears exactly once in the path. We refer to [2] for further details on digraphs.

3 Complexity with Unbounded Qualifications

In this section, we study PQI and NQI where each individual can qualify as many as up to n individuals. Our main results are polynomial-time algorithms for PQI and NQI for consent rules, liberal-start-respecting rule and consensus-start-respecting rule.

Consider first consent rules. Our polynomial-time algorithms are based on the observation that if an individual a is socially qualified and someone else disqualifying a changes her opinion to qualifying a, then a is still qualified. In particular, this observation enables us to safely reset the values of many $\varphi(a, b)$ with $\varphi(a, b) = *$ in advance to 1. We omit the proof due to space restrictions.

Theorem 1. *PQI and NQI for consent rules $f^{(s,t)}$ can be solved in $O(n^2)$ time, for all integers s and t.*

Now we turn our attention to the procedural rules f^{CSR} and f^{LSR}. Based on a similar observation as the one for consent rules, we develop polynomial-time algorithms for PQI and NQI for these two rules. We omit the proof due to space restrictions.

Theorem 2. *PQI and NQI for f^{LSR} and f^{CSR} can be solved in $O(n^2)$ time.*

4 Complexity with r-Partial Profiles

In this section, we study two variants of PQI and NQI, namely r-PQI and r-NQI. In particular, in these two variants every individual is allowed to qualify exactly r individuals in the extensions of the given partial profile. We have seen in the previous section that PQI and NQI are polynomial-time solvable for all social rules considered in this paper. The intuition is that in all cases, maximizing the number of individuals who qualify an individual a increases the possibility of a to be socially qualified in the extensions, and maximizing the individuals who disqualify a decreases the possibility of a to be socially qualified. As such, in PQI/NQI we generally replace $*$ with $1/0$. However, if every individual is allowed to qualify exactly r individuals, we have to carefully replace $*$ with 1 or 0.

We prove that r-PQI and r-NQI for consent rules are polynomial-time solvable too. However, the polynomial-time algorithms studied in this section are not trivial generalizations of the ones studied in the previous section. In fact, we derive different algorithms for consent rules with different consent quotes s and t. Moreover, some of the polynomial-time algorithms are derived only when r and t are both constants. However, the polynomial-time algorithms studied in the previous section hold for all integers s and t. For the two procedural social rules f^{CSR} and f^{LSR}, we prove that 1-PQI and 1-NQI are polynomial-time solvable. However, if r increases just by one, we show that r-PQI for both procedural rules becomes NP-hard. Hence, we obtain a complexity dichotomy result for r-PQI for f^{CSR} and f^{LSR} with respect to the values of r.

An observation that is useful in deriving the polynomial-time algorithms is as follows: if there is an $a \in N$ such that $|1(a, \varphi)| = r$, then in each r-extension ϕ of φ it must be that $\phi(a, b) = 0$ for every $b \in *(a, \varphi)$. In addition, if $|*(a, \varphi)| = r - |1(a, \varphi)|$ for an individual $a \in N$, in every r-extension ϕ of φ it must be that $\phi(a, b) = 1$ for every $b \in *(a, \varphi)$.

4.1 Consent Rules

We consider first r-PQI and r-NQI for consent rules. Our first result is a polynomial-time algorithm for r-NQI for consent rules $f^{(s,t)}$. Moreover, the polynomial-time solvability holds regardless of the values of s and t.

Theorem 3. *r-NQI for all consent rules $f^{(s,t)}$ can be solved in $O(n^2)$ time.*

Proof. We develop a polynomial-time algorithm as follows. First, the algorithm calculates $|1(a, \varphi)|$ and $|*(a, \varphi)|$ for all $a \in N$. This can be done in $O(n^2)$ time. Then, the algorithm breaks down I into $|S|$ subinstances, each of which takes as input I and an individual $a \in S$, and asks if there is an r-extension ϕ of φ such that $a \notin f^{(s,t)}(\phi)$. Clearly, I is a NO-instance if and only if at least one of the subinstances is a YES-instance. Let $I' = (I, a \in S)$ be a subinstance. We show how to solve I' in polynomial time, by distinguishing between the following cases.

Case 1. $\varphi(a, a) \in \{0, 1\}$.

In this case, we do the following. For every $b \in N$ such that $a \in *(b, \varphi)$, if $|*(b, \varphi)| > r - |1(b, \varphi)|$, reset $\varphi(b, a) = 0$; otherwise, reset $\varphi(b, a) = 1$. This can be done in $O(n)$ time. If $\varphi(a, a) = 1$ and $|\{b \in N \mid \varphi(b, a) = 1\}| < s$, or $\varphi(a, a) = 0$ and $|\{b \in N \mid \varphi(b, a) = 0\}| \geq t$ after doing so, I' is a YES-instance. Otherwise, I' is a NO-instance. We can check this in $O(n)$ time.

Case 2. $\varphi(a, a) = *$.

Assume that $|*(a, \varphi)| > r - |1(a, \varphi)|$ (and also $|1(a, \varphi)| < r$), since otherwise, a qualifies (disqualifies) herself in all r-extensions of φ. Hence, we can reset $\varphi(a, a) = 1$ ($\varphi(a, a) = 0$) and solve I' by calling the procedure in Case 1. We deal with Case 2 with this assumption as follows. First, for every $b \in N \setminus \{a\}$ such that $a \in *(b, \varphi)$, if $|*(b, \varphi)| > r - |1(b, \varphi)|$, reset $\varphi(b, a) = 0$; otherwise, reset $\varphi(b, a) = 1$. After doing so, we compare $s - 2$ with the number of individuals qualifying a. In particular, if $|\{b \in N \mid \varphi(b, a) = 1\}| \leq s - 2$, a is not socially qualified in every r-extension of φ after resetting $\varphi(a, a) = 1$. Hence, we can conclude that I' is a YES-instance. If, however,

$|\{b \in N \mid \varphi(b,a) = 1\}| \geq s - 1$, we reset $\varphi(a,a) = 0$. Then, if $|\{b \in N \mid \varphi(b,a) = 0\}| \geq t$, a is not socially qualified in every r-extension of φ, implying that I' is a YES-instance as well. In all other cases, I' is a NO-instance. The above procedure can be done in $O(n)$ time.

As we have at most $|S| \leq n$ subinstances, the whole running time of the algorithm is bounded by $O(n^2) + n \cdot O(n) = O(n^2)$, where the first $O(n^2)$ is the time for calculating $|1(a,\varphi)|$ and $|*(a,\varphi)|$ for all $a \in N$. $\qquad\square$

Now we consider r-PQI for consent rules. We start with a special case of r-PQI where in the input profile φ it holds that $\varphi(a,a) \in \{0,1\}$ for every $a \in S$. We denote this special case by r-PQI-S and show that this problem for all consent rules $f^{(s,t)}$ is polynomial-time solvable even for r,s,t being non-constants. This algorithm will be used to develop polynomial-time algorithms for r-PQI for some consent rules later.

For an individual $a \in N$, let $1^{-1}(a,\varphi)$ be the set of individuals qualifying a in φ, i.e., $1^{-1}(a,\varphi) = \{b \in N \mid \varphi(b,a) = 1\}$.

Lemma 1. *r-PQI-S for consent rules can be solved in $O(n^3)$ time.*

Proof. Let $I = (N, \varphi, S \subseteq N)$ be a given instance of r-PQI-S for $f^{(s,t)}$, where φ is an r-partial profile and $\varphi(a,a) \in \{0,1\}$ for every $a \in S$. We solve the problem in polynomial-time by reducing it to the MAXIMUM FLOW problem. We create the following network.

For every $a \in N$, we create one vertex $v(a)$. Moreover, for every $a \in S$, we further create one vertex $u(a)$. Finally, we create a source vertex x and a sink vertex y. The arcs and capacities of the arcs are as follows. First, there is an arc from the source x to every $v(a)$ with capacity $c(x,v(a)) = r - |1(a,\varphi)|$, indicating that a can further qualify at most $r - |1(a,\varphi)|$ individuals in S. Second, there is an arc from a vertex $v(a), a \in N$ to a vertex $u(b), b \in S$ with capacity 1 if and only if $\varphi(a,b) = *$, indicating that it is possible to let a qualify b in an r-extension of φ. Finally, there is an arc from every vertex $u(a), a \in S$ to the sink y with capacity $c(u(a),y) = \max\{0, s' - |1^{-1}(a,\varphi)|\}$, where $s' = s$ if $\varphi(a,a) = 1$ and $s' = |N| - t + 1$ if $\varphi(a,a) = 0$. The capacity of the arc from $u(a)$ to y indicates how many qualifications are still needed to make a socially qualified. See Fig. 1 for an illustration of the network.

We argue that G has a flow of size $\sum_{a \in S} c(u(a),y)$ if and only if I is a YES-instance. Assume that there is an r-extension ϕ of φ under which all individuals in S are socially qualified with respect to $f^{(s,t)}$. Consider the following flow. First, the flow on each arc from $u(a), a \in S$ to y is $c(u(a),y)$. Consider now the flows on the arcs in the middle. As each individual $a \in S$ is socially qualified in ϕ, the number of individuals b such that $\varphi(b,a) = *$ and $\phi(b,a) = 1$ is at least $\max\{0, s' - |1^{-1}(a,\varphi)|\} = c(u(a),y)$, where s' is defined as above. Then, from these individuals, we select any arbitrary $c(u(a),y)$ individuals. Moreover, for each selected individual $b \in N$, the flow on the arc from $v(b)$ to $u(a)$ is 1. Finally, the flow on each arc from x to every $v(a), a \in N$ is the sum of flows leaving $v(a)$. Note that there can be at most $r - |1(a,\varphi)|$ individuals $b \in S$ such that $\varphi(a,b) = *$ and $\phi(a,b) = 1$. Hence, the flow on each arc from x to $v(a), a \in N$ does not exceed the capacity of the arc. The flows on all remaining arcs are 0. Clearly, the size of the flow is $\sum_{a \in S} c(u(a),y)$. It remains to prove the other direction. Assume that G has a flow of size $\sum_{a \in S} c(u(a),y)$. Due to the integrality theorem [9], there is an integer flow F of the same size. We can find an r-extension of φ under which all individuals

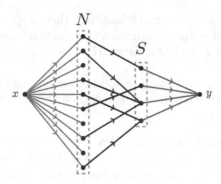

Fig. 1. An illustration of the construction of the network in the proof of Lemma 1. Each red arc from x to a vertex $v(a)$ corresponding to an individual $a \in N$ has capacity $r - |1(a, \varphi)|$. Each blue arc from a vertex $u(a)$ corresponding to an individual $a \in S$ to y has capacity $\max\{0, s' - |1^{-1}(a, \varphi)|\}$. A black arc from $v(a), a \in N$ to $u(a'), a' \in S$ means that $\varphi(a, a') = *$. Moreover, each black arc in the middle has capacity 1. (Color figure online)

in S are socially qualified as follows. Consider the r-partial profile ϕ obtained from φ by resetting $\varphi(b, a) = 1$ for every arc from a vertex $v(b), b \in N$ to a vertex $u(a), a \in S$ with flow 1. Hence, for each $a \in S$ with $\varphi(a, a) = 1$, there are at least $|1^{-1}(a, \varphi)| + c(u(a), y) \geq s$ individuals qualifying a with respect to ϕ. In addition, for each $a \in S$ with $\varphi(a, a) = 0$, there are at least $|1^{-1}(a, \varphi)| + c(u(a), y) \geq |N| - t + 1$ individuals qualifying a. This means that there are at most $t - 1$ individuals disqualifying a with respect to ϕ. Then, due to the definition of $f^{(s,t)}$, any r-extension of ϕ is a solution of I.

It remains to analyze the running time of the algorithm. It was recently shown by Orlin [17] that the MAXIMUM FLOW problem is solvable in $O(p \cdot q)$ time, where p is the number of arcs and q is the number of vertices in the network. As the network constructed above has $O(n^2)$ arcs and $O(n)$ vertices, the running time of our algorithm is bounded by $O(n^3)$. □

Armed with Lemma 1, we are ready to show our polynomial-time algorithms for r-PQI for consent rules. We start with some results for r-PQI for consent rules with small constant consent quotas s and t but unbounded values of r.

Theorem 4. r-PQI for $f^{(1,1)}$ and for $f^{(1,2)}$ can be solved in $O(n^2)$ time, and for $f^{(2,1)}$ can be solved in $O(n^3)$ time.

Proof. We develop polynomial-time algorithms with the corresponding running times as stated in the theorem as follows.

$f^{(1,1)}$: If there is an individual $a \in S$ such that $\varphi(a, a) = 0$, I is a NO-instance. This can be checked in $O(|S|) = O(n)$ time. In addition, if there is an $a \in S$ such that $\varphi(a, a) = *$ and $|1(a, \varphi)| = r$, the instance is a NO-instance as well, as in all r-extensions of φ the individual a disqualifies herself. If no such an individual a as described above exists, the instance must be a YES-instance. As it takes $O(n)$ time to calculate $|1(a, \varphi)|$ for each $a \in S$, the algorithm terminates in $O(n) + |S| \cdot O(n) = O(n^2)$ time.

$f^{(1,2)}$: If there exist $a \in S$ and $b \in N \setminus \{a\}$ such that $\varphi(a,a) = \varphi(b,a) = 0$, we conclude that I is a NO-instance. In addition, if there exist $a \in S$ and $b \in N \setminus \{a\}$ such that $\varphi(a,a) = *$ and $\varphi(b,a) = 0$, we reset $\varphi(a,a) = 1$. If after the reset $|1(a,\varphi)| > r$, we conclude that I is a NO-instance. We deal with the remaining cases as follows. It is clear that if $\varphi(a,a) = 1$ for some $a \in S$, or $\varphi(a,a) = 0$ and $\varphi(b,a) = 1$ for all $b \in N \setminus \{a\}$, then a is socially qualified in every r-extension of φ. We need only to focus on other individuals in S. Let S_0^* be the set of individuals $a \in S$ such that $\varphi(a,a) = 0$ and there exists $b \in N \setminus \{a\}$ such that $a \in *(b,\varphi)$. We maintain the set S_0^* throughout the algorithm. Clearly, for every $a \in S_0^*$, we have to reset $\varphi(b,a) = 1$ for all $b \in N$ such that $\varphi(b,a) = *$, since otherwise a would not be socially qualified. Due to this, we derive a procedure to deal with the remaining cases as follows. While $S_0^* \neq \emptyset$, for every $a \in S_0^*$ and $b \in N$ such that $\varphi(b,a) = *$, reset $\varphi(b,a) = 1$. If $r < |1(b,\varphi)|$ after resetting $\varphi(b,a) = 1$, the procedure immediately returns "NO".

After the while loop, for all $b \in N$ such that $r = |1(b,\varphi)|$, we do the following: (1) reset $\varphi(b,a) = 0$ for all $a \in *(b,\varphi)$; (2) update S_0^*; and (3) go back to the while loop if $S_0^* \neq \emptyset$. If no conclusion is drawn after the while loops terminate, we can conclude that I is a YES-instance. The reason is as follows. For every $a \in S$ with $\varphi(a,a) = *$, if $|1(a,\varphi)| < r$, we can reset $\varphi(a,a) = 1$ to make a socially qualified. If $|1(a,\varphi)| = r$, then due to the above algorithm, it must be that all the other individuals qualify a. Hence, we can safely reset $\varphi(a,a) = 0$ (in this case, a is socially qualified in every r-extension). The above algorithm can be implemented in $O(n^2)$ time.

$f^{(2,1)}$: If there is an $a \in S$ such that $\varphi(a,a) = 0$, or $\varphi(a,a) = *$ and $r = |1(a,\varphi)|$, then I is a NO-instance, since a is not socially qualified in every r-extension of φ. As $|1(a,\varphi)|$ can be calculated in $O(n)$ time, this can be done in $O(n^2)$ time. Assume that no such individuals a discussed above exist. Then, if there is an $a \in S$ such that $\varphi(a,a) = *$, we can safely reset $\varphi(a,a) = 1$, since if there is an r-extension of φ under which all individuals in S are socially qualified, then a must qualify herself in the r-extension. This can be clearly done in $O(|S|) = O(n)$ time. Hence, we have now that $\varphi(a,a) = 1$ for every $a \in S$. Then, due to Lemma 1, we can solve the instance in $O(n^3)$ time. The whole running time of the algorithm is clearly bounded by $O(n) + O(n^3) = O(n^3)$. □

We have shown that if $(s,t) \in \{(1,1),(1,2),(2,1)\}$, r-PQI for $f^{(s,t)}$ is polynomial-time solvable even when r is not a constant. In the following, we continue to show some polynomial-time algorithms based on Lemma 1. However, these algorithms take the assumption that r and t are constants. Before showing the result, let's first study a property, as summarized in the following lemma. In general, it states that in any r-profile the number of socially qualified individuals disqualifying themselves is bounded by $r + t - 1$. Hence, due to this property, we can enumerate all individuals $a \in S$ with $\varphi(a,a) = *$ who disqualify themselves in a solution in $O(n^{r+t-1})$ time. If both r and t are constants, this can be done in polynomial time. Moreover, we can solve r-PQI in polynomial-time based on Lemma 1.

Lemma 2. *Let φ be an r-profile over N. Let A be the set of socially qualified individuals with respect to the consent rule $f^{(s,t)}$ who disqualify themselves, i.e., $A = \{a \in f^{(s,t)}(\varphi) \mid \varphi(a,a) = 0\}$. Then, $|A| \leq r + t - 1$.*

Proof. Consider the profile φ restricted to A. Let x be the number of qualifications in the subprofile, i.e., $x = |\{\varphi(a,b) \mid a,b \in A, \varphi(a,b) = 1\}|$. As each individual qualifies r individuals, we have that $x \leq r \cdot |A|$. On the other hand, as each individual in A is socially qualified, for each $a \in A$ there are at least $|A| - t + 1$ individuals in A qualifying a. It follows that $x \geq |A| \cdot (|A| - t + 1)$. In summary, we have that $|A| \cdot (|A| - t + 1) \leq r \cdot |A|$. It directly follows that $|A| \leq r + t - 1$. □

Now we are ready to show an algorithm to solve r-PQI for all consent rules.

Theorem 5. *r-PQI for $f^{(s,t)}$ can be solved in $O(n^{(t+r+2)})$ time.*

Proof. Let $S_* = \{a \in S \mid \varphi(a,a) = *\}$. According to Lemma 2, if I is a YES-instance, then for any solution ϕ of I, it must be that $|\{a \in S \mid \phi(a,a) = 0\}| \leq r + t - 1$. Due to this, we can enumerate all subsets A of S_* such that all individuals in A disqualify themselves and all individuals in $S_* \setminus A$ qualify themselves in a solution ϕ of I, i.e., $A = \{a \in S_* \mid \phi(a,a) = 0\}$. Then, for each enumerated A, we extend φ by resetting $\varphi(a,a) = 0$ for all individuals $a \in A$ and resetting $\varphi(a,a) = 1$ for all individuals $a \in S_* \setminus A$ (if φ is not an r-partial profile after the resetting, we discard this enumeration). So, each enumeration corresponds to an instance of r-PQI-S. Clearly, I is a YES-instance if and only if at least one of the enumerations corresponds to a YES-instance. Due to Lemma 1, r-PQI-S can be solved in $O(n^3)$ time. As we have in total at most $\binom{|S|}{r+t-1} \leq \binom{n}{r+t-1} = O(n^{(r+t-1)})$ enumerations, we can solve I in $O(n^3) \cdot O(n^{(r+t-1)}) = O(n^{(r+t+2)})$ time. □

According to the above theorem, if both r and t are constants, we can solve r-PQI for $f^{(s,t)}$ in polynomial-time. One may wonder if we could get a similar result but with the assumption that s is a constant. As a matter of fact, if we assume that $n - r$ is a constant, i.e., each individual is allowed to disqualify a constant number of individuals, we could obtain a similar lemma as Lemma 2. Based on the lemma, we could obtain an algorithm with running time $O(n^{(n-r+s+2)})$. In addition, by utilizing Lemma 1 we could derive an algorithm with running time $O(2^{|S|} \cdot n^3)$: enumerate all possible values of $\varphi(a,a)$ for all $a \in S$ with $\varphi(a,a) = *$. Clearly, there are at most $2^{|S|}$ enumerations and we can solve each enumerated case in $O(n^3)$ time according to Lemma 1.

4.2 Procedural Rules

Now we study r-PQI and r-NQI for the two procedural rules f^{CSR} and f^{LSR}. We prove that the values of r have significant impact on the complexity of these two problems. First, we show that 1-PQI and 1-NQI for f^{LSR} and f^{CSR} are polynomial-time solvable, based on the following lemma.

Lemma 3. *Let $X \in \{CSR, LSR\}$ and ϕ a 1-extension of φ. Then, an individual $a \in N$ is socially qualified in ϕ with respect to f^X if and only if a is in the initial set of socially qualified individuals in ϕ, i.e., $a \in K_0^X(\phi)$.*

Proof. As ϕ is a 1-extension of φ, every individual in N qualifies exactly one individual. Then, due to the definition of f^X, an individual in the initial set of ϕ with respect to f^X does not qualify anyone not in the initial set. It directly follows that $f^X(\phi) = K_0^X(\phi)$, i.e., the socially qualified individuals with respect to f^X and ϕ are exactly the individuals in the initial set of socially qualified individuals. The lemma follows. □

Based on the above lemma, we are able to show our results.

Theorem 6. *1-PQI and 1-NQI for f^{CSR} and f^{LSR} can be solved in $O(n^2)$ time.*

Proof. Due to Lemma 3, 1-PQI for f^{CSR} (f^{LSR}) consists in determining whether there is a 1-extension of φ where every $a \in S$ is in the initial set K_0^{CSR} (K_0^{LSR}) of socially qualified individuals. We develop a polynomial-time algorithm for 1-PQI as follows.

f^{LSR}: Return "NO" if and only if there is an $a \in S$ such that $\varphi(a,a) = 0$, or $\varphi(a,a) = *$ and $r = |1(a, \varphi)|$. This clearly can be done in $O(n^2)$ time. Notice that if $\varphi(a,a) = *$ for some $a \in S$ and $r > |1(a, \varphi)|$, we can simply reset $\varphi(a,a) = 1$ without changing the answer to the instance.

f^{CSR}: As in every 1-extension, there can be at most 1 individual in the initial set of socially qualified individual with respect to f^{CSR}, due to Lemma 3, if $|S| \geq 2$, we can immediately conclude that I is a NO-instance. Assume now that $S = \{a\}$. Again, due to Lemma 3, if there is an individual $b \in N$ such that $\varphi(b,a) = 0$, or $\varphi(b,a) = *$ and $r = |1(b, \varphi)|$, I is a NO-instance; otherwise, I is a YES-instance, as we can find a solution obtained from φ by resetting $\varphi(b,a) = 1$ for every $\varphi(b,a) = *$ in advance. In the worst case, we need to calculate $1(b, \varphi)$ for all $b \in N$. As $1(b, \varphi)$ can be calculated in $O(n)$ time, the algorithm terminates in $n \cdot O(n) = O(n^2)$ time.

Now we turn our attention to 1-NQI for f^{CSR} (resp. f^{LSR}). To solve the problem, it suffices to determine if there is an $a \in S$ and a 1-extension ϕ of φ such that $a \notin f^{CSR}(\phi)$ (resp. $f^{LSR}(\phi)$). Due to Lemma 3, this is equivalent to determining if there is a 1-extension of φ such that a is in the initial set of socially qualified individuals. Hence, we can use similar algorithms to the ones for 1-PQI to solve the instance here. □

In contrast to the polynomial-time solvability of 1-PQI and 1-NQI, we show that if the individuals are allowed to qualify one more individual, r-PQI for f^{CSR} and f^{LSR} becomes NP-hard, i.e., r-PQI for f^{LSR} and f^{CSR} is NP-hard for every constant $r \geq 2$. Our proof is based on a reduction from the HAMILTONIAN PATH problem, which is known to be NP-hard [8].

HAMILTONIAN PATH

Given: A digraph $G = (V, A)$.

Question: Is there a Hamiltonian path in G?

We assume that in the given digraph of a HAMILTONIAN PATH instance, each vertex has at least one outneighbor. This assumption does not change the complexity of the problem. We first show the NP-hardness of r-PQI for f^{LSR}.

Theorem 7. *r-PQI for f^{LSR} is NP-hard for every constant $r \geq 2$.*

Proof. Let $G = (V, A)$ be a given instance of the HAMILTONIAN PATH problem. We create an instance $I = (N, \varphi, S \subseteq N)$ of r-PQI for f^{LSR} as follows.

For each vertex $v \in V(G)$, we create an individual $a(v)$. In addition, we have r individuals $b_0, \ldots, b_{r-1}$. We define $S = \{a(v) \mid v \in V(G)\}$. Hence, $N = S \cup \{b_0, \ldots, b_{r-1}\}$. The r-partial profile φ is defined as follows.

- $\varphi(a,a) = 0$ for every $a \in S$.
- $\varphi(a,b_i) = 1$ for every $a \in N$ and $0 \le i \le r-2$.
- $\varphi(b_i,b_{r-1}) = 1$ for every $1 \le i \le r-1$ and $\varphi(a,b_{r-1}) = 0$ for every $a \in S \cup \{b_0\}$.
- $\varphi(b_0,a(v)) = *$ for every $a(v) \in S$.
- $\varphi(b_i,a(v)) = 0$ for every $1 \le i \le r-1$ and $a(v) \in S$.
- For $a(v),a(u) \in S$, if $(v,u) \in A(G)$, then $\varphi(a(v),a(u)) = *$; otherwise, $\varphi(a(v),a(u)) = 0$.

It remains to prove the correctness of the reduction. We claim that there exists a Hamiltonian path in G if and only if there is an r-extension of φ with respect to which all individuals in S are socially qualified.

($\Rightarrow$) Assume that there is a Hamiltonian path $(v_1,\ldots,v_q)$ in G, where $q = |V(G)|$. Consider the r-extension ϕ of φ obtained from φ by resetting the values of $*$ entries as follows.

- reset $\varphi(b_0,a(v_1)) = 1$.
- For every $1 \le i \le q-1$, reset $\varphi(a(v_i),a(v_{i+1})) = 1$.
- Reset $\varphi(a(v_q),a(v_i)) = 1$, where v_i is any arbitrary outneighbor of v_q in G.
- Reset $\varphi(a(v_i),a(v_j)) = 0$ for all entries not defined above.

In this extension, b_0 is socially qualified. According to the definition of f^{LSR}, $a(v_1)$ is also socially qualified as b_0 qualifies $a(v_1)$ in ϕ. Moreover, if $a(v_i)$ is socially qualified, so is $a(v_{i+1})$ for every $i \in \{1,2,\ldots,q-1\}$. This means all individuals in S are socially qualified in ϕ. Hence, I is a YES-instance.

($\Leftarrow$) Let ϕ be an r-extension of φ with respect to which all individuals in S are socially qualified. As each individual $a(v) \in S$ has already qualified the $r-1$ individuals $b_0,b_1,\ldots,b_{r-2}$, $a(v)$ qualifies exactly one more individual $a(u) \in S$. Moreover, b_0 qualifies exactly one individual in S. Hence, from b_0, we can find a sequence of individuals $a(v_1),a(v_2),\ldots,a(v_q)$ such that b_0 qualifies $a(v_1)$ and $a(v_i)$ qualifies $a(v_{i+1})$ for every $1 \le i \le q-1$. Due to the construction, $(v_1,v_2,\ldots,v_q)$ is a Hamiltonian path in G.   □

We can prove the NP-hardness of r-PQI for f^{CSR} by a similar reduction as in the above proof. Due to space limitation, we omit the proof.

Theorem 8. *r-PQI for f^{CSR} is NP-hard for every constant $r \ge 2$.*

In the r-partial profiles constructed in the proofs of Theorems 7 and 8, all individuals in S disqualify themselves. Hence, we in fact proved that r-PQI-S for both f^{LSR} and f^{CSR} with $r \ge 2$ is NP-hard.

Table 1. A summary of our results for r-PQI and r-NQI. Here, n is the number of individuals.

	Consent rule $f^{(s,t)}$			f^{CSR}	f^{LSR}
	$s+t=2$	$s+t=3$	$s+t \ge 4$		
r-PQI	$O(n^2)$	$s=1:O(n^2)$	$O(n^{(r+t+2)})$	$r \ge 2$: NP-hard	$r \ge 2$: NP-hard
		$t=1:O(n^3)$		$r=1:O(n^2)$	$r=1:O(n^2)$
r-NQI	$O(n^2)$			$r=1:O(n^2)$	$r=1:O(n^2)$

5   Conclusion

We have studied PQI and NQI under the framework of group identification. In these problems, we are given a partial profile and a subset S of individuals. The question is whether the individuals in S are socially qualified in an extension of the given partial profile (PQI), or in every extension of the given partial profile (NQI). In addition, we considered r-PQI and r-NQI which differ from PQI and NQI in that in the searched extensions we restrict each individual to qualify exactly r individuals.

We studied the complexity of these problems for the prevalent consent rules and two procedural rules f^{LSR} and f^{CSR}. We derived both polynomial-time algorithms as well as NP-hardness results. In general, our results reveal that most of these problems are polynomial-time solvable. Moreover, for consent rules, the complexity of our polynomial-time algorithms increases slightly as the consent quotes increase. Furthermore, the consent quotes s and t play different roles in the complexity of these problems. For instance, for consent rules $f^{(s,t)}$, r-PQI can be solved in $O(n^2)$ time if $s = 1$ and $t \leq 2$. For $t = 1$ and $s = 2$, the running time of the algorithm for r-PQI increases to $O(n^3)$. When one of s and t further increases, the running time of our algorithm for r-PQI increases to $O(n^{(r+t+2)})$, leaving the algorithm to be polynomial-time only when $r + t$ is a constant. However, for r-NQI, the running time of our algorithms is $O(n^2)$ for all consent quotes. We obtained a complexity dichotomy result for r-PQI for f^{LSR} and f^{CSR} with respect to the values of r. In particular, if $r = 1$, r-PQI for f^{CSR} and f^{LSR} can be solved in $O(n^2)$ time; otherwise, it becomes NP-hard. For r-NQI for f^{CSR} and f^{LSR}, we have an $O(n^2)$-time algorithm for $r = 1$, and leave the complexity of the remaining cases open. The intuition that r-PQI and r-NQI for consent rules are generally easy to solve is that whether an individual a is socially qualified can be independently determined by the number of individuals qualifying or disqualifying a. However, in f^{CSR} and f^{LSR}, if an individual a is socially qualified depends not only on who qualify a but also on the connectivity between the individuals qualifying a and the initially socially qualified individuals. Between r-PQI and r-NQI, our results reveal that r-NQI is easier to solve. The reason is that in r-NQI, we need only to determine if there is just one individual $a \in S$ and one r-extension of φ with respect to which a is not socially qualified. We refer to Table 1 for a summary of our results.

There remain several open questions. For instance, we do not know the complexity of r-NQI for both f^{LSR} and f^{CSR} for $r > 1$. In addition, investigating the NP-hard problems studied in this paper from the parameterized complexity point of view would be an interesting research topic. A natural parameter would be $|S|$. As we pointed out in Subsect. 4.1, r-PQI for consent rules $f^{(s,t)}$ is fixed-parameter tractable (FPT) with respect to $|S|$. In fact, we had a single-exponential time algorithm for the problem with respect to $|S|$ (see the discussion after Theorem 5). However, whether r-PQI/r-NQI for the two procedural rules f^{CSR} and f^{LSR} is FPT with respect to $|S|$ remains open.

References

1. Aziz, H., Gaspers, S., Gudmundsson, J., Mackenzie, S., Mattei, N., Walsh, T.: Computational aspects of multi-winner approval voting. In: AAMAS, pp. 107–115 (2015)
2. Bang-Jensen, J., Gutin, G.: Digraphs - Theory, Algorithms and Applications. Springer, London (2008)
3. Briskorn, D., Erdélyi, G., Reger, C.: Bribery in k-Approval and k-Veto under partial information: (extended abstract). In: AAMAS, pp. 1299–1300 (2016)
4. Dey, P., Misra, N., Narahari, Y.: Complexity of manipulation with partial information in voting. In: IJCAI, pp. 229–235 (2016)
5. Dimitrov, D.: The Social Choice Approach to Group Identification. In: Herrera-Viedma, E., García-Lapresta, J.L., Kacprzyk, J., Fedrizzi, M., Nurmi, H., Zadrożny, S. (eds.) Consensual Processes. Studies in Fuzziness and Soft Computing, vol. 267, pp. 123–134. Springer, Heidelberg (2011)
6. Dimitrov, D., Sung, S.C., Xu, Y.: Procedural group identification. Math. Soc. Sci. **54**(2), 137–146 (2007)
7. Erdélyi, G., Reger, C., Yang, Y.: The complexity of bribery and control in group identification. In: AAMAS, pp. 1142–1150 (2017)
8. Garey, M.R., Johnson, D.S., Stockmeyer, L.: Some simplified NP-complete graph problems. Theor. Comput. Sci. **1**(3), 237–267 (1976)
9. Ford, L.R., Fulkerson, D.R.: Flows in Networks. Princeton Unviersity Press, Princeton (1962)
10. Kasher, A., Rubinstein, A.: On the question "Who is a J?": a social choice approach. Log. Anal. **160**, 385–395 (1997)
11. Konczak, K., Lang, J.: Voting procedures with incomplete preferences. In: MPREF, pp. 124–129 (2005)
12. Meir, R., Procaccia, A.D., Rosenschein, J.S.: A broader picture of the complexity of strategic behavior in multi-winner elections. In: AAMAS, pp. 991–998 (2008)
13. Meir, R., Procaccia, A.D., Rosenschein, J.S., Zohar, A.: Complexity of strategic behavior in multi-winner elections. J. Artif. Intell. Res. (JAIR) **33**, 149–178 (2008)
14. Miller, A.D.: Group identification. Games. Econ. Behav. **63**(1), 188–202 (2008)
15. Nicolas, H.: "I want to be a J!": liberalism in group identification problems. Math. Soc. Sci. **54**(1), 59–70 (2007)
16. Obraztsova, S., Zick, Y., Elkind, E.: On manipulation in multi-winner elections based on scoring rules. In: AAMAS, pp. 359–366 (2013)
17. Orlin, J.B.: Max flows in $O(nm)$ time, or better. In: STOC, pp. 765–774 (2013)
18. Procaccia, A.D., Rosenschein, J.S., Zohar, A.: Multi-winner elections: complexity of manipulation, control and winner-determination. In: IJCAI, pp. 1476–1481 (2007)
19. Samet, D., Schmeidler, D.: Between liberalism and democracy. J. Econ. Theor. **110**(2), 213–233 (2003)
20. Xia, L., Conitzer, V.: Stackelberg voting games: computational aspects and paradoxes. In: AAAI, pp. 921–926 (2010)
21. Yang, Y., Dimitrov, D.: How hard is it to control a group? In: COMSOC 2016. https://arxiv.org/abs/1606.03366
22. Yang, Y., Guo, J.: Possible winner problems on partial tournaments: a parameterized study. J. Comb. Optim. **33**(3), 882–896 (2017)

Multi-criteria Coalition Formation Games

Ayumi Igarashi[1]($\boxtimes$) and Diederik M. Roijers[1,2]

[1] Department of Computer Science, University of Oxford, Oxford, UK
ayumi.igarashi@cs.ox.ac.uk
[2] Artificial Intelligence Laboratory, Vrije Universiteit Brussel, Brussels, Belgium
droijers@ai.vub.ac.be

Abstract. When forming coalitions, agents have different utilities per coalition. Game-theoretic approaches typically assume that the scalar utility for each agent for each coalition is public information. However, we argue that this is not a realistic assumption, as agents may not want to divulge this information or are even incapable of expressing it. To mitigate this, we propose the *multi-criteria coalition formation game* model, in which there are different publicly available quality metrics (corresponding to different criteria) for which a value is publicly available for each coalition. The agents have private utility functions that determine their preferences with respect to these criteria, and thus also with respect to the different coalitions. Assuming that we can ask agents to compare two coalitions, we propose a heuristic (best response) algorithm for finding stable partitions in MC2FGs: *local stability search (LSS)*. We show that while theoretically individually stable partitions need not exist in MC2FGs in general, empirically stable partitions can be found. Furthermore, we show that we can find individually stable partitions after asking only a small number of comparisons, which is highly important for applying this model in practice.

Keywords: Multi-criteria · Hedonic games · Coalition formation games · Preference elicitation · Local search

1 Introduction

Coalitions are an essential part of life; typically we are not able to achieve our goals, or that of an organisation we are part of, by ourselves. Therefore, we need to cooperate in order to achieve these goals. For example, this paper has been written by a coalition of authors, who could not have written this paper individually, without the help of the other authors. *Hedonic games*, initiated by Banerjee et al. [4] and Bogomolnaia and Jackson [5], offer a versatile framework to model how coalitions are formed by multiple autonomous agents. An important consideration in this respect is *coalitional/individual stability*, i.e., no individuals wish to deviate from their current situation.

In order to determine which coalition structures are stable, we need to know the *preferences* over different possible coalitions for each agent. The standard

© Springer International Publishing AG 2017
J. Rothe (Ed.): ADT 2017, LNAI 10576, pp. 197–213, 2017.
DOI: 10.1007/978-3-319-67504-6_14

setting of hedonic games [6] assumes that we know, or can compute, the exact preferences of each agent over possible coalitions. However, this is not a realistic assumption; not only may agents not want to divulge this information for social or privacy reasons, but the agents might not even be able to specify their utilities a priori to begin with. For example, when working for a company, it may not be socially acceptable to order different possible teams you may be on.

1.1   Our Contributions

In this paper, we aim to mitigate this lack of information in two important ways. First, we take the information we do have into account, in the form of different quality metrics for coalitions. E.g., a manager who wants to assign teams, may rely on statistics from previous teams, such as a previous productivity metric, a number of occurred conflicts, and reports by previous team leaders on how creative teams have been. Combining this with a predictor for new possible coalitions, we may assign an expectation of how well such a team will do with respect to these criteria. This leads us to the formulation of a *multi-criteria coalition formation game (MC2FG)*, i.e., we have access to the quality of a coalition with respect to multiple quality criteria.

Contrary to previous work on related (vector-valued) cooperative games [20], we take a *utility-based approach* [16]: we assume that each agent, i, has a private utility function, $f^{(i)}$, that takes the multi-criteria value of a coalition, and produces a scalar utility. Because these functions are private, we cannot use these to check whether a proposed coalition is stable. We are however able to make some assumptions about $f^{(i)}$, e.g., that it is a monotonically increasing function w.r.t. all criteria. Furthermore, we assume that we can increase our knowledge about $f^{(i)}$, i.e., impose extra constraints, by asking agent i to compare two coalitions.

We make the following contributions. First, in Sect. 2, we propose our model: the *multi-criteria coalition formation game (MC2FG)*. We prove that while for single-criterion coalition formations individually stable partitions exist, this is not always the case for MC2FGs (Sect. 3). Therefore, we define a new heuristic search algorithm called *local stability search (LSS)* to investigate whether we can find stable partitions in practice (Sect. 4). We test this on random MC2FGs, as well as a class of MC2FGs we call `Author × Author`, inspired on forming teams of scientists to work together. We show empirically that it is possible to find individually stable, and sometimes even Nash-stable, partitions. Further, we show that it is possible to limit the number of questions we need to ask to an agent to a number that is likely to be feasible to ask from human decision makers. We therefore conclude that this type of model, with individual utility functions but public quality metrics, is a promising area that warrants further research.

1.2   Related Work

In the existing literature of cooperative games, Fernández et al. [8] were the first to incorporate multi-criteria characteristic functions. In [8], the classical solution concepts, such as the core, have been adapted to the multi-criteria setting (see also the survey by Tanino [20]). Faliszewski et al. [7] also extended weighted voting games to the multi-criteria cases in which coalitions are defined to be winning if they are winning in several levels of games specified by the propositional formula. We note that in these settings, the players' utilities are *transferable*; hence, the goal of these models is to divide the resulting values among players in a reasonable way. In contrast, our work assumes that the utilities are *non-transferable*, and focuses on coalition formation among players.

There is a rich body of the literature on preference elicitation in computational social choice. In particular, Balcan et al. [2] studied the PAC (probably approximately correct) learnability of cooperative transferable utility games. Specifically, they investigate whether given random samples of coalitions, one can efficiently predict the unknown values of coalitions as well as an allocation that is likely to be stable. Although there are some classes of games that are computationally intractable to learn, it has been shown that one can always stabilize a game by finding an allocation that is likely to be in the core. The most closely related to our approach is perhaps the paper by Benabbou and Perny [3], who designed an incremental preference elicitation procedure for the knapsack problem with multiple decision-makers; in their work, players have individual utilities for each item, and each item corresponds to its own criterion (in our framework).

2   The General Framework

For $k \in \mathbb{N}$, let $[k] = \{1, 2, \ldots, k\}$. We define multi-criteria coalition formation games.

Definition 1. *A* multi-criteria coalition formation game *(MC2FG) is a triple* $(N, q, (f^{(i)})_{i \in N})$ *where* $N = [n]$ *is a finite set of players,* $q : 2^N \to \mathbb{R}^m$ *is a multi-valued set function that represents the quality* $q_k(S)$ *of a subset* S *for each criterion* $k \in [m]$, *and each* $f^{(i)} : \mathbb{R}^m \to \mathbb{R}$ *is a private utility function for each player* $i \in N$. *We assume that each* $f^{(i)}$ *is strictly monotone, namely,* $f^{(i)}(\boldsymbol{x}) > f^{(i)}(\boldsymbol{y})$ *whenever* $\boldsymbol{x} > \boldsymbol{y}$.

Throughout the paper, we assume that $q(\emptyset) = \mathbf{0}$. We refer the subsets of N as *coalitions*. We let $\mathcal{N}_i$ denote the collection of all possible coalitions containing i. Preference relations derived from scalarisations can be naturally defined as follows. Let $i \in N$ and coalitions $S, T \in \mathcal{N}_i$. We say that player i *weakly prefers* S to T (denoted by $S \succeq_{f^{(i)}} T$) if $f^{(i)}(q(S)) \geq f^{(i)}(q(T))$; player i *strictly prefers* S to T (denoted by $S \succ_{f^{(i)}} T$) if $f^{(i)}(q(S)) > f^{(i)}(q(T))$; player i is *indifferent* between S and T (denoted by $S \sim_{f^{(i)}} T$) if $f^{(i)}(q(S)) = f^{(i)}(q(T))$.

An MC2FG $(N, q, (f^{(i)})_{i \in N})$ is said to be a *single-criterion coalition formation game* if forming a coalition derives a single non-transferable value (e.g., the expected number of citations for authors working together on a paper), and all players rank coalitions simply according to these values, irrespective of their scalarisation functions. We use the notation (N, q) to denote a single-criterion coalition formation game.

We say that $f : \mathbb{R}^m \to \mathbb{R}$ is a *linear scalarisation function* if there exists a non-negative vector $\boldsymbol{w} \in \mathbb{R}^m_+$ such that $\sum_{k=1}^m w_k = 1$ and $f(\boldsymbol{x}) = \boldsymbol{w} \cdot \boldsymbol{x} = \sum_{k=1}^m w_k x_k$ for any $\boldsymbol{x} \in \mathbb{R}^m$. We use the notation $(N, q, (\boldsymbol{w}^{(i)})_{i \in N})$ to denote an MC2FG with scalarisation vectors $(\boldsymbol{w}^{(i)})_{i \in N}$. In this paper, we focus on linear scalarisations as a model for user utility.

Solution Concepts. An *outcome* of an MC2FG is a partition π of players into disjoint coalitions. Given a partition π of N and a player $i \in N$, let $\pi(i)$ denote the coalition in π containing i. We adapt stability concepts in hedonic games [4,5] as follows. A partition π of N is said to be *individually rational* if no player strictly prefers staying alone to their own coalitions, i.e., each player $i \in N$ weakly prefers $\pi(i)$ to $\{i\}$. A coalition $S \subseteq N$ *blocks* a partition π of N if every player in $i \in S$ strictly prefers S to $\pi(i)$.

Definition 2. *A partition π of N is said to be* core stable (CR) *if no coalition $S \subseteq N$ where $S \neq \emptyset$ blocks π.*

We now introduce stability concepts that are immune to deviations by individual players. Consider a player $i \in N$ and a pair of coalitions $S \in 2^N \setminus \mathcal{N}_i$ (possibly the empty set) and $T \in \mathcal{N}_i$. A player i *wants to deviate* from T to S if i prefers $S \cup \{i\}$ to T. A player $j \in S$ *accepts* a deviation of i to S if j weakly prefers $S \cup \{i\}$ to S.

Definition 3. *A deviation of i from T to S is*

- *an NS-deviation if i wants to deviate from T to S.*
- *an IS-deviation if it is an NS-deviation and all players in S accept it.*

Definition 4. *A partition π of N is called* Nash stable (NS) *(respectively, individually stable (IS)) if no player $i \in N$ has an NS-deviation (respectively, an IS-deviation) from $\pi(i)$ to another coalition $S \in \pi$ or to $\emptyset$.*

Intuitively, the easier players can deviate, the more stringent the corresponding solution concept is. It is thus easy to see that any Nash stable partition is individually stable. Similarly to the standard hedonic games [4,5], stable partitions may not necessarily exist as can be seen in the following example.

Example 1. Consider three researchers who can potentially form a research team and write a paper together. They have access to the quality of each group with respect to the three criteria: productivity, creativity, and timeliness. Any pair of researchers can produce some positive correlations, whereas the other coalitions produce nothing or some negative correlations. One can formulate this scenario as an MC2FG $(N, q, (\boldsymbol{w}^{(i)})_{i \in N})$ as follows. The player set is $N = \{1, 2, 3\}$, and $q : 2^N \to \mathbb{R}^3$ is given by

$$q(\{1\}) = q(\{2\}) = q(\{3\}) = (0, 0, 0)^\top,$$
$$q(\{1, 2\}) = (2, 1, 1)^\top, q(\{2, 3\}) = (1, 2, 1)^\top, q(\{1, 3\}) = (1, 1, 2)^\top,$$
$$q(\{1, 2, 3\}) = (-1, -1, -1)^\top.$$

Player 1 (respectively, 2 and 3) finds the first (respectively, the second and the third) quality measure very important, so the scalarisation functions are given by $\boldsymbol{w}^{(1)} = (1,0,0)^\top$ $\boldsymbol{w}^{(2)} = (0,1,0)^\top$, and $\boldsymbol{w}^{(3)} = (0,0,1)^\top$. The resulting preference profile is as follows:

$$1 \; : \; \{1,2\} \succ_{\boldsymbol{w}^{(1)}} \{1,3\} \succ_{\boldsymbol{w}^{(1)}} \{1\} \succ_{\boldsymbol{w}^{(1)}} \{1,2,3\},$$
$$2 \; : \; \{2,3\} \succ_{\boldsymbol{w}^{(2)}} \{1,2\} \succ_{\boldsymbol{w}^{(2)}} \{2\} \succ_{\boldsymbol{w}^{(2)}} \{1,2,3\},$$
$$3 \; : \; \{1,3\} \succ_{\boldsymbol{w}^{(3)}} \{2,3\} \succ_{\boldsymbol{w}^{(3)}} \{3\} \succ_{\boldsymbol{w}^{(3)}} \{1,2,3\}.$$

This game admits four individually rational partitions: three partitions that consists of a singleton and a pair of the others, $\pi_1 = \{\{1\},\{2,3\}\}$, $\pi_2 = \{\{2\},\{1,3\}\}$, $\pi_3 = \{\{1\},\{2,3\}\}$, and the partition of singletons $\pi_4 = \{\{1\},\{2\},\{3\}\}$. It is not difficult to see that none of them is a core stable partition or an individually stable partition.

3   Existence of Stable Outcomes

As we have seen in the previous section, the set of stable outcomes can be empty in general. Nonetheless, it turns out that for single-criterion coalition formation games core stability and individual stability can be simultaneously achieved: one can find such an outcome by detecting a sequence of undominated coalitions. A similar construction can be found in [15] for dichotomous hedonic games.

Theorem 1. *Every single-criterion coalition formation game admits a partition that is both core and individually stable.*

Proof. We iteratively find a maximal coalition $S \subseteq N$ of the highest quality $q(S)$, and add S to π. Then, the resulting partition π is both core and individually stable. Observe first that π is core stable, since if there exists a blocking coalition $T \subseteq N$, T would be added to π before any $S \in \pi$ such that $S \cap T \neq \emptyset$. Second, π is individually stable. Notice that no player wants to deviate to a later formed coalition. Moreover, if there exists a player who can IS-deviate to a former formed coalition, this would contradict the maximality of the S.

Moreover, in single-criterion cases, any dynamics under individual stability always converges. Specifically, we define *IS dynamics* to be a procedure by which while the current partition π is not individually stable, we choose an arbitrary player i and a coalition $S \in \pi \cup \{\emptyset\}$ such that i has an IS-deviation to S, and move to the partition $\pi' = (\pi \setminus \{\pi(i), S\}) \cup \{S \cup \{i\}, \pi(i) \setminus \{i\}\}$.

Theorem 2. *In a single-criterion coalition formation game, from an arbitrary initial partition, IS dynamics converges to an individually stable partition.*

Proof. We prove this by an induction on the number of players $|N|$. When $|N| = 1$, our claim clearly holds. Assume that for any $|N| \leq k - 1$, IS dynamics converges, and consider the case when $|N| = k$. We will construct a digraph $D = (V, A)$ where V is given by the set of partitions of N, and $(\pi, \pi') \in A$ if and only if π' is *reachable* from π, i.e., there is a player i and a coalition $S \in \pi$ such that i has an IS-deviation to S and $\pi' = \pi \setminus \{S, \pi(i)\} \cup \{S \cup \{i\}, \pi(i) \setminus \{i\}\}$. Suppose towards a contradiction that there is a directed cycle $\mathcal{C} = \{\pi_1, \pi_2, \ldots, \pi_s\}$ in D where $(\pi_t, \pi_{t+1}) \in A$ for all $t \in [s]$; here, we let $\pi_{s+1} = \pi_1$. Now let $S_t \in \operatorname{argmax}_{S \in \pi_t} q(S)$, $q_t = q(S_t)$ for $t \in [s]$, and $q_{s+1} = q_1$. We start by proving the following lemma.

Lemma 1. $q_t \leq q_{t+1}$ *for all* $t \in [s]$.

Proof. Take any $t \in [s]$. Consider the transition from π_t to π_{t+1}. First, if no player moves from or to the coalition S_t, then it is clear that $q_t \leq q_{t+1}$. Second, if there is a player $i \in S_t$ who deviates from S_t to $T \in \pi \cup \{\emptyset\}$, then $q_t < q(T \cup \{i\}) \leq q_{t+1}$. Third, if there is a player $i \in N \setminus S_t$ who deviates from $\pi_t(i)$ to S_t, then it must be the case that $q(S_t) \leq q(S_t \cup \{i\})$ in order for i to be accepted by the players in S_t, and hence $q_t \leq q_{t+1}$. In all cases, we have that $q_t \leq q_{t+1}$.

By Lemma 1, we have that $q_1 \leq \ldots \leq q_s \leq q_1$, implying that the quality of the best coalition does not change along the cycle, i.e., $q_t = q_{t+1}$ for all $t \in [s]$. Now let us focus on the coalition S_1. If there is a player who deviates from S_1 at some point, then the coalition to which the player deviates would produce a higher value than S_1 and hence $q_1 < q_t$ for some $t \in [s]$, a contradiction. Thus, no player moves from S_1 along the cycle $\mathcal{C}$. If there is a player who deviates to S_1 at some point in the cycle, then the size of the coalition S_1 strictly increases, and hence the dynamics cannot come back to π_1, a contradiction. Therefore, the coalition S_1 remains the same in all the partitions in $\mathcal{C}$, namely, $S_1 \in \pi_t$ for all $t \in [s]$. Now, we again construct a directed graph $D' = (V', A')$ where V' is the set of partitions of $N \setminus S_1$ and $(\pi, \pi') \in A'$ if and only if π' is reachable from π. By the induction hypothesis, D' must be acyclic. However, $\mathcal{C}' = \{\pi_t \setminus \{S_1\} \mid t \in [s]\}$ forms a directed cycle in D' by the facts that $\mathcal{C}$ forms a directed cycle and that S_1 does not change along $\mathcal{C}$. Hence, we obtain a contradiction, and conclude that from any initial partition the IS dynamics converges.

Due to Theorems 1 and 2, if all players have the same scalarisation function, there always exists a partition that is core and individually stable; moreover, IS dynamics always converges to individual stability.

Corollary 1. *Every multi-criteria coalition formation game admits a core and individually stable partition if all the scalarisation functions are the same. Moreover, IS dynamics always converges to an individually stable partition.*

Proof. Given the scalarisation function $f : \mathbb{R}^m \to \mathbb{R}$, define a single-criterion coalition formation game (N, q') where $q'(S) = f(q(S))$ for each $S \subseteq 2^N$. It is not difficult to see that core or individually stable partitions of the resulting game are also stable partitions of the original game.

We note that Nash stable outcomes may not exist even in single-criterion cases. Consider for instance the two-player game (N, q) where $q(\{1\}) > q(\{1, 2\}) > q(\{2\})$; if player 1 is alone, then player 2 would deviate to his coalition, which would again cause the deviation by player 1. Now, it is natural to wonder what can be said if we have "similar" scalarisation functions. Even in such cases, however, there always exists an MC2FG whose stable partitions are empty.

Theorem 3. *For any positive integer n and for any $0 < \varepsilon < \frac{1}{2}$, there exists an MC2FG $(N, q, \{w^{(i)}\}_{i \in N})$ which admits neither a core nor individually stable partition, where the number of players $|N| = n$, the number of criteria $m = 2$, and $|w_k^{(i)} - w_k^{(j)}| \leq \varepsilon$ for any $i, j \in N$ and any $k \in [m]$.*

Proof. Take any $0 < \varepsilon < \frac{1}{2}$. We choose ε' such that $0 < \varepsilon' < \varepsilon$. Let $c = \frac{1+\varepsilon-\varepsilon'}{2+\varepsilon-\varepsilon'}$. Observe that $\min\{c, 1-\varepsilon\} > \frac{1}{2}$ and hence there exists $\alpha \in \mathbb{R}$ such that $\frac{1}{2} < \alpha < \min\{c, 1-\varepsilon\}$.

Now we construct a two-criteria coalition formation game where $N = [n]$,

$$w_1^{(1)} = \alpha + \varepsilon, w_2^{(1)} = 1 - \alpha - \varepsilon,$$

$$w_1^{(i)} = w_1^{(j)} = \alpha, \text{ and } w_2^{(i)} = w_2^{(i)} = 1 - \alpha, \text{ for all } i \in N \setminus \{i\},$$

and $q : 2^N \to \mathbb{R}^2$ is given as follows:

$$q(\{i\}) = (0, 0) \text{ for all } i \in N,$$
$$q(\{1,2\}) = (1, 0), q(\{2,3\}) = (1, \varepsilon'), q(\{3,1\}) = (0, 1 + \varepsilon),$$
$$q(\{1,2,3\}) = (-1, -1)$$
$$q(S) = (-1, -1) \text{ for all } S \not\subseteq \{1,2,3\} : |S| \neq 1.$$

Clearly, all players except for $1, 2, 3$ strictly prefer being alone to being together with somebody; hence, these players stay alone at any individually rational partition. The players $1, 2, 3$ strictly prefer pairs to their singletons, and strictly prefer the singletons to the coalition $\{1, 2, 3\}$ and any coalition $S \not\subseteq \{1, 2, 3\}$. Thus, for any individually rational partition π and for all $i = 1, 2, 3$, we have $\pi(i) \subsetneq \{1, 2, 3\}$. Also, it is not difficult to see that $\{2,3\} \succ_{w^{(2)}} \{1,2\}$ as the vector $q(\{2,3\}) = (1, \varepsilon')$ Pareto-dominates $q(\{1,2\}) = (1, 0)$. Further, we have that $\{3,1\} \succ_{w^{(3)}} \{2,3\}$ and $\{1,2\} \succ_{w^{(1)}} \{3,1\}$, since

$$w_1^{(3)} q_1(\{1,3\}) + w_2^{(3)} q_2(\{1,3\}) - w_1^{(3)} q_1(\{2,3\}) - w_2^{(3)} q_2(\{2,3\})$$
$$= (1+\varepsilon-\varepsilon') - \alpha \cdot (2+\varepsilon-\varepsilon') > (1+\varepsilon-\varepsilon') - c \cdot (2+\varepsilon-\varepsilon') = 0,$$

and

$$w_1^{(1)} q_1(\{1,2\}) + w_2^{(1)} q_2(\{1,2\}) - w_1^{(1)} q_1(\{3,1\}) - w_2^{(1)} q_2(\{3,1\})$$
$$= \alpha \cdot (2 + \varepsilon) + (\varepsilon^2 + \varepsilon - 1) > \frac{1}{2} \cdot (2 + \varepsilon) + (\varepsilon^2 + \varepsilon - 1) > 0.$$

The resulting preferences restricted to $\{1, 2, 3\}$ are the same as in Example 1, meaning that the instance has neither a core nor an individually stable partition.

Another implication of Theorem 3 is that even if the number of criteria is much smaller than the number of players there exists a two-criteria MC2FG which does not admit a stable partition.

4   Algorithms

Because none of stable partitions necessarily exists in an MC2FG, there is no algorithm that can guarantee a stable partition as an outcome. However, because we know (Theorem 1 and Corollary 1) that if there is only one criterion or if all the agents have the same scalarisation function, individually stable partitions do exist, we *expect* the chances of stable partitions existing in a random MC2FG to increase as the number of criteria decreases. In order to test this hypothesis, we devise heuristic algorithms for constructing stable partitions. Here, we do not focus on the core since checking core stability is computationally intractable (see e.g. [19]): we need to iterate through all subsets of players to see whether it is a blocking coalition.

We aim for our algorithms to minimise the number of questions that need to be asked to each agent. This is essential, as asking such questions to people can be time-consuming — both in terms of time required by the humans, and the time the system needs to wait until an answer is received — and experienced as hindrance by these humans.

We define a so-called *local search (LS)*, or *best-response*, algorithm for MC2FGs called *local stability search (LSS)*. LSS starts from a partition, π. At each time-step, the algorithm selects an agent, i, computes whether there exists a deviation (Definition 3) from $\pi(i)$ to any other coalition $T \in \pi \setminus \pi(i)$, and if it does performs the deviation. When there are no more deviations for any agent, the partition is stable.

A key aspect of LSS is that at any given iteration, LSS may not be able to decide whether an agent prefers an alternative coalition over another before explicitly asking that agent. For example, imagine that LSS is currently considering a partition π, and knows nothing about the $\boldsymbol{w}^{(i)}$ of an agent, i. When considering whether i wants to deviate from $\pi(i)$ to say, a coalition $T \in \pi$, we must know whether, $\boldsymbol{w}^{(i)} \cdot \boldsymbol{q}(\pi(i)) < \boldsymbol{w}^{(i)} \cdot \boldsymbol{q}(T \cup \{i\})$, where "." denotes the inner product. When for example, $\boldsymbol{q}(\pi(i)) = (0,3)$ and $\boldsymbol{q}(T \cup \{i\}) = (1,4)$, i will always prefer to deviate, as there is no $\boldsymbol{w}^{(i)}$ for which $\boldsymbol{w}^{(i)} \cdot \boldsymbol{q}(\pi(i)) \geq \boldsymbol{w}^{(i)} \cdot \boldsymbol{q}(T \cup \{i\})$. However, if $\boldsymbol{q}(\pi(i)) = (2,3)$ and $\boldsymbol{q}(T \cup \{i\}) = (1,4)$, there are possible values for $\boldsymbol{w}^{(i)}$ that would make i prefer $\pi(i)$. In such cases we have to *elicit* the preferences of agent i with respect to these two vectors.

LSS is provided in Algorithm 1 and is parameterised by the agents and quality function of an MC2FG, i.e., N and $\mathbf{q}$. However, we assume that we have no *direct* access to $f^{(i)}$ (i.e., $\boldsymbol{w}^{(i)}$). In fact, LSS only knows that each $\boldsymbol{w}^{(i)}$ adheres to the simplex constraints, i.e., $\sum_x w_x^{(i)} = 1$ and $\forall x : 0 \leq w_x^{(i)} \leq 1$. Therefore, it creates a set of sets of constraints on line 1 containing the simplex constraints for each $\boldsymbol{w}^{(i)}$.

LSS is also parameterised by a starting coalition π, which can e.g., be initialised randomly. Finally LSS is parameterised by a function checkDeviation. This function checks whether a deviation exists, and thus requires different implementations for NS-deviations (Algorithm 2), and IS-deviations (Algorithm 3). It is also this function that will elicit preferences from the agents.

Algorithm 1. LSS($N, \boldsymbol{q}, \pi, \texttt{checkDeviation}$)

```
1  C ← a set of simplex constraints on w^(i), C^(i), for each i ∈ N
2  stable ← false
3  while ¬stable ∨ ¬timeout() do
4  |   stable ← true
5  |   foreach i ∈ N do
6  |   |   foreach T ∈ (π \ π(i)) that i could join do
7  |   |   |   if (T ∪ {i} ≻_P T) ∧ (T ∪ {i} ≻_P π(i)) then
8  |   |   |   |   π ← (π \ {π(i), T}) ∪ {π(i) \ {i}, T ∪ {i}}
9  |   |   |   |   stable ← false
10 |   |   |   |   continue from top while-loop (line 3)
11 |   |   |   end
12 |   |   end
13 |   end
14 |   foreach i ∈ N do
15 |   |   // for all agents (in random order), check for and perform deviations:
16 |   |   foreach T ∈ (π \ π(i)) that i could join do
17 |   |   |   possibleDeviation ← checkDeviation(i, π(i), T, C)
18 |   |   |   if possibleDeviation then
19 |   |   |   |   π ← (π \ {π(i), T}) ∪ {π(i) \ {i}, T ∪ {i}}
20 |   |   |   |   stable ← false
21 |   |   |   |   continue from top while-loop (line 3)
22 |   |   |   end
23 |   |   end
24 |   end
25 end
26 if stable then return π ;
27 else return No stable partitioning was found ;
```

Algorithm 2. checkNSDeviation($i, \pi(i), T, \mathcal{C}^{(i)}$)

```
1  maxDiffNew ← max_{w^(i)} w^(i) · (q(T ∪ {i}) − q(π(i))) s.t. C^(i)
2  maxDiffOld ← max_{w^(i)} w^(i) · (q(π(i)) − q(T ∪ {i})) s.t. C^(i)
3  if maxDiffOld ≥ 0 ∧ maxDiffNew > 0 then
4  |   // not enough information, ask agent i:
   |   prefNew ← askAgent(T ∪ {i} ≻_{f(i)} π(i))
5  |   if prefNew then
6  |   |   C^(i) ← C^(i) ∪ {w^(i) · (q(T ∪ {i}) − q(π(i))) > 0}
7  |   |   return true
8  |   else
9  |   |   C^(i) ← C^(i) ∪ {w^(i) · (q(T ∪ {i}) − q(π(i))) ≤ 0}
10 |   |   return false
11 |   end
12 else return maxDiffNew > maxDiffOld ;
```

Algorithm 3. checkISDeviation($i, \pi(i), T, \mathcal{C}$)

1 nsDeviation $\leftarrow$ checkNSDeviation($i, \pi(i), T, \mathcal{C}^{(i)}$)
2 **if** nsDeviation **then**
3   **foreach** $j \in T$ **do**
4   maxDiffNew $\leftarrow$ $\max_{w^{(j)}} w^{(j)} \cdot (q(T \cup \{i\}) - q(T))$ s.t. $\mathcal{C}^{(j)}$
5   maxDiffOld $\leftarrow$ $\max_{w^{(j)}} w^{(j)} \cdot (q(T) - q(T \cup \{i\}))$ s.t. $\mathcal{C}^{(j)}$
6   **if** maxDiffOld $> 0 \wedge$ maxDiffNew ≥ 0 **then**
7   //not enough information, ask agent j:
8   prefOld $\leftarrow$ askAgent($T \succ_{f(j)} T \cup \{i\}$)
9   **if** prefOld **then**
10                 $\mathcal{C}^{(j)} \leftarrow \mathcal{C}^{(j)} \cup \{w^{(j)} \cdot (q(T) - q(T \cup \{i\}) > 0\}$
11   **return false**
12   **else**
13                 $\mathcal{C}^{(j)} \leftarrow \mathcal{C}^{(j)} \cup \{w^{(j)} \cdot (q(T) - q(T \cup \{i\}) \leq 0\}$
14   **end**
15   **else if** maxDiffOld > 0 **then**
16   **return false**
17   **end**
18   **end**
19   **return true**
20 **else return false** ;

In the main loop (lines 3–25), LSS iterates over all agents two times, and checks whether it has a deviation it wants to perform (lines 7 and 17). The first time LSS loops over all agents, it checks whether i can deviate to a coalition T for which $q(T \cup \{i\})$ Pareto-dominates, i.e., is better or equal in all criteria and better in at least one criterion than, both $q(T)$ and $q(\pi(i))$; if that is the case, both i and the agents in T will prefer that definition, will thus allow both an NS- and IS-deviation. If such a deviation exists, it is performed (line 8). The second time LSS loops over the agents, it checks for each agent i whether there is an NS-deviation or an IS-deviation using an NS- or IS-specific subroutine. If such a deviation exists, the deviation is performed (line 19). When none of the agents have a deviation LSS terminates. In order to check whether an agent has a deviation, we need a specific algorithm for each type of deviation. For NS-deviations, the algorithm is given in Algorithm 2. The algorithm is called with an agent i that may want to deviate from $\pi(i)$ to T, given the known constraints on $w^{(i)}$, $\mathcal{C}^{(i)}$. On the first two lines, *linear programs (LPs)* are run to calculate the maximal possible difference in utility $w^{(i)} \cdot q(S)$ if the new coalition, i.e., $S = T \cup \{i\}$, is preferred over the old coalition, i.e., $\pi(i)$, resp. if the old coalition is preferred over the new one, given the known constraints $\mathcal{C}^{(i)}$. When both these values are positive, it is both possible that the old coalition has a higher utility for i and that the new coalition has a higher utility for i. In other words, LSS cannot determine which coalition is preferred by i and it must thus ask the agent directly (line 3). We denote this asking the agent as `askAgent`$(T \cup \{i\} \succ_{f^{(i)}} \pi(i))$. We assume the agent will always answer truthfully with either `true` or `false`. When an agent answers a comparison question, for example, it states that `askAgent`$((2,3) \succ_{f^{(i)}} (1,4)) \rightarrow$ `true`, this imposes a constraint on $\boldsymbol{w}^{(i)}$. In this case, it imposes the constraint $\boldsymbol{w}^{(i)} \cdot (2-1, 3-4) > 0$. In general, the imposed constraints are:

$$\boldsymbol{q}(S) \succ_{f^{(i)}} \boldsymbol{q}(T) \implies \boldsymbol{w}^{(i)} \cdot (\boldsymbol{q}(S) - \boldsymbol{q}(T)) > 0, and$$

$$\neg(\boldsymbol{q}(S) \succ_{f^{(i)}} \boldsymbol{q}(T)) \implies \boldsymbol{w}^{(i)} \cdot (\boldsymbol{q}(S) - \boldsymbol{q}(T)) \leq 0.$$

Therefore, by eliciting such constraints through asking agents for comparisons between quality vectors, LSS can learn the relevant preference information of the agents. LSS adds the constraints to $\mathcal{C}^{(i)}$ (lines 6 and 9).

When LSS is run with `checkNSDeviation` (Algorithm 2) as `checkDeviation`, it will try to find a Nash-stable partitioning. However, we may want to consider weaker stability concepts; for individual stability, we need to use `checkISDeviation` (Algorithm 3) instead. Because individual stability imposes extra constraints on deviations over Nash-stability, `checkISDeviation` first calls `checkNSDeviation` (line 1), to check whether i wants to deviate. Then, if that is indeed the case, it loops over all the agents of the coalition i wants to deviate to, T, to check whether none of these agents lose utility by i joining T (lines 3–18). This is done using similar LPs as for the agent that wants to deviate (lines 4 and 5). Again, the algorithm might not be able to determine this from the current constraints, and thus elicits comparisons from the appropriate agent. Note that when one agent in T loses utility, the `checkISDeviation` terminates immediately (line 11). This is because we want to minimise the number of questions asked.

We note that on lines 6 and 16 of Algorithm 1 (LSS), we may not allow i to deviate to arbitrary $T \in \pi \setminus \pi(i)$. For example, we may impose social network constraints [11], i.e., that the agents are embedded in a graph representing which agents know each other, and only allow coalitions that are connected subgraphs.

5   Experiments

In this section we empirically test the LSS algorithm for both Nash stability and individual stability. We initialise the partition at the start of each run as the *partition of singletons*, i.e., initially each agent is in a separate coalition. As a baseline, we compare to an algorithm that does not employ linear programs, but only tests for Pareto-dominance instead, i.e., Algorithm 1 is the same, but in the `checkDeviation` subroutines (Algorithms 2 and 3) the linear programming steps (e.g., lines 1–2 in Algorithm 2) are skipped, and instead the agent is always asked for a comparison. We refer to this baseline algorithm as *always ask stability search (AASS)*.

We implemented all algorithms in Python 3, making use of (the default solver of) the PuLP library (version 1.6.1) for linear programs. We ran the experiments on a MacBook Pro, with a 2.9 GHz Intel Core i5 processor and 16GB memory, running macOS Sierra (version 10.12.1).

5.1 Test Problems

In order to test the performance of LSS and AASS we make use of two test classes of MC2FGs. A Random instance, is one in which every possible coalition $S \subseteq N$ has a randomly drawn quality vector $q(S)$, from a uniform distribution on the unit hypercube of m dimensions (i.e., between the origin, $\mathbf{0}$, and the vector containing only ones, $\mathbf{1}$). The scalarisation function for each agent, i, is linear, with a weight vector $\boldsymbol{w}^{(i)}$ that is drawn independently from a uniform distribution on the weight simplex.

The second problem, Author × Author, is inspired on the example of scientists writing papers together. Imagine we have n authors (with scalarisation functions generated in the same way as for the Random instances) whom will be advised to work together in coalitions by a recommender system. This system interacts with the agents by asking them whether they would prefer to be in one of two proposed coalitions. For each coalition, the recommender system presents the expected quality q in m dimensions (e.g., expected impact and expected novelty when writing a paper together), of the coalitions to the agents. It is therefore essential that the number of questions posed to the agents is minimised, as interaction with human decision makers is the slowest part of the process. The quality vectors are formed by summing over randomly drawn agent quality vectors $\mathbf{v}_i$ for each agent (containing values between 1 and 4 drawn independently from a uniform distribution), and subtracting a group size penalty:

$$q(S) = (-2^{|S|-1})\mathbf{1} + \sum_{i \in S} \mathbf{v}_i.$$

Because of the group size penalty, stable partitions will typically consist of coalitions of 3 or 4 agents. Furthermore, because of this structure, we have empirically found that Nash stable partitions typically do exist in Author × Author instances.

5.2 Random

To test whether we can find stable partitions with LSS, we first run LSS on Random MC2FG instances, as defined above. Note that we do not test AASS separately, as they have the same number of iterations; the only difference between the two is when they ask the agents to compare quality vectors.

In Fig. 1, proportion of instances (of 50 in total) for which LSS found a Nash or individually stable partition within 1000 iterations for varying numbers of agents (top left) and criteria (top right).

We observe that for the individual stability criterion, stable partitions were nearly always found (in only 2 out of the 800 Random instances an individually stable coalition was not found by LSS), while for Nash stability this is not the case. This is according to expectation, as it is more difficult to reach a Nash stable partition (as Nash stability is a stronger stability concept than individual stability). Furthermore, we observe that the proportion of stable partitions found goes down as a function of the number of agents (as expected), but does not change significantly as a function of the number of criteria. This is surprising

as more criteria make the likelihood of agents disagreeing more
likely. We thus conclude that
MC2FGs become significantly
harder as the number of agents
in the problem increases.

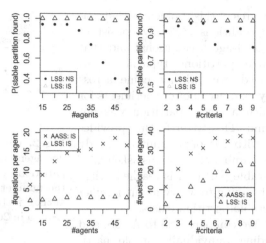

Secondly, we measure how
many questions we need to ask
per agent in order to find stable partitions. This is important, as this may correspond to
asking humans for their preferences, which can be time-
consuming and experienced as
hindrance by these humans.
Because we have not always
found Nash stable partitions,
we focus on individual stability only for the **Random** MC2FG
instances. From Fig. 1 (bottom
left) we observe that for LSS
leads to significantly less questions asked per agents than
AASS across all numbers of

Fig. 1. The proportion of instances for which a
stable partition was found by LSS using the Nash
stability criterion and Individual stability criterion within 1000 iterations (top) and average number of questions asked per agent until an IS-stable
partitioning was found (bottom), as a function of
(left) the number of agents in 50 2-criteria **Random**
instances, and (right) the number of criteria in 50
20-agent **Random** instances.

agents. The highest number of
questions per agents (at 40 agents) was 3.26 for LSS and 18.65 for AASS. LSS
scales better in the number of agents than AASS; when we fit a line for the
number of questions per agent as a function of the number of agents, we obtain
a slope of 0.017 questions per agent per additional agent for LSS, and a slope of
0.22 for AASS.

LSS needs less questions per agent than AASS across different numbers of
criteria (Fig. 1; bottom right). However, the difference in slope, i.e., the number
of questions per agent per additional criterion is not significant (2.96 for LSS,
versus 3.36 for AASS).

We conclude that LSS can be used to find stable partitions in **Random** MC2FG
instances, but that Nash stable partitions are harder to find than individually
stable partitions. Furthermore, LSS can significantly decrease the number of
comparisons that need to be made by the agents when looking for individually
stable partitions, and LSS scales better in the number of agents than a naive
question asking scheme like AASS.

5.3   Author × Author

In order to test the performance of LSS with respect to AASS, we test the
algorithms on our real-world inspired Author × Author problem. Important

features of this problem is that stable coalitions typically consist of 3- or 4-agent coalitions, and that stable partitions typically exist. Indeed, we did not come across any instance in our experiments for which a stable partition was not found. This latter feature provides us with the opportunity to study the number of questions asked *until* a stable partition is reached, without having to worry about *whether* a stable partition exists.

We compare LSS to AASS, and finding individually stable partitions to finding Nash stable partitions on `Author × Author` instances. For all

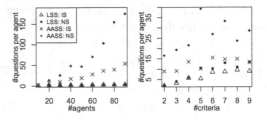

Fig. 2. The average number of questions asked per agent until an individual or Nash stable partitioning was found as a function of (left) the number of agents in 2-criteria `Author × Author` instances, and (right) the number of criteria in 20-agent `Author × Author` instances.

`Author × Author` instances Nash and individually stable partitions were found within 1000 iterations of the main loop of LSS/AASS (Algorithm 1, lines 3–25). While only singleton initialisation is displayed in Fig. 2, we also compared to undominated initialisation, but like for `Random` instances found no significant difference.

When comparing finding Nash stable partitions to individually stable partitions for varying numbers of agents (Fig. 2 (left)), we observe that AASS performs significantly worse for individual stability than for Nash stability in terms of the number of questions, there is no significant difference between the number of questions until a Nash stable partition and an individually stable partition is found by LSS. This is a surprising result, as it is typically much harder to find Nash stable partitions (as can be seen from the AASS curve for the number of questions asked), and it does take more iterations to find a Nash stable partition than an individually stable partition (on average 253 versus 144 iterations for 90 agent instances). Furthermore, both for individual stability and for Nash stability, LSS scales better than AASS in terms of the number of questions per agent.

For 80-agent 2-criteria `Author × Author` instances, LSS required 3.7 questions on average per agent. When we consider the example use case of matching small groups of authors for an event, asking authors to compare 3 or 4 groups is probably feasible. On the other hand, if we employ a more naive question-asking scheme, i.e., AASS, the numbers are 40 for individual stability, and 154, which probably would not be feasible. We thus conclude that LSS can keep the number of questions that need to be asked to agents can be kept at feasible numbers even for higher number of agents.

When we compare `Author × Author` instances of 20 agents for varying numbers of criteria (Fig. 2 (right)), LSS also outperforms AASS by a large margin for both Nash stability and individual stability. For higher criteria, LSS using individual stability is slightly more efficient than LSS using Nash stability. For

9 criteria, 20 instances required 10.6 questions per agents for finding an individually stable partition with LSS, and 14.4 for finding Nash stable partitions. We conclude that for this real-world inspired problem, LSS can reduce the number of questions that need to be asked to agents can be kept at reasonable numbers, even for higher numbers of criteria.

6   Discussion

In this paper, we proposed the *multi-criteria coalition formation game (MC2FG)* for forming stable partitions while having limited access to the preferences over different possible coalitions for each agent. This is important as agents may either not want to divulge their complete preference profiles for social or privacy reasons, but more importantly might not even be able to specify their utilities a priori to begin with. Instead, MC2FGs model the quality of coalitions as vectors containing different agent-independent metrics corresponding to different criteria that agents may have. However, the agents may have different preferences with respect to these criteria. To model this, in MC2FGs each agent, i, has a private utility function, $f^{(i)}$, that takes the multi-criteria quality vector of a coalition, and produces a scalar utility. Because these functions are private, we cannot use these to check whether a proposed coalition is stable. However, some a priori assumptions about $f^{(i)}$ can be made, and we can increase our knowledge about $f^{(i)}$, by asking agent i to compare two coalitions. In this paper we made the a priori assumption that each $f^{(i)}$ is a linear function.

Because (Nash and individually) stable partitions need not exist in MC2FGs, we proposed a local search algorithm that we call *local stability search (LSS)* to find stable partitions where possible. We showed empirically that LSS is able to discover stable partitions. By exploiting the additional knowledge gained by asking agents for comparisons between quality vectors for different coalitions (via LPs), we show empirically that the number of comparisons that need to be asked from agents can be kept to a minimum.

Because stable partitions need not exist in general, in future research we aim to find subclasses of MC2FGs in stable partitions are guaranteed to exist. Specifically, we aim to find realistic subclasses corresponding to use cases like the Author × Author problem. Furthermore, we aim to improve the Author × Author problem, using studies on scientific cooperation [1,12] to redefine the utility functions.

More generally, we have applied a utility-based approach to multi-criteria multi-agent decision making. In our model, all agents in a coalition share a value vector, but may have different private utilities w.r.t. this value vector, as their preferences with respect to the criteria may vary. We believe that this kind of model is realistic for many real-world decision-making problems in which there are heterogeneous agents, e.g., colleagues in different phases of their careers cooperating in a project. We aim to investigate how this perspective can be applied to other multi-agent decision-making models, such as *(multi-objective) coordination graphs* [9,17], *(cooperative) Bayesian games* [14] and *(partially observable) stochastic games* [10,13]. Finally, we want to investigate possibility of noisy pairwise comparisons [18].

Acknowledgements. We are grateful for very useful comments by the anonymous reviewers at ADT and CoopMAS, and thank Edith Elkind for her helpful suggestions. Diederik M. Roijers is a postdoctoral fellow of the Research Foundation – Flanders (FWO). This work is in part supported by the EC-FP7 under grant agreement no. #611153 (TERESA).

References

1. Ackerman, M., Brânzei, S.: The authorship dilemma: alphabetical or contribution? In: AAMAS 2014, pp. 1487–1488 (2014)
2. Balcan, M., Procaccia, A.D., Zick, Y.: Learning cooperative games. In: IJCAI 2015, pp. 475–481 (2015)
3. Benabbou, N., Perny, P.: Solving multi-agent knapsack problems. In: ECAI 2016, pp. 1318–1326 (2016)
4. Banerjee, S., Konishi, H., Sönmez, T.: Core in a simple coalition formation game. Soc. Choice Welf. **18**(1), 135–153 (2001)
5. Bogomolnaia, A., Jackson, M.O.: The stability of hedonic coalition structures. Games Econ. Behav. **38**(2), 201–230 (2002)
6. Chalkiadakis, G., Elkind, E., Wooldridge, M.: Computational aspects of cooperative game theory. Synth. Lect. Artif. Intell. Mach. Learn. **5**(6), 1–168 (2011)
7. Faliszewski, P., Elkind, E., Wooldridge, M.: Boolean combinations of weighted voting games. In: AAMAS 2009, pp. 185–192 (2009)
8. Fernández, F.R., Hinojosa, M.A., Puerto, J.: Core concepts in vector-valued games. J. Optim. Theor. Appl. **112**(2), 331–360 (2002)
9. Guestrin, C., Koller, D., Parr, R.: Multiagent planning with factored MDPs. In: NIPS 2002, pp. 1523–1530 (2002)
10. Hansen, E.A., Bernstein, D.S., Zilberstein, S.: Dynamic programming for partially observable stochastic games. In: AAAI 2004, pp. 709–715 (2004)
11. Igarashi, A., Elkind, E.: Hedonic games with graph-restricted communication. In: AAMAS 2016, pp. 242–250 (2016)
12. Kleinberg, J., Oren, S.: Mechanisms for (mis) allocating scientific credit. In: STOC 2011, pp. 529–538 (2011)
13. Littman, M.L.: Markov games as a framework for multi-agent reinforcement learning. In: ICML 1994, pp. 157–163 (1994)
14. Oliehoek, F.A., Whiteson, S., Spaan, M.T.J.: Exploiting structure in cooperative Bayesian games. In: UAI 2012, pp. 654–664 (2012)
15. Peters, D.: Complexity of hedonic games with dichotomous preferences. In: AAAI 2016, pp. 579–585 (2016)
16. Roijers, D.M., Vamplew, P., Whiteson, S., Dazeley, R.: A survey of multi-objective sequential decision-making. J. Artif. Intell. Res. **47**, 67–113 (2013)
17. Roijers, D.M., Whiteson, S., Oliehoek, F.A.: Computing convex coverage sets for faster multi-objective coordination. J. Artif. Intell. Res. **52**, 399–443 (2015)
18. Roijers, D.M., Zintgraf,L.M., Nowé, A.: Interactive thompson sampling for multi-objective multi-armed bandits. In: ADT 2017 (2017, To appear)

19. Sung, S.C., Dimitrov, D.: On core membership testing for hedonic coalition formation games. Oper. Res. Lett. **35**(2), 155–158 (2007)
20. Tanino, T.: Multiobjective cooperative games with restrictions on coalitions. In: Barichard, V., Ehrgott, M., Gandibleux, X., T'Kindt, V. (eds.) Multiobjective Programming and Goal Programming. Lecture Notes in Economics and Mathematical Systems, vol. 618, pp. 167–174. Springer, Heidelberg (2009)

Precise Complexity of the Core in Dichotomous and Additive Hedonic Games

Dominik Peters[(⊠)]

Department of Computer Science, University of Oxford, Oxford, UK
dominik.peters@cs.ox.ac.uk

Abstract. Hedonic games provide a general model of coalition formation, in which a set of agents is partitioned into coalitions, with each agent having preferences over which other players are in her coalition. We prove that with additively separable preferences, it is Σ_2^p-complete to decide whether a core- or strict-core-stable partition exists, extending a result of Woeginger (2013). Our result holds even if valuations are symmetric and non-zero only for a constant number of other agents. We also establish Σ_2^p-completeness of deciding non-emptiness of the strict core for hedonic games with dichotomous preferences. Such results establish that the core is much less tractable than solution concepts such as individual stability.

1 Introduction

Suppose agents wish to form coalitions, perhaps to jointly achieve some task or common goal, with the payoff to an agent depending on the make-up of the coalition the agent is joining. In many situations, it makes sense to model agents' preferences to only depend on the *identity* of the players in a group. Such games are called *hedonic games*, because agents in a hedonic game can be seen as deriving pleasure from each other's presence.

An agent in a hedonic game specifies a preference ordering over all sets (*coalitions*) of agents. An outcome of the game is a partition of the agent set into disjoint coalitions. A player prefers those partitions in which she is part of a preferred coalition. The main focus of the literature on hedonic games is studying outcomes that are *stable* [7,9].

Of the many notions of stability discussed in the literature, the most prominent is the concept of the *core*. A partition π of the agent set is core-stable if there is no non-empty set S of agents all of which strictly prefer S to where they are in π. Intuitively, if the state of affairs were π, then the members of S would find each other and decide to defect together because π does not offer them enough utility. In this case, we say that S *blocks* π. A related concept is the *strict core*. The partition π is strict-core-stable if there is no set S of agents all of which *weakly* prefer S to π, but with at least one agent i in S having a strict preference in favour of S. Intuitively, i can offer some of his profit of the deviation to the other players, inducing S to deviate from π; and even if utility

© Springer International Publishing AG 2017
J. Rothe (Ed.): ADT 2017, LNAI 10576, pp. 214–227, 2017.
DOI: 10.1007/978-3-319-67504-6_15

is entirely non-transferable, the players who are indifferent between π and S will be easily convinced to join the deviation.

Computationally speaking, hedonic games are large objects: every player is described by a preference ordering over exponentially many sets. However, there are attractive concise representations of such preferences, many of which have been studied in the computer science literature [3,6,10,12,22]. Given a concise representation (which usually is not universally expressive), it makes sense to pose the computational problem of finding a stable outcome given a hedonic game. Since many games do not admit stable outcomes, we can conveniently consider the decision problem whether one exists. This problem turns out to be NP-hard for most cases (see Peters and Elkind [18] for a systematic study of representations inducing NP-hard CORE-EXISTENCE problems, and Woeginger [23] for a survey). In particular, the case of *additively separable hedonic games* is an example of a class of succinctly represented hedonic games where it is hard to distinguish games admitting stable outcomes from those which do not [2,22]. In this model, agents assign numeric values $v_i(j)$ to other players, and the utility of a coalition is the sum $\sum_{j \in S} v_i(j)$ of the values of the players in it.

In general, CORE-EXISTENCE, the problem of deciding whether a given hedonic game admits a core-stable partition, is contained in the complexity class Σ_2^p. This is because the question under consideration is characterised by a single alternation of quantifiers: "does there *exist* a partition π such that *for all* coalitions S, S does not block π?". Here, we have assumed to be able to efficiently decide whether a given coalition S blocks, which is trivially the case for all commonly considered representations. Alternatively, since $\Sigma_2^p = \mathrm{NP}^{\mathrm{NP}}$, we can see containment of CORE-EXISTENCE through the following non-deterministic algorithm: guess a partition π, and use the NP-oracle to check whether π is core-stable.

In the case of additively separable games, it is coNP-complete to verify that a given partition π is core-stable [11]. Thus, it is unlikely that the CORE-EXISTENCE problem is contained in NP. Indeed, Woeginger [24] proved that the problem is Σ_2^p-complete. The problem thus encapsulates the full hardness of the second level of the polynomial hierarchy, making it much harder to decide than NP-complete problems. (Woeginger likens NP-complete problems to "rotten eggs" while Σ_2^p-complete problems are at the level of "radioactive thallium".)

Recent decades have shown impressive advances in general-purpose tools to handle NP-complete problems, such as through SAT and ILP solvers. Thus, for solution concepts such as Nash stability, whose EXISTENCE problem is NP-complete, we should expect tractability in many practically relevant cases. On the other hand, because of Σ_2^p-hardness, finding a core-stable partition will likely require exponentially many calls to solvers. This suggests that the core will remain computationally elusive for some time to come.

In this paper, we extend Woeginger's result to the *strict core*. In addition, our reduction—which works for both the core and the strict core—produces additive hedonic games that are *symmetric* and *sparse*, showing that imposing these additional restrictions do not lead to a drop in complexity. Before we turn to additive games, however, we first consider hedonic games with *dichotomous preferences*, or *Boolean hedonic games* [4]. In these games, players only distinguish between *approved* and *non-approved* coalitions. Preferences in these games can be specified by giving each agent a goal formula of propositional logic; the approved coalitions are those which satisfy the goal. We show that deciding the existence of a strict-core-stable outcome in a Boolean hedonic game is Σ_2^p-complete. For the framework of Aziz et al. [4], this is interesting because it suggests that no polynomial-sized formula in their logic will be able to characterize strict-core-stable outcomes. Our hardness result for *additive* games is then obtained by implementing the reduction for the dichotomous case using additive valuations.

2   Preliminaries

Given a finite set of agents N, a *hedonic game* is a pair $G = \langle N, (\succcurlyeq_i)_{i \in N} \rangle$, where each agent $i \in N$ possesses a complete and transitive preference relation $\succcurlyeq_i$ over $\mathcal{N}_i = \{ S \subseteq N : i \in S \}$, the set of coalitions containing i. If $S \succcurlyeq_i T$, we say that i *weakly prefers* S to T. If $S \succcurlyeq_i T$ but $T \not\succcurlyeq_i S$, we say that the preference is *strict* and write $S \succ_i T$. An agent is *indifferent* between S and T whenever both $S \succcurlyeq_i T$ and $T \succcurlyeq_i S$.

An *outcome* of a hedonic game is a partition π of N into disjoint coalitions. We write $\pi(i)$ for the coalition of π that contains i. If $\pi(i) \succcurlyeq_i \{i\}$ for each $i \in N$, then π is *individually rational*. We say that a non-empty coalition $S \subseteq N$ blocks π if $S \succ_i \pi(i)$ for all $i \in S$. Thus, all members of a blocking coalition strictly prefer that coalition to where they are in π. The partition π is *core-stable* if there is no blocking coalition. A non-empty coalition $S \subseteq N$ *weakly blocks* π if $S \succcurlyeq_i \pi(i)$ for all $i \in S$, and $S \succ_i \pi(i)$ for some $i \in S$. The partition π is *strict-core-stable* if there is no weakly blocking coalition. Clearly, if π is strict-core-stable, then it is also core-stable. There are many other stability concepts for hedonic games that we do not consider here, see Aziz and Savani [5] for a survey.

A hedonic game has *dichotomous preferences*, and is called a *Boolean hedonic game*, if for each agent $i \in N$ the coalitions $\mathcal{N}_i = \{ S \subseteq N : i \in S \}$ can be partitioned into *approved* coalitions $\mathcal{N}_i^+$ and non-approved coalitions $\mathcal{N}_i^-$ such that i strictly prefers approved coalitions to non-approved coalitions, but is indifferent within the two groups: so $S \succ_i T$ iff $S \in \mathcal{N}_i^+$ and $T \in \mathcal{N}_i^-$. We can specify a Boolean hedonic game by assigning to each agent $i \in N$ a formula ϕ_i (*i*'s *goal*) of propositional logic with the propositional atoms given by the agent set N. A coalition $S \ni i$ is then approved by i if and only if $S \models \phi_i$, that is the formula ϕ_i is satisfied by the truth assignment that sets variable $j \in N$ true iff $j \in S$. For example, if i's goal formula ϕ_i is $(j \vee k) \wedge \neg \ell$, then i approves coalitions containing agent j or k as long as they do not contain agent ℓ.

A hedonic game is *additively separable*, and is called an *additive hedonic game*, if there are valuation functions $v_i : N \to \mathbb{Z}$ for each agent $i \in N$ such that $S \succcurlyeq_i T$ if and only if $\sum_{j \in S} v_i(j) \geqslant \sum_{j \in T} v_i(j)$. An additive hedonic game is *symmetric* if $v_i(j) = v_j(i)$ for all $i, j \in N$.

The complexity class Σ_2^p, the *second level of the polynomial hierarchy*, is $\mathrm{NP}^{\mathrm{NP}}$, the class of problems solvable by a polynomial-time non-deterministic

Turing machine when given an NP-oracle. It can also be seen as the class of problems polynomial-time reducible to the language TRUE $\exists\forall$-QBF, which consists of true quantified Boolean formulas with only 1 alternation of quantifiers.

3 Related Work

Boolean hedonic games were introduced by Aziz et al. [4] who study them from a mainly logical point of view. (Notice that they use a different choice of propositional atoms—*pairs* of agents—to allow future generalisations to games played on general coalition structures. Our choice is more natural for the hedonic setting.) In particular, Aziz et al. [4] show that every Boolean hedonic game admits a core-stable partition. Thus, only the complexity of the existence of the *strict* core needs to be settled. Peters [16] shows that *finding* a core-stable partition, while it is guaranteed to exist, is FNP-hard.

Elkind and Wooldridge [12] introduce a representation formalism for hedonic games called *hedonic coalition nets* (HC-nets), which can be seen as a powerful mixture of the additive and Boolean representations introduced above. Here, agents provide several goals ϕ_i weighted by real numbers, and an agent obtains as utility the sum of the weights of the formulas that are satisfied by the coalition. Thus, additively separable games are given by HC-nets in which every formula is just a single positive literal, and Boolean hedonic games are given by HC-nets in which every agent has only a single formula. It follows from a general result of Malizia et al. [14] that core-existence is Σ_2^p-complete to decide for games given by HC-nets (see also Elkind and Wooldridge [12]). We strengthen this to also apply to the strict core, and to hold even if either every agent only has a single formula, or if every formula is given by a single literal.

This paper follows the work of Woeginger [24] who proves that deciding core-existence is Σ_2^p-complete for additive hedonic games. His reduction (from the same problem that we reduce from) does not work for the strict core, and in his survey [23] he poses the problem to establish Σ_2^p-hardness for this solution concept. Doing this is the main contribution of this paper. We also strengthen Woeginger's [24] result for the core to hold even for symmetric valuations, and even if $v_i(j)$ is non-zero for at most 10 agents j, so that the game is "sparse". This closes off two avenues for potential avoidance of Σ_2^p-hardness.

Since a preprint of this paper appeared on arXiv, some additional Σ_2^p-hardness results for hedonic games have been obtained. Ohta et al. [15] show that in hedonic games based on aversion to enemies (introduced by Dimitrov et al. [11]), if one allows 'neutral' players, then deciding the existence of the (strict) core is Σ_2^p-hard. The games studied by Ohta et al. [15] are in fact additively separable; thus, they imply Σ_2^p-hardness of the (strict) core in additive hedonic games. Hardness holds even if we restrict players' valuations to take at most three values, so that $v_i(j) \in \{-n, 0, 1\}$ for all $i, j \in N$. However, in contrast to our reduction, the games produced in their reduction are not symmetric and not sparse. Ohta et al. [15] also show Σ_2^p-hardness for the strict core for games based on friend appreciation in the presence of neutral players. Aziz et al. [1]

consider *fractional hedonic games*, where players care about the *average* value of their coalition partners, rather than the sum. They show that deciding the existence of the core is Σ_2^p-hard, even if valuations are symmetric and simple, so that $v_i(j) \in \{0, 1\}$ for all $i, j \in N$. The reductions in both of these recent papers is from the complement of the MINMAX CLIQUE problem [13], which seems to be well-suited as a starting point for reductions for hedonic games. In this paper, like in Woeginger [24], we instead use a problem based on quantified Boolean formulas.

Peters [17] introduces *graphical hedonic games*, which are hedonic games equipped with an underlying graph on the agent set. This is a direct analogue of non-cooperative graphical games. In this language, our result for additive hedonic games implies that deciding the existence of the (strict) core is Σ_2^p-hard even for graphical hedonic games of bounded degree—that is, games whose underlying graph has a max-degree of at most 10.

4   A Useful Restricted Hard Problem

Stockmeyer [21] proved that the following basic problem is Σ_2^p-complete:

TRUE ∃∀-3DNF

Instance: A quantified Boolean formula of form
$$\exists x_1, \ldots, x_m \; \forall y_1, \ldots, y_n \; \phi(x_1, \ldots, x_m, y_1, \ldots, y_n),$$
where ϕ is in disjunctive normal form with each disjunct containing 2 or 3 literals.

Question: Is the formula true?

Here we show that the problem remains Σ_2^p-complete even if we place restrictions on the number of occurrences of the variables, like is standard practice when proving NP-completeness.

RESTRICTED TRUE ∃∀-3DNF

Instance: A quantified Boolean formula of form
$$\exists x_1, \ldots, x_m \forall y_1, \ldots, y_n \; \phi(x_1, \ldots, x_m, y_1, \ldots, y_n),$$
where ϕ is in disjunctive normal form with

- each disjunct containing 2 or 3 literals,
- each x-variable occurring exactly once positive and once negative
- each y-variable occurring exactly three times, and at least once positively and at least once negatively.

Question: Is the formula true?

We may further insist that every disjunct contain at most 2 x-literals, because a disjunct containing only x-literals makes the formula trivially true.

Proposition 1. *The problem* RESTRICTED TRUE $\exists\forall$-3DNF *is Σ_2^p-complete.*

Proof. Membership in Σ_2^p is clear.

Let us note that a true $\exists\forall$-3DNF-formula is the same thing as a *false* $\forall\exists-$3CNF-formula; we will use this latter view since CNF formulas are more familiar. Thus, we may reduce from the unrestricted problem FALSE-$\forall\exists$-3CNF. We will in polynomial time transform a given such formula

$$\forall x_1,\ldots,x_m \exists y_1,\ldots,y_n \, \phi(x_1,\ldots,x_m,y_1,\ldots,y_n)$$

into a formula of equal truth value which is restricted as above, establishing hardness of the restricted variant.

First, by unit propagation, we may assume that no clause contains only a single literal. Next, for each x-variable x_i, relabel all its occurrences as $y_i^1,\ldots,y_i^{n_i}$ where n_i is the number of occurrences of x_i, and the y_i^r are new variables. Existentially quantify over these new variables, keeping x_i universally quantified. Add clauses $(x_i \to y_i^1) \wedge (y_i^1 \to y_i^2) \wedge \cdots \wedge (y_i^{n_i} \to x_i)$ to force all copies to have the same truth value. Similarly, for each old y-variable, we relabel all its occurrences (existentially quantifying) and add a 'wheel of implications' for them as well (here we may discard the old y-variable in the process). The resulting formula satisfies the restrictions and has the same truth value as the original formula. $\square$

By using the techniques of Berman et al. [8], we can similarly prove that the problem remains Σ_2^p-complete if disjuncts are required to contain exactly 3 distinct literals, each x-literal occurs exactly once, and each y-literal occurs exactly twice. One can also show that the problem remains Σ_2^p-complete if every clause contains at most one x-literal [13, Theorem 10]. For the reductions in this paper, we do not need these other restrictions.

5 Strict Core for Boolean Hedonic Games

Our first hardness result concerns Boolean hedonic games, as introduced by Aziz et al. [4]. While for this type of hedonic game, the core is always guaranteed to exist, the *strict* core is more difficult to handle.

Theorem 1. *The problem "does a given Boolean hedonic game admit a strict-core-stable partition?" is Σ_2^p-complete.*

Proof. Membership in Σ_2^p is clear, since we are asking: does there *exist* a partition such that *for all* coalitions S, S does not block?

For hardness, we reduce from RESTRICTED TRUE $\exists\forall$-3DNF. Let $\varphi = \exists x \forall y \phi$ be an instance of this problem, where $x = (x_i)$ and $y = (y_j)$ denote vectors of variables. We rewrite φ as $\exists x(\neg\exists y\neg\phi)$. Note that $\neg\phi$ is a 3CNF formula, and when below we talk about clauses, we are always referring to clauses of $\neg\phi$. In the hedonic game which we construct below, a strict-core-stable partition corresponds to an assignment to the x-variables. If there is a y-assignment satisfying $\neg\phi$ (so that φ is false), this will form a weakly blocking coalition, and conversely, such a coalition induces a satisfying assignment. If the formula is true, such a blocking coalition cannot exist.

For our construction, we take the following agents:

- For each x_i, four agents x_i, $\overline{x}_i$, t_i, f_i.
- For each y_j, two agents y_j, $\overline{y}_j$.
- For each clause c_k in $\neg\phi$, one agent c_k.
- A single player φ representing the formula.

We now specify agents' goals. For a clause c_k of $\neg\phi$, we let $\ell_1^k, \ell_2^k, \ell_3^k$ denote the agents corresponding to the literals occurring in it. For example, if clause c_k is $(x_1 \vee \neg y_2 \vee y_3)$, then ℓ_1^k refers to agent x_1, ℓ_2^k refers to $\overline{y}_2$, and ℓ_3^k refers to y_3. If a clause only contains 2 literals, just let $\ell_2^k = \ell_3^k$.

- $x_i : f_i \wedge \neg t_i \wedge \overline{x}_i$
- $\overline{x}_i : t_i \wedge \neg f_i \wedge x_i$
- $t_i : \neg\varphi$
- $f_i : \neg\varphi$
- $y_j : \neg\overline{y}_j$
- $\overline{y}_j : \neg y_j$
- $c_k : \neg\varphi \vee ((\ell_1^k \vee \ell_2^k \vee \ell_3^k) \wedge c_{k+1})$, or
  $c_k : \neg\varphi \vee (\ell_1^k \vee \ell_2^k \vee \ell_3^k)$ if c_{k+1} does not exist
- $\varphi : c_1 \wedge (x_1 \vee \overline{x}_1)$

This hedonic game has a strict-core-stable outcome if and only if φ is true.

$\Longleftarrow$: Suppose φ is true. Take an assignment $\mathcal{A}$ to the x-variables certifying truth of φ. Then take the partition π with coalitions $\{t_i, x_i, \overline{x}_i\}$ for true x_i, with coalitions $\{f_i, x_i, \overline{x}_i\}$ for false x_i, and singleton coalitions for all other players. We show that π is strict-core-stable.

Most agents' goals are satisfied in π, except for true x-literals and the player φ. If π is not stable, then there is a weakly blocking coalition S including a player whose goal is satisfied in S but not in π; also, no other player in S can be worse off in S than in π. Now the profiting player cannot be a true x-literal, for if this player were to gain then its complementary literal must be part of S and this literal would lose, because complementary literals have incompatible goals. Hence any weakly blocking coalition S must include the φ-player and must satisfy it. Looking at φ's goal, this means that $c_1 \in S$. Indeed, by induction, $c_k \in S$ for all c_k, since $c_k \in S$ cannot be worse off in S, and thus c_k's goal must be satisfied, which requires $c_{k+1} \in S$.

Since every c_k is satisfied in π, the goal of every c_k is also satisfied in S. This means that the literals present in S must satisfy each clause of $\neg\phi$. Thus, the literals present in S satisfy $\neg\phi$. This contradicts truth of φ under the assignment $\mathcal{A}$ once we can show that S does not contain complementary y-literals, and only contains x-literals that are true in $\mathcal{A}$. But both of these requirements are easy to see: since all y-literals have their goal satisfied in π, they must have their goal satisfied in S, which means their complementary literal is not part of S. Also, a false x-literal is happy in π but would be unhappy in S since $t_i, f_i \notin S$ (because they hate φ), and thus false x-literals are not part of the weakly blocking S. Hence π is strict-core-stable.

$\Longrightarrow$: Suppose the game has a strict-core-stable outcome π. We show that φ is true. First we will find an appropriate assignment to the x-variables. Fix some variable x_i. Since the goals of x_i and $\overline{x}_i$ are incompatible, at most one of them is happy in π. If both are unhappy, then $\{t_i, x_i, \overline{x}_i\}$ weakly blocks. Hence for each x_i, exactly one of x_i and $\overline{x}_i$ has their goal satisfied. Define the assignment that sets that literal true which is *not* satisfied.

Soon we will need to know that the φ player does not have its goal satisfied in π. For a contradiction suppose it does. Then $x_1 \in \pi(\varphi)$ or $\overline{x}_1 \in \pi(\varphi)$. Since x_1 or $\overline{x}_1$ has their goal satisfied in π, both of them are together with either t_1 or f_1. So either $t_1 \in \pi(\varphi)$ or $f_1 \in \pi(\varphi)$; but then $\{t_1\}$ or $\{f_1\}$ blocks π, a contradiction.

Now take an arbitrary assignment to the y-variables, and suppose for a contradiction that under these assignments to the x- and y-variables the formula ϕ becomes false, so that $\neg\phi$ becomes true so every clause is true. Let S be the coalition consisting of player φ, all clauses c_k, all true x-literals, and all true y-literals. In S, every player except for the true x-literals is satisfied, so no player is worse off. However φ did not have its goal satisfied in π, so is strictly better off in S, and hence S weakly blocks π, a contradiction. Thus, φ must be true. $\square$

Hardness holds even if every agent mentions at most 5 other agents in their goal. By rewriting the formulas, we can see that hardness also holds even if goals are given in 3-DNF or 4-CNF.

6   Core and Strict Core for Additive Hedonic Games

The structure of our reduction for Boolean hedonic games can be adapted to work in the additive case. The resulting reduction is necessarily less straightforward, because we have to simulate the clausal structure using additive valuations. On the other hand, the resulting reduction works for *both* the core and the strict core. Further, this is the first hardness reduction for additive hedonic games that applies even to "sparse" games, where players assign non-zero valuations to only at most a fixed number of other players.

Theorem 2. *The problem "does a given additive hedonic game admit a strict-core-stable partition?" is Σ_2^p-complete. The same question for the core is also Σ_2^p-complete, and both problems remain hard even for symmetric utilities that only assign non-zero values to at most 10 other players.*

Proof. Membership in Σ_2^p is clear, since we are asking: does there *exist* a partition such that *for all* coalitions S, S does not block?

For hardness, we reduce from RESTRICTED TRUE $\exists\forall$-3DNF. Let $\varphi = \exists x \forall y \phi$ be an instance of this problem, where $x = (x_i)$ and $y = (y_j)$ denote vectors of variables. We rewrite φ as $\exists x (\neg \exists y \neg \phi)$. Note that $\neg\phi$ is a 3CNF formula, and when below we talk about clauses, we are always referring to clauses of $\neg\phi$. In the hedonic game which we construct below, a (strict-)core-stable partition corresponds to an assignment to the x-variables. If there is a y-assignment satisfying $\neg\phi$, this will form a blocking coalition, and conversely, such a coalition

Table 1. The agent valuations $v_a(b)$. All values not specified are 0. The value "$-\infty$" denotes any sufficiently large negative number; -100 will do. Notice that the valuations are symmetric ($v_a(b) = v_b(a)$) and every agent specifies at most 10 non-zero values, noting that we can ensure that no clause contains more than 2 x-literals.

$a\ b$	$v_a(b)$	$a\ b$	$v_a(b)$	$a\ b$	$v_a(b)$	$a\ b$	$v_a(b)$
$x_i\ \bar{x}_i$	-10	$\bar{x}_i\ x_i$	-10	$y_j\ \bar{y}_j$	$-\infty$	$\bar{y}_j\ y_j$	$-\infty$
f_i	20	f_i	14	$c(y_j)$	5	$c(\bar{y}_j)$	5
t_i	14	t_i	20	$c'(y_j)$	$-\infty$	$c'(\bar{y}_j)$	$-\infty$
$c(x_i)$	5	$c(\bar{x}_i)$	5				
f'_i	$-\infty$	f'_i	$-\infty$	$c_k\ c_{k-1}$	13	$c'_k\ c_k$	30
t'_i	$-\infty$	t'_i	$-\infty$	c_{k+1}	13	$\ell(c_k)$	$-\infty$
$c'(x_i)$	$-\infty$	$c'(\bar{x}_i)$	$-\infty$	$\ell(c)$	5	c_{k-1}	$-\infty$
				c'_k	30	c_{k+1}	$-\infty$
$t_i\ x_i$	14	$f_i\ x_i$	20	t_i/f_i	$-\infty$		
$\bar{x}_i$	20	$\bar{x}_i$	14	c'_{k-1}	$-\infty$	$t'_i\ t_i$	30
t'_i	30	f'_i	30	c'_{k+1}	$-\infty$	$x_i,\ \bar{x}_i$	$-\infty$
f_i	$-\infty$	t_i	$-\infty$				
$c(x_i)$	$-\infty$	$c(\bar{x}_i)$	$-\infty$			$f'_i\ f_i$	30
						$x_i,\ \bar{x}_i$	$-\infty$

induces a satisfying assignment. If the formula φ is true under the assignment to the x-variables, such a blocking coalition cannot exist.

Before we start, let us add a new x-variable (call it x_*) to the formula and add the clause $c_1 := (x_* \vee \bar{x}_*)$ to $\neg\phi$. This preserves the truth value of the formula, and preserves the restrictions of the input problem. This will help in the proof of implication (iii) $\Rightarrow$ (i) later.

We take the following agents:

- For each x_i, four agents $x_i, \bar{x}_i, t_i, f_i$.
- For each y_j, two agents $y_j, \bar{y}_j$.
- For each clause c_k in $\neg\phi$, one agent c_k.
- A helper player c'_k for each c_k, and helper players t'_i and f'_i for each t_i and f_i.

We say that a player p who has a helper player p' is *supported*. The purpose of the helper players will be to guarantee that supported players obtain utility at least 30 in stable outcomes.

In Table 1, we specify agents' symmetric utilities, see also Figs. 1 and 2. All utilities not specified in the table are 0; the figures also omit $-\infty$ valuations for clarity. As notation, for a literal ℓ, we let $c(\ell)$ denote the clauses that contain ℓ; for x-literals there is just one such clause, but there are up to two for y-literals. We also let $\ell(c)$ denote the literals occurring in c. We take arithmetic in subscripts of clause names (as in 'c_{k+1}') to be modulo the number of clauses in $\neg\phi$. We show that the following are equivalent:

(i) The input formula is true.
(ii) The game admits a strict-core-stable partition.
(iii) The game admits a core-stable partition.

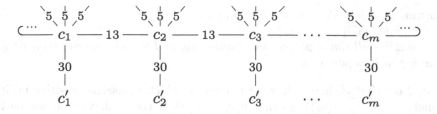

Fig. 1. The clause gadgets. Clauses are arranged in a cycle. We will consider a partition π where each clause agent c_k is in a pair with its helper c'_k. If the formula is false, though, all the clauses join forces and can deviate together with a falsifying selection of y-literals (and of true x-literals).

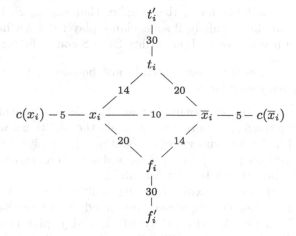

Fig. 2. The x-variable gadget. If the variable is set true, the upper triangle forms a coalition. If it is false, the lower triangle forms a coalition. Note that in this configuration, the true (but not the false) literal is willing to deviate with the connected clause agent.

Call a coalition S *feasible* if it is individually rational (and in particular does not contain players who evaluate each other as $-\infty$), and if each 'supported' player (those that have a helper player: $c_k/t_i/f_i$) obtains utility ≥ 30 in S. Call a coalition *infeasible* otherwise. Observe that all coalitions in a core-stable partition must be feasible (as otherwise it is not individually rational or a coalition like $\{c_k, c'_k\}$ blocks). Observe further that if a partition π is weakly blocked by some coalition S, then it is weakly blocked by a feasible coalition. This is because if S is not individually rational for player i, then π is also weakly blocked by the feasible coalition $\{i\}$ or $\{i, i'\}$, and if S gives less than utility 30 to a supported player i, then the feasible coalition $\{i, i'\}$ also weakly blocks π. These properties of feasible coalitions formalise the intuitive purpose of the helper players: they guarantee a minimum payoff to the supported player in any stable outcome.

Lemma. *In a feasible coalition $S \ni c_k$, either*
(a) $c_k' \in S$ but $c_{k-1}, c_{k+1}, \ell(c_k) \notin S$, or
(b) S contains all clause players simultaneously, and for each clause, some literal occurring in it is part of S.

Proof. Note that c_k' hates all other players to which c_k assigns positive utility (namely $c_{k-1}, c_{k+1}, \ell(c_k)$), so that if $c_k' \in S$ then those players cannot be in S. So suppose $c_k' \notin S$. We will prove by induction that we are in case (b). By feasibility of S, c_k obtains utility at least 30. This is only possible if both $c_{k-1} \in S$ and $c_{k+1} \in S$ (subscripts mod the number of clauses), since the literals connected to c_k only give total utility $3 \cdot 5 = 15$. Thus, $c_{k+1} \in S$. Assume now that $c_k, c_{k+1}, \ldots, c_{s-1}, c_s \in S$ for some s. As c_{s-1} and c_s' hate each other, we have $c_s' \notin S$. But feasibility for c_s then implies that $c_{s+1} \in S$. By induction, all clause players are in S. Finally, if some clause player did not have any of its literals in S, then it would only obtain utility 26 in S contradicting feasibility. □

Let us note that in case (b), some x-literal must be part of S; namely at least one x_*-literal by feasibility for c_1.

(i) $\Rightarrow$ (ii): Suppose the input formula is true. Take an assignment $\mathcal{A}$ of the x-variables certifying truth of the formula. Then take the partition π with coalitions $\{t_i, x_i, \overline{x}_i\}, \{t_i'\}, \{f_i, f_i'\}$ for true x_i, with coalitions $\{f_i, x_i, \overline{x}_i\}, \{f_i'\}, \{t_i, t_i'\}$ for false x_i, with coalitions $\{c_k, c_k'\}$ for each c_k, and singleton coalitions $\{y_j\}, \{\overline{y}_j\}$ for y-literals. We show that π is strict-core-stable.

Suppose not, and there is a weakly blocking coalition S which we may assume to be feasible. In π, the following players are already in a best coalition among feasible ones: false x-literals (with utility 10), t_i and f_i players together with literals (with utility 34), c_k' (with utility 30), and paired t_i', f_i' (with utility 30). Since S is weakly blocking, it contains a player p who *strictly* prefers S to π. This p cannot be on the preceding list.

Now p also cannot be a t_i or f_i player together with its helper for it cannot offer its associated literals a feasible coalition as good as they have in π (namely, the false literal involved would move from utility 10 to 4). Also, p cannot be a singleton t_i' or f_i' as t_i or f_i (currently with utility 34) will be less happy in feasible coalitions together with t_i' or f_i' (which only offers 30). Hence, p is either a true x-literal, a y-literal, or a c_k player. In the former two cases, in order that p strictly gains utility, S must also include a clause player associated with the literal p. So in either case S includes a clause player, and by the lemma, since we cannot be in case (a), S must contain *all* clause players, and enough literals to satisfy every clause. This means that the literals present in S satisfy $\neg\phi$. This contradicts truth of φ under assignment $\mathcal{A}$ once we can show that the literals in S form a valid assignment, i.e., S does not contain complementary y-literals, and only contains true x-literals. But both of these requirements are easy to see: by feasibility of S no y-literal has their complementary literal in S (since they hate each other). Also, a false x-literal (currently with utility 10) would be worse off in S since $t_i, f_i \notin S$ (by feasibility since t_i, f_i hate a clause) and so the false

x-literal would obtain at most 5 in S. Thus, false x-literals are not part of the weakly blocking S. Hence π is strict-core-stable.

(ii) $\Rightarrow$ (iii): Strict-core-stability implies core-stability.

(iii) $\Rightarrow$ (i): Suppose the game has a core-stable outcome π. We show that the formula is true. Recall that every coalition in π must be feasible. Write $i \sim j$ for agents i, j that appear in the same coalition in π (so $\sim$ is the equivalence relation associated with π).

First we will find an appropriate assignment for the x-variables. Consider variable x_i. It cannot be that both $x_i \sim c(x_i)$ and $\overline{x}_i \sim c(\overline{x}_i)$ for then by the lemma all clause players are together in π, which implies $x_i \not\sim t_i, f_i$ (since they hate a clause) and so x_i obtains utility at most $5 - 10 = -5$, contradicting feasibility. If exactly one of x_i and $\overline{x}_i$ is together with clause players, say $x_i \sim c(x_i)$, then $\overline{x}_i$ must obtain utility 0, because feasibility for f_i and t_i implies that they obtain utility at least 30, which they can only obtain if they are either together with *both* x_i and $\overline{x}_i$ (but this is impossible because $x_i \not\sim \overline{x}_i$ by the lemma and our assumption that $\overline{x}_i \not\sim c(\overline{x}_i)$) or if they are together with their helper player but not with either x_i or $\overline{x}_i$. Thus $\{x_i, \overline{x}_i, f_i\}$ blocks, a contradiction. The only remaining possibility is that both $x_i \not\sim c(x_i)$ and $\overline{x}_i \not\sim c(\overline{x}_i)$. In this case, we can see that core-stability implies either $x_i \sim \overline{x}_i \sim t_i$ or $x_i \sim \overline{x}_i \sim f_i$ but not both: having both is infeasible because t_i and f_i hate each other; having neither is unstable because $\{x_i, \overline{x}_i, f_i\}$ would block. Define the assignment that sets those variables x_i true for which $x_i \sim t_i$, and sets those variables x_i false for which $x_i \sim f_i$.

Soon we will need to know that π makes most players not very happy. Indeed, we claim that in π, clause players c_k obtain utility 30, and y-literals obtain utility 0. This is because if any two of these players would obtain strictly more in π, then case (b) of the lemma applies, so that all clause players are together in π, and they are together with literals which include an x_*-literal. But we already know that x-literals are not together with their clause player, a contradiction.

Now take an arbitrary assignment to the y-variables, and suppose for a contradiction that under these assignments to the x- and y-variables the formula φ becomes false, so that $\neg\phi$ becomes true meaning each clause of $\neg\phi$ is true. Let S be the coalition consisting of all clauses c_k, all true x-literals, and all true y-literals. In S, true x-literals obtain utility $5 > 4$, y-literals obtain positive utility, and clauses obtain utility $\geq 31 = 13 + 13 + 5$. Hence everyone in S is strictly better off in S than in π. Thus π is not core-stable, contradiction. Thus, φ must be true. $\square$

7   Conclusions and Open Problems

We have shown that deciding the existence of a core- or strict-core-stable partition in a given additive hedonic game is Σ_2^p-complete, and similarly that deciding the existence of a strict-core-stable partition in a Boolean hedonic game is Σ_2^p-complete. This answers the complexity status of these questions conclusively, and implies that solving them is even harder than solving NP-complete problems (unless the polynomial hierarchy collapses).

The root cause for Σ_2^p-hardness is that blocking coalitions may be very large. If we consider a solution concept according to which only coalitions of bounded size may block, the EXISTENCE problem is contained in NP. If the size bound is 2, the problem is identical to the *stable roommates* setting, and in some cases there are known tractability results. For larger size bounds, it is likely that hardness holds. It would be interesting to further explore this variation of the concept of the core.

Our reduction for additive games produced a hardness result that holds even for sparse game that have maximum degree bounded by 10. It would be interesting to decide the complexity of cases where other parameters are small. For example, Ohta et al. [15] show that the core remains Σ_2^p-hard even if agents only assign three different valuation numbers to the other players. Another parameter that may be interesting is the number of agent *types*, that is, the number of different valuation functions appearing in the game (see, e.g., Shrot et al. [20]). In the framework of graphical hedonic games [17], it would be interesting to decide the complexity of cases where the underlying graph is planar, or bipartite, or satisfying some other topological constraint. Looking in another direction, our hardness result appears to be the first that applies to sparse games, i.e., games in which agents only assign non-zero valuations to at most a constant number of other agents. It would be interesting to see whether deciding the existence of other solution concepts, such as Nash or individually stable partitions, remains hard for sparse games. More technically, it is likely possible to improve our bound of 10 to a smaller maximum degree.

We also reiterate here a problem posed by Woeginger [23]: is strict-core-existence Σ_2^p-hard even for games based on aversion to enemies? These are additive hedonic games in which $v_i(j) \in \{-\infty, 1\}$ for all agents $i, j \in N$. It appears that an entirely different approach is necessary for this setting. Partial progress on this problem is presented by Rey et al. [19] and by Ohta et al. [15].

Acknowledgements. I thank the anonymous reviewers for helpful feedback that improved the clarity of presentation, and Lena Schend for useful discussions. I am supported by EPSRC, by ERC under grant number 639945 (ACCORD), and by COST Action IC1205.

References

1. Aziz, H., Brandl, F., Brandt, F., Harrenstein, P., Olsen, M., Peters, D.: Fractional hedonic games, [CS.GT] (2017). arXiv:1705.10116
2. Aziz, H., Brandt, F., Seedig, H.G.: Computing desirable partitions in additively separable hedonic games. Artif. Intell. **195**, 316–334 (2013)
3. Aziz, H., Brandt, F., Harrenstein, P.: Fractional hedonic games. In: Proceedings of the 13th International Conference on Autonomous Agents and Multiagent Systems (AAMAS), pp. 5–12 (2014)
4. Aziz, H., Harrenstein, P., Lang, J., Wooldridge, M.: Boolean hedonic games. In: Proceedings of the 15th International Conference on Principles of Knowledge Representation and Reasoning (KR), pp. 166–175 (2016)

5. Aziz, H., Savani, R.: Hedonic games. In: Brandt, F., Conitzer, V., Endriss, U., Lang, J., Procaccia, A.D. (eds.) Handbook of Computational Social Choice. Cambridge University Press (2016). Chapter 15
6. Ballester, C.: NP-completeness in hedonic games. Games Econ. Behav. **49**(1), 1–30 (2004)
7. Banerjee, S., Konishi, H., Sönmez, T.: Core in a simple coalition formation game. Soc. Choice Welfare **18**(1), 135–153 (2001)
8. Berman, P., Karpinski, M., Scott, A.D.: Approximation hardness of short symmetric instances of MAX-3SAT. Technical report. ECCC TR03-049 (2003)
9. Bogomolnaia, A., Jackson, M.O.: The stability of hedonic coalition structures. Games Econ. Behav. **38**(2), 201–230 (2002)
10. Cechlárová, K., Hajduková, J.: Stable partitions with $\mathcal{W}$-preferences. Discrete Appl. Mathe. **138**(3), 333–347 (2004)
11. Dimitrov, D., Borm, P., Hendrickx, R., Sung, S.C.: Simple priorities and core stability in hedonic games. Soc. Choice Welfare **26**(2), 421–433 (2006)
12. Elkind, E., Wooldridge, M.: Hedonic coalition nets. In: Proceedings of the 8th International Conference on Autonomous Agents and Multiagent Systems (AAMAS), pp. 417–424 (2009)
13. Ko, K.I., Lin, C.L.: On the complexity of min-max optimization problems and their approximation. In: Du, D.Z., Pardalos, P.M. (eds.) Minimax and Applications, vol. 4, pp. 219–239. Springer, Boston (1995)
14. Malizia, E., Palopoli, L., Scarcello, F.: Infeasibility certificates and the complexity of the core in coalitional games. In: Proceedings of the 20th International Joint Conference on Artificial Intelligence (IJCAI), pp. 1402–1407 (2007)
15. Ohta, K., Barrot, N., Ismaili, A., Sakurai, Y., Yokoo, M.: Core stability in hedonic games among friends and enemies: impact of neutrals. In: Proceedings of the 26th International Joint Conference on Artificial Intelligence (IJCAI) (2017)
16. Peters, D.: Complexity of hedonic games with dichotomous preferences. In: Proceedings of the 30th AAAI Conference on Artificial Intelligence (AAAI), pp. 579–585 (2016)
17. Peters, D.: Graphical hedonic games of bounded treewidth. In: Proceedings of the 30th AAAI Conference on Artificial Intelligence (AAAI), pp. 586–593 (2016)
18. Peters, D., Elkind, E.: Simple causes of complexity in hedonic games. In: Proceedings of the 24th International Joint Conference on Artificial Intelligence (IJCAI), pp. 617–623 (2015)
19. Rey, A., Rothe, J., Schadrack, H., Schend, L.: Toward the complexity of the existence of wonderfully stable partitions and strictly core stable coalition structures in enemy-oriented hedonic games. Ann. Mathe. Artif. Intell., 1–17 (2015)
20. Shrot, T., Aumann, Y., Kraus, S.: On agent types in coalition formation problems. In: Proceedings of the 9th International Conference on Autonomous Agents and Multiagent Systems (AAMAS), pp. 757–764 (2010)
21. Stockmeyer, L.J.: The polynomial-time hierarchy. Theoret. Comput. Sci. **3**(1), 1–22 (1976)
22. Sung, S.C., Dimitrov, D.: Computational complexity in additive hedonic games. Eur. J. Oper. Res. **203**(3), 635–639 (2010)
23. Woeginger, G.J.: Core stability in hedonic coalition formation. In: van Emde Boas, P., Groen, F.C.A., Italiano, G.F., Nawrocki, J., Sack, H. (eds.) SOFSEM 2013. LNCS, vol. 7741, pp. 33–50. Springer, Heidelberg (2013). doi:10.1007/978-3-642-35843-2_4
24. Woeginger, G.J.: A hardness result for core stability in additive hedonic games. Mathe. Soc. Sci. **65**(2), 101–104 (2013)

The Subset Sum Game Revisited

Astrid Pieterse[1] and Gerhard J. Woeginger[2(✉)]

[1] Department of Computer Science, Eindhoven University of Technology,
Eindhoven, Netherlands
[2] Department of Computer Science, RWTH Aachen, Aachen, Germany
woeginger@cs.rwth-aachen.de

Abstract. We discuss a game theoretic variant of the subset sum problem, in which two players compete for a common resource represented by a knapsack. Each player owns a private set of items, players pack items alternately, and each player either wants to maximize the total weight of his own items packed into the knapsack or to minimize the total weight of the items of the other player.

We show that finding the best packing strategy against a hostile or a selfish adversary is PSPACE-complete, and that against these adversaries the optimal reachable item weight for a player cannot be approximated within any constant factor (unless P=NP). The game becomes easier when the adversary is short-sighted and plays greedily: finding the best packing strategy against a greedy adversary is NP-complete in the weak sense. This variant forms one of the rare examples of pseudo-polynomially solvable problems that have a PTAS, but do not allow an FPTAS.

1 Introduction

The subset sum game is a combinatorial game for two players A and B with perfect information. An instance of the game consists of $m + n$ items and a knapsack of capacity c. The A-items have weights $a_1, a_2, \ldots, a_m$ and belong to player A, while the B-items have weights $b_1, b_2, \ldots, b_n$ and belong to player B. Throughout we assume that every item weight is bounded by the knapsack capacity c. The players move alternately, and the instance specifies whether player A or player B makes the first move. In every move, the active player picks one of his items (which has not been picked in any earlier move) and puts it into the knapsack. As usual, an item can only be added to the knapsack, if the overall weight of all packed items does not exceed the knapsack capacity c. A player may pass on a move, but only in case none of his items fits. The game ends as soon as none of the remaining unpacked items fits into the knapsack.

We will always look at this game through the eyes of player A, whose goal is simply to maximize the total weight of A-items in the knapsack. Player B will be considered our adversary and enemy, who behaves in one of the following ways.

– **Hostile:** The objective of adversary B is to hurt player A as much as possible, and to minimize the total weight of A-items in the knapsack.

© Springer International Publishing AG 2017
J. Rothe (Ed.): ADT 2017, LNAI 10576, pp. 228–240, 2017.
DOI: 10.1007/978-3-319-67504-6_16

- **Selfish:** The objective of adversary B is to get as much profit for himself as possible, and hence to maximize the total weight of B-items in the knapsack.
- **Greedy:** The (short-sighted) objective of adversary B is to pack in every single move a B-item of largest possible weight.

While the behavior of the greedy adversary is easy to understand (and easy to predict), the behavior of the two other adversaries needs a more precise mathematical definition that considers the game in extensive form. The hostile adversary and the selfish adversary are defined via the underlying game tree; this tree is an acyclic directed graph whose vertices correspond to the possible game situations. A situation is fully specified by the current contents of the knapsack. For every possible move in the game, the game tree contains a corresponding arc between the two corresponding situations. The initial situation (with empty knapsack) is denoted p_0. Final situations (where none of the remaining unpacked items fits into the knapsack) have no out-going arcs.

Let us first specify the hostile adversary against some fixed (deterministic) strategy σ of player A. For evaluating a final situation p, we look at the contents of the knapsack and use $a(p)$ to denote the total weight of packed A-items. For evaluating a situation q somewhere in the middle of the game, we enumerate all situations $q_1, \ldots, q_k$ that can be reached from q in a single move. If it is player A's turn then his strategy σ will lead him to a well-defined situation q_j, and we define $a(q) = a(q_j)$. If it is player B's turn then $a(q) = \min_i a(q_i)$. When the game terminates, player A will end up with a total weight of $a(p_0)$.

Next let us specify the selfish adversary against a fixed (deterministic) strategy σ of player A. For evaluating a final situation p, we denote by $a(p)$ the total weight of packed A-items and by $b(p)$ the total weight of packed B-items. For evaluating a situation q in the middle of the game, let $q_1, \ldots, q_k$ denote the situations that can be reached from q in a single move.

- If it is player A's turn, then strategy σ leads him from situation q to a well-defined situation q_j. We set $a(q) = a(q_j)$ and $b(q) = b(q_j)$.
- If it is player B's turn, then $b(q) = \max_i b(q_i)$. To make the game determinate, we furthermore set $a(q) = \max_k a(q_k)$ with $k \in \arg\max_i b(q_i)$.

This means that whenever the adversary may choose between several moves that yield the same profit for himself, he will always pick the move that is best for player A. We stress that all our results can be carried over to the variant where the selfish adversary picks the move that is worst for player A. When the game terminates, player A will reach a weight of $a(p_0)$ and player B will reach a weight of $b(p_0)$.

In this paper, we study certain algorithmic questions centered around subset sum games. The central algorithmic decision problem is defined as follows:

Instance: A knapsack of capacity c; positive integer weights $a_1, a_2, \ldots, a_m$ and $b_1, b_2, \ldots, b_n$; a starting player (A or B); a positive integer bound α; an adversary type (hostile, selfish, greedy).

Question: Does player A have a deterministic strategy that allows him to pack A-items of total weight at least α into the knapsack, if he plays the game against a player B of the given adversary type?

The resulting three variants of the subset sum game will be denoted SSG-hostile, SSG-selfish, and SSG-greedy. The respective optimization versions of the game ask to find the largest possible weight α^* that player A can pack, and the respective approximation versions ask to find an approximation of α^*.

Known and related results. Motivated by certain applications in the area of operations research, Darmann, Nicosia, Pferschy and Schauer [2] introduced the subset sum game variant against the selfish adversary. They analyze a number of intuitive strategies for the game, that are either pure greedy approaches or greedy-based strategies that use some kind of bounded look-ahead. Among other results, they show that a certain greedy strategy reaches a worst case ratio of 2 when it is applied against a selfish adversary. We stress that this result assumes an oracle-access to the selfish adversary; it works move by move through the entire game, and guarantees that at the very end the weight of the packed item set is at least 50% of the weight reached by an optimal strategy. As the selfish adversary has high computational complexity, this approach does not yield a polynomial time algorithm in the classical sense, but just a policy that can be applied while playing the game. In strong contrast to this, in the current paper we will analyze these packing games by purely looking at the given instance, and we do not assume cheap oracle-access to the expensive adversary.

The combinatorics of the subset sum game is far from trivial and sometimes shows a quite counter-intuitive behavior. For instance, our intuition tells us that it should always be better to pack large items before small items. However, in [2] it is demonstrated that for certain instances of SSG-selfish it might be optimal for player A to first pack some smaller items and only later on pack large items.

Caprara et al. [1] study three packing games that are centered around bilevel variants of the knapsack problem. These games consist of only two rounds; the first player (called leader) packs some items in the first round, and then the second player (called follower) reacts by packing some items in the second round. The objective value of the leader depends on the profits of all items in the final packing. All bilevel packing games considered in [1] are Σ_2^p-complete, most of them are in-approximable, and only one has a PTAS.

Our results. We provide a complete picture of the computational complexity and the approximability landscape of the subset sum game against the three adversaries types.

- The games against the hostile and selfish adversaries are PSPACE-complete. Unless $P = NP$, these games do not allow any polynomial time approximation algorithm with constant worst case guarantee.
- The game against the greedy adversary is weakly NP-hard and pseudo-polynomially solvable. This game yields one of the rare pseudo-polynomially solvable problems that have a PTAS, but do not allow an FPTAS.

The rest of the paper is organized as follows. Section 2 states several technical observations. Section 3 proves the in-approximability results for SSG-hostile and SSG-selfish (no constant factor approximation) and SSG-greedy (no

FPTAS). Section 4 pinpoints the computational complexity of SSG-greedy, and Sect. 5 derives a PTAS for SSG-greedy. Finally Sect. 6 derives the PSPACE-completeness of SSG-hostile and SSG-selfish.

2   Technical Preliminaries

When we introduced the subset sum game in the first paragraph of this paper, we stated that every instance of the game explicitly specifies whether player A or player B makes the first move. The following lemma shows that from the computational complexity point of view, there is no difference whether player A or player B starts the game. In other words, this lemma allows us to shift the first move from player A to player B and vice versa.

Lemma 1. *For the games SSG-hostile, SSG-selfish, SSG-greedy, the computational complexity of the variant where player A has the first move coincides with the complexity of the variant where the adversary B has the first move.*

The definitions of the two games SSG-hostile and SSG-selfish look very similar to each other, and it might not be clear at first sight that these two definitions actually yield two *different* games. The following instance illustrates that these two games indeed are different.

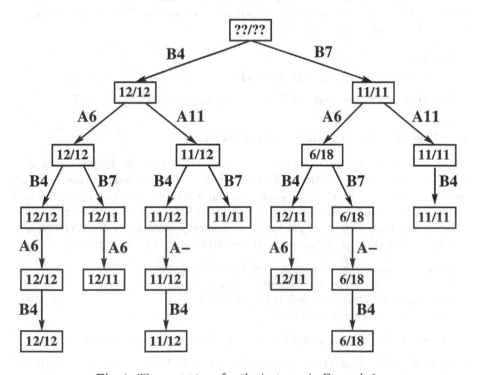

Fig. 1. The game tree for the instance in Example 1.

Example 1. Consider the subset sum game with B-items $4, 4, 4, 7, 7$, with A-items $6, 6, 11$, and with a knapsack of capacity $c = 24$. The first move belongs to the adversary B.

Figure 1 lists the full game tree for the game in Example 1. Every directed arc describes a possible move. The label of the arc states the active player (A or B), followed by the weight of the packed item; a dash indicates that the player passes (as none of his remaining items can be packed). The number pairs in the rectangular boxes indicate the values of the current situation for the two players. The first number gives the highest reachable packed weight for player A. The second number gives the packed weight for player B; it turns out (and the reader may want to verify this) that with the sole exception of the root situation, the hostile adversary and the selfish adversary assign equal values to every situation.

In the root situation, the adversary B has to choose between packing an item of weight 4 (which eventually leads to profits of 12 for both players) and packing an item of weight 7 (which eventually leads to profits of 11 for both players). Hence player A will make profit 11 against the hostile adversary and profit 12 against the selfish adversary.

3   Inapproximability Results

In this section, we derive two inapproximability results for the subset sum game. Both results are derived by means of reductions from the NP-complete PARTITION problem; see Garey and Johnson [3].

Problem: PARTITION

Instance: Positive integers $U \geq 3$ and $u_1, \ldots, u_t$ with $\sum_{i=1}^{t} u_i = 2U$.

Question: Does there exist $J \subseteq \{1, \ldots, t\}$ such that $\sum_{i \in J} u_i = U$?

3.1   The Hostile and the Selfish Adversary

Our first reduction from PARTITION simultaneously settles the inapproximability for the hostile and for the selfish adversary. Suppose for the sake of contradiction that the optimal value α^* in SSG-hostile or in SSG-selfish can be approximated in polynomial time within a factor of r for some fixed real r with $0 < r < 1$. Fix an integer R with $R > 1/r$. We take an arbitrary instance of PARTITION, and construct the following instance of SSG-hostile and SSG-selfish from it.

- For $i = 1, \ldots, tR$ there is an A-item of weight 1.
- For $i = 1, \ldots, t$, there is a B-item of weight $tR\,u_i$.
- The capacity of the knapsack is $c = tRU + t$.
- Player A has the first move.

The proofs of the following two lemmas can be found in the full version.

Lemma 2. *If the* PARTITION *instance has answer YES, then player A packs a total weight of at most t against a hostile or selfish adversary.*   □

Lemma 3. *If the* PARTITION *instance has answer NO, then player A can pack a total weight of* tR. □

According to Lemmas 2 and 3, the approximation algorithm with worst case guarantee $r > 1/R$ can distinguish in polynomial time between YES-instances and NO-instances of the PARTITION problem.

Theorem 1. *Unless* $P = NP$, *the problems SSG-hostile and SSG-selfish do not allow any polynomial time approximation algorithm with finite worst case guarantee.* □

3.2 The Greedy Adversary

We turn to our second reduction, to prove that the game against the greedy adversary has no FPTAS assuming $P \neq NP$. We take an arbitrary instance of PARTITION, and construct the following instance of SSG-greedy from it.

- For $i = 1, \ldots, t$, there is one corresponding dummy A-item of weight $4U$ and one corresponding standard A-item of weight $4U + u_i$. Furthermore, there is one special A-item of weight $3U - 1$.
- For $i = 1, \ldots, t$ there is a B-item of weight $3U$.
- The capacity of the knapsack is $c = (7t + 1)U - 1$.
- Player A has the first move.

The proofs of the following two lemmas can be found in the full version.

Lemma 4. *If the* PARTITION *instance has answer YES, then player A can pack a total weight of* $(4t + 4)U - 1$. □

Lemma 5. *If the* PARTITION *instance has answer NO, then player A can never pack a total weight above* $(4t + 2)U$. □

Now suppose for the sake of contradiction that SSG-greedy allows an FPTAS. For any instance of SSG-greedy and for any $\varepsilon > 0$, the FPTAS takes a running time that is polynomially bounded in the instance size and in $1/\varepsilon$, and then outputs an approximation value α that is at least $1 - \varepsilon$ times the optimal value α^*. We execute this FPTAS with precision $\varepsilon = 1/(4t+4)$ on the instance constructed above. The resulting running time is polynomially bounded in the instance size. If the PARTITION instance has answer YES, then by Lemma 4 we have

$$\alpha \geq (1 - \varepsilon)\alpha^* \geq \frac{4t + 3}{4t + 4}((4t + 4)U - 1) > (4t + 3)U - 1.$$

If the PARTITION instance has answer NO, then by Lemma 5 we have $\alpha \leq \alpha^* \leq (4t + 2)U$. Hence, by analyzing the value α generated by the FPTAS, we would be able to separate in polynomial time the YES-instances from the NO-instances of the PARTITION problem.

Theorem 2. *Unless* $P = NP$, *problem SSG-greedy does not possess an FPTAS.*

4 Complexity of the Game Against the greedy adversary

As problem PARTITION is weakly NP-hard, the reduction in Sect. 3.2 implies that also problem SSG-greedy is weakly NP-hard. Our main goal in this section is to show that SSG-greedy is pseudo-polynomially solvable. Note that for SSG-greedy, a strategy of player A is fully specified by the ordered list of packed A-items. The following lemma will be useful.

Lemma 6 *(Darmann et al. [2])*. *Against the greedy adversary, there exists an optimal strategy for player A that packs the items in non-increasing order of weight.*

Proof. Consider an optimal strategy of player A that (without loss of generality) packs the items in order $a_1, a_2, \ldots, a_s$. If the item weights in this list are not non-increasing, consider the smallest index k with $a_k < a_{k+1}$. If we swap items k and $k+1$ in the list, the reactions of the greedy adversary will not change. We may repeat this swapping step until the weights are non-increasing.   □

Let us assume that the A-items are ordered as $a_1 \geq a_2 \geq \cdots \geq a_m$ and that the B-items are ordered as $b_1 \geq b_2 \geq \cdots \geq b_n$. By Lemma 6 there exists an optimal strategy for player A that packs items in order of non-increasing weight. For $i = 0, \ldots, m+1$, for $j = 0, \ldots, n+1$, and for $W_A = 0, \ldots, c$ and $W_B = 0, \ldots, c$, we introduce a corresponding state $[i, j, W_A, W_B]$ that encodes the following situation that might potentially arise in some run of the game:

> In situation $[i, j, W_A, W_B]$ player A is to move next. In his last move, player A has packed the i'th A-item (where $i = 0$ means that no A-item has been packed yet, and where $i = m+1$ means that no further A-item fits). Similarly, the greedy adversary B has packed the j'th B-item in his last move. The total weight of the packed A-items equals W_A and the total weight of the packed B-items equals W_B.

For two situations $S' = [i', j', W'_A, W'_B]$ and $S = [i, j, W_A, W_B]$, it is easy to check whether S can be reached from S' by means of a move of player A and a following counter-move by player B. In this case we must have $i > i'$, $j > j'$, $W_A = W'_A + a_i$ and $W_B = W'_B + b_j$. Furthermore $W_A + W_B \leq c$, and b_j must be the largest available B-item of size $\leq c - (W'_A + W'_B + a_i)$.

This suggests the following approach. We generate all possible states $[i, j, W_A, W_B]$ where the variables i, j, W_A, W_B may take all possible values from their ranges as listed above. Altogether this yields $O(mnc^2)$ states. Then we compute for every pair S and S' of states, whether S can be reached from S'. Next, we determine (by a standard depth-first-search traversal) all states that are reachable from the starting situation of the game. The maximum value W_A among all reachable states yields the largest possible weight α^* that player A can pack. We summarize the findings of this section.

Theorem 3. *Problem SSG-greedy is weakly NP-complete and solvable in pseudo-polynomial time $O(m^2 n^2 c^4)$.*   □

The literature contains certain routine approaches that automatically translate certain types of pseudo-polynomial algorithms into an FPTAS; see for instance Woeginger [5]. Unfortunately, the above pseudo-polynomial algorithm for SSG-greedy does not fall into that category and does not allow such an automatic translation. Theorem 2 explicitly excludes the existence of an FPTAS (unless $P = NP$). The following section derives the strongest approximation result possible under these circumstances.

5 The Approximation Scheme

In this section, we derive a polynomial time approximation scheme (PTAS) for problem SSG-greedy. The PTAS uses an enumeration approach (for the large A-items) together with a greedy strategy (for the smaller A-items) against the greedy adversary. Throughout this section we will assume without loss of generality that player A has the first move. The following lemma (whose proof can be found in the full version of the paper) will be crucial.

Lemma 7. *Consider an instance of SSG-greedy, and let $a_{\max}$ denote the weight of the largest A-item. Let α^* denote the total weight player A packs under an optimal strategy, and let α^G denote the total weight player A packs when he applies the greedy strategy. Then*

$$\alpha^* - a_{\max} < \alpha^G.$$

We turn to the description of the PTAS. We assume that the A-items are ordered as $a_1 \geq a_2 \geq \cdots \geq a_m$ and the B-items are ordered as $b_1 \geq b_2 \geq \cdots \geq b_n$. We will furthermore assume by Lemma 6 that both players pack their items in order of non-decreasing weight. Let ε with $0 < \varepsilon < 1$ be a small positive real number; for the sake of simplicity we will assume that the reciprocal value $1/\varepsilon$ is integer.

Let S be a subset of A-items with $|S| \leq 1/\varepsilon$, and let a_k denote the lowest weight A-item in S (that is, the item with largest index in S). We run the following two-phase procedure on S by simulating a game of player A against the greedy adversary.

- The first phase consists of the first $|S|$ moves. Player A packs the items in S in non-increasing order of weight.
- The second phase consists of the remaining moves. Player A ignores all A-items with indices up to k, and plays greedily with the A-items with indices at least $k + 1$.

In case we cannot pack all the items of S in the first phase, we call set S infeasible and ignore it. Otherwise, set S is feasible and the procedure yields a certain total packed weight for player A that we denote $\alpha(S)$. The PTAS outputs the maximum value $\alpha(S)$ over all feasible sets S.

Now consider an optimal strategy for player A that packs a sequence of items with weights $a_1^* \geq a_2^* \geq \cdots \geq a_t^*$. If $t \leq 1/\varepsilon$, then the items in the sequence form a feasible set S^*. In this case our PTAS analyzes S^* in the first phase, and eventually outputs the optimal objective value $\alpha^* = \alpha(S^*)$. If $t > 1/\varepsilon$, then define S^* as the set of the $1/\varepsilon$ largest items in the sequence; note that S^* is a feasible set. What does our two-phase procedure do with S^*?

- The first phase simply follows the first $1/\varepsilon$ moves of the optimal strategy: player A packs the first $1/\varepsilon$ items from the optimal sequence, and the greedy adversary picks the largest B-items that fit. As player A altogether packs $1/\varepsilon$ items during the first phase, the smallest item weight in S^* is at most $\varepsilon\alpha^*$.
- The second phase is only played with the A-items that are smaller than the smallest item in S^*, and hence have weight at most $\varepsilon\alpha^*$. The knapsack capacity is the remaining capacity that has not been used in the first phase.

Let α_1^+ denote the total A-weight packed during the first phase, and let α_2^+ denote the total A-weight packed during the second phase. Let α_1^* denote the weight of the items in S^*, and let $\alpha_2^* = \alpha^* - \alpha_1^*$ denote the weight of the remaining items in the optimal packing for A. Clearly $\alpha_1^* = \alpha_1^+$, and Lemma 7 yields that $\alpha_2^* - \varepsilon\alpha^* \leq \alpha_2^+$. These two inequalities imply

$$\alpha(S^*) = \alpha_1^+ + \alpha_2^+ \geq \alpha_1^* + (\alpha_2^* - \varepsilon\alpha^*) = (1 - \varepsilon)\alpha^*.$$

As the PTAS yields a strategy with profit at least $\alpha(S^*)$, this yields the desired approximation guarantee. Up to polynomial factors, the time complexity of the PTAS is proportional to the number of analyzed sets S, which is bounded by $m^{1/\varepsilon}$. This completes our analysis and yields the following theorem.

Theorem 4. *Problem SSG-greedy has a polynomial time approximation scheme.*

6   The PSPACE Completeness Result

In this section we determine the computational complexity of the problem variants with a hostile or selfish adversary. Both problems are PSPACE-complete, and the proof is done by means of a polynomial time reduction from the following variant of the quantified satisfiability problem.

Problem: QUANTIFIED 1-IN-3-SAT

Instance: Two sets $X = \{x_1, \ldots, x_s\}$ and $Y = \{y_1, \ldots, y_s\}$ of Boolean variables. A Boolean formula ϕ over $X \cup Y$ in conjunctive normal form with clauses $C_1, \ldots, C_t$; every (disjunctive) clause C_j consists of exactly three literals.

Question: Assuming that a clause in ϕ is satisfied if and only if it contains *exactly* one true literal, is $\forall x_1 \exists y_1 \forall x_2 \exists y_2 \ldots \forall x_s \exists y_s \ \phi$ true?

As usual, we interpret a quantified formula as a game between a universal player (controlling the universal quantifiers) and an existential player (controlling the existential quantifiers). The goal of the existential player is that in the end formula ϕ evaluates to true, whereas the universal player wants to prevent this.

Lemma 8. QUANTIFIED 1-IN-3-SAT *is PSPACE-complete.*

Proof. There is a reduction from the classic Quantified 3-Satisfiability problem, whose PSPACE-completeness was established by Schaefer [4] reduction is routine and omitted in this extended abstract.   □

Now we turn to the PSPACE-hardness proofs for SSG-hostile and SSG-selfish. We present a single reduction that simultaneously settles both cases. We start from an arbitrary instance of QUANTIFIED 1-IN-3-SAT with $2s$ variables $x_1, \ldots, x_s$ and $y_1, \ldots, y_s$ and with t clauses, and we construct the following subset sum game from it.

All item weights will be specified in decimal representation, and will all have at most $2s + t + 3$ digits. The first $2s$ digits (in the high positions) form the so-called *variable piece*; the $(2i - 1)$th such digit from the left corresponds to the Boolean variable x_i, and the $(2i)$th digit from the left corresponds to the Boolean variable y_i. The three digits after the variable piece form the so-called *middle piece*. The last t digits (in the low positions) form the so-called *clause piece*; the jth such digit from the left in the clause piece corresponds to clause C_j. The item weights are defined as follows.

- For every literal $\ell \in \{x_i, \overline{x_i}\}$ with $1 \leq i \leq s$ there is a corresponding item $B(\ell)$. The decimal representation of its weight has a 1-digit in the position corresponding to x_i in the variable piece. The middle piece digits and all other digits in the variable piece are 0. Furthermore, the clause piece has a digit 1 in the position corresponding to clause C_j if ℓ occurs in C_j; otherwise, the digit is 0.
- Symmetrically, for every literal $\ell \in \{y_i, \overline{y_i}\}$ with $1 \leq i \leq s$ there is a corresponding item $A(\ell)$. The decimal representation of its weight has a 1-digit in the position corresponding to y_i in the variable piece. The middle piece digits and all other digits in the variable piece are 0. Furthermore, the clause piece has a digit 1 in the position corresponding to clause C_j if ℓ occurs in C_j; otherwise, the digit is 0.
- For every variable x_i and every variable y_i, there is a corresponding threat item $T(x_i)$ respectively $T(y_i)$. The decimal representation of the weight has a 1-digit in the position corresponding to that variable in the variable piece. All other digits are 0.
- Furthermore, there are four verification items V_1, V_2, V_3, U with weights $w(V_1)$, $w(V_2)$, $w(V_3)$, $w(U)$. All digits in the variable pieces of these item weights are 0. The middle pieces of $w(V_1), w(V_2), w(V_3), w(U)$ respectively are 111, 100, 010, 011. All other digits in the four decimal representations are 0, with the sole exception of the lowest digit in the weight of V_1, which is set to 1.

We stress the following property of our construction: if we add up the weights of any subset of items in decimal, then there will be no carry overs from one position to the next one. The knapsack capacity c has digits 1 in all $2s + t + 3$ positions. The goal weight α of player A is defined as follows: The decimal representation of α has a digit 1 in the even positions in the variable piece (that is, in the positions

	Owner	Variable piece $x_1 y_1 \ldots x_i y_i \ldots x_s y_s$	Middle piece	Clause piece $C_1 \ldots C_j \ldots C_t$
$B(x_i)$	B	0 0 ... 1 0 ... 0 0	000	0...010...00
$B(\overline{x_i})$	B	0 0 ... 1 0 ... 0 0	000	0...010...00
$A(y_i)$	A	0 0 ... 0 1 ... 0 0	000	0...010...00
$A(\overline{y_i})$	A	0 0 ... 0 1 ... 0 0	000	0...010...00
$T(x_i)$	A	0 0 ... 1 0 ... 0 0	000	0...000...00
$T(y_i)$	B	0 0 ... 0 1 ... 0 0	000	0...000...00
V_1	B	0 0 ... 0 0 ... 0 0	111	0...000...01
V_2	B	0 0 ... 0 0 ... 0 0	100	0...000...00
V_3	B	0 0 ... 0 0 ... 0 0	010	0...000...00
U	A	0 0 ... 0 0 ... 0 0	011	0...000...00
c	−	1 1 ... 1 1 ... 1 1	111	1...111...11
α	−	0 1 0 1 ... 0 1 0 1	011	0...000...00
β	−	0 0 ... 0 0 ... 0 0	111	1...111...11

Fig. 2. Summary of the decimal representation of the item weights.

corresponding to variables y_i), a middle piece 011, and digits 0 in the remaining positions. Finally, we define an auxiliary number β (with digits 1 in middle piece and clause piece, and 0s in the variable piece). The decimal representations of all the introduced numbers are summarized in Fig. 2.

All items $A(y_i)$ and $A(\overline{y_i})$, all threat items $T(x_i)$, and the verification item U belong to player A. All items $B(x_i)$ and $B(\overline{x_i})$, all threat items $T(y_i)$, and the verification items V_1, V_2, V_3 belong to player B. Player B has the first move. The question is to decide whether player A can pack a total weight of at least α (in which case we say that player A wins the game).

Lemma 9. *Assume that player A and the (hostile or selfish) adversary B both apply optimal strategies. Then in round $2i − 1$ (with $1 \leq i \leq s$), the adversary B either packs item $B(x_i)$ or item $B(\overline{x_i})$. In round $2i$ (with $1 \leq i \leq s$), player A either packs item $A(y_i)$ or item $A(\overline{y_i})$.*

Proof. We assume inductively that up to round $2k$ (with $0 \leq k \leq s − 1$), the statement holds and both players have followed the described moves. Let d denote the current contents of the knapsack at th beginning of round $2k$. Then the decimal representation of d has digits 1 in the first $2k$ positions of the variable piece. This implies that none of the remaining items $A(y_i)$, $A(\overline{y_i})$, $T(y_i)$, $B(x_i)$, $B(\overline{x_i})$, $T(x_i)$ with $1 \leq i \leq k$ can be picked anymore: their weight is so large that they do not fit into the remaining empty part of the knapsack. All other items (the four verification items and the items corresponding to variables x_i or y_i with $i \geq k + 1$) are small enough to fit.

First we consider a hostile adversary B. If B neither packs item $B(x_k)$ nor $B(\overline{x_k})$ in round $2k+1$, then player A can react in the next round by packing the threat item $T(x_k)$. This immediately brings A's weight above the threshold α, so that A wins the game. Hence the hostile B must pack $B(x_k)$ or $B(\overline{x_k})$ in round $2k+1$. If A then does neither pack $A(y_1)$ nor $A(\overline{y_1})$ in round $2k+2$, the adversary will pack the threat item $T(y_k)$ in his next move. In this case A will never be able to pack $A(y_k)$ or $A(\overline{y_k})$ or $T(x_k)$ in the future, and his weight will permanently stay below α. Hence the claimed statement also holds up to round $2k+2$.   □

Next we consider a selfish adversary B. If B neither packs item $B(x_k)$ nor $B(\overline{x_k})$ in round $2k+1$, then player A can react in the next round by packing the threat item $T(x_k)$. Exactly as in the hostile case, the weight of player A then exceeds the threshold α, and A wins the game. On the other hand, the selfish player B will never be able to compensate for the loss of $B(x_k)$ and $B(\overline{x_k})$. All in all, this means that the selfish B must pack item $B(x_k)$ or $B(\overline{x_k})$ in round $2k+1$. Finally, we may argue exactly as in the hostile case that player A then must pack item $A(y_1)$ or $A(\overline{y_1})$ in round $2k+2$, Hence, also in the selfish case the claimed statement holds up to round $2k+2$.

Lemma 10. *Let c' denote the remaining knapsack capacity at the end of round $2s$. Let $w(A)$ and $w(B)$ denote the weight packed by player A and the (hostile or selfish) adversary B in rounds $2s+1, 2s+2, 2s+3$.*

(S1)   If $c' \geq w(V_1)$ then $w(A) = 0$ and $w(B) = w(V_1)$.
(S2)   If $c' = w(V_1)-1$ then $w(A) = w(U)$ and $w(B) = w(V_2)$ or $w(B) = w(V_3)$.
(S3)   If $c' \leq w(V_1) - 2$ then $w(A) = 0$ and $w(B) = w(V_2) + w(V_3)$.

The proof of this lemma can be found in the full version of the paper.

By Lemmas 9 and 10, player A will win the game if situation (S2) occurs after round $2s$, and he will lose the game under situations (S1) and (S3). If player B is hostile, then his goal will be to avoid situation (S2). If player B is selfish, then he will also avoid situation (S2), as both (S1) and (S3) yield a better profit for him. Hence in either case, the objective of player A is to reach situation (S2) and the objective of the adversary is exactly to avoid this situation (S2).

Now let us finally connect our analysis to the considered instance of QUAN-TIFIED 1-IN-3-SAT. By Lemma 9, after the first $2s$ rounds the knapsack contains exactly one of items $B(x_i)$ and $B(\overline{x_i})$ and exactly one of items $A(y_i)$ and $A(\overline{y_i})$ for $1 \leq i \leq s$. We construct the following truth setting T^* from this packing: If the knapsack contains $B(x_i)$ (respectively $B(\overline{x_i})$) then variable x_i is set to true (respectively false), and if the knapsack contains $A(y_i)$ (respectively $A(\overline{y_i})$) then variable y_i is set to true (respectively false). The remaining knapsack capacity c' at the end of round $2s$ equals $w(V_1) - 1$, if and only if under truth setting T^* every clause in formula ϕ contains exactly one true literal. In other words, for player A reaching situation (S2) in the subset sum game is equivalent to reaching a satisfying truth setting T^* for the QUANTIFIED 1-IN-3-SAT instance.

This yields a natural bijection between the moves of the players in the QUAN-TIFIED 1-IN-3-SAT instance and the moves of the players in the constructed instance of SSG-hostile and SSG-selfish. This implies PSPACE-hardness of these games. Furthermore, these games can be fully analyzed with polynomial space by a depth-first-search traversal of the underlying game tree; this yields containment in PSPACE. We summarize our findings in the following theorem.

Theorem 5. *The games SSG-hostile and SSG-selfish are PSPACE-complete.*

7   Final Remarks

We have analyzed the three variants SSG-hostile, SSG-selfish, and SSG-greedy of the subset sum game. Our analysis fully describes the complexity and the approximability landscape of these three games.

The two games SSG-hostile and SSG-selfish look and behave very similarly. In fact, a single proof (for Theorem 5) suffices to settle the complexity status of both problems, and a single proof (for Theorem 1) settles their approximability status. We do not have a good understanding of the actual difference between these two games. In particular, the following question (motivated by the game instance in Example 1) remains open: What is the computational complexity of deciding for a given instance of the subset sum game, whether player A can enforce a strictly larger profit against a selfish adversary than against a hostile adversary? We suspect this problem to be computationally intractable.

References

1. Caprara, A., Carvalho, M., Lodi, A., Woeginger, G.J.: A complexity and approximability study of the bilevel knapsack problem. SIAM J. Optim. **24**, 823–838 (2014)
2. Darmann, A., Nicosia, G., Pferschy, U., Schauer, J.: The subset sum game. Euro. J. Oper. Res. **233**, 539–549 (2014)
3. Garey, M.R., Johnson, D.S.: Computers and Intractability: A Guide to the Theory of NP-Completeness. Freeman, San Francisco (1979)
4. Schaefer, T.J.: On the complexity of some two-person perfect-information games. J. Comput. Syst. Sci. **16**, 185–225 (1978)
5. Woeginger, G.J.: When does a dynamic programming formulation guarantee the existence of a fully polynomial time approximation scheme (FPTAS)? INFORMS J. Comput. **12**, 57–75 (2000)

Higher-Order Decision Theory

Jules Hedges[1], Paulo Oliva[2], Evguenia Shprits[3], Viktor Winschel[4],
and Philipp Zahn[5(✉)]

[1] Department of Computer Science, University of Oxford, Oxford, UK
julian.hedges@cs.ox.ac.uk
[2] Department of Electronic Engineering and Computer Science, Queen Mary
University of London, London, UK
p.oliva@qmul.ac.uk
[3] Department of Economics, University of Mannheim, Mannheim, Germany
eugeniaw@rumms.uni-mannheim.de
[4] OICOS GmbH, Mannheim, Germany
viktor.winschel@gmail.com
[5] Department of Economics, University of St. Gallen, St. Gallen, Switzerland
philipp.zahn@unsig.ch

Abstract. This paper investigates a surprising relationship between
decision theory and proof theory. Using constructions originating in proof
theory based on higher-order functions, so called *quantifiers* and *selection
functions*, we show that these functionals model choice behavior of individual agents. Our framework is expressive, it captures classical theories
such as utility functions and preference relations but it can also be used
to faithfully model abstract goals such as coordination. It is directly
implementable in functional programming languages. Lastly, modeling
an agent with selection functions and quantifiers is modular and thereby
allows to seamlessly combine agents bridging decision theory and game
theory.

Keywords: Decision theory · Utility functions · Preferences · Higher-
order functions · Quantifiers

1 Introduction

A *higher-order function* (or *functional*) is a function whose domain is itself a
set of functions. In this paper we investigate a surprising and deep connection between decision theory on one side and a higher-order construction that
originated in proof theory [3,4] on the other side: We use a particular class of
higher-order functions as a way of describing the choice behavior of individual
agents.

Assume X is the overall set of alternatives, and R a set of observable outcomes
or measures. For instance, X could be the set of all books, and $R = \mathbb{R}^+$ the non-
negative reals representing prices. We then see functions of type $X \to R$ as
describing the agent's *decision context*, e.g. $X \to \mathbb{R}^+$ as the mapping from books
to prices.

© Springer International Publishing AG 2017
J. Rothe (Ed.): ADT 2017, LNAI 10576, pp. 241–254, 2017.
DOI: 10.1007/978-3-319-67504-6_17

The core idea we explore here is to model agents' decision goals as higher-order functions of type $(X \to R) \to R$. Functionals of this type have been called *quantifiers* [14], since the standard $\exists$ and $\forall$ logical quantifiers are particular cases of these when $R = \mathbb{B}$ is the type of booleans. Going back to the example of books and prices, an agent who prefers the cheapest book will be modeled by the quantifier

$$\min : (X \to \mathbb{R}^+) \to \mathbb{R}^+$$

saying that given any catalogue of prices $p : X \to \mathbb{R}^+$ the agent will choose to pay the cheapest price on the catalogue $\min(p)$.

Therefore, quantifiers describe the outcome (e.g. price paid for the choice of book) an agent considers to be good in any given decision context. A corresponding notion is that of a *selection function*, i.e. a higher-order function of type $(X \to R) \to X$ which calculates a concrete choice that meets the desired goal. In the example above, the corresponding selection function would pick one of the cheapest books according to the prices given in $p : X \to \mathbb{R}^+$.

Since the $\max : (X \to \mathbb{R}) \to \mathbb{R}$ operator is also a quantifier and the corresponding $\mathrm{argmax} : (X \to \mathbb{R}) \to X$ operator is a selection function, we have that the standard approach of modeling preferences via utility functions or preference relations are instances of our modeling framework. But, quantifiers and selection functions can capture alternative decision criteria as well as decision heuristics. We demonstrate this through several simple examples.

As we see it, there are three crucial advantages of adopting a higher-order modeling of choice behavior: *expressivity*, *modularity*, and *computability*:

- *Expressivity*: We show that describing agent's choices through higher-order functionals captures existing standard approaches such as utility maximization. Moreover, we can directly model at the level of higher-order functionals and thereby faithfully describe the agent's behavior. As an example, we introduce a fixed-point operator which captures an agent's goal to coordinate with other agents.
- *Modularity*: By taking into account the decision context $p : X \to R$ when describing an agent's goal, we can fully describe the agent "locally", without having to refer to "global" utility functions or similar constructs. This allows us to seamlessly combine individual agents into strategic games, bridging the gap between decision theory and game theory.
- *Computability*: Finally, such higher-order function, as abstract as they might appear, can be directly expressed and coded into functional programming languages such as Haskell and Ocaml. In fact, most of the theory we have developed here and in [8] has been implemented and tested in Haskell. This Haskell code has also served to guide us in discovering new forms of game equilibria which are available in this setting of higher-order decisions and games.

The plan for the paper is as follows. We give a brief introduction to higher-order functions in the next section. We then show how quantifiers can be used to model behavior directly at the level of these higher-order functions. In Sect. 3, we also show how to instantiate preference relations and utility maximization as special cases. In Sects. 4 and 4.3 we introduce a series of non-optimizing examples and show how they can be represented by higher-order functions. These include coordination goals as fixpoint operators. We conclude in Sect. 5.

Elsewhere we show that selection functions are very powerful building blocks for game theory and implementations thereof [8]. Moreover, in [3–5,7,10] selection functions are a building block for a compositional approach to game theory. These papers rely on simple instances of selection functions. All instances of selection functions and quantifiers introduced in this paper, can be directly integrated into these game theoretic models.

2   Agents as Quantifiers

A *higher-order function* (or *functional*) is a function whose domain is itself a set of functions. Given sets X and Y we denote by $X \to Y$ the set of all functions with domain X and codomain Y. A higher-order function is therefore a function $f : (X \to Y) \to Z$ where X, Y and Z are sets.

Here are some well known examples of higher-order functions. In case of the maximization of a utility function $u \colon X \to \mathbb{R}$

$$\max_{x \in X} u(x)$$

the max operator takes the utility function $u \colon X \to \mathbb{R}$ as its input and returns a real number $\max_{x \in X} u(x)$ as the output. Therefore, the max operator has type

$$\max \colon (X \to \mathbb{R}) \to \mathbb{R}$$

In a similar vein, the argmax operator is also a higher-order function of a particular type:

$$\mathrm{argmax} \colon (X \to \mathbb{R}) \to \mathcal{P}(X)$$

where $\mathcal{P}(X)$ is the set of subsets of X. For a given function $u \colon X \to \mathbb{R}$ we have that $\mathrm{argmax}(u)$ is the set of points where u attains its maximum value.

2.1   Agent Context

We want to model an agent $\mathcal{A}$ in a choice *situation* or *context* and formulate his motivations and his choices. We shall model such contexts as mappings $X \to R$ that encode for choices in X their effects on the outcomes in R.

Definition 1 (Agent context). *We call any function $p : X \to R$ a possible context for the agent $\mathcal{A}$ who is choosing a move from a set X, having in sight a final outcome in a set R,*

For instance, X could be the set of available flights between two cities, and $R = \mathbb{R}^+$ could be the set of positive real numbers that represent prices. An agent who is interested in choosing a flight having in mind only the cost of the flight will consider the price list $X \to \mathbb{R}^+$ as a sufficient context for his decision. If, however, the number of stops (or changes) is an important factor in the decision of the agent, we could take $R = \mathbb{R}^+ \times \mathbb{N}$ and the agent's context would then be $X \to \mathbb{R}^+ \times \mathbb{N}$.

2.2   Quantifiers

Suppose the agent $\mathcal{A}$ has to make a decision in the context $p\colon X \to R$. The agent will consider some of the possible outcomes to be *good* (or *acceptable*), and others to be bad (or *unacceptable*). Such choices define a higher-order function of the following type:

Definition 2. (Quantifier, [3,4]). *Mappings*

$$\varphi : (X \to R) \to \mathcal{P}(R)$$

from contexts $p : X \to R$ to sets of outcomes $\varphi(p) \subseteq R$ are called quantifiers.

The terminology comes from the observation that the usual existential $\exists$ and universal $\forall$ quantifiers of logic can be seen as operations of type $(X \to \mathbb{B}) \to \mathbb{B}$, where $\mathbb{B}$ is the type of booleans. Mostowski [14] has called arbitrary functionals of type $(X \to \mathbb{B}) \to \mathbb{B}$ *generalized quantifiers*. This was generalized further in [3] to the type given here.

We model agents $\mathcal{A}$ as quantifiers $\varphi_{\mathcal{A}}$ and take $\varphi_{\mathcal{A}}(p)$ as the set of outcomes that the agent $\mathcal{A}$ considers preferable in each context $p\colon X \to R$. For instance, consider again the example where X is the set of flights, and $R = \mathbb{R}^+ \times \mathbb{N}$ indicates prices together with the number of stops. An agent that wishes to minimize the cost of the flight, but does not wish to make more than two stops will be modeled by the quantifier:

$$\varphi(p) = \{p(x) \, : \, x \in X, p_0(x) = \min p_0, p_1(x) \le 2\}$$

where $p_0\colon X \to \mathbb{R}^+$ and $p_1\colon X \to \mathbb{N}$ are the two projections of $p\colon X \to \mathbb{R}^+ \times \mathbb{N}$.

Our main objective in this paper is to convince the reader that this is a general, modular, and highly flexible way of describing an agent's goal or objective.

The classical example of a quantifier is utility maximization. Suppose an agent has a utility function $u'\colon R \to \mathbb{R}$ mapping outcomes into utilities. Composing the context $p\colon X \to R$ and $u'\colon R \to \mathbb{R}$ we get a new context that maps actions directly into utility $u\colon X \to \mathbb{R}$. Given this new context, the good outcomes for the player are precisely those for which his utility function is maximal. This quantifier is given by

$$\max(u) = \{r \in \mathrm{Im}(u) \mid r \ge u(x') \text{ for all } x' \in X\}$$

where $\mathrm{Im}(u)$ denotes the image of the utility function $u\colon X \to R$.

2.3 Context-Dependence

In general, we are going to allow the set of outcomes that the agent considers good to be arbitrary. It is reasonable, however, to assume that for each context $p\colon X \to R$ we have $\varphi(p) \neq \varnothing$. This is to say that in any given context the agent must have a preferred outcome (even if this would be the least bad one). We will call such quantifiers *total*. Another more interesting class of quantifiers consists of those we call *context-independent*:

Definition 3 (Context-independence). *A quantifier* $\varphi\colon (X \to R) \to \mathcal{P}(R)$ *is said to be* context-independent *if the value* $\varphi(p)$ *only dependents on* $\mathrm{Im}(p)$, *i.e.*
$$\mathrm{Im}(p) = \mathrm{Im}(p') \implies \varphi(p) = \varphi(p').$$

Dually, a quantifier φ *will be called* context-dependent *if for some contexts* p *and* p', *with* $\mathrm{Im}(p) = \mathrm{Im}(p')$, *the sets of preferred outcomes* $\varphi(p)$ *and* $\varphi(p')$ *are different.*

Intuitively, a context-dependent quantifier will select good outcomes not only based on which outcomes are possible, but will also take into account how the outcomes are actually achieved. It is easy to see that the quantifier max is context-independent, since it can be written as a function of $\mathrm{Im}(p)$ only.

Our prototypical example of a context-dependent quantifier is the fixpoint operator
$$\mathrm{fix} : (X \to X) \to \mathcal{P}(X)$$

Recall that a fixpoint of a function $f : X \to X$ is a point $x \in X$ satisfying $f(x) = x$. If the set of moves is equal to the set of outcomes then there is a quantifier whose good outcomes are precisely the fixpoints of the context. If the context has no fixpoints we shall assume that the agent will be equally satisfied with any outcome. Such a quantifier is given by

$$\mathrm{fix}(p) = \begin{cases} \{x \in X \mid p(x) = x\} & \text{if nonempty} \\ X & \text{otherwise.} \end{cases}$$

Clearly $\mathrm{fix}(\cdot)$ is context-dependent, since we could have different contexts $p, p'\colon X \to X$ having the same image set $\mathrm{Im}(p) = \mathrm{Im}(p')$ but with p and p' having different sets of fixpoints. For example, if we take $p, p' : \mathbb{R} \to \mathbb{R}$ to be given by $p(x) = x$ and $p'(x) = -x$ then $\mathrm{Im}(p) = \mathrm{Im}(p') = \mathbb{R}$, but $\mathrm{fix}(p) = \mathbb{R}$ and $\mathrm{fix}(p') = \{0\}$. We will discuss the relevance of this quantifier in Sect. 4.3.

2.4 Attainability

Another important property of quantifiers that we shall consider is that of *attainability*:

Definition 4 (Attainability). *A quantifier* $\varphi : (X \to R) \to \mathcal{P}(R)$ *is called* attainable *if, for every context* $p : X \to R$, *for some* $r \in \varphi(p)$ *there exists an* x *such that* $p(x) = r$. *(In particular, attainable quantifiers are total.)*

In other words, an agent modeled by an attainable quantifier will select at least one preferred outcome r that is actually *achievable* by some move x. An equivalent definition is that $\varphi \colon (X \to R) \to \mathcal{P}(R)$ is attainable if and only if

$$\varphi(p) \cap \operatorname{Im}(p) \neq \emptyset.$$

Remark 1. We could also define a *strong attainability* notion whereby all $r \in \varphi(p)$ need to be achievable by some $x \in X$, i.e.

$$\varphi(p) \subseteq \operatorname{Im}(p).$$

For our purposes the weaker notion of Definition 4 has been sufficient and reasonably well-behaved.

Attainable quantifiers bring out the relevance of moves in the decision making process. Sometimes an agent might actually wish to spell out the preferred *moves* instead of the preferred *outcomes*. This leads to the definition of another class of higher-order functions:

Definition 5 (Selection functions). *A selection function is any function of type*

$$\varepsilon : (X \to R) \to \mathcal{P}(X)$$

Remark 2. In the computer science literature, where quantifiers and selection functions have been considered previously, the initial focus was on single-valued ones [3]. Multi-valued quantifiers were first considered in [4]. As the definition above shows, we also make use of multi-valued selection functions, as these are extremely important in our examples and also in game theoretic approaches that take selection functions as a building block [5,7,10].

Similarly to quantifiers, the canonical example of a selection function is maximizing $\mathbb{R}$, defined by

$$\operatorname{argmax}(p) = \{x \in X \mid p(x) \geq p(x') \text{ for all } x' \in X\}$$

The argmax selection function is naturally multi-valued: a function may attain its maximum value at several different points.

Proposition 1. *A quantifier* $\varphi \colon (X \to R) \to \mathcal{P}(R)$ *is attainable if and only if there exists a total selection function* $\varepsilon \colon (X \to R) \to \mathcal{P}(X)$ *such that, for all* $p \colon X \to R$,

$$x \in \varepsilon(p) \implies p(x) \in \varphi(p)$$

If such a relationship between a quantifier $\varphi : (X \to R) \to \mathcal{P}(R)$ and a selection function $\varepsilon : (X \to R) \to \mathcal{P}(X)$ holds then we shall say that ε *attains* φ. The attainability relation holds between the quantifier max and the selection function argmax. The fixpoint quantifier is also a selection function, and it attains itself since

$$x \in \operatorname{fix}(p) \implies p(x) \in \operatorname{fix}(p).$$

3 Preference Relations and Utility Maximization

In this section we relate the concepts of quantifiers and selection functions to standard concepts of classical (economic) decision theory: utility functions and preference relations. In particular, we show that both correspond to context-independent quantifiers that have the same structure. We now want to characterize the relationship between preference relations and context-independent quantifiers.

Suppose R is the set of possible outcomes, and an agent has a partial order relation $\succeq$ on R as preferences, so that $x \succeq y$ means that the agent prefers the outcome x to y. These partial orders lead to choice functions $f : \mathcal{P}(R) \to \mathcal{P}(R)$ where $f(S)$ are the maximal elements in the set of possible outcomes S with respect to the order $\succeq$. Note that these f satisfy $f(S) \subseteq S$, and $f(S) \neq \varnothing$ for non-empty S.

Every such f can be turned into a quantifier φ in a generic way, using the fact that the image operator is a higher-order function $\mathrm{Im} : (X \to R) \to \mathcal{P}(R)$:

$$(X \to R) \xrightarrow{\mathrm{Im}} \mathcal{P}(R) \xrightarrow{f} \mathcal{P}(R)$$

so that $f \circ \mathrm{Im} : (X \to R) \to \mathcal{P}(R)$ are quantifiers.

Proposition 2. *Assume $|X| \geq |R|$, i.e. the number of choices is bigger than the number of possible outcomes. Then a quantifier $\varphi : (X \to R) \to \mathcal{P}(R)$ is context-independent if and only if $\varphi = f \circ \mathrm{Im}$, for some choice function $f : \mathcal{P}(R) \to \mathcal{P}(R)$.*

Proof. If $\varphi = f \circ \mathrm{Im}$ then φ is context-independent. For the other direction, note that since $|X| \geq |R|$ we have for any subset $S \subseteq R$ a map $u_S : X \to R$ such that $\mathrm{Im}(u_S) = S$. Assume φ is context-independent and let us define $f(S) = \varphi(u_S)$. Clearly,

$$\varphi(p) = \varphi(u_{\mathrm{Im}(p)}) = f(\mathrm{Im}(p))$$

where the first step uses that φ is context-independent and that $\mathrm{Im}(p) = \mathrm{Im}(u_{\mathrm{Im}(p)})$ by the assumption on the family of maps u_S, while the second steps simply uses the definition of f.

Agents who are defined by context-independent quantifiers are choosing the set of good outcomes simply by ranking the set of outcomes that can be achieved in a given context but are ignoring all the information about how each of the outcomes arise from particular choices of moves.

For instance, we might have a set of actions that will lead us to earn some large sums of money. Some of these, however, might be illicit. A classical agent who cares only about the direct consequences of his decision and is defined in a context-independent way would choose the outcome that gives himself the maximum sum of money, regardless of the nature of action. If however the agent also cares about the actions themselves and their indirect consequences, he might not consider the largest amount of money as preferable.

The following proposition guarantees the attainability of context independent quantifiers arising from preference relations:

Proposition 3. *Whenever $f_{\succeq}$ is a choice function arising from a partial order $\succeq$, then the context-independent quantifier $\varphi = f_{\succeq} \circ \mathrm{Im}$ is attainable.*

Proof. By the definition of φ we have that if $r \in \varphi(p)$ then r is a maximal element in $\mathrm{Im}(p)$. Hence we must have an $x \in X$ be such that $p(x) = r$. $\qquad \blacksquare$

Another example of a context-independent quantifier is the maximization of a utility function. A utility function can be characterized as the context $p \colon X \to \mathbb{R}$ that attaches a real number to each element of the set of choices X with the quantifier defined as

$$\varphi(p) = \max_{x \in X} p(x).$$

Moreover, this quantifier is attained by the selection function

$$\varepsilon(p) = \arg\max p$$

Note the types $\varphi \colon (X \to \mathbb{R}) \to \mathcal{P}(\mathbb{R})$ and $\varepsilon \colon (X \to \mathbb{R}) \to \mathcal{P}(X)$ respectively. And indeed we have that

$$x \in \varepsilon(p) \implies p(x) \in \varphi(p).$$

Thus, max and arg max operators, are the prototypical examples of a context-independent quantifier and a selection function attaining it.

4   Alternatives to Optimization

We have seen how the higher-order notion of a context-independent quantifiers is able to model choices based on rational preferences (or equivalently on utility maximization). For simplicity, we consider decisions under certainty but it is straightforward to consider uncertainty and consider expected utility theory. Other decision criteria such as regret minimization [13,15] or maximin choices [16] are also captured by our framework.

Instead of going in this direction, in the following we consider another direction: Utility functions as well as preference relations are intimately linked to the assumption that the agent fully optimizes. The behavioral economic literature as well as the psychological literature have documented deviations from optimizing behavior [2,11]. Quantifiers provide a direct way to model such deviations. Here we give a few examples by allowing for a different structure on the set of outcomes R or by allowing for a different mapping $f \colon \mathcal{P}(R) \to \mathcal{P}(R)$, or by relaxing both.

4.1   Context-Independent Agents

Example 1 (Averaging Agent). Consider an agent who prefers the outcome to be as close as possible to the average of all achievable outcomes. Given a decision context $p: X \to \mathbb{R}$, the average amongst the possible outcomes can be calculated as

$$A_p = \frac{\Sigma_{r \in \mathrm{Im}(p)} r}{|\mathrm{Im}(p)|}$$

Therefore, such agent can be directly modeled via the averaging quantifier $\varphi^A: (X \to \mathbb{R}) \to \mathcal{P}(\mathbb{R})$ as

$$\varphi^A(p) = \{r \in \mathrm{Im}(p) \mid |r - A_p| \text{ is minimal}\}$$

The next example represents the second best decision problem discussed in [12].

Example 2 (Second-best Agent). Consider a simple heuristic of a person ordering wine in a restaurant whereby he always chooses the second most-expensive wine. In terms of quantifiers, let X be the set of wines available in a restaurant, and $p : X \to \mathbb{R}$ the price attached to each wine x_i ($i = 1, ..., N$) on the menu, so that $r_i = p(x_i)$ denotes the price of wine x_i. Given a maximal strict chain $r_n > r_{n-1} > ... > r_1$ in $\mathbb{R}$, let us call r_{n-1} a sub-maximal element. The goal of the agent can be described by the quantifier

$$\varphi_>(p^{X \to \mathbb{R}}) = \{\text{sub-maximal elements with respect to} > \text{within} \mathrm{Im}(p)\}.$$

A crucial point of the above examples is the additional degree of freedom of modeling as it is possible to vary the choice operator itself and not being automatically restricted to the max operator.

4.2 Context-Dependent Agents

So far, we have focused only on context-independent quantifiers. Yet, we can do more. As we have discussed in Sect. 2, we can allow for quantifiers that do not only take the image of p as input but the complete function. Again, we consider several examples.

Example 3 (Ideal-move Agent [6]). Let $r > 0$ be a fixed real number. For a point $v \in \mathbb{R}^n$ we define the closed ball with centre v and radius r by

$$B(v; r) = \{w \in \mathbb{R}^n \mid d(v, w) \le r\}$$

where d is the Euclidean distance. Let the set of choices X have a distinguished element $x_0 \in X$. Define the quantifier $\varphi : (X \to \mathbb{R}^n) \to \mathcal{P}(\mathbb{R}^n)$ by

$$\varphi(p) = B(p(x_0); r)$$

This quantifier is attained by the constant selection function $\varepsilon(p) = \{x_0\}$.

The last example illustrates Simon's satisficing behavior. The value $r > 0$ can be considered as a satisficing threshold around outcomes that are close to the outcome of an ideal point. Such an agent is equally satisfied with all outcomes which are close enough to the outcome of the ideal choice.

Example 4 (Averaging – revised). Consider again an agent who prefers the outcome to be as close as possible to the average outcome. But this time we assume that he takes into account the number of possible ways an outcome may be attained. Given a decision context $p: X \to \mathbb{R}$, the weighted average in this case can be calculated as

$$A_p = \frac{\Sigma_{x \in X} p(x)}{|X|}$$

Such an agent can be modeled via the weighted averaging quantifier $\varphi: (X \to \mathbb{R}) \to \mathcal{P}(\mathbb{R})$ as

$$\varphi(p) = \{r \in \mathrm{Im}(p) \mid |r - A_p| \text{ is minimal}\}$$

It easy to check that this is a *context-dependent* quantifier.

Now, consider the example where the set of actions allows an agent to earn some money but some actions are illicit and hence not considered to be a permissible behavior. If we care about the actions themselves, we might not necessarily consider the largest sum of money as preferable.

Example 5 (Honest Agents). Consider an agent with a set of possible actions X leading to monetary outcomes $M \subseteq \mathbb{R}$. Assume some of these actions $I \subset X$ are illegal or dishonest. Hence, the set $L = X \backslash I$ consists of the legal, or honest, actions. In the first instance consider an honest agent who maximizes over the outcomes which follows from honest actions. Such a honest agent can be modeled by the quantifier:

$$\varphi^h(p) = \{r \mid r \text{ a maximal element in the set } p(L)\}$$

where $p(L)$ is the image of L under p. Consider, however, a more complicated case where the agent is prepared to consider dishonest or illegal actions when the reward associated with some of these actions is above a threshold T. This subtler preference can be directly modeled as

$$\varphi^d(p) = \begin{cases} \{r \mid r \text{ is maximal in } \mathrm{Im}(p)\} & \text{if } \max_{x \in I} p(x) > T \\ \varphi^h(p) & \text{otherwise} \end{cases}$$

so that the dishonest agent will behave as the honest one if the maximal reward for a dishonest action is low, but he will consider any action to be acceptable if the gain from a dishonest or illegal action is high enough.

In the next example we introduce an extreme case of an agent who decides on preferred outcomes solely based on the set of moves that lead to that outcome.

Example 6 (Safe Agents). Given a decision context $p: X \to R$ and an outcome $r \in \mathrm{Im}(p)$, we can calculate the number of different ways r can be attained by

$$n_r^p = |\{x \in X \mid p(x) = r\}| \,.$$

We say that an outcome r is most unavoidable if n_r^p is maximal over the set of possible outcomes $\mathrm{Im}(p)$. We say that an agent is safe if he prefers most unavoidable outcomes. Such agents are modeled by the quantifier

$$\phi(p) = \{r \in \mathrm{Im}(p) \mid n_r^p \text{ maximal}\}$$

In order to illustrate this quantifier, suppose there are three beaches, and the agent is indifferent between them. The first can be reached by one highway, the second by two highways and the third by three highways. The agent has to choose which highway to take, and the outcome is the beach that the agent goes to. The safe agent decides to visit the beach which can be reached by the most different routes, which is the third, in order to avoid the risk of being stuck in a traffic jam.

4.3   Fixed Points as Coordination

We now discuss the specific situations where the set of actions X and outcomes R are the same $X = R$. In this case elements of the type

$$(X \to X) \to \mathcal{P}(X)$$

can be either viewed as quantifiers or selection functions. Agents of this type are common in voting contests:

Example 7 (Voting Agent). Consider three judges $J = \{J_1, J_2, J_3\}$ voting for two contestants $X = \{A, B\}$. The winner is determined by the simple majority rule of type $\mathrm{maj} : X \times X \times X \to X$. The set X denotes both the set of choices and the set of possible outcomes of the contest. We first assume that the judges rank the contestants according to a preference ordering. For example, suppose judges 1 and 2 prefer A and judge 3 prefers B. Consider the decision problem of the first judge. He has an ordering on the set X, namely $A \succeq_1 B$, and his goal is to maximize the outcome with respect to this ordering. Hence, he is modeled via the quantifier:

$$\varphi_1^J(p) = \max_{x \in (X, \succeq_1)} p(x)$$

The set X is equipped with a partial order and the max operator $(X \to X) \to \mathcal{P}(X)$ describes the agent.

Another very interesting example of an agent with an important economic interpretation, is the fixpoint operator, that we have already mentioned in Sect. 2.3.

Example 8 (Keynesian Agent). Consider the last example but now assume that judge 1 has different preferences: he prefers to support the winner of the contest. He is only interested in voting for the winner of the contest and he has no preferences for the contestants per se. The selection function of such a Keynesian agent can be described by a fixpoint operator as

$$\varepsilon_1^K(p) = \mathrm{fix}(p) = \{x \in X \mid p(x) = x\}.$$

Interestingly, such an agent is best described by a selection function, rather than via the corresponding quantifier

$$\varphi_1^K(p) = \{p(x) \mid p(x) = x\}.$$

We note, it is perfectly possible to model such a Keynesian agent via standard utility functions, attaching say utility 1 to good outcomes and 0 to the bad ones, so that the judges maximize over the set of monetary payoffs. In this process of attaching utilities to the decision, however, one has to compute the outcome of the votes, then to check for the second and the third judges whether their vote is in line with the outcome, and finally to attach the utilities. If, instead, we use the fixpoint operator to represent the agent's goals, no such calculation is necessary.

As briefly discussed above, most functions $p\colon X \to X$ do not have a fixpoint and the fixpoint operator will often give the empty set. For the purposes of modeling a particular situation we might want to totalize the fixpoint operator in different ways and describe what an agent might do in case that no fixpoint exists. The fixpoint goals are far more interesting when we consider a game with several agents with different concerns, for instance some with usual preferences and some with fixpoint goals. We analyze such a game in [8].[1]

Let us conclude with another example of a reflexive agent.

Example 9 (Coordinating Agent). Consider two players, $\{0, 1\}$, who want to coordinate, for instance, about the restaurant where to meet for lunch. The set of actions $X_0 = X_1 = \{A, B\}$ denotes the different restaurants at choice. The set of outcomes $R = X_0 \times X_1$ denotes the two restaurants where the agents might end up. The fact that these two agents want to meet in the same restaurant can be directly described by another sort of fixpoint operator:

$$\varepsilon_i(p) = \{x \in X_i \mid x = (\pi_{1-i} \circ p)(x)\}$$

where $\pi_i\colon X_0 \times X_1 \to X_i$ are the projection functions. The preferred move of agent i is the one which leads him to the same place as the other agent $1 - i$.

These two examples above show that the overall goal of the Keynesian and the coordinating agent are similar, and can be captured by some variants of the same fixpoint operators. Even though it is possible to use utility functions in order to model these concerns in the particular examples, it is not obvious that this commonality can be made explicit when modeling with utility functions. In our more abstract formalization via higher-order functions, it is possible to detect patterns across problems that are hard to find when one only looks at the compiled level of utility maximization.

5   Conclusions

We introduced quantifiers and selection functions to model agents' choices. We illustrated that classical and standard approaches such as utility functions can

---

[1] See also the working paper version [9] for more details.

be instantiated in our framework as examples. Alternatives to optimization can be similarly captured. Lastly, one can directly model at the level of these higher-order functions. Overall, higher-order functions provide a possibility to abstraction of lower-level instantiations and by that realize commonality between seemingly different approaches.

In this paper, we limit ourselves to show that quantifiers and selection functions do capture different deterministic approaches. We already noted above that decision-making under uncertainty such as expected utility theory can also be represented in our framework. In fact, analogous to the deterministic case, different theories can be dealt with. What is more, there exist non-deterministic and probabilistic extensions of the quantifiers and selection functions based on monads. In future research we will explore how these constructs can be used to model decisions under uncertainty and under risk more generally.

Also not explored in this paper but on our agenda for the future is the composition of selection functions which can be naturally defined. By that one can consider the aggregation of different individuals, as for instance in the literature of social choice, or model individuals as "multiple selves", as for instance in [1,12], where different dimensions of an agent are aggregated and the different dimensions taken together determine his final choices.

Lastly, and probably most importantly, selection functions are a building block for game theoretic approaches built on high-order functions. Thus, the ability to express various goals as discussed here scales to strategic interactions. In [8] we show that goal functions such as the fixed point player introduced above can be fruitfully applied to model voting contests. More generally, we show that the Nash equilibrium extends to games based on quantifiers and selection functions; we show that selection functions and quantifiers yield the same set of equilibria in the case of max and argmax operators; but we also show that for other classes of goal functions, such as the fixed point agent, quantifiers and selection functions yield different equilibria. In fact, we show that the equilibria induced by selection functions are a refinement of the equilbria induced by quantifiers. As a last point, we also explore selection functions as a building block to *open games* in [5,7,10]. Open games are a further abstraction based on category theory. Selection functions are an essential component because as in this paper they represent the individual agent's goal function.

References

1. Ambrus, A., Rozen, K.: Rationalising choice with multi-self models. Econ. J. **125**(585), 1136–1156 (2015)
2. Camerer, C.F.: Behavioral Game Theory: Experiments in Strategic Interaction. The Roundtable Series in Behavioral Economics. Princeton University Press, Princeton (2011)
3. Escardó, M., Oliva, P.: Selection functions, bar recursion and backward induction. Math. Struct. Comput. Sci. **20**(2), 127–168 (2010)
4. Escardo, M., Oliva, P.: Sequential games and optimal strategies. Proc. R. Soc. A Math. Phys. Eng. Sci. **467**(2130), 1519–1545 (2011)

5. Ghani, N., Hedges, J.: A compositional approach to economic game theory. CoRR, abs/1603.04641 (2016)
6. Hedges, J.: A generalisation of Nash's theorem with higher-order functionals. Proc. R. Soc. A **469**(2154), 1–18 (2013)
7. Hedges, J.: String diagrams for game theory. CoRR, abs/1503.06072 (2015)
8. Hedges, J., Oliva, P., Shprits, E., Winschel, V., Zahn, P.: Selection equilibria of higher-order games. In: Lierler, Y., Taha, W. (eds.) PADL 2017. LNCS, vol. 10137, pp. 136–151. Springer, Cham (2017). doi:10.1007/978-3-319-51676-9_9
9. Hedges, J., Oliva, P., Sprits, E., Winschel, V., Zahn, P.: Higher-order game theory. CoRR, abs/1506.01002 (2015)
10. Hedges, J., Shprits, E., Winschel, V., Zahn, P.: Compositionality and string diagrams for game theory. CoRR, abs/1604.06061 (2016)
11. Kahneman, D.: Thinking, Fast and Slow. Farrar, Straus and Giroux, New York (2011)
12. Kalai, G., Rubinstein, A., Spiegler, R.: Rationalizing choice functions by multiple rationales. Econometrica **70**(6), 2481–2488 (2002)
13. Loomes, G., Sugden, R.: Regret theory: An alternative theory of rational choice under uncertainty. Econ. J. **92**(368), 805–824 (1982)
14. Mostowski, A.: On a generalization of quantifiers. Fund. Math. **44**(1), 12–36 (1957)
15. Savage, L.J.: The theory of statistical decision. J. Am. Stat. Assoc. **46**(253), 55–67 (1951)
16. Wald, A.: Statistical Decision Functions. Wiley, Oxford (1950)

On Simplified Group Activity Selection

Andreas Darmann[1]([✉]), Janosch Döcker[2], Britta Dorn[2], Jérôme Lang[3],
and Sebastian Schneckenburger[2]

[1] University of Graz, Graz, Austria
andreas.darmann@uni-graz.at
[2] University of Tübingen, Tübingen, Germany
[3] Université Paris-Dauphine, Paris, France

Abstract. Several real-world situations can be represented in terms of
agents that have preferences over activities in which they may participate. Often, the agents can take part in at most one activity (for instance,
since these take place simultaneously), and there are additional constraints on the number of agents that can participate in an activity. In
such a setting we consider the task of assigning agents to activities in
a reasonable way. We introduce the simplified group activity selection
problem providing a general yet simple model for a broad variety of settings, and start investigating the case where upper and lower bounds of
the groups have to be taken into account. We apply different solution
concepts such as envy-freeness and core stability to our setting and provide a computational complexity study for the problem of finding such
solutions.

1 Introduction

Several real-world situations can be represented in terms of agents that have
preferences over activities in which they may participate, subject to some feasibility constraints on the way they are assigned to the different activities. Here
'activity' should be taken in a wide sense; here are a few examples, each with its
specificities which we will discuss further:

1. a group of co-workers may have to decide in which project to work, given
   that each project needs a fixed number of participants;
2. the participants to a big workshop, who are too numerous to fit all in a single
   restaurant, want to select a small number of restaurants (say, between two
   and four) out of a wider selection, with different capacities, and that serve
   different types of food, and to assign each participant to one of them;
3. a group of pensioners have to select two movies out of a wide selection, to be
   played simultaneously in two different rooms, and each of them will be able
   to see at most one of them;
4. a group of students have to choose one course each to follow out of a selection,
   given that each course opens only if it has a minimum number of registrants
   and has also an upper bound;

© Springer International Publishing AG 2017
J. Rothe (Ed.): ADT 2017, LNAI 10576, pp. 255–269, 2017.
DOI: 10.1007/978-3-319-67504-6_18

5. a set of voters want to select a committee of k representatives, given that each voter will be represented by one of the committee members.

While these examples seem to vary in several aspects, they share the same general structure: there is a set of *agents*, a set of available *activities*; each agent has preferences over the possible activities; there are constraints bearing on the selection of activities and the way agents are assigned to them; the goal is to assign each agent to one activity, respecting the constraints, and respecting as much as possible the agents' preferences.

Sometimes the set of selected activities is fixed (as Example 1), sometimes it will be determined by the agents' preferences. The nature of the constraints can vary: sometimes there are constraints that are *local* to each activity (typically, bounds on the number of participants, although we might imagine more complex constraints), as Examples 1, 2, 4, 5, and also 3 if the rooms have a capacity smaller than the number of pensioners; sometimes there are *global* constraints, that bear on the whole assignment (typically, bounds on the number of activities that can be selected; once again, we may consider more complex constraints), as in Examples 2, 3. Sometimes each agent *must* be assigned to an activity (as in Examples 1 and 5), sometimes she has the option of not being assigned to any activity.

This class of problems can be seen as a simplified version of the *group activity selection problem* (GASP), which asks how to assign agents to activities in a "good" way. In the original form introduced by Darmann et al. [5], agents express their preferences both on the activities and on the number of participants for the latter; in general, these preferences are expressed by means of weak orders over pairs "(activity, group size)". Darmann [4] considers the variant of GASP in which the agents' preferences are strict orders over such pairs and analyzes the computational complexity of finding assignments that are stable or maximize the number of agents assigned to activities.

Our model considers a simplified version of the group activity selection problem, called s-GASP. Here, agents only express their preferences over the set of activities. However, the activities come with certain constraints, such as restrictions on the number of participants, concepts like balancedness, or more global restrictions. The goal is again to find a "good" assignment of agents to activities, respecting both the agents' preferences as well as the constraints.

But what is a good assignment? Clearly, this essentially depends on the application on hand, but there are several concepts in the social choice and game theory literature that propose for an evaluative solution. We consider two classes of criteria for assessing the quality of an assignment:

– *solution concepts* that mainly come from game theory and that aim at telling whether an assignment is stable enough (that is, immune to some types of deviations) to be implemented. First, *individual rationality* requires that each agent is assigned to an activity she likes better than not being assigned to any activity at all. Then, a solution concept considered both in hedonic games, where coalition building is studied, and in matching theory, is the notion of *stability*. It asks whether the assignment is stable in the sense that no agent would want to or be able to deviate from her coalition, her match, or in our case, her assigned activity. Besides considering different variants of *core stability*, it also makes sense in our setting to investigate variations of *virtual stability*, meaning that it is not possible that an agent deviates from her assigned activity due to the given constraints.

– *criteria* that mainly come from social choice and that measure, qualitatively or quantitatively, the welfare of agents. A common quality measure in terms of efficiency of an assignment is the notion of *Pareto optimality*: there should be no feasible assignment in which there is an agent that is strictly better off, while the remaining agents do not change for the worse. More generally, one may wish to *optimize social welfare*, for some notion of utility derived from the agents' preferences: for instance, one may simply be willing to maximize the number of agents assigned to an activity. If fairness is important in the design, the notion of *envy-freeness* makes sense: an assignment respecting the constraints is envy-free if no agent strictly prefers the group another agent is assigned to.

Related Work. Apart from GASP, our model is related to various streams of work:

Course allocation, e.g. [2,6,10,14]. Students bear preferences over courses they would like to be enrolled in (these preferences are typically strict orders), and there are typically constraints given on the size of the courses. Courses will only be offered if a minimum number of participants is found, and there are upper bounds due to space or capacity limitations. In particular, Cechlárová and Fleiner [2] consider a course-allocation framework, so for them it makes sense that one agent can be matched to more than one activity (course), while [10,14] consider the case in which an agent can be assigned to at most one activity (project). The latter works are very close to our setting with constraints over group sizes. In contrast to above works however, our setting contains a dedicated outside option (the *void activity*), and agents' preferences are represented by weak orders over activities instead of strict rankings.

Hedonic games (see the recent survey by Aziz and Savani [1]) are coalition formation games where each agent has preferences over coalitions containing her. The stability notions we will focus on are derived from those for hedonic games. However, in our model, agents do not care about who else is assigned to the same activity as them, but only on the activity to which they are assigned to.[1]

---

[1] Still, it is possible to express simplified group activity selection within the setting of hedonic games, by adding special agents corresponding to activities, who are indifferent between all locally feasible coalitions. See the work by Darmann et al. [5] for such a translation for the more general group activity selection problem. But it is a rather artificial, and overly complex, representation of our model, which moreover does not help characterizing and computing solution concepts.

In *multiwinner elections*, there is a set of candidates, voters have preferences over single candidates, and a subset of k candidates has to be elected. In some approaches to multiwinner elections, each voter is assigned to one of the members of the elected committee, who is supposed to represent her. Sometimes there are no constraints on the number of voters assigned to a given committee member (as is the case for the *Chamberlin-Courant* rule [3]), in which case each voter is assigned to her most preferred committee member; on the other hand, for the *Monroe* rule [13], the assignment has to be balanced. A more general setting, with more general constraints, has been defined by Skowron et al. [16]. Note also that multiwinner elections can also be interpreted as *resource allocation* with items that come in several units (see again [16]) and as *group recommendation* [12]. While assignment-based multiwinner elections problems are similar to simplified group activity, an important difference is that for the former, stability notions play no role, as the voters are not assumed to be able to deviate from their assigned representatives.

Contents and Outline. In this work, we will take into account various solution concepts and ask two questions: First, do "good" assignments exist? Can we decide this efficiently? And if they exist, can we find them efficiently? Our second concern is optimization: we are looking for desirable assignments that maximize the number of agents which can be assigned to an activity. Again, we may ask whether an assignment that is optimal in this sense exists, and we can try to find it.

We will focus on one family of constraints concerning the size of the groups—we assume that each activity comes with a lower and an upper bound on the number of participants—and give a detailed analysis of the described problems for this class.

Our results for this class are twofold. First, we show that it is often possible to find assignments with desirable properties in an efficient way: we propose several polynomial time algorithms to find good assignments or to optimize them. We complement these findings with NP- and coNP-completeness results for certain solution concepts. Whenever we encounter computational hardness, we identify tractable special cases: we will see that all our problems can be solved in polynomial time if there is no restriction on the minimum number of participants for the activities to take place. An overview of our computational complexity results is given in Table 1 in Sect. 3; due to space constraints, we do not elaborate all proofs. Second, we show that also in this class of problems considered, there is a certain tension between the concepts of envy-freeness and Pareto-optimality, even for small instances.

The remainder of this work is organized as follows. In Sect. 2, we formally introduce the simplified model as well as possible constraints and several solution concepts. Section 3 is the main part of the paper and provides an analysis of the computational complexity of the questions described above. Section 4 deals with the tension between envy-freeness and Pareto optimality. In Sect. 5, we conclude and discuss future directions of research connected to s-GASP.

2   Model, Constraints, and Solution Concepts

We start with defining our model and with introducing the solution concepts we want to consider.

Simplified Group Activity Selection, Constraints. An instance (N, A, P, R) of the *simplified group activity selection problem* (s-GASP) is given as follows. The set $N = \{1, \ldots, n\}$ denotes a set of agents and $A = A^* \cup \{a_\emptyset\}$ a set of activities with $A^* = \{a_1, \ldots, a_m\}$, where $a_\emptyset$ stands for the *void activity*. An agent who is assigned to $a_\emptyset$ can be thought of as not participating in any activity. The preference profile $P = \langle \succsim_1, \ldots, \succsim_n \rangle$ consists of n votes (one for each agent), where $\succsim_i$ is a weak order over A for each $i \in N$. The set R is a set of side constraints that restricts the set of assignments.

A mapping $\pi : N \to A$ is called an *assignment*. Given assignment π, $\#(\pi) = |\{i \in N : \pi(i) \neq a_\emptyset\}|$ denotes the number of agents π assigns to a non-void activity; for activity $a \in A$, $\pi^a := \{i \in N : \pi(i) = a\}$ is the set of agents π assigns to a.

The goal will be to find "good" assignments that satisfy the constraints in R. The structure of the set R depends on the application. Some typical kinds of constraints are (combinations of) the following cases:

1. each activity comes with a lower and/or upper bound on the number of participants;
2. no more than k activities can have some agent assigned to them;
3. the number of voters per activity should be balanced in some way;

Intuitively, if there are no constraints or the constraints are flexible enough, then agents go where they want and the problem becomes trivial. If the constraints are tight enough (e.g., perfect balancedness, provided $|A|$ and $|V|$ allow it), then some agents are generally not happy, but they are unable to deviate because most deviations violate the constraints. The interesting cases can therefore be in between these two extreme cases.

In this work, we will start investigations for s-GASP for the first class of constraints: We assume that each activity $a \in A^*$ comes with a lower bound $\ell(a)$ and an upper bound $u(a)$, and all constraints in R are of the following type: for each $a \in A^*$, $|\pi^a| \in \{0\} \cup [\ell(a), u(a)]$.

Feasible Assignments, Solution Concepts. Let an instance (N, A, P, R) of s-GASP be given. A *feasible assignment* is an assignment meeting the constraints in R. We will consider the following properties. A feasible assignment π is

- *envy-free* if there is no pair of agents $(i, j) \in N \times N$ with $\pi(j) \in A^*$ such that $\pi(j) \succ_i \pi(i)$ holds;
- *individually rational* if for each $i \in N$ we have $\pi(i) \succsim_i a_\emptyset$;

- *individually stable* if there is no agent i and no activity $a \in A$ such that (i) $a \succ_i \pi(i)$ and (ii) the mapping π' defined by $\pi'(i) = a$ and $\pi'(k) = \pi(k)$ for $k \in N \setminus \{i\}$ is a feasible assignment;
- *core stable* if there is no set $E \subseteq N$ and no activity $a \in A$ such that (i) $a \succ_i \pi(i)$ for all $i \in E$, (ii) $\pi^a \subset E$ holds if $a \in A^*$, and (iii) the mapping π' defined by $\pi'(i) = a$ for $i \in E$ and $\pi'(k) = \pi(k)$ for $k \in N \setminus E$ is a feasible assignment; (Note that the respective activity a to which the set E of agents wishes to deviate must be either $a_\emptyset$ or currently unused.)
- *strictly core stable* if there is no set $E \subseteq N$ and no activity $a \in A$ such that (i) $a \succsim_i \pi(i)$ for all $i \in E$ where $a \succ_i \pi(i)$ for at least one $i \in E$, (ii) $\pi^a \subset E$ holds if $a \in A^*$, and (iii) the mapping π' defined by $\pi'(i) = a$ for all $i \in E$ and $\pi'(k) = \pi(k)$ for $k \in N \setminus E$ is a feasible assignment;
- *Pareto optimal* if there is no feasible assignment $\pi' \neq \pi$ such that $\pi'(i) \succsim_i \pi(i)$ for all $i \in N$ and $\pi'(i) \succ_i \pi(i)$ for at least one $i \in N$;

Finally, an individually rational assignment π is *maximum individually rational* if for all individually rational assignments π' we have $\#(\pi) \geq \#(\pi')$. Analogously, maximum feasible/envy-free/.../Pareto optimal assignments are defined.

For the class of constraints we consider, the notion of *virtual stability* is interesting. It requires that any deviation from the assigned towards a more preferred activity $a \in A^*$ violates the capacity constraints of a. Formally, we define the following stability concepts.

A feasible assignment π is

- *virtually individually stable* if there is no agent i and no activity $a \in A$ with $\ell(a) \leq |\pi^a| + 1 \leq u(a)$ such that $a \succ_i \pi(i)$ holds;
- *virtually core stable* if there is no set $E \subseteq N$ and no activity $a \in A$ with $\ell(a) \leq |E| \leq u(a)$ such that $a \succ_i \pi(i)$ for all $i \in E$, and (ii) $\pi^a \subset E$ holds if $a \in A^*$;
- *virtually strictly core stable* if there is no set $E \subseteq N$ and no activity $a \in A$ with $\ell(a) \leq |E| \leq u(a)$ such that (i) $a \succsim_i \pi(i)$ for all $i \in E$ where $a \succ_i \pi(i)$ for at least one $i \in E$, and (ii) $\pi^a \subset E$ holds if $a \in A^*$.

Note that as in the definition of core stability, also in virtual core stability the respective activity a to which the set E of agents wishes to deviate must be either $a_\emptyset$ or currently unused.

The relationships between the solution concepts is shown in Fig. 1 (for an overview of the relationships between solution concepts in hedonic games we refer to [1]).

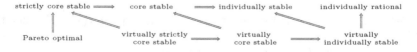

Fig. 1. Relations between the solution concepts we consider.

3 Computational Complexity for s-GASP with Group Size Constraints

We will now consider the computational complexity of s-GASP for various solution concepts. An overview of our results is given in Table 1.

Table 1. Overview of results for constraints $|\pi^a| \in \{0\} \cup [\ell(a), u(a)]$, $a \in A^*$.

Find assignment that is	general	$u(a) = n$	$\ell(a) = 1$
Feasible	in P (Proposition 1)	in P (Proposition 1)	in P (Proposition 1)
Individually rational	in P (Theorem 2)	in P (Theorem 2)	in P (Corollary 2)
Envy-free	in P (trivial)	in P (trivial)	in P (trivial)
Individually stable	in P (Theorem 1)	in P (Theorem 1)	in P (Corollary 2)
Core stable	in P (Theorem 1)	in P (Theorem 1)	in P (Corollary 2)
Strictly core stable	in P (Theorem 1)	in P (Theorem 1)	in P (Corollary 2)
Virtually individually stable	in P (Theorem 2)	in P (Theorem 2)	in P (Corollary 2)
Virtually core stable	NP-c (Corollary 1)	NP-c (Corollary 1)	in P (Corollary 2)
Virtually strictly core stable	NP-c (Theorem 3)	NP-c (Theorem 3)	in P (Corollary 2)
Pareto optimal	?	?	in P (Theorem 7)
Is there an assignment π with $\#(\pi) \geq k$ ($k \in \mathbb{N}$) that is	general	$u(a) = n$	$\ell(a) = 1$
Feasible	in P (Proposition 1)	in P (Proposition 1)	in P (Proposition 1)
Individually rational	NP-c (Theorem 4)	NP-c (Theorem 4)	in P (Theorem 5)
Envy-free	NP-c (Theorem 6)	in P (trivial)	?
Virtually individually stable	NP-complete	NP-complete	in P (Corollary 2)
Virtually core stable	NP-c (Corollary 1)	NP-c (Corollary 1)	in P (Corollary 2)
Virtually strictly core stable	NP-c (Theorem 3)	NP-c (Theorem 3)	in P (Corollary 2)
Pareto optimal	?	?	in P (Theorem 7)
Given assignment π, is π PO?	coNP-c (Theorem 8)	coNP-c (Theorem 8)	in P (Theorem 9)

3.1 Finding "Good" Assignments

The first interesting question is whether "good" assignments exist and how to find them. Obviously, assigning the void activity to every agent results is a feasible, individually rational and envy-free assignment. However, this is not a satisfying solution in terms of stability because agents will want to deviate. The good news is that for several stability concepts, a corresponding assignment always exists and can efficiently be found, as shown in the following theorem.

Theorem 1. *A strictly core stable assignment always exists and can be found in polynomial time.*

Proof. We sketch the basic algorithmic idea. Starting with a feasible assignment π, for each agent i and each activity b which i prefers to $\pi(i)$ we check whether there is a subset of agents including agent i that wants to deviate to b such that the resulting assignment is feasible. That is, we check whether there is a subset $E \supset \pi^b$ such that (i) for all $j \in E$ we have that $b \succsim_j \pi(j)$ holds (recall that for agent i $b \succ_i \pi(i)$ holds) and (ii) π' with $\pi'(i) = b$ for $i \in E$ and $\pi'(j) = \pi(j)$ for $j \in N \setminus E$ is a feasible assignment. In order to do so, for each activity $c \in A \setminus \{b\}$, we compute the possible numbers of agents in the set π^c that agree with joining b and can be removed from π^c while still enabling a feasible assignment. Finally, given these numbers, we need to verify if—including i and the agents in π^b—these add up to an integer contained in $[\ell(b), u(b)]$ by taking exactly one number from each activity. The latter problem reduces to the MULTIPLE-CHOICE SUBSET-SUM problem (see Pisinger [15]), which, in our case, allows for an overall polynomial time algorithm for finding a strictly core stable assignment. ∎

Recall that a strictly core stable assignment is also core stable and individually stable. Hence, as a consequence of the above theorem, also a core stable and an individually stable assignment always exist.

Theorem 2. *A virtually individually stable assignment always exists and can be found in polynomial time.*

Proof. In an instance (N, A, P, R) of s-GASP, we initially assign each agent to $a_\emptyset$, i.e., set $\pi(i) := a_\emptyset$ for $i \in N$. For $a \in A^*$ with $\ell(a) \geq 2$, if no agent is assigned to such a, then $\ell(a) \leq |\pi^a| + 1$ cannot hold. Hence, in what follows, we only consider activities $a \in A^*$ with $\ell(a) = 1$. For $1 \leq i \leq n$, assign agent i to the best ranked such activity $a \succ_i a_\emptyset$ with $|\pi^a| < u(a)$ and update π (i.e., set $\pi(i) := a$ while $\pi(j)$ remains unchanged for $j \in N \setminus \{i\}$). It is easy to see that the resulting assignment π is virtually individually stable. ∎

In contrast, a virtually core stable (and thus a virtually strictly core stable) assignment does not always exist, as the following example shows.

Example 1. Let $N = \{1, 2, 3\}$ and $A^* = \{a, b, c\}$, with $a \succ_1 b \succ_1 c \succ a_\emptyset$, $b \succ_2 c \succ_2 a \succ a_\emptyset$, and $c \succ_3 a \succ_3 b \succ a_\emptyset$. The restrictions on the activities are given by $|\pi^x| \in \{0\} \cup [2, 3]$, for each $x \in A^*$. By the restrictions given, there is at most one non-void activity to which agents can be assigned. Clearly, for any activity $z \in A$ there is a $y \in A^*$ such that two agents prefer y to z. As a consequence, there can be no virtually core stable assignment.

In addition, the problem to decide whether or not a virtually strictly core stable assignment exists turns out to be computationally difficult.

Theorem 3. *It is NP-complete to decide if there is a virtually strictly core stable assignment, even when for each activity $a \in A^*$ we have $u(a) = n$.*

Proof. Membership in NP is not difficult to verify. The proof proceeds by a reduction from EXACT COVER BY 3-SETS (X3C). The input of an instance of X3C consists of a pair $\langle X, \mathcal{Z} \rangle$, where $X = \{1, \ldots, 3q\}$ and $\mathcal{Z} = \{Z_1, \ldots, Z_p\}$ is a collection of 3-element subsets of X; the question is whether we can cover X with exactly q sets of $\mathcal{Z}$. X3C is known to be NP-complete even when each element of X is contained in exactly three sets of $\mathcal{Z}$ (see [7,8]); note that in such a case $p = 3q$ holds. For each $i \in X$, let the sets containing i be denoted by $Z_{i_1}, Z_{i_2}, Z_{i_3}$ with $i_1 < i_2 < i_3$.

Define instance $\mathcal{I} = (N, A, P, R)$ of s-GASP as follows. Let $N = \{V_{i,1}, V_{i,2}, V_{i,3} \mid 1 \leq i \leq p\}$ and $A^* = \{y_i, a_i, b_i, c_i \mid 1 \leq i \leq p\}$. For $1 \leq i \leq p$, let $\ell(a_i) = \ell(b_i) = \ell(c_i) = 2$, and $\ell(y_i) = 9$. For each $a \in A^*$, let $u(a) = |N|$. Since any virtually strictly core stable assignment is individually rational, in the profile P we omit the activities ranked below $a_\emptyset$; for each $i \in \{1, \ldots, p\}$, let the ranking of the agents $V_{i,1}, V_{i,2}, V_{i,3}$ (each of which represents element $i \in X$) be given as follows:

$$V_{i,1}: \quad y_{i_1} \succ_{i,1} y_{i_2} \succ_{i,1} y_{i_3} \succ_{i,1} a_i \succ_{i,1} b_i \succ_{i,1} c_i \succ_{i,1} a_\emptyset$$
$$V_{i,2}: \quad y_{i_2} \succ_{i,2} y_{i_3} \succ_{i,2} y_{i_1} \succ_{i,2} b_i \succ_{i,2} c_i \succ_{i,2} a_i \succ_{i,2} a_\emptyset$$
$$V_{i,3}: \quad y_{i_3} \succ_{i,3} y_{i_1} \succ_{i,3} y_{i_2} \succ_{i,3} c_i \succ_{i,3} a_i \succ_{i,3} b_i \succ_{i,3} a_\emptyset$$

Note that each set Z contains three elements, and hence each y_i, $1 \leq 1 \leq p$, is preferred to $a_\emptyset$ by exactly 9 agents. We show that there is an exact cover in instance $\langle X, \mathcal{Z} \rangle$ if and only if there is a virtually strictly core stable assignment in instance $\mathcal{I}$.

Assume there is an exact cover C. Consider the assignment π defined by $\pi(V_{i,h}) = y_j$ if $i \in Z_j$ and $Z_j \in C$, for $i \in \{1, \ldots, p\}$ and $h \in \{1, 2, 3\}$. Since C is an exact cover, assignment π is well-defined and feasible; note that each agent is assigned to an activity she ranks first, second or third. In addition, note that for $Z_j \in C$, each agent that prefers y_j to $a_\emptyset$ is assigned to y_j. Assume a set of agents E wishes to deviate to another activity d, such that at least one member $i \in E$ prefers d over $\pi(i)$ while there is no $j \in E$ with $\pi(j) \succ_j d$. By the definition of π, $d \in \{y_i \mid 1 \leq i \leq p\}$ holds. Observe that $\pi^d = \emptyset$ holds because C is an exact cover. Due to $\ell(d) = 9$, it hence follows that each agent of those who prefer d to $a_\emptyset$ must prefer d to the assigned activity, which is impossible since, by construction of the instance, for at least one of these agents j the assigned activity is top-ranked, i.e., $\pi(j) \succ_j d$ holds. Therewith, π is virtually strictly core stable.

Conversely, assume there is a virtually strictly core stable assignment π. Assume that there is an agent $V_{i,h}$ who is not assigned to one of the activities $y_{i_1}, y_{i_2}, y_{i_3}$. Then, by $\ell(y_i) = 9$ and the fact that exactly 9 agents prefer y_i to $a_\emptyset$ for each $i \in \{1, \ldots, p\}$, it follows that no agent is assigned to one of $y_{i_1}, y_{i_2}, y_{i_3}$; in particular none of $V_{i,1}, V_{i,2}, V_{i,3}$ is assigned to one of these activities. Analogously to Example 1 it then follows that there is no virtually strictly core stable assignment, in contradiction with our assumption.

Thus, π assigns each agent $V_{i,h}$ to one of the activities $y_{i_1}, y_{i_2}, y_{i_3}$. For each $i \in \{1, \ldots, p\}$, by $\ell(y_i) = 9$ and the fact that exactly 9 agents prefer y_i to $a_\emptyset$

it follows that to exactly one of $y_{i_1}, y_{i_2}, y_{i_3}$ exactly 9 agents are assigned, while no agent is assigned to the remaining two activities. As a consequence, the set $C = \{Z_i \mid |\pi^{y_i}| = 9, 1 \leq i \leq p\}$ is an exact cover in instance $\langle X, \mathcal{Z} \rangle$.   ∎

In the instance considered in the above proof, an assignment is virtually strictly core stable if and only if it is virtually core stable. As a consequence, we get the following corollary.

Corollary 1. *It is* NP-*complete to decide if there is a virtually core stable assignment, even if for each activity $a \in A^*$ we have $u(a) = n$.*

However, for the case of $\ell(a) = 1$ for each $a \in A^*$, we get a positive complexity result (see Sect. 3.2). In particular, we can show that in this case a virtually strictly core stable assignment that maximizes the number of agents assigned to a non-void activity can be found in polynomial time.

Turning to Pareto optimality, in the special case of $\ell(a) = 1$ for each $a \in A^*$, there is a simple algorithm to compute a Pareto optimal assignment. In that case, it is easy to see that a Pareto optimal assignment is always individually rational. Thus, neglecting activities ranked below $a_\emptyset$, we start with the assignment $\pi(i) = a_\emptyset$ for each $i \in N$ and iteratively assign an agent to the best-ranked among the activities a with $|\pi^a| < u(a)$. However, in the case of $\ell(a) = 1$ for each $a \in A^*$ we can even find a Pareto optimal assignment that maximizes the number of agents assigned to a non-void activity in polynomial time (see Sect. 3.2).

3.2   Maximizing the Number of Agents Assigned to a Non-void Activity

We now turn to an optimization problem: Among all feasible assignments that feature a certain property, one is usually interested in finding one that maximizes the number of agents that are assigned to a non-void activity, thus keeping the number of agents who cannot be enrolled in any activity low.

Proposition 1. *In polynomial time we can find a feasible assignment that maximizes the number of agents assigned to a non-void activity.*

But already for individual rational assignments, it is hard to decide whether all agents can be assigned to a non-void activity, as the following theorem shows. We omit its proof which is again a reduction from the EXACT COVER BY 3-SETS problem.

Theorem 4. *It is* NP-*complete to decide if there is an individually rational assignment that assigns each agent to some $a \in A^*$, even if for each activity $a \in A^*$ we have $u(a) = n$.*

However, if we assume that each activity admits a group size of 1, then we can find an optimal individually rational assignment efficiently.

Theorem 5. *If for each activity $a \in A^*$ we have $\ell(a) = 1$, then in polynomial time we can find a maximum individually rational assignment.*

Proof. Reduction to max integer flow with upper bounds. Given an instance $\mathcal{I} = (N, A, P, R)$ of s-GASP with $\ell(a) = 1$ for all $a \in A^*$, we construct an instance $\mathcal{M}$ of max integer flow with directed graph $G = (V, E)$. Set $V := \{s, t\} \cup N \cup A^*$, and let the edges and their capacities be given as follows: for each $i \in N$, introduce edge (s, i) with capacity 1; for each $a \in A^*$ and $i \in N$ introduce an edge (i, a) of capacity 1 if $a \succsim_i a_\emptyset$ holds; for each $a \in A^*$, introduce edge (a, t) of capacity $u(a)$. It is easy to see that a max integer flow from s to t induces a maximum individually rational assignment in $\mathcal{I}$ and vice versa. ∎

For envy-freeness, optimizing the number of "active" agents turns again out to be a hard problem which can be shown by a reduction from EXACT COVER BY 3-SETS as well.

Theorem 6. *It is NP-complete to decide if there is an envy-free assignment that assigns each agent to some $a \in A^*$.*

We obtain tractability for envy-freeness if we loosen the constraints on the upper bounds of the group sizes: Clearly, if there is an activity with "unlimited" capacity (i.e., its upper bound equals n), we can assign all agents to it and obtain envy-freeness.

3.3   Pareto Optimality

In this subsection, we consider the computational complexity involved in Pareto optimal assignments.

In the framework of course allocation, if all agents have strict preferences it is known that a Pareto optimal matching—that assigns an agent to an activity (course) only if the activity is acceptable for the agent—can be found in polynomial time (see [2,10]). Since in our setting (i) the agents' preferences are represented by weak orders and (ii) Pareto optimality does not require individual rationality, these results do not immediately translate. For the latter reason, the computational intractability result of [2] for finding a Pareto optimal matching maximizing the number of agents assigned to a non-void activity if each agent can be assigned to at most one activity does not immediately translate to our setting either. In particular, in general we do not know the computational complexity status of finding a Pareto optimal assignment (or of finding one that maximizes the number of agents assigned to non-void activities) in s-GASP. As the following theorem shows, the latter issue is computationally tractable if we relax the constraint on the lower bound of the group sizes.

Theorem 7. *If for each activity $a \in A^*$ we have $\ell(a) = 1$, then in polynomial time we can find a Pareto optimal assignment that maximizes the number of agents assigned to a non-void activity.*

Proof. In that case, any Pareto optimal assignment is individually rational. Let k be the maximum number of agents assigned to non-void activities by an individually rational assignment. Hence, it is sufficient to find a Pareto optimal assignment π with $\#(\pi) = k$. Given an instance $\mathcal{I} = (N, A, P, R)$ of s-GASP with $\ell(a) = 1$ for all $a \in A^*$, we construct an instance $\mathcal{F}$ of the minimum cost flow problem. Instance $\mathcal{F}$ corresponds to instance $\mathcal{M}$ of the proof of Theorem 5 except that we add the following edge costs: for each $a \in A^*$ and $i \in N$ edge (i, a) has cost $-(1 + |\{b \in A^* | a \succ_i b, b \succ_i a_\emptyset\}|)$, all remaining edges have zero cost. Let f be a minimum integer cost flow of size k in instance $\mathcal{F}$. Then f induces the assignment π by setting $\pi(i) = a$ iff f sends a unit of flow through edge (i, a). Clearly, π is Pareto optimal since otherwise a flow f' of lower total cost than the total cost of f could be induced. ∎

Note that in the case $\ell(a) = 1$ for each $a \in A^*$, also any strictly core stable, core stable, or individually stable assignment is individually rational. In addition, in this case virtually (strict) core stability coincides with (strict) core stability, and virtually individually stability coincides with individual stability. Hence we can state the following corollary.

Corollary 2. *If for each activity $a \in A^*$ we have $\ell(a) = 1$, then in polynomial time we can find a maximum individually rational assignment that is Pareto optimal, (virtually) individually stable, (virtually) core stable and (virtually) strictly core stable.*

However, checking whether a given assignment is Pareto optimal turns out to be coNP-complete, as Theorem 8 shows. We omit the proof which makes use of the NP-completeness of X3C.

Theorem 8. *It is coNP-complete to decide if a given assignment is Pareto optimal, even if for each activity $a \in A^*$ we have $u(a) = n$.*

Again, if there are no restrictions on the minimum number of participants of each activity, the latter problem becomes tractable.

Theorem 9. *If for each activity $a \in A^*$ we have $\ell(a) = 1$, then in polynomial time we can decide if a given assignment is Pareto optimal.*

Proof. Given instance $\mathcal{I} = (N, A, P, R)$ of s-GASP with $\ell(a) = 1$ for all $a \in A^*$ and assignment π, we construct instance $\mathcal{C}$ of the minimum cost flow problem as follows with lower and upper edge capacities. Note that π must be individually rational. In instance $\mathcal{C}$, the directed graph $G = (V, E)$, edge costs and capacities are given as follows. $G = (V, E)$ has vertex set $V := \{s, t\} \cup N \cup A^*$, the edge set E consists of the following edges:

- for $i \in N$, edge (s, i) of zero cost, and, for $a \in A^*$ with $a \succsim_i \pi(i)$, edge (i, a) of cost -1 if $a \succ_i \pi(i)$ and of cost 0 if $a \sim_i \pi(i)$;
- for $a \in A^*$ edge (a, t) of upper capacity bound $u(a)$.

The lower and upper capacity bound of edge (s, i) is 1 iff $\pi(i) \succ a_\emptyset$ holds. Unless otherwise specified, the lower capacity bound of edge $e \in E$ is 0 and the upper capacity bound is 1, and its cost is 0.

Assume there is an integer flow f of negative total cost. Consider the assignment π' defined by $\pi'(i) = a$ iff f sends flow through edge (i, a). Then, by construction we must have $\pi'(i) \sim_i \pi(i)$ or $\pi'(i) \succ_i \pi(i)$ for each $i \in N$, where the latter holds for at least one agent $i \in N$ by the negative total cost of f. Thus, π is not Pareto optimal.

If, on the other hand, π is not Pareto optimal, then there is an assignment π' with $\pi'(i) \sim_i \pi(i)$ or $\pi'(i) \succ_i \pi(i)$ for each $i \in N$, where the latter holds for at least one $i \in N$. The integer flow f' that sends flow along the edges $(s, i), (i, a), (a, t)$ iff $\pi'(i) = a$ holds, has negative total cost.

Therewith, for verifying if π is Pareto optimal it is sufficient to find an integer minimum cost flow in instance $\mathcal{C}$. ∎

4   Envy-Freeness vs. Pareto Optimality

In many social choice settings, there is a tension between envy-freeness and Pareto optimality. This is also the case for our simplified group activity selection problem, as the following proposition and the subsequent corollary show.

Proposition 2. *For any $k \geq 2$, there is an instance (N, A, P, R) of s-GASP with $|N| = k$ and $\ell(a) = 1$ for each $a \in A^*$, for which there does not exist an assignment π which is both Pareto optimal and envy-free.*

Proof. We provide a proof for $k = 2$, which easily extends to $n = k$ for any $k > 2$. Consider the instance with $N = \{1, 2\}$, $A^* = \{a\}$, with the rankings $a \succ_1 a_\emptyset$ and $a \succ_2 a_\emptyset$, and the restrictions given by $\ell(a) = u(a) = 1$. Any Pareto optimal assignment assigns exactly one agent to a, which is clearly not envy-free. ∎

Corollary 3. *There is no mechanism that determines an assignment that is both Pareto optimal and envy-free for each given instance (N, A, P, R) of s-GASP, even if $\ell(a) = 1$ holds for each $a \in A^*$.*

Interestingly, this tension also holds if the only relevant constraint is the lower bound of the activities (i.e., $u(a) = n$ for all a).

Proposition 3. *For any $k \geq 6$, there is an instance (N, A, P, R) of s-GASP with $|N| = k$ and $u(a) = k$ for each $a \in A^*$, for which there does not exist an assignment π which is both Pareto optimal and envy-free.*

Proof. We provide a proof for $k = 6$, which easily extends to $k = n$ for any $n > 6$. Consider the instance of s-GASP with $N = \{1, 2, 3, 4, 5, 6\}$, $A^* = \{a, b, c\}$ and for any $x \in A^*$ we have $\ell(x) = 3, u(x) = 6$. The rankings are

$$\succsim_1: a \succ_1 b \succ_1 c \succ_1 a_\emptyset \quad \succsim_4: a \succ_4 b \succ_4 c \succ_4 a_\emptyset$$
$$\succsim_2: b \succ_2 c \succ_2 a \succ_2 a_\emptyset \quad \succsim_5: b \succ_5 c \succ_5 a \succ_5 a_\emptyset$$
$$\succsim_3: c \succ_3 a \succ_3 b \succ_3 a_\emptyset \quad \succsim_6: c \succ_6 a \succ_6 b \succ_6 a_\emptyset$$

Due to the feasibility constraints, there are only 4 types of feasible assignments:

(i) 3–5 agents are assigned to the same activity $x \neq a_\emptyset$, and the rest to $a_\emptyset$.
(ii) All agents are assigned to the void activity.
(iii) All agents are assigned to the same activity $x \neq a_\emptyset$.
(iv) 3 agents are assigned to the same activity $x \neq a_\emptyset$ and the other 3 agents are assigned to another activity $y \notin \{x, a_\emptyset\}$.

The assignments of type (i) and (ii) are Pareto dominated by some assignment of type (iii). An assignment π_1 of type (iii) is envy-free but not Pareto optimal. Due to the symmetrical construction of the preferences profiles, we can assume without loss of generality $\pi_1^a = N$. But then the assignment is Pareto dominated by the assignment π_2 with $\pi_2^a = \{1, 3, 4\}$ and $\pi_2^c = \{2, 5, 6\}$. An assignment of type (iv) cannot be envy-free. Without loss of generality we can assume $x = a$ and $y = b$. Assume, for the sake of contradiction, that there is an envy-free assignment. Agents 1 and 4 must be assigned to activity a and agents 2 and 5 to activity b. As the preference profiles of the remaining agents both rank a strictly better than b, the assignment cannot be an envy-free assignment. ∎

Corollary 4. *There is no mechanism that determines an assignment that is both Pareto optimal and envy-free for each given instance (N, A, P, R) of s-GASP, even if $u(a) = n$ holds for each $a \in A^*$.*

5   Conclusion

We have formulated a simplified version of GASP where the assignment of agents to activities depends on the agents' preferences as well as on exogenous constraints. This model is powerful enough to capture many real world applications. We have made a first step by analyzing one family of constraints and have studied several solution concepts for this family.

An obvious next step is to drive a similar analysis for other interesting classes of constraints as described in Sect. 2. In particular, it would be interesting to characterize families of constraints guaranteeing or not guaranteeing existence of a stable solution for the different solution concepts we considered, or exploring forbidden structures that prevent stability. Also, it would be nice to provide a detailed analysis of the parameterized complexity of the hard cases, as done by Lee and Williams [11] for the stable invitation problem and by Igarashi et al. [9] for GASP on social networks. Another variant would be to consider typed agents as in the paper by Spradling and Goldsmith [17].

Acknowledgement. This work was partly supported by COST Action IC1205 on Computational Social Choice.

References

1. Aziz, H., Savani, R.: Hedonic games. In: Brandt, F., Conitzer, V., Endriss, U., Lang, J., Procaccia, A.D. (eds.) Handbook of Computational Social Choice. Cambridge University Press (2015). Chap. 15

2. Cechlárová, K., Fleiner, T.: Pareto optimal matchings with lower quotas. Math. Soc. Sci. **88**, 3–10 (2017)
3. Chamberlin, J.R., Courant, P.N.: Representative deliberations and representative decisions: proportional representation and the Borda rule. Am. Polit. Sci. Rev. **77**(3), 718–733 (1983)
4. Darmann, A.: Group activity selection from ordinal preferences. In: Walsh, T. (ed.) ADT 2015. LNCS (LNAI), vol. 9346, pp. 35–51. Springer, Cham (2015). doi:10.1007/978-3-319-23114-3_3
5. Darmann, A., Elkind, E., Kurz, S., Lang, J., Schauer, J., Woeginger, G.: Group activity selection problem. In: Goldberg, P.W. (ed.) WINE 2012. LNCS, vol. 7695, pp. 156–169. Springer, Heidelberg (2012). doi:10.1007/978-3-642-35311-6_12
6. Goto, M., Hashimoto, N., Iwasaki, A., Kawasaki, Y., Ueda, S., Yasuda, Y., Yokoo, M.: Strategy-proof matching with regional minimum quotas. In: Proceedings of the 2014 International Conference on Autonomous Agents and Multi-agent Systems, pp. 1225–1232. International Foundation for Autonomous Agents and Multiagent Systems (2014)
7. Hein, J., Jiang, T., Wang, L., Zhang, K.: On the complexity of comparing evolutionary trees. Discrete Appl. Math. **71**, 153–169 (1996)
8. Hickey, G., Dehne, F., Rau-Chaplin, A., Blouin, C.: SPR distance computation for unrooted trees. Evol. Bioinform. **4**, 17–27 (2008)
9. Igarashi, A., Bredereck, R., Elkind, E.: On parameterized complexity of group activity selection problems on social networks (2017). CoRR abs/1703.01121
10. Kamiyama, N.: A note on the serial dictatorship with project closures. Oper. Res. Lett. **41**(5), 559–561 (2013)
11. Lee, H., Williams, V.V.: Parameterized complexity of group activity selection. In: Proceedings of the 16th International Conference on Autonomous Agents and Multiagent Systems (AAMAS-2017), IFAAMAS, May 2017
12. Lu, T., Boutilier, C.: Budgeted social choice: From consensus to personalized decision making. In: IJCAI (2011)
13. Monroe, B.L.: Fully proportional representation. Am. Polit. Sci. Rev. **89**, 925–940 (1995)
14. Monte, D., Tumennasan, N.: Matching with quorums. Econ. Lett. **120**(1), 14–17 (2013)
15. Pisinger, D.: An O(nr) Algorithm for the Subset Sum Problem. DIKU-Rapport, Datalogisk Institut, Københavns Universitet, Datalogisk Institut København (1995)
16. Skowron, P., Faliszewski, P., Slinko, A.M.: Achieving fully proportional representation: Approximability results. Artif. Intell. **222**, 67–103 (2015)
17. Spradling, M., Goldsmith, J.: Stability in role based hedonic games. In: Proceedings of the Twenty-Eighth International Florida Artificial Intelligence Research Society Conference, FLAIRS 2015, Hollywood, Florida, 18–20 May 2015, pp. 85–90 (2015)

Equilibria in Sequential Allocation

Haris Aziz[1,2]($\boxtimes$), Paul Goldberg[3], and Toby Walsh[1,2]

[1] Data61, CSIRO, Sydney, Australia
toby.walsh@data61.csiro.au
[2] UNSW, Sydney, Australia
haris.aziz@data61.csiro.au
[3] University of Oxford, Oxford, UK
paul.goldberg@cs.ox.ac.uk

Abstract. Sequential allocation is a simple mechanism for sharing multiple indivisible items. We study strategic behavior in sequential allocation. In particular, we consider Nash dynamics, as well as the computation and Pareto optimality of pure equilibria, and Stackelberg strategies. We first demonstrate that, even for two agents, better responses can cycle. We then present a linear-time algorithm that returns a profile (which we call the "bluff profile") that is in pure Nash equilibrium. Interestingly, the outcome of the bluff profile is the same as that of the truthful profile and the profile is in pure Nash equilibrium for *all* cardinal utilities consistent with the ordinal preferences. We show that the outcome of the bluff profile is Pareto optimal with respect to pairwise comparisons. In contrast, we show that an assignment may not be Pareto optimal with respect to pairwise comparisons even if it is a result of a preference profile that is in pure Nash equilibrium for all utilities consistent with ordinal preferences. Finally, we present a dynamic program to compute an optimal Stackelberg strategy for two agents, where the second agent has a constant number of distinct values for the items.

1 Introduction

A simple but popular mechanism to allocate indivisible items is *sequential allocation* [3,5,8,9,12,14,15]. Sequential allocation is used, for example, by the Harvard Business School to allocate courses to students [10] as well as multimillion dollar sports drafts [8]. In a sequential allocation mechanism, a *picking sequence* specifies the turns of the agents. For example, for sequence 1212, agents 1 and 2 alternate with agent 1 taking the first turn. Agents report their preferences over the items. Then the items are allocated to the agents in the following manner. In each turn, the agent in that turn is given the most preferred item that has not yet been allocated. In this paper we focus on the "direct revelation" version where agents submit their complete rankings at the same time (and are committed to them), as opposed to the "extensive form" version where agents take turns choosing and are only committed to items chosen previously. Sequential allocation is an ordinal mechanism since the outcome only depends on the ordinal preferences of agents over items. Although the agents are asked to report

© Springer International Publishing AG 2017
J. Rothe (Ed.): ADT 2017, LNAI 10576, pp. 270–283, 2017.
DOI: 10.1007/978-3-319-67504-6_19

ordinal preferences, we will assume a standard assumption in the literature that agents have underlying additive utilities for the items.

It has long been known that sequential allocation is not strategy-proof when agents do not have consecutive turns. An agent may not pick their most preferred item remaining if they expect this item to remain till a later turn. Instead, the agent may pick a slightly less preferred item that they would not otherwise get. Of course, this requires reasoning about how the agents may behave strategically at the same time. Since the sequential allocation mechanism is not strategy-proof, how precisely should agents behave? There has already been some work on strategic behavior in the setting where sequential allocation is viewed as a repeated game. Kohler and Chandrasekaran [14] presented a linear-time algorithm to compute a subgame perfect Nash equilibrium (SPNE) when there are two agents and the picking sequence is alternating (121212...). The result was generalized to the case of any sequence [13]. Brams and Straffin [8] stated that "no algorithm is known which will produce optimal play more efficiently than by checking many branches of the game tree." Recently, it was proved that there can be an exponential number of subgame perfect Nash equilibria and finding even one of them is PSPACE-hard for an unbounded number of agents [13].

However, it is also natural to view sequential allocation as a *one shot* game rather than a repeated game. At the Harvard Business School, students submit a single ranked list of courses to a central organization that runs the sequential allocation mechanism on these fixed preferences. This is essentially then a one shot game. This suggests considering the more general solution concept of pure Nash equilibrium rather than that of subgame perfect Nash equilibrium. In this paper, *we will view sequential allocation as a one shot strategic game in which the possible actions of the agents are possible ordinal preferences over the items, and the agents know each others' true ordinal preferences, as well as the picking sequence.* Surprisingly no algorithm to date has yet been proposed in the literature for efficiently computing a pure Nash equilibrium (PNE). We therefore propose a simple linear time method to compute a PNE even for an unbounded number of agents. We also consider Pareto optimality of pure Nash equilibria. This issue is similar to previous work on price of anarchy/stability of equilibria in other strategic domains. Finally, we consider Stackelberg strategies in sequential allocation where an agent announces the preference he or she intends to report.

Results. We study the computational problems of finding the equilibria of sequential allocation when viewed as a one shot game. No algorithm to date has been proposed in the literature for efficiently computing a pure Nash equilibrium (PNE) of sequential allocation.

One general method to compute a PNE is to compute a sequence of better responses. Indeed, for any finite potential game, this is guaranteed to find a PNE. We first show better responses need not converge to a pure Nash equilibrium. Even for two agents, better responses can cycle. Instead, we propose a simple linear time method to compute the preference profile of a PNE even for an unbounded number of agents. We refer to the output of this algorithm as the *bluff profile*. Interestingly, the allocation generated by the bluff profile is the same as that of the truthful profile, and this profile is in equilibrium for *all* cardinal utilities consistent with the ordinal preferences. The fact that this equilibrium can be computed in linear time is perhaps a little surprising because computing just a single best response with the sequential allocation mechanism has been recently shown to be NP-hard [4]. In addition, computing a subgame perfect Nash equilibrium of the repeated game is PSPACE-hard [13], and this is a PNE of the one shot game. Our result that there exists a linear-time algorithm to compute a PNE profile in the one shot game also contrasts with the fact that computing a PNE profile is NP-hard under the related probabilistic serial (PS) random assignment mechanism for fair division of indivisible goods [1].

We also consider Pareto optimality and other fairness properties of the pure Nash equilibria (Sect. 6). This is in line with work on the price of anarchy/stability of equilibria in other strategic domains. We show that the outcome of the bluff profile is Pareto optimal with respect to pairwise comparisons (defined in Sect. 6). Hence, in sequential allocation, pure Nash equilibrium is not incompatible with ordinal Pareto optimality. On the other hand, we also prove that an assignment may not be Pareto optimal with respect to pairwise comparisons even if it is a result of a preference profile that is PNE for all utilities consistent with ordinal preferences.

Finally, in Sect. 7 we show that an agent may have an advantage from committing and declaring his preference and that committing to the truthful report may not be optimal. For 2 players we present a polynomial-time algorithm to compute an optimal strategy to commit to in the case that the other agent has a small number of utility values.

2   Preliminaries

We consider the setting in which we have $N = \{1, \ldots, n\}$ a set of agents, $O = \{o_1, \ldots, o_m\}$ a set of items, and the preference profile $\succ = (\succ_1, \ldots, \succ_n)$ specifies for each agent i his complete, strict, and transitive preference $\succ_i$ over O.

Each agent may additionally express a cardinal utility function u_i consistent with $\succ_i$: $u_i(o) > u_i(o')$ iff $o \succ_i o'$. We will assume that each item is positively valued, i.e., $u_i(o) > 0$ for all $i \in N$ and $o \in O$. The set of all utility functions consistent with $\succ_i$ is denoted by $\mathcal{U}(\succ_i)$. We will denote by $\mathcal{U}(\succ)$ the set of all utility profiles $u = (u_1, \ldots, u_n)$ such that $u_i \in \mathcal{U}(\succ_i)$ for each $i \in N$. When we consider agents' valuations according to their cardinal utilities, we will assume additivity, that is $u_i(O') = \sum_{o \in O'} u_i(o)$ for each $i \in N$ and $O' \subseteq O$.

An *assignment* is an allocation of items to agents, represented as an $n \times m$ matrix $[p(i)(o_j)]_{1 \leq i \leq n, 1 \leq j \leq m}$ such that for all $i \in N$, and $o_j \in O$, $p(i)(o_j) \in \{0, 1\}$; and for all $j \in \{1, \ldots, m\}$, $\sum_{i \in N} p(i)(o_j) = 1$. An agent i gets item o_j if and only if $p(i)(o_j) = 1$. Each row $p(i) = (p(i)(o_1), \ldots, p(i)(o_m))$ represents the *allocation* of agent i.

We will also present the cardinal utilities in matrix form. A utility matrix U is an $n \times m$ matrix $[U(i)(j)]_{1 \leq i \leq n, 1 \leq j \leq m}$ such that for all $i \in N$, and $j \in O$, the entry $U(i)(j)$ in the i-th row and j-th column is $u_i(o_j)$. We say that utilities are *lexicographic* if for each agent $i \in N$, $o \in O$, $u_i(o) > \sum_{o' \prec_i o} u_i(o')$. By $S \succ_i T$, we will mean $u_i(S) > u_i(T)$.

Example 1. Consider the setting in which $N = \{1, 2\}$, $O = \{o_1, o_2, o_3, o_4\}$, the preferences of agents are

$$1: \quad o_1, o_2, o_3, o_4 \qquad\qquad 2: \quad o_1, o_3, o_2, o_4$$

Then for the picking sequence 1221, agent 1 gets $\{o_1, o_4\}$ while 2 gets $\{o_2, o_3\}$. The assignment resulting from sequential allocation (SA) can be represented as follows.

$$SA(\succ_1, \succ_2) = \begin{pmatrix} 1 & 0 & 0 & 1 \\ 0 & 1 & 1 & 0 \end{pmatrix}.$$

The allocation of agent 1 is denoted by $SA(\succ_1, \succ_2)(1)$.

For a reported preference profile $(\succ'_1, \ldots, \succ'_n)$, an agent i's *best response* is a preference report $\succ''_i$ that maximizes utility $u_i(SA(\succ''_i, \succ'_{-i})(i))$. We say that a reported preference profile $(\succ'_1, \ldots, \succ'_n)$ is in *pure Nash equilibrium (PNE)* if no agent i can report a preference $\succ''_i$ such that $u_i(SA(\succ''_i, \succ'_{-i})(i)) > u_i(SA(\succ')(i))$.

3   Nash Dynamics

Since we are interested in computing a PNE, a natural approach is to simulate better responses and hope they converge. For finite potential games, such an approach is guaranteed to find a PNE. However, we show that even for two agents, computing better responses will not always terminate and is thus not a method that is guaranteed to find a pure Nash equilibrium.

Theorem 1. *For two agents, better responses can cycle.*

Proof. Let the sequence be the alternating one: 121212..... The following 5 step sequence of better responses leads to a cycle.

The ordinal preferences corresponding to the utility functions are as follows.

$$\succ_1: \quad o_3, o_4, o_5, o_6, o_9, o_{10}, o_7, o_8, o_1, o_2$$
$$\succ_2: \quad o_9, o_{10}, o_5, o_6, o_7, o_8, o_1, o_2, o_3, o_4$$

It is sufficient to consider the agents having lexicographic utilities although the argument works for any utilities consistent with the ordinal preferences.

This yields the following assignment and utilities at the start:

$$SA(\succ_1, \succ_2) = \begin{pmatrix} 1 & 0 & 1 & 1 & 1 & 0 & 1 & 0 & 0 & 0 \\ 0 & 1 & 0 & 0 & 0 & 1 & 0 & 1 & 1 & 1 \end{pmatrix}$$

In Step 1, Agent 1 misreports to increase his utility.

$$\succ_1^1: \quad o_5, o_6, o_7, o_8, o_3, o_4, o_1, o_2, o_9, o_{10}$$
$$\succ_2^1: \quad o_9, o_{10}, o_5, o_6, o_7, o_8, o_1, o_2, o_3, o_4$$

$$SA(\succ^1) = \begin{pmatrix} 0\,0\,1\,1\,1\,1\,1\,0\,0\,0 \\ 1\,1\,0\,0\,0\,0\,0\,1\,1\,1 \end{pmatrix}$$

In Step 2, Agent 2 changes his report in response.

$\succ_1^2$: $o_5, o_6, o_7, o_8, o_3, o_4, o_1, o_2, o_9, o_{10}$

$\succ_2^2$: $o_5, o_6, o_7, o_8, o_9, o_{10}, o_1, o_2, o_3, o_4$

$$SA(\succ^2) = \begin{pmatrix} 1\,0\,1\,1\,1\,0\,1\,0\,0\,0 \\ 0\,1\,0\,0\,0\,1\,0\,1\,1\,1 \end{pmatrix}$$

In Step 3, Agent 1 changes his report in response.

$\succ_1^3$: $o_5, o_6, o_9, o_{10}, o_3, o_4, o_1, o_2, o_7, o_8$

$\succ_2^3$: $o_5, o_6, o_7, o_8, o_9, o_{10}, o_1, o_2, o_3, o_4$

$$SA(\succ^3) = \begin{pmatrix} 0\,0\,1\,1\,1\,0\,0\,0\,1\,1 \\ 1\,1\,0\,0\,0\,1\,1\,1\,0\,0 \end{pmatrix}$$

In Step 4, Agent 2 changes his report in response.

$\succ_1^4$: $o_5, o_6, o_9, o_{10}, o_3, o_4, o_1, o_2, o_7, o_8$

$\succ_2^4$: $o_9, o_{10}, o_5, o_6, o_7, o_8, o_1, o_2, o_3, o_4$

$$SA(\succ^4) = \begin{pmatrix} 1\,0\,1\,1\,1\,1\,0\,0\,0\,0 \\ 0\,1\,0\,0\,0\,0\,1\,1\,1\,1 \end{pmatrix}$$

In Step 5, Agent 1 changes his report in response.

$\succ_1^5$: $o_5, o_6, o_7, o_8, o_3, o_4, o_1, o_2, o_9, o_{10}$

$\succ_2^5$: $o_9, o_{10}, o_5, o_6, o_7, o_8, o_1, o_2, o_3, o_4$

$$SA(\succ^5) = \begin{pmatrix} 0\,0\,1\,1\,1\,1\,1\,0\,0\,0 \\ 1\,1\,0\,0\,0\,0\,0\,1\,1\,1 \end{pmatrix}$$

Since $\succ^1 = \succ^5$, we have cycled. $\square$

4   The Bluff Profile

In this section, we outline a linear-time algorithm to compute a pure Nash equilibrium preference profile. Surprisingly, we will show that the preference profile constructed is in pure Nash equilibrium for *all* utilities consistent with the ordinal preferences.

> *Simulate sequential allocation with the truthful preferences. Set the preferences of each agent to the order in which the items are picked when simulating sequential allocation under truthful preferences.*

We refer to the profile constructed as the bluff profile since the idea behind the profile is that an agent wants to get the most preferred item immediately because if he does not, some other agent will take it. We observe the following characteristics of the bluff profile.

Lemma 1. *In the bluff profile, (i) all agents have the same preferences; (ii) the order in which items are picked is the same as the order in which items are picked under the truthful profile; and (iii) the allocations of agents are the same as in the truthful profile.*

We show that the bluff profile is in pure Nash equilibrium if the utilities are lexicographic.

Lemma 2. *The bluff profile is in pure Nash equilibrium if the utilities are lexicographic.*

Proof. We prove by induction on the number of picks that no agent has an incentive to pick some other item when his turn comes which means that he picks the same item that he picks in the bluff profile which is also the most preferred item among the available items. This is equivalent to proving that no agent has an incentive to change his report from that in the bluff profile.

For the base case, let us consider the first agent who takes the first turn. If he does not take his most-preferred item, the next agent will take it. Since utilities are lexicographic, the first agent gains most by getting his most-preferred item. Regarding the other agents, they are not disadvantaged by placing that item first in their preferences lists, since it is taken by the first agent. It does not affect their ability to express their preferences amongst the remaining items.

Similarly, let us assume that agents in the first k turns did not have an incentive to misreport and pick some item other than the most preferred available item. Then we show that agent j in the $k + 1$-st turn does not have an incentive to change his report. Note that the item that j picks according to the bluff profile is his most preferred item o amongst those still available. This is because the order in which items are picked and allocations that are made exactly coincide with the truthful profile. Now if j does not make the consistent pick, he will not be able to recover the loss of not getting o because the utilities are lexicographic. Again, for the other agents (who do not get o) it does not disadvantage them to put o in the $k + 1$-st place in their preference list.   □

We now prove the following lemma.

Lemma 3. *Consider a profile in which all agents in $N \setminus \{i\}$ report the same preferences. Then agent i's best response results in the same allocation for him for all utilities consistent with the ordinal preferences.*

Proof. When all agents in $N \setminus \{i\}$ report the same preferences, then for agent i, from the perspective of agent i, all the turns of agents in $N \setminus \{i\}$ can be replaced by a single agent, representative of $N \setminus \{i\}$ who has the same preferences as agents in $N \setminus \{i\}$. Thus, computing a best response for agent i when all agents in

$N \setminus \{i\}$ report the same preferences is equivalent to computing a best response for agent i when there is only one other agent (with the same preference as the agents in $N \setminus \{i\}$) and each turn of agents in $N \setminus \{i\}$ is replaced by the representative agent. When there is one other agent, [6] proved that the best response results in the same allocation for the agent for all utilities consistent with the ordinal preferences.[1]   □

Combining Lemmas 2 and 3, we are in a position to prove the following:

Theorem 2. *The bluff profile is in pure Nash equilibrium under all utilities consistent with the ordinal preferences.*

Proof. From Lemma 2, we know that the bluff profile is in pure Nash equilibrium if the utilities are lexicographic. From Lemma 1, we know that all agents have the same preferences in the bluff profile. This immediately implies that for any agent i, all agents in $N \setminus \{i\}$ report the same preferences. From Lemma 3, each agent i's best response to the bluff profile results in the same unique allocation for all utilities consistent with the ordinal preferences. This allocation should be the same as allocation achieved by i when he reports the bluff preferences because they yield the best allocation under lexicographic utilities. Hence the bluff profile is in pure Nash equilibrium under all utilities consistent with the ordinal preferences.   □

5   The Crossout Profile

Since sequential allocation can also be viewed as a perfect information extensive form game, it admits a SPNE (Subgame-Perfect Nash Equilibrium) and hence a pure Nash equilibrium for the *game tree*.[2] Computing a SPNE of the game tree is PSPACE-complete [13]. On the other hand, the optimal play for the extensive form game can be computed in polynomial time for the case of two agents. The strategy corresponding to the SPNE is to play so that the last agent gets their least preferred item, the second from last, their next least preferred item, and so on. We first show that for the case of two agents, similar ideas can also be used to construct a PNE preference profile for the one-shot game.

We use the expression *crossout profile* to refer to the preference profile in which both agents have the preferences which are the same as the item picking ordering in the optimal play of perfect information extensive form game. The crossout preference profile can be computed as follows:

> *Reverse and then invert (exchange 1s with 2s and vice versa) the picking sequence. Reverse the preferences of the two agents. Find the order*

---

[1] This argument does not work when the number of other agents is more than one and they have different preferences. It can be shown that for three or more agents, best responses need not result in the same allocation.

[2] For readers not familiar with extensive form games and Subgame-Perfect Nash Equilibrium), we refer them to [16].

L in which items are allocated to the agents according to the new picking sequence and preferences. Return reverse of L as the preference of each agent.

We now show that the crossout profile is in PNE for certain utilities consistent with the ordinal preferences. We say that utilities are *upward lexicographic* if for each agent $i \in N$ and two allocations with equal number of items, if the agent prefers the allocation with the better least-preferred item and in case of equality, the one with a better second-least-preferred item and so on. Such a preference relation can be captured by cardinal utilities as follows. If agent i has ordinal preferences $o_1, o_2, \ldots, o_m$, then utilities are as follows: $u_i(o_j) = 1 - (1/2^{m+1-j})$ for all $j \in \{1, \ldots, m\}$.

Lemma 4. *For two agents and for upward lexicographic utilities, the crossout profile is in PNE.*

Proof (Proof Sketch). Consider agent $i \in \{1, 2\}$ and denote by $-i$ the other agent. Let π be the sequence of turns of the agents so that $\pi(j)$ is the agent with the j-th turn. Now if agent $-i = \pi(m)$ has the last turn, then in the first $m - 1$ turns, whenever agent i's turn comes, he has an option to get an item better than i's least preferred item o_m. Hence i can guarantee to not get his least preferred item o_m and hence guarantee $-i$ to get o_m. This will always be the best response for agent i if he has upward lexicographic utilities. Since $-i$ gets o_m in any case, $-i$ may as well rank o_m last, and use his higher slots to prioritise amongst the other items. We can now consider a situation in which o_m does not exist in O and it is fixed as the least preferred item in both agents' preferences. Then the same argument can be applied recursively.   □

Lemma 4 can be used to prove the following theorem.

Theorem 3. *For two agents and for all utilities consistent with the ordinal preferences, the crossout profile is in PNE.*

Proof. From Lemma 4, we know that for two agents and for upward lexicographic utilities, the crossout profile is in PNE. When there is one other agent, Bouveret and Lang [6] proved that the best response results in the same allocation for the agent under all utilities consistent with the ordinal preferences.   □

Next we show that even for two agents, the outcome of a crossout profile may not be the same as the truthful assignment.

Example 2. Even for two agents, the outcome of the crossout profile (and hence the SPNE assignment) may not be the same as the truthful assignment. Consider the sequence 1212 and profile:

$$\succ_1: \quad a, b, c, d \qquad\qquad \succ_2: \quad b, c, a, d$$

The picking sequence obtained after reversing and inverting the picking sequence is again 1212. The modified preferences are as follows.

$$\succ_1'': \quad d, c, b, a \qquad\qquad\qquad \succ_2'': \quad d, a, c, b$$

Under picking sequence 1212 and profile $\succ''$, the items are picked as follows: d, a, c, b. We reverse this ordering to obtain the following crossout profile:

$$\succ_1'': \quad b, c, a, d \qquad\qquad\qquad \succ_2'': \quad b, c, a, d.$$

Under this profile and original picking sequence 1212, 1 gets $\{b, a\}$ and 2 gets $\{c, d\}$. Also note that the SPNE path is as follows: 1 gets b, 2 gets c, 1 gets a and then 2 gets d. In contrast, in the truthful assignment, 1 gets a and c.

6 Pareto Optimality of Pure Nash Equilibria

We next consider the Pareto optimality of equilibria. An allocation S is *at least as preferred with respect to pairwise comparisons by a given agent i* as allocation T, if there exists an injection f from T to S such that for each item $o \in T$, i prefers $f(o)$ at least as much as o. We note that an agent strictly prefers S over T with respect to pairwise comparisons if S results from T by a sequence of replacements of an item in T with a strictly more preferred item. Note that the pairwise comparison relation is transitive but not necessarily complete. We will focus on Pareto optimality with respect to pairwise comparisons.

We first show there exists a PNE whose outcome is Pareto optimal with respect to pairwise comparisons. Hence, unlike some other games, Pareto optimality is not incompatible with Nash equilibria in sequential allocation.

Theorem 4. *The outcome of the bluff profile is Pareto optimal with respect to pairwise comparisons.*

The argument is as follows. Since the outcome of the bluff profile is the same as the outcome of the truthful profile and since the outcome of each truthful profile is Pareto optimal with respect to pairwise comparisons [7], the outcome of the bluff profile is Pareto optimal with respect to pairwise comparisons as well. Although the argument for the theorem is simple, it shows the following: if the truthful outcome satisfies some normative properties such as envy-freeness or other fairness properties [2], we know that there exists at least one PNE which results in an assignment with the same normative properties. The theorem above is in sharp contrast with the result in [8] that there exist utilities under which no SPNE assignment is Pareto optimal with respect to pairwise comparisons. Other relevant papers that deal with implementing Pareto optimal outcomes in other settings include [11,17].

Next, we show that there may exist a PNE whose outcome is not Pareto optimal with respect to pairwise comparisons. The statement holds even if the PNE in question is in PNE with respect to all utilities consistent with the ordinal preferences!

Theorem 5. *An assignment may not be Pareto optimal with respect to pairwise comparisons even if it is a result of a preference profile that is in PNE for all utilities consistent with ordinal preferences.*

Proof. Consider the preference profile:

$$\succ_1: \quad a, b, c, d, e, f \qquad\qquad \succ_2: \quad e, f, b, a, d, c$$
$$\succ_3: \quad c, f, e, d, a, b$$

Let the sequence be 123123. Then the outcome of the truthful preference profile can be summarized as

$$1: \ \{a, b\} \qquad\qquad 2: \ \{e, f\} \qquad\qquad 3: \ \{c, d\}$$

Consider the following profile $\succ'$:

$$\succ'_1: \quad c, f, a, b, d, e \qquad\qquad \succ'_2: \quad b, a, e, c, d, f$$
$$\succ'_3: \quad f, e, d, a, b, c$$

Then the outcome of the profile $\succ'$ can be summarized as

$$1: \ \{c, a\} \qquad\qquad 2: \ \{b, e\} \qquad\qquad 3: \ \{f, d\}$$

We argue that the profile $\succ'$ is in PNE. In his reported preference $\succ'_1$, agent 1 gets $\{a, c\}$. The only better outcome agent 1 can get is $\{a, b\}$. If he goes for a first, he does not get c or b. If he goes for b first, he does not get a. So agent 1 plays his best response for all utilities consistent with his ordinal preferences.

In his reported preference $\succ'_2$, agent 2 gets $\{e, b\}$. The only better outcome agent 2 can get is $\{e, f\}$. Now agent 2 in his best response will try to get $\{e, f\}$. If agent 2 tries to pick f first, he will not get e. Hence agent 2 plays his best response for all utilities consistent with his ordinal preferences.

Finally, for agent 3, he cannot get c. The best he can get is $\{e, f\}$. If 3 goes for e first, then he does not get f. If 3 goes for f first, he can only get d. The best he can get is $\{f, d\}$ so his reported preference is his best response for all utilities consistent with the ordinal preferences.   □

7   Advantage of Commitment

In prior work on strategic aspects of sequential allocation, the focus has been on computing manipulations or equilibria. We now consider another strategic aspect: Stackelberg strategies to commit to in order to obtain outcomes that are better for the individual agent. In this setting, agent 1 (the leader) announces a preference R of all the items, and commits to selecting, whenever it is his turn, the highest-ranked item in R that is not yet taken. The following example illustrates a leadership advantage.

Example 3. There are 2 agents and 4 items denoted *a,b,c,d*. Suppose the agents choose items in order 1212. The ordinal preferences are

$$\succ_1: \quad a, d, c, b \qquad\qquad\qquad \succ_2: \quad a, b, d, c$$

Then in an SPNE, agent 1 takes item *a*, then agent 2 takes item *d* (since agent 1 will not take item *b*, it is okay for agent 2 to take *d*, ending up with *b* and *d*). Then agent 1 takes *c*. Also if agent 1 reports the truth a, d, c, b, then agent 2 is guaranteed to get *b* so he can report d, b, a, c and get $\{b, d\}$ which means that 1 gets $\{a, c\}$.

However, consider the case where agent 1 is leader, and announces the preference list $\succ'_1$: $\quad a, b, d, c$. Then agent 2 must use a preference list that results in agent 2 taking item *b* first. Agent 1 has a credible threat to take item *b*, if agent 2 does not take it next (despite the fact that agent 1 doesn't value item *b*). So, agent 1 gets items *a* and *d*.

This raises the following question. For two agents, what is the complexity of finding the best preference report for the leader, assuming that the follower will best-respond. Next, we consider an interesting special case in which the problem can be solved in polynomial time.

Theorem 6. *For $n = 2$ and any fixed picking sequence, there is an algorithm whose runtime is polynomial in the number of items m, to compute an optimal Stackelberg strategy for agent 1 when agent 2 has a constant number of distinct values for items.*

We make the assumption, standard in the study of optimal Stackelberg strategies, that if agent 2 has more than one best response, then agent 1 breaks the tie in his (agent 1's) favour. Let k (constant) be the number of distinct values that agent 2 has for items. Agent 1 has to identify a ranking of the items such that if agent 2 best-responds, agent 1's total value is maximised.

It is convenient to proceed by solving the following slight generalisation of the problem. Given a picking sequence P (a sequence of 1's and 2's of length m), we add a parameter ℓ, where ℓ is at most the number of 2's in P, and agent 2 may receive only ℓ items. We make the following observation:

Observation 1. *Suppose (for picking sequence P) agent 2 is allowed to receive ℓ items. We can regard agent 2's selection of items as working as follows. Given agent 1's preference ranking $\succ_1$, agent 2 places a token on the ℓ items in $\succ_1$ whose positions correspond to the positions of 2 in P. Agent 2 is allowed to move any token from any item x to an item x', provided that $x \succ_1 x'$, subject to the constraint that tokens lie at distinct items. Finally, items marked with tokens are the ones that agent 2 receives. Agent 2 chooses the most valuable set that can be obtained in this way.*

Proposition 1. *We may assume that in an optimal (for agent 1) ranking of the items, if items x and x' have the same value for agent 2, and agent 1 values x higher than x', then agent 1 ranks x higher than x'.*

Proof. We claim first that since x and x' have the same value to agent 2, then given any ranking by agent 1, any best response by agent 2 can be modified to avoid an outcome where agent 2 takes the higher-ranked of $\{x, x'\}$, but not the lower-ranked of $\{x, x'\}$.

Noting Observation 1, if the higher-ranked of $\{x, x'\}$ has a token, but not the lower-ranked of them, the token can be moved to the lower-ranked of $\{x, x'\}$ without loss of utility to agent 2. If agent 1 ranks the lower-valued of $\{x, x'\}$ higher in $\succ_1$, they can be exchanged, and the new ranking (with the right best response by agent 2) is at least as good for agent 1.   □

Notation: Recall that m denotes the number of items. Let $S_1, \ldots, S_k$ be the partition of the items into subsets that agent 2 values equally. For $1 \leq i \leq k$ let $m_i = |S_i|$. Let $o_{i,j} \in S_i$ be the member of S_i that has the j-th highest value to agent 1. Let $S_i(j) \subseteq S_i$ be the set $\{o_1, \ldots, o_j\}$, that is, the j highest value (to agent 1) members of S_i. Let $U(j_1, \ldots, j_k; \ell)$ be the highest utility that agent 1 can get, assuming that items $S_1(j_1) \cup \cdots \cup S_k(j_k)$ are being shared, and agent 2 is allowed to take ℓ of them, where $\ell \leq m$. In words, we consider subsets of the S_i obtained by taking the best items in S_i, and consider various numbers of items that we limit agent 2 may to receive.

Proof. (of Theorem 6) If picking sequence P contains m' occurrences of "2", we are interested in computing $U(m_1, \ldots, m_k; m')$ and its associated ranking. We express the solution recursively be expressing $U(j_1, \ldots, j_k; \ell)$ in terms of various values of $U(j_1', \ldots, j_k'; \ell')$, where $j_i' \leq j_i$ and $\ell' \leq \ell$, and at least one inequality is strict. Furthermore, for any values $j_1, \ldots, j_k, \ell$ we also evaluate and remember agent 2's best response. This can be seen to be achievable in polynomial time via dynamic programming, since there are $O(m^{k+1})$ sets of values that can be taken by these parameters.

We compute $U(j_1, \ldots, j_k; \ell)$ as follows. Let $j = j_1 + \ldots + j_k$ and assume that there are at least ℓ occurrences of "2" in the first j entries of the picking sequence.

By Proposition 1, in an (agent 1)-optimal ranking of items in $S_1(j_1) \cup \cdots \cup S_k(j_k)$, the lowest ranked item must be one of $o_{1,j_1}, \cdots, o_{k,j_k}$. We consider two cases, according to whether or not agent 2 takes that lowest-ranked item.

Suppose agent 2 takes that item. Then $U(j_1, \ldots, j_k; \ell)$ is given by:

$$\max_{i \in [k]} \Big(U(j_1, \ldots, j_{i-1}, j_i - 1, j_{i+1}, \ldots, j_k; \ell - 1) \Big) \tag{1}$$

Alternatively agent 2 may fail to take that item, in which case $U(j_1, \ldots, j_k; \ell)$ is given by:

$$\max_{i \in [k]} \Big(u_1(o_{i,j_i}) + U(j_1, \ldots, j_{i-1}, j_i - 1, j_{i+1}, \ldots, j_k; \ell) \Big) \tag{2}$$

where $u_1(o_{i,j_i})$ is agent 1's value for item o_{i,j_i} lowest-ranked.

By way of explanation of (1) and (2), in the case of (1) where agent 2 takes o_{i,j_i}, agent 1's utility will be that of the optimal ranking of the other $j - 1$ items under the constraint that agent 2 only gets to take $\ell - 1$ of them. If agent 2 does not take this lowest-ranked item o_{i,j_i}, then (2) gives agent 1's utility as his value $u_1(o_{i,j_i})$ for that item, plus the best outcome for agent 1 assuming agent 2 may take ℓ items from amongst the other $j - 1$ items.

In checking which case applies for a given choice of i and corresponding item o_{i,j_i}, we check whether the optimal ranking of the other items for agent 2 taking $\ell - 1$ of them, when extended to agent 2's selection of that additional item, is indeed a best-response for agent 2 given that he gets ℓ of all the items. This can be done efficiently, since best responses can be efficiently computed [6].   □

8   Conclusion

Sequential allocation is a simple and frequently used mechanism for resource allocation. Its strategic aspects have been formally studied for the last forty years. To our surprise, some fundamental questions have been unaddressed in the literature about sequential allocation when viewed as an one shot game. This is despite the fact that in many settings, it is essentially played as an one shot game. We have therefore studied in detail the pure Nash equilibrium of sequential allocation mechanisms. We presented a number of results on Nash dynamics, as well as on the computation of pure Nash equilibrium, and the Pareto optimality of equilibria. In particular, we presented the first polynomial-time algorithm to compute a PNE that applies to all utilities consistent with the ordinal preferences. We have also explored some other new directions such as Stackelberg strategies that have so far not been examined in sequential allocation.

References

1. Aziz, H., Gaspers, S., Mackenzie, S., Mattei, N., Narodytska, N., Walsh, T.: Equilibria under the probabilistic serial rule. In: Proceedings of the 24th International Joint Conference on Artificial Intelligence (IJCAI), pp. 1105–1112 (2015)
2. Aziz, H., Gaspers, S., Mackenzie, S., Walsh, T.: Fair assignment of indivisible objects under ordinal preferences. Artif. Intell. **227**, 71–92 (2015)
3. Aziz, H., Walsh, T., Xia, L.: Possible and necessary allocations via sequential mechanisms. In: Proceedings of the 24th International Joint Conference on Artificial Intelligence (IJCAI), pp. 468–474 (2015)
4. Aziz, H., Bouveret, S., Lang, J., Mackenzie, S.: Complexity of manipulating sequential allocation. In: Proceedings of the 31st AAAI Conference on Artificial Intelligence (AAAI), pp. 328–334 (2017)
5. Bouveret, S., Lang, J.: A general elicitation-free protocol for allocating indivisible goods. In: Proceedings of the 22nd International Joint Conference on Artificial Intelligence (IJCAI), pp. 73–78. AAAI Press (2011)
6. Bouveret, S., Lang, J.: Manipulating picking sequences. In: Proceedings of the 21st European Conference on Artificial Intelligence (ECAI), pp. 141–146 (2014)
7. Brams, S.J., King, D.L.: Efficient fair division: help the worst off or avoid envy? Ration. Soc. **17**(4), 387–421 (2005)
8. Brams, S.J., Straffin, P.D.: Prisoners' dilemma and professional sports drafts. Am. Math. Mon. **86**(2), 80–88 (1979)

9. Brams, S.J., Taylor, A.D., Division, F.: From Cake-Cutting to Dispute Resolution. Cambridge University Press, Cambridge (1996)
10. Budish, E., Cantillion, E.: The multi-unit assignment problem: theory and evidence from course allocation at Harvard. Am. Econ. Rev. **102**(5), 2237–2271 (2012)
11. Kalai, G., Meir, R., Tennenholtz, M.: Bidding games and efficient allocations. In: Proceedings of the 16th ACM Conference on Economics and Computation (ACM-EC), pp. 113–130 (2015)
12. Kalinowski, T., Narodytska, N., Walsh, T.: A social welfare optimal sequential allocation procedure. In: Proceedings of the 22nd International Joint Conference on Artificial Intelligence (IJCAI), pp. 227–233. AAAI Press (2013)
13. Kalinowski, T., Narodytska, N., Walsh, T., Xia, L.: Strategic behavior when allocating indivisible goods sequentially. In: Proceedings of the 27th AAAI Conference on Artificial Intelligence (AAAI), pp. 452–458. AAAI Press (2013)
14. Kohler, D.A., Chandrasekaran, R.: A class of sequential games. Oper. Res. **19**(2), 270–277 (1971)
15. Levine, L., Stange, K.E.: How to make the most of a shared meal: plan the last bite first. Am. Math. Mon. **119**(7), 550–565 (2012)
16. Leyton-Brown, K., Shoham, Y.: Essentials of Game Theory: A Concise, Multidisciplinary Introduction. Morgan & Claypool, San Rafael (2008)
17. Moulin, H.: Implementing the Kalai-Smorodinsky bargaining solution. J. Econ. Theory **33**(1), 32–45 (1984)

Obtaining a Proportional Allocation
by Deleting Items

Britta Dorn[1], Ronald de Haan[2], and Ildikó Schlotter[3(✉)]

[1] University of Tübingen, Tübingen, Germany
britta.dorn@uni-tuebingen.de
[2] University of Amsterdam, Amsterdam, The Netherlands
R.deHaan@uva.nl
[3] Budapest University of Technology and Economics, Budapest, Hungary
ildi@cs.bme.hu

Abstract. We consider the following control problem on fair allocation of indivisible goods. Given a set I of items and a set of agents, each having strict linear preference over the items, we ask for a minimum subset of the items whose deletion guarantees the existence of a proportional allocation in the remaining instance; we call this problem PROPORTIONALITY BY ITEM DELETION (PID). Our main result is a polynomial-time algorithm that solves PID for three agents. By contrast, we prove that PID is computationally intractable when the number of agents is unbounded, even if the number k of item deletions allowed is small, since the problem turns out to be W[3]-hard with respect to the parameter k. Additionally, we provide some tight lower and upper bounds on the complexity of PID when regarded as a function of $|I|$ and k.

1 Introduction

We consider a situation where a set I of indivisible items needs to be allocated to a set N of agents in a way that is perceived as *fair*. Unfortunately, it may happen that a fair allocation does not exist in a setting. In such situations, we might be interested in the question how our instance can be modified in order to achieve a fair outcome. Naturally, we seek for a modification that is as small as possible. This can be thought of as a *control action* carried out by a central agency whose task is to find a fair allocation. The computational study of such control problems was first proposed by Bartholdi, III et al. [3] for voting systems; our paper follows the work of Aziz et al. [2] who have recently initiated the systematic study of control problems in the area of fair division.

The idea of fairness can be formalized in various different ways such as proportionality, envy-freeness, or max-min fair share. Here we focus on *proportionality*, a notion originally defined in a model where agents use utility functions to represent their preferences over items. In that context, an allocation is called proportional if each agent obtains a set of items whose utility is at least $1/|N|$ of their total utility of all items. One way to adapt this notion to a model with linear preferences (not using explicit utilities) is to look for an allocation that is proportional with respect to *any* choice of utility functions for the agents that

© Springer International Publishing AG 2017
J. Rothe (Ed.): ADT 2017, LNAI 10576, pp. 284–299, 2017.
DOI: 10.1007/978-3-319-67504-6_20

is compatible with the given linear preferences (see Aziz et al. [1] for a survey of other possible notions of proportionality and fairness under linear preferences). Aziz et al. [1] referred to this property as "necessary proportionality"; for simplicity, we use the shorter term "proportionality."

We have two reasons for considering linear preferences. First, an important advantage of this setting is the easier elicitation of agents' preferences, which enables for more practical applications. Second, this simpler model is more tractable in a computational sense: under linear preferences, the existence of a proportional allocation can be decided in polynomial time [1], whereas the same question for cardinal utilities is NP-hard already for two agents [9]. Clearly, if already the existence of a proportional allocation is computationally hard to decide, then we have no hope to solve the corresponding control problem efficiently.

Control actions can take various forms. Aziz et al. [2] mention several possibilities: control by adding/deleting/replacing agents or items in the given instance, or by partitioning the set of agents or items. In this paper we concentrate only on control by *item deletion*, where the task is to find a subset of the items, as small as possible, whose removal from the instance guarantees the existence of a proportional allocation. In other words, we ask for the maximum number of items that can be allocated to the agents in a proportional way.

1.1   Related Work

We follow the research direction proposed by Aziz et al. [2] who initiated the study of control problems in the area of fair division. As an example, Aziz et al. [2] consider the complexity of obtaining envy-freeness by adding or deleting items or agents, assuming linear preferences. They show that adding/deleting a minimum number of items to ensure envy-freeness can be done in polynomial time for two agents, while for three agents it is NP-hard even to decide if an envy-free allocation exists. As a consequence, they obtain NP-hardness also for the control problems where we want to ensure envy-freeness by adding/deleting items in case there are more than two agents, or by adding/deleting agents.

The problem of deleting a minimum number of items to obtain envy-freeness was first studied by Brams et al. [4] who gave a polynomial-time algorithm for the case of two agents.[1] In the context of cake cutting, Segal-Halevi et al. [13] proposed the idea of distributing only a portion of the entire cake in order to obtain an envy-free allocation efficiently. For the Hospitals/Residents with Couples problem, Nguyen and Vohra [11] considered another type of control action: they obtained stability by slightly perturbing the capacities of hospitals.

1.2   Our Contribution

We first consider the case where the number of agents is unbounded (see Sect. 3). We show that the problem of deciding whether there exist at most k items whose

---

[1] For a complete proof of the correctness of their algorithm, see also [2].

deletion allows for a proportional allocation is NP-complete, and also W[3]-hard with parameter k (see Theorem 2). This latter result shows that even if we allow only a few items to be deleted, we cannot expect an efficient algorithm, since the problem is not fixed-parameter tractable with respect to the parameter k (unless FPT = W[3]).

Additionally, we provide tight upper and lower bounds on the complexity of the problem. In Theorem 3 we prove that the trivial $|I|^{O(k)}$ time algorithm—that, in a brute force manner, checks for each subset of I of size at most k whether it is a solution—is essentially optimal (under the assumption FPT $\neq$ W[1]). We provide another simple algorithm in Theorem 4 that has optimal running time, assuming the Exponential Time Hypothesis.

In Sect. 4, we turn our attention to the case with only three agents. In Theorem 5 we propose a polynomial-time algorithm for this case, which can be viewed as our main result. This algorithm is based on dynamic programming, but relies heavily on a non-trivial insight into the structure of solutions.

For lack of space, proofs marked by an asterix are deferred to a detailed technical report [12].

2   Preliminaries

We assume the reader to be familiar with basic complexity theory, in particular with parameterized complexity [6].

Preferences. Let N be a set of agents and I a set of indivisible items that we wish to allocate to the agents in some way. We assume that each agent $a \in N$ has strict preferences over the items, expressed by a preference list L^a that is a linear ordering of I, and set $L = \{L^x \mid x \in N\}$. We call the triple (N, I, L) a *(preference) profile*. We denote by $L^a[i : j]$ the subsequence of L^a containing the items ranked by agent a between the positions i and j, inclusively, for any $1 \leq i \leq j \leq |I|$. Also, for a subset $X \subseteq I$ of items we denote by L^a_X the restriction of L^a to the items in X.

Proportionality. Interestingly, the concept of proportionality (as described in Sect. 1) has an equivalent definition that is more direct and practical: we say that an allocation $\pi : I \to N$ mapping items to agents is *proportional* if for any integer $i \in \{1, \ldots, |I|\}$ and any agent $a \in N$, the number of items from $L^a[1 : i]$ allocated to a by π is at least $i/|N|$. Note that, in particular, this means that in a proportional allocation, each agent needs to get his or her first choice. Another important observation is that a proportional allocation can only exist if the number of items is a multiple of $|N|$, since each agent needs to obtain at least $|I|/|N|$ items.

Control by deleting items. Given a profile $\mathcal{P} = (N, I, L)$ and a subset U of items, we can define the preference profile $\mathcal{P} - U$ obtained by removing all items in U from I and from all preference lists in L. Let us define the PROPORTION-ALITY BY ITEM DELETION (PID) problem as follows. Its input is a pair $(\mathcal{P}, k)$ where $\mathcal{P} = (N, I, L)$ is a preference profile and k is an integer. We call a set

$U \subseteq I$ of items a *solution* for $\mathcal{P}$ if its removal from I allows for proportionality, that is, if there exists a proportional allocation $\pi : I \setminus U \to N$ for $\mathcal{P} - U$. The task in PID is to decide if there exists a solution of size at most k.

3   Unbounded Number of Agents

Since the existence of a proportional allocation can be decided in polynomial time by standard techniques in matching theory [1], the PROPORTIONAL ITEM DELETION problem is solvable in $|I|^{O(k)}$ time by the brute force algorithm that checks for each subset of I of size at most k whether it is a solution. In terms of parameterized complexity, this means that PID is in XP when parameterized by the solution size.

Clearly, such a brute force approach may only be feasible if the number k of items we are allowed to delete is very small. Searching for a more efficient algorithm, one might ask whether the problem becomes fixed-parameter tractable with k as the parameter, i.e., whether there exists an algorithm for PID that, for an instance $(\mathcal{P}, k)$ runs in time $f(k)|\mathcal{P}|^{O(1)}$ for some computable function f. Such an algorithm could be much faster in practice compared to the brute force approach described above.

Unfortunately, the next theorem shows that finding such a fixed-parameter tractable algorithm seems unlikely, as PID is W[2]-hard with parameter k. Hence, deciding whether the deletion of k items can result in a profile admitting a proportional allocation is computationally intractable even for small values of k.

Theorem 1. PROPORTIONALITY BY ITEM DELETION *is* NP-*complete and* W[2]-*hard when parameterized by the size* k *of the desired solution.*

Proof. We are going to present an FPT-reduction from the W[2]-hard problem k-DOMINATING SET, where we are given a graph $G = (V, E)$ and an integer k and the task is to decide if G contains a dominating set of size at most k; a vertex set $D \subseteq V$ is *dominating* in G if each vertex in V is either in D or has a neighbor in D. We denote by $N(v)$ the set of neighbors of some vertex $v \in V$, and we let $N[v] = N(v) \cup \{v\}$. Thus, a vertex set D is dominating if $N[v] \cap D \neq \emptyset$ holds for each $v \in V$.

Let us construct an instance $I_{\text{PID}} = (\mathcal{P}, k)$ of PID with $\mathcal{P} = (N, I, L)$ as follows. We let N contain $3n + 2m + 1$ agents where $n = |V|$ and $m = |E|$: we create $n + 1$ so-called *selection agents* $s_1, \ldots, s_{n+1}$, and for each $v \in V$ we create a set $A_v = \{a_v^j \mid 1 \leq j \leq |N[v]| + 1\}$ of *vertex agents*. Next we let I contain $2|N| + k$ items: we create distinct first-choice items $f(a)$ for each agent $a \in N$, a *vertex item* i_v for each $v \in V$, a dummy item d_v^j for each vertex agent $a_v^j \in N$, and $k + 1$ additional dummy items $c_1, \ldots, c_{k+1}$.

Let F denote the set of all first-choice items, i.e., $F = \{f(a) \mid a \in N\}$. For any set $U \subseteq V$ of vertices in G, let $I_U = \{i_v \mid v \in U\}$; in particular, I_V denotes the set of all vertex items.

Before defining the preferences of agents, we need some additional notation. We fix an arbitrary ordering $\prec$ over the items, and for any set X of items we let $[X]$ denote the ordering of X according to $\prec$. Also, for any $a \in N$, we define the set F_i^a as the first i elements of $F \setminus \{f(a)\}$, for any $i \in \{1, \ldots, |N| - 1\}$. We end preference lists below with the symbol '...' meaning all remaining items not listed explicitly, ordered according to $\prec$.

Now we are ready to define the preference list L^a for each agent a.

- If a is a selection agent $a = s_i$ with $1 \leq i \leq n - k$, then let

$$L^a : f(a), \underbrace{[F_{|N|-n}^a], [I_V], [F_{|N|-n+k}^a \setminus F_{|N|-n}^a]}_{|N| \text{ items}}, \underbrace{\phantom{[F_{|N|-n+k}^a \setminus F_{|N|-n}^a]}}_{k \text{ items}}, \ldots$$

- If a is a selection agent $a = s_i$ with $n - k < i \leq n + 1$, then let

$$L^a : f(a), \underbrace{[F_{|N|-n}^a], [I_V], [F_{|N|-n+k-1}^a \setminus F_{|N|-n}^a]}_{|N| \text{items}}, \underbrace{\phantom{[F_{|N|-n+k-1}^a \setminus F_{|N|-n}^a]}}_{k-1 \text{ items}}, c_{i-(n-k)}, \ldots$$

- If a is a vertex agent $a = a_v^j$ with $1 \leq j \leq |N[v]| + 1$, then let

$$L^a : f(a), \underbrace{[F_{|N|-|N[v]|}^a], [I_{N[v]}]}_{|N| \text{ items}}, d_v^j, \ldots$$

This finishes the definition of our PID instance I_{PID}.

Suppose that there exists a solution S of size at most k to I_{PID} and a proportional allocation π mapping the items of $I \setminus S$ to the agents in N. Observe that by $|I| = 2|N| + k$, we know that S must contain exactly k items.

First, we show that S cannot contain any item from F. For contradiction, assume that $f(a) \in S$ for some agent a. Since the preference list of a starts with more than k items from F (by $N - n > k$), the first item in $L_{I \setminus S}^a$ must be an item $f(b)$ for some $b \in N$, $b \neq a$. The first item in $L_{I \setminus S}^b$ is exactly $f(b)$, and thus any proportional allocation should allocate $f(b)$ to both a and b, a contradiction.

Next, we prove that $S \subseteq I_V$. For contradiction, assume that S contains less than k items from I_V. Then, after the removal of S, the top $|N| + 1$ items in the preference list $L_{I \setminus S}^{s_i}$ of any selection agent s_i are all contained in $I_V \cup F$. Hence, π must allocate at least two items from $I_V \cup F$ to s_i, by the definition of proportionality. Recall that for any agent a, π allocates $f(a)$ to a, meaning that π would need to distribute the n items in I_V among the $n + 1$ selection agents, a contradiction. Hence, we have $S \subseteq I_V$.

We claim that the k vertices $D = \{v \mid i_v \in S\}$ form a dominating set in S. Let us fix a vertex $v \in V$. For sake of contradiction, assume that $N[v] \cap D = \emptyset$, and consider any vertex agent a in A_v. Then the top $|N| + 1$ items in $L_{I \setminus S}^a$ are the same as the top $|N| + 1$ items in $L^a = L_I^a$ (using that $S \cap F = \emptyset$), and these items form a subset of $I_{N[v]} \cup F$ for every $a \in A_v$. But then arguing as above, we get that π would need to allocate an item of $I_{N[v]}$ to each of the $|N[v]| + 1$ vertex agents in A_v; again a contradiction. Hence, we get that $N[v] \cap D \neq \emptyset$ for each $v \in V$, showing that D is indeed a dominating set of size k.

For the other direction, let D be a dominating set of size k in G, and let S denote the set of k vertex items $\{i_v \mid v \in D\}$. To prove that S is a solution for I_{PID}, we define a proportional allocation π in the instance obtained by removing S. First, for each selection agent s_i with $1 \leq i \leq n-k$, we let π allocate $f(s_i)$ and the ith item from $I_{V \setminus D}$ to s_i . Second, for each selection agent s_{n-k+i} with $1 \leq i \leq k+1$, we let π allocate $f(s_{n-k+i})$ and the dummy item c_i to s_{n-k+i}. Third, π allocates the items $f(a_v^j)$ and d_v^j to each vertex agent $a_v^j \in N$.

It is straightforward to check that π is indeed proportional.

For proving NP-completeness, observe that the presented FPT-reduction is a polynomial reduction as well, so the NP-hardness of DOMINATING SET implies that PID is NP-hard as well; since for any subset of the items we can verify in polynomial time whether it yields a solution, containment in NP follows.   □

In fact, we can strengthen the W[2]-hardness result of Theorem 1 and show that PID is even W[3]-hard with respect to parameter k.[2]

Theorem 2 ($\star$). PROPORTIONALITY BY ITEM DELETION *is* W[3]-*hard when parameterized by the size k of the desired solution.*

Theorem 2 implies that we cannot expect an FPT-algorithm for PID with respect to the parameter k, the number of item deletions allowed, unless FPT $\neq$ W[3]. Next we show that the brute force algorithm that runs in $|I|^{O(k)}$ time is optimal, assuming the slightly stronger assumption FPT $\neq$ W[1].

Theorem 3. *There is no algorithm for PID that on an instance $(\mathcal{P}, k)$ with item set I runs in $f(k)|I|^{o(k)}|\mathcal{P}|^{O(1)}$ time for some function f, unless* FPT $\neq$ W[1].[3]

Proof. Chen et al. [5] introduced the class of $W_l[2]$-hard problems based on the notion of *linear FPT-reductions*. They proved that DOMINATING SET is $W_l[2]$-hard, and that this implies a strong lower bound on its complexity: unless FPT $\neq$ W[1], DOMINATING SET cannot be solved in $f(k)|V|^{o(k)}(|V| + |E|)^{O(1)}$ time for any function f.

Observe that in the FPT-reduction presented in the proof of Theorem 1 the new parameter has linear dependence on the original parameter (in fact they coincide). Therefore, this reduction is a linear FPT-reduction, and consequentially, PID is $W_l[2]$-hard. Hence, as proved by Chen et al. [5], PID on an instance $(\mathcal{P}, k)$ with item set I cannot be solved in time $f(k)|I|^{o(k)}|\mathcal{P}|^{O(1)}$ time for any function f, unless FPT $\neq$ W[1].   □

If we want to optimize the running time not with respect to the number k of allowed deletions but rather in terms of the total number of items, then we can also give the following tight complexity result, under the Exponential Time Hypothesis (ETH). This hypothesis, formulated in the seminal paper by Impagliazzo, Paturi, and Zane [8] says that 3-SAT cannot be solved in $2^{o(n)}$ time, where n is the number of variables in the 3-CNF fomular given as input.

---

[2] We present Theorem 1 so that we can re-use its proof for Theorems 3 and 4.

[3] Here, we use an effective variant of "little o" (see, e.g. [7, Definition 3.22]).

Theorem 4. *PID can be solved in $O^*(2^{|I|})$ time, but unless the ETH fails, it cannot be solved in $2^{o(|I|)}$ time, where I is the set of items in the input.*

Proof. The so-called Sparsification Lemma proved by Impagliazzo et al. [8] implies that assuming the ETH, 3-SAT cannot be solved in $2^{o(m)}$ time, where m is the number of clauses in the 3-CNF formula given as input. Since the standard reduction from 3-SAT to DOMINATING SET transforms a 3-CNF formula with n variables and m clauses into an instance (G, n) of DOMINATING SET such that the graph G has $O(m)$ vertices and maximum degree 3 (see, e.g., [14]), it follows that DOMINATING SET on a graph (V, E) cannot be solved in $2^{o(|V|)}$ time even on graphs having maximum degree 3, unless the ETH fails.

Recall that the reduction presented in the proof of Theorem 1 computes from each instance (G, k) of DOMINATING SET with $G = (V, E)$ an instance $(\mathcal{P}, k)$ of PID where the number of items is $3|V| + 2|E| + 1$. Hence, assuming that our input graph G has maximum degree 3, we obtain $|I| = O(|V|)$ for the set I of items in $\mathcal{P}$. Therefore, an algorithm for PID running in $2^{o(|I|)}$ time would yield an algorithm for DOMINATING SET running in $2^{o(|V|)}$ time on graphs of maximum degree 3, contradicting the ETH. □

4   Three Agents

It is known that PID for two agents is solvable in polynomial-time: the problem of obtaining an envy-free allocation by item deletion is polynomial-time solvable if there are only two agents [2,4]; since for two agents an allocation is proportional if and only it is envy-free [1], this proves tractability of PID for $|N| = 2$ immediately. In this section, we generalize this result by proving that PID is polynomial-time solvable for three agents.

Let us define the underlying graph G of our profile $\mathcal{P}$ of PID as the following bipartite graph. The vertex set of G consists of the set I of items on the one side, and a set S on the other side, containing all pairs of the form (x, i) where $x \in N$ is an agent and $i \in \{1, \ldots, \lceil |I|/|N| \rceil\}$. Such pairs are called *slots*. We can think of the slot (x, i) as the place for the ith item that agent x receives in some allocation. We say that an item is *eligible* for a slot (x, i), if it is contained in $L^x[1 : |N|(i-1)+1]$. In the graph G, we connect each slot with the items that are eligible for it. Observe that any proportional allocation corresponds to a perfect matching in G; see Lemma 1 for a proof.

In what follows, we suppose that our profile $\mathcal{P}$ contains three agents, so let $N = \{a, b, c\}$.

4.1   Basic Concepts: Prefixes and Minimum Obstructions

Since our approach to solve PID with three agents is to apply dynamic programming, we need to handle partial instances of PID. Let us define now the basic necessary concepts.

Prefixes. For any triple (i_a, i_b, i_c) with $1 \leq i_a, i_b, i_c \leq |I|$ we define a *prefix* $\mathcal{Q} = \mathcal{P}[i_a, i_b, i_c]$ of $\mathcal{P}$ as the triple $(L^a[1 : i_a], L^b[1 : i_b], L^c[1 : i_c])$, listing only the first i_a, i_b, i_c items in the preference list of agents a, b, and c, respectively. We call (i_a, i_b, i_c) the *size* of Q and denote it by size(Q). We also define the *suffix* $P - Q$ as the triple $(L^a[i_a + 1 : |I|], L^b[i_b + 1 : |I|], L^c[i_c + 1 : |I|])$, which can be thought of as the remainder of P after deleting Q from it.

We say that a prefix $P_i = P[i_a, i_b, i_c]$ is *contained in* another prefix $P_j = P[j_a, j_b, j_c]$ if $j_x \leq i_x$ for each $x \in N$; the containment is *strict* if $j_x < i_x$ for some $x \in N$. We say that P_i and P_j are *intersecting* if none of them contains the other; we call the unique largest prefix contained both in P_i and in P_j, i.e., the prefix $P[\min(i_a, j_a), \min(i_b, j_b), \min(i_c, j_c)]$, their *intersection*, and denote it by $P_i \cap P_j$.

For some prefix $Q = P[i_a, i_b, i_c]$, let $I(Q)$ denote the set of all items appearing in Q, and let $S(Q)$ denote the set of all slots appearing in Q, i.e., $S(Q) = \{(x, i) \mid 1 \leq i \leq \lceil (i_x + 2)/3 \rceil, x \in N\}$. We also define the graph $G(Q)$ underlying Q as the subgraph of G induced by all slots and items appearing in Q, that is, $G(Q) = G[S(Q) \cup I(Q)]$. We say that a slot is *complete* in Q, if it is connected to the same items in $G(Q)$ as in G; clearly the only slots which may be incomplete are the last slots in Q, that is, the slots $(x, \lceil (i_x + 2)/3 \rceil)$, $x \in N$.

Solvability. We say that a prefix Q is *solvable*, if the underlying graph $G(Q)$ has a matching that covers all its complete slots. Hence, a prefix is solvable exactly if there exists an allocation π from $I(Q)$ to N that satisfies the condition of proportionality restricted to all complete slots in Q: for any agent $x \in N$ and any index $i \in \{1, \ldots, i'_x\}$, the number of items from $L^x[1 : i_x]$ allocated by π to x is at least $i_x/3$; here $i'_x = 3(\lfloor (i_x + 2)/3 \rfloor) - 2$ is the last position in Q that is contained in a complete slot for agent x.

Minimal obstructions. We say that a prefix Q is a *minimal obstruction*, if it is not solvable, but all prefixes strictly contained in Q are solvable. See Fig. 1 for an illustration. The next lemmas claim some useful observations about minimal obstructions.

Profile P: Min. obstruction Q: Graph $G(Q)$:

a: $1, 3, 2, 4, 6, 5, 7$.   a: $1, 3, 2, 4$.

b: $3, 1, 5, 2, 7, 4, 6$.   b: $3, 1, 5, 2$.

c: $2, 4, 5, 3, 6, 7, 1$.   c: $2, 4, 5, 3$.

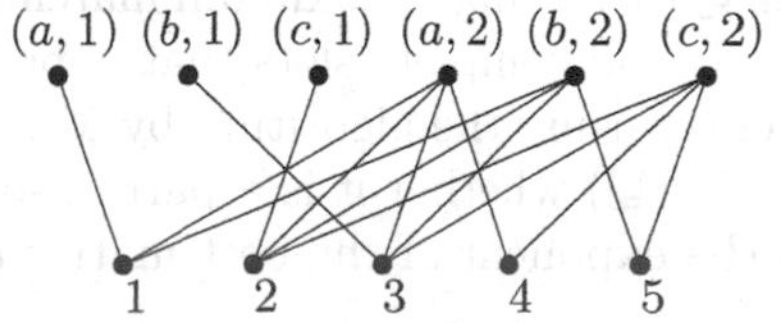

$Q - \{2\}$: $P - \{2\}$:

a: $\underline{1}, 3, 4$.   a: $\underline{1}, 3, 4, \underline{6}, 5, 7$.

b: $\underline{3}, 1, 5$.   b: $\underline{3}, 1, 5, \underline{7}, 4, 6$.

c: $\underline{4}, 5, 3$.   c: $\underline{4}, \underline{5}, 3, 6, 7, 1$.

Fig. 1. An example profile P with item set $I = \{1, 2, \ldots, 7\}$, a minimal obstruction Q of size $(4, 4, 4)$ in P and its associated graph $G(Q)$. Note that the partial solution $\{2\}$ for Q is a solution for P as well. We depicted a proportional allocation for $Q - \{2\}$ and $P - \{2\}$ by underlining in each agent's preference list the items allocated to her.

Lemma 1 ($\star$). *Profile $\mathcal{P}$ admits a proportional allocation if and only if the underlying graph G contains a perfect matching. Also, in $O(|I|^3)$ time we can find either a proportional allocation for $\mathcal{P}$, or a minimal obstruction $\mathcal{Q}$ in $\mathcal{P}$.*

Lemma 2 ($\star$). *Let $\mathcal{Q} = \mathcal{P}[i_a, i_b, i_c]$ be a prefix of $\mathcal{P}$ that is a minimal obstruction. Then $i_a \equiv i_b \equiv i_c \equiv 1 \mod 3$, and either*

(i) $i_a = i_b = i_c$, or
(ii) $i_x = i_y = i_z + 3$ for some choice of agents x, y, and z with $\{x, y, z\} = \{a, b, c\}$.

Moreover, if (ii) holds, then $L^x[1 : i_x]$ and $L^y[1 : i_y]$ contain exactly the same item set, namely $I(\mathcal{Q})$.

Based on Lemma 2, we define the *shape* of a minimal obstruction $\mathcal{Q}$ as either *straight* or *slant*, depending on whether $\mathcal{Q}$ fulfills the conditions (i) or (ii), respectively. More generally, we also say that a prefix has straight or slant shape if it fulfills the respective condition. Furthermore, we define the *boundary items* of $\mathcal{Q}$, denoted by $\delta(\mathcal{Q})$, as the set of all items that appear once or twice (but not three times) in $\mathcal{Q}$.

Lemma 3 ($\star$). *Let $\mathcal{Q}$ be a prefix of $\mathcal{P}$ that is a minimal obstruction. Then the boundary of $\mathcal{Q}$ contains at most three items: $|\delta(\mathcal{Q})| \leq 3$.*

4.2 Partial Solutions and Branching Sets

Partial solutions. For a prefix $\mathcal{Q}$ and a set U of items, we define $\mathcal{Q} - U$ in the natural way: by deleting all items of U from the (partial) preference lists of the profile (note that the total length of the preference lists constituting the profile may decrease). We say that an item set $Y \subseteq I(\mathcal{Q})$ is a *partial solution for $\mathcal{Q}$* if $\mathcal{Q} - Y$ is solvable. See again Fig. 1 for an example.

Observe that for any item set Y we can check whether it is a partial solution for $\mathcal{Q}$ by finding a maximum matching in the corresponding graph (containing all items and complete slots that appear in $\mathcal{Q} - Y$), which has at most $2|I|$ vertices. Hence, using the algorithm by Mucha and Sankowski [10], we can check for any $Y \subseteq I(\mathcal{Q})$ whether it is a partial solution for $\mathcal{Q}$ in $O(|I|^\omega)$ time where $\omega < 2.38$ is the exponent of the best matrix multiplication algorithm.

Branching set. To solve PID we will repeatedly apply a branching step: whenever we encounter a minimal obstruction $\mathcal{Q}$, we shall consider several possible partial solutions for $\mathcal{Q}$, and for each partial solution Y we try to find a solution U that contains Y. To formalize this idea, we say that a family $\mathcal{Y}$ containing partial solutions for a minimal obstruction $\mathcal{Q}$ is a *branching set for $\mathcal{Q}$*, if there exists a solution U of minimum size for the profile $\mathcal{P}$ such that $U \cap I(\mathcal{Q}) \in \mathcal{Y}$. Such a set is exactly what we need to build a search tree algorithm for PID.

Lemma 4 shows that we never need to delete more than two items from any minimal obstruction. This will be highly useful for constructing a branching set.

Lemma 4 (⋆). *Let Q be a minimal obstruction in a profile P, and let U denote an inclusion-wise minimal solution for P. Then $|U \cap I(Q)| \leq 2$.*

Lemma 4 implies that simply taking all partial solutions of $I(Q)$ of size 1 or 2 yields a branching set for Q.

Corollary 1. *For any minimal obstruction Q in a profile, a branching set $\mathcal{Y}$ for Q of cardinality at most $|I(Q)| + \binom{|I(Q)|}{2} = O(|I|^2)$ and with $\max_{Y \in \mathcal{Y}} |Y| \leq 2$ can be constructed in polynomial time.*

4.3   Domination: Obtaining a Smaller Branching Set

To exploit Lemma 4 in a more efficient manner, we will rely on an observation about the equivalence of certain item deletions, which can be used to reduce the number of possibilities that we have to explore when encountering a minimal obstruction, i.e., the size of our branching set. To this end, we need some additional notation. Given a prefix $Q = P[i_a, i_b, i_c]$, we define its *tail* as the set $T(Q)$ of items as follows, depending on the shape of Q.

- If Q has straight shape, then $T(Q)$ contains the last three items contained in Q for each agent, that is, all items in $L^a[i_a - 2 : i_a]$, $L^b[i_b - 2 : i_b]$, and $L^c[i_c - 2 : i_c]$.
- If Q has slant shape with $i_z = i_x - 3 = i_y - 3$ for some choice of agents x, y, and z with $\{x, y, z\} = \{a, b, c\}$, then $T(Q)$ contains the last six items in Q listed by agents x and y, that is, all items in $L^x[i_x - 5 : i_x]$ and $L^y[i_y - 5 : i_y]$.

Let us state the main property of the tail which motivates its definition.

Lemma 5 (⋆). *Suppose Q is a minimum obstruction in P, and R is a prefix of P intersecting Q such that $R - X$ is a minimum obstruction for some item set X with $|X \cap I(Q)| \leq 2$. Then any item that occurs more times in Q than in R must be contained in the tail of Q.*

Next, we give a condition that guarantees that some partial solution for a minimum obstruction Q is "not worse" than some other. Given two sets of items $Y, Y' \subseteq I(Q)$, we say that Y' *dominates* Y with respect to the prefix Q, if

(1) $|Y| = |Y'|$,
(2) Y' only contains an item from the boundary or the tail of Q if Y also contains that item, i.e., $Y' \cap (\delta(Q) \cup T(Q)) \subseteq Y \cap (\delta(Q) \cup T(Q))$.

Lemma 6 (⋆). *If U is an inclusion-wise minimal solution for the profile P, Q is a minimal obstruction in P, $Y = U \cap I(Q)$ and $Y' \subseteq I(Q)$ is a partial solution for Q that dominates Y, then $U \setminus Y \cup Y'$ is a solution for P.*

Lemma 6 means that if a branching set $\mathcal{Y}$ contains two different partial solutions Y and Y' for a minimum obstruction such that Y' dominates Y, then removing Y from $\mathcal{Y}$ still results in a branching set. Using this idea, we can construct a branching set of constant size.

Lemma 7. *There is a polynomial-time algorithm that, given a minimal obstruction Q in the profile $\mathcal{P}$, produces a branching set $\mathcal{Y}$ with $\max_{Y \in \mathcal{Y}} |Y| \leq 2$ and $|\mathcal{Y}| = O(1)$.*

Proof. First observe that for any two item sets Y and Y' in Q, we can decide whether Y dominates Y' in $O(\min(|Y|, |Y'|))$ time. Hence, we can simply start from the branching set $\mathcal{Y}$ guaranteed by Corollary 1, and check for each $Y \in \mathcal{Y}$ whether there exists some $Y' \in \mathcal{Y}$ that dominates Y; if so, then we remove Y. By Lemma 6, at the end of this process the set family $\mathcal{Y}$ obtained is a branching set.

We claim that $\mathcal{Y}$ has constant size. To see this, observe that if Y_1 and Y_2 are both in $\mathcal{Y}$ and have the same size, then both $Y_1 \setminus Y_2$ and $Y_2 \setminus Y_1$ contain an element from $T(Q) \cup \delta(Q)$. Thus, we can bound $|\mathcal{Y}|$ using the pigeon-hole principle: first, $\mathcal{Y}$ may contain at most $|T(Q) \cup \delta(Q)|$ partial solutions of size 1, and second, it may contain at most $\binom{|T(Q) \cup \delta(Q)|}{2}$ partial solutions of size 2. Recall that $|T(Q)| \leq 9$ by definition, and we also have $|\delta(Q)| \leq 3$ by Lemma 3, proving our claim. $\square$

4.4   Polynomial-Time Algorithm for PID for Three Agents

Let us now present our algorithm for solving PID on our profile $\mathcal{P} = (N, I, L)$.

We are going to build the desired solution step-by-step, iteratively extending an already found partial solution. Namely, we propose an algorithm $\mathrm{MinDel}(\mathcal{T}, U)$ that, given a prefix $\mathcal{T}$ of $\mathcal{P}$ and a partial solution U for $\mathcal{T}$, returns a solution S for $\mathcal{P}$ for which $S \cap I(\mathcal{T}) = U$, and has minimum size among all such solutions. We refer to the set $S \setminus U$ as an *extension* for $(\mathcal{T}, U)$; note that an extension for $(\mathcal{T}, U)$ only contains items from $I \setminus I(\mathcal{T})$. We will refer to the set of items in $I(\mathcal{T}) \setminus U$ as *forbidden* w.r.t. $(\mathcal{T}, U)$.

Branching set with forbidden items. To address the problem of finding an extension for $(\mathcal{T}, U)$, we modify the notion of a branching set accordingly. Given a minimal obstruction Q and a set $F \subseteq I(Q)$ of items, we say that a family $\mathcal{Y}$ of partial solutions for Q is a *branching set for Q forbidding F*, if the following holds: either there exists a solution U for the profile $\mathcal{P}$ that is disjoint from F and has minimum size among all such solutions, and moreover, fulfills $U \cap I(Q) \in \mathcal{Y}$, or $\mathcal{P}$ does not admit any solution disjoint from F.

Lemma 8. *There is a polynomial-time algorithm that, given a minimal obstrucion Q in a profile and a set $F \subseteq I(Q)$ of forbidden items, produces a branching set $\mathcal{Y}$ forbidding F with $\max_{Y \in \mathcal{Y}} |Y| \leq 2$ and $|\mathcal{Y}| = O(1)$.*

Proof. The algorithm given in Lemma 7 can be adapted in a straightforward fashion to take forbidden items into account: it suffices to simply discard in the first place any subset $Y \subseteq I(Q)$ that is not disjoint from F. It is easy to verify that this modification indeed yields an algorithm as desired. $\square$

Equivalent partial solutions. We will describe MinDel as a recursive algorithm, but in order to ensure that it runs in polynomial time, we need to apply dynamic programming. For this, we need a notion of equivalence: we say that two partial solutions U_1 and U_2 for $\mathcal{T}$ are *equivalent* if (1) $|U_1| = |U_2|$, and (2) $(\mathcal{T}, U_1)$ and $(\mathcal{T}, U_2)$ admit the same extensions.

Ideally, whenever we perform a call to MinDel with a given input $(\mathcal{T}, U)$, we would like to first check whether an equivalent call has already been performed, i.e., whether MinDel has been called with an input $(\mathcal{T}, U')$ for which U and U' are equivalent. However, the above definition of equivalence is computationally hard to handle: there is no easy way to check whether two partial solutions admit the same extensions or not. To overcome this difficulty, we will use a stronger condition that implies equivalence.

Deficiency and strong equivalence. Consider a solvable prefix $\mathcal{Q}$ of $\mathcal{P}$. We let the *deficiency* of $\mathcal{Q}$, denoted by $\mathrm{def}(\mathcal{Q})$, be the value $|S(\mathcal{Q})| - |I(\mathcal{Q})|$. Note that due to possibly incomplete slots in $\mathcal{Q}$, the deficiency of $\mathcal{Q}$ may be positive even though $\mathcal{Q}$ is solvable. However, if $\mathcal{Q}$ contains only complete slots, then its solvability implies $\mathrm{def}(\mathcal{Q}) \leq 0$. We define the *deficiency pattern* of $\mathcal{Q}$ as the set of all triples

$$(\mathrm{size}(\mathcal{Q} \cap \mathcal{R}), \mathrm{def}(\mathcal{Q} \cap \mathcal{R}), I(\mathcal{Q} \cap \mathcal{R}) \cap \delta(Q))$$

where $\mathcal{R}$ can be any prefix with a straight or a slant shape that intersects $\mathcal{Q}$. Roughly speaking, the deficiency pattern captures all the information about $\mathcal{Q}$ that is relevant for determining whether a given prefix intersecting $\mathcal{Q}$ is a minimal obstruction or not.

Now, we call the partial solutions U_1 and U_2 for $\mathcal{T}$ *strongly equivalent*, if

1. $|U_1| = |U_2|$,
2. $U_1 \cap \delta(\mathcal{T}) = U_2 \cap \delta(\mathcal{T})$, and
3. $\mathcal{T} - U_1$ and $\mathcal{T} - U_2$ have the same deficiency pattern.

As the name suggests, strong equivalence is a sufficient condition for equivalence.

Lemma 9 ($\star$). *If U_1 and U_2 are strongly equivalent partial solutions for $\mathcal{T}$, then they are equivalent as well.*

Now, we are ready to describe the MinDel algorithm in detail. Let $(\mathcal{T}, U)$ be the input for MinDel. Throughout the run of the algorithm, we will store all inputs with which MinDel has been computed in a table SolTable, keeping track of the corresponding solutions for $\mathcal{P}$ as well. Initially, SolTable is empty.

Step 0: Check for strongly equivalent inputs. For each $(\mathcal{T}, U')$ in SolTable, check whether U' and U are strongly equivalent, and if so, return MinDel$(\mathcal{T}, U')$.

Step 1: Check for trivial solution. Check if $\mathcal{P} - U$ is solvable. If so, then store the entry $(\mathcal{T}, U)$ together with the solution U in SolTable, and return U.

Step 2: Find a minimal obstruction. Find a minimal obstruction $\mathcal{Q}$ in $\mathcal{P} - U$; recall that $\mathcal{P} - U$ is not solvable in this step. Let $\mathcal{T}'$ be the prefix of $\mathcal{P}$ for which $\mathcal{T}' - U = \mathcal{Q}$.

Step 3: Compute a branching set. Using Lemma 8, determine a branching set $\mathcal{Y}$ for $\mathcal{Q}$ forbidding $I(\mathcal{T}) \setminus U$. If $\mathcal{Y} = \emptyset$, then stop and reject.

Step 4: Branch. For each $Y \subseteq \mathcal{Y}$, compute $S_Y := \mathrm{MinDel}(\mathcal{T}', U \cup Y)$.

Step 5: Find a smallest solution. Compute a set S_{Y*} for which $|S_{Y*}| = \min_{Y \in \mathcal{Y}} |S_Y|$. Store the entry $(\mathcal{T}, U)$ together with the solution S_{Y*} in SolTable, and return S_{Y*}.

Lemma 10 ($\star$). *Algorithm MinDel is correct, i.e., for any prefix $\mathcal{T}$ of $\mathcal{P}$ and any partial solution U for $\mathcal{T}$, $\mathrm{MinDel}(\mathcal{T}, U)$ returns a solution S for $\mathcal{P}$ with $S \cap I(\mathcal{T}) = U$, having minimum size among all such solutions (if existent).*

Lemma 10 immediately gives us an algorithm to solve PID. Let $\mathcal{T}_\emptyset$ denote the empty prefix of our input profile $\mathcal{P}$, i.e. $\mathcal{P}[0, 0, 0]$; then $\mathrm{MinDel}(\mathcal{T}_\emptyset, \emptyset)$ returns a solution S for $\mathcal{P}$ of minimum size; we only have to compare $|S|$ with the desired solution size k.

The next lemma states that MinDel gets called polynomially many times.

Lemma 11. *Throughout the run of algorithm MinDel initially called with input $(\mathcal{T}_\emptyset, \emptyset)$, the table SolTable contains $O(|I|^7)$ entries.*

Proof. Let us consider table SolTable at a given moment during the course of algorithm MinDel, initially called with the input $(\mathcal{T}_\emptyset, \emptyset)$ (and having possibly performed several recursive calls since then). Let us fix a prefix $\mathcal{T}$. We are going to give an upper bound on the maximum size of the family $\mathcal{U}_\mathcal{T}$ of partial solutions U for $\mathcal{T}$ for which SolTable contains the entry $(\mathcal{T}, U)$.

By Step 0 of algorithm MinDel, no two sets in $\mathcal{U}_\mathcal{T}$ are strongly equivalent. Recall that if U_1 and U_2, both in $\mathcal{U}_\mathcal{T}$, are not strongly equivalent, then either $|U_1| \neq |U_2|$, or $\delta(\mathcal{T}) \cap U_1 \neq \delta(\mathcal{T}) \cap U_2$, or $\mathcal{T} - U_1$ and $\mathcal{T} - U_2$ have different deficiency patterns. Let us partition the sets in $\mathcal{U}_\mathcal{T}$ into *groups*: we put U_1 and U_2 in the same group, if $|U_1| = |U_2|$ and $\delta(\mathcal{T}) \cap U_1 = \delta(\mathcal{T}) \cap U_2$.

Examining Steps 2–4 of algorithm MinDel, we can observe that if $U \neq \emptyset$, then for some $Y_U \subseteq U$ of size 1 or 2, the prefix $\mathcal{T} - (U \setminus Y_U)$ is a minimal obstruction $\mathcal{Q}_U$. Since removing items from a prefix cannot increase the size of its boundary, Lemma 3 implies that the boundary of $\mathcal{T} - U$ contains at most 3 items. We get $|\delta(\mathcal{T}) \setminus U| \leq |\delta(\mathcal{T} - U)| \leq 3$, from which it follows that $\delta(\mathcal{T}) \cap U$ is a subset of $\delta(\mathcal{T})$ of size at least $|\delta(\mathcal{T})| - 3$. Therefore, the number of different values that $\delta(\mathcal{T}) \cap U$ can take is $O(|I|^3)$. Since any $U \in \mathcal{U}_\mathcal{T}$ has size at most $|I|$, we get that there are $O(|I|^4)$ groups in $\mathcal{U}_\mathcal{T}$. Let us fix some group $\mathcal{U}_g$ of $\mathcal{U}_\mathcal{T}$. We are going to show that the number of different deficiency patterns for $\mathcal{T} - U$ where $U \in \mathcal{U}_g$ is constant.

Recall that the deficiency pattern of $\mathcal{T} - U$ contains triples of the form $(\mathrm{size}(\mathcal{R}^\cap), \mathrm{def}(\mathcal{R}^\cap), I(\mathcal{R}^\cap) \cap \delta(\mathcal{T} - U))$, where $\mathcal{R}^\cap$ is the intersection of $\mathcal{T} - U$ and some prefix $\mathcal{R}$ of $\mathcal{P} - U$ with a slant or a straight shape.

First observe that by the definition of a group, $\mathrm{size}(\mathcal{T} - U_1) = \mathrm{size}(\mathcal{T} - U_2)$ holds for any $U_1, U_2 \in \mathcal{U}_g$. Let us fix an arbitrary $U \in \mathcal{U}_g$. Since $\mathcal{T} - U$ can be obtained by deleting 1 or 2 items from a minimal obstruction, Lemma 2 implies that there can only be a constant number of prefixes $\mathcal{R}$ of $\mathcal{P} - U$ which intersect

$T - U$ and have a slant or a straight shape; in fact, it is not hard to check that the number of such prefixes R is at most 5 for any given $T - U$. Therefore, the number of values taken by the first coordinate size($\mathcal{R}^{\cap}$) of any triple in the deficiency pattern of $T - U$ is constant. Since $T - U$ has the same size for any $U \in \mathcal{U}_g$, we also get that these values coincide for any $U \in \mathcal{U}_g$. Hence, we obtain that (A) the total number of values the first coordinate of any triple in the deficiency pattern of $T - U$ for any $U \in \mathcal{U}_g$ can take is constant.

Let $\mathcal{R}_{\cap}$ be the intersection of $T - U$ and some prefix of straight or slant shape. By definition, $\mathcal{R}_{\cap}$ is contained in $\mathcal{Q}_U$. By $|Y_U| \leq 2$, there are only a constant number of positions which are contained in $\mathcal{Q}_U$ but not in $\mathcal{R}_{\cap}$. From this both $||I(\mathcal{R}_{\cap})| - |I(\mathcal{Q}_U)|| = O(1)$ and $||S(\mathcal{R}_{\cap})| - |S(\mathcal{Q}_U)|| = O(1)$ follow. As $\mathcal{Q}_U$ is a minimal obstruction, we also have $|I(\mathcal{Q}_U)| = |S(\mathcal{Q}_U)| - 1$, implying that (B) the deficiency def($\mathcal{R}_{\cap}$) = $|S(\mathcal{R}_{\cap})| - |I(\mathcal{R}_{\cap})|$ can only take a constant number of values too; note that we have an upper bound on $|\text{def}(\mathcal{R}_{\cap})|$ that holds for any $U \in \mathcal{U}_g$. Considering that $I(\mathcal{R}_{\cap}) \cap \delta(T - U)$ is the subset of $\delta(T - U)$, and we also know $|\delta(T - U)| \leq 3$, we obtain that (C) the set $I(\mathcal{R}_{\cap}) \cap \delta(T - U)$ can take at most 2^3 values (again, for all $U \in \mathcal{U}_g$).

Putting together the observations (A), (B), and (C), it follows that the number of different deficiency patterns of $T - U$ taken over all $U \in \mathcal{U}_g$ is constant. This implies $|\mathcal{U}_T| = O(|I|^4)$. Since there are $O(|I|^3)$ prefixes T of $\mathcal{P}$, we arrive at the conclusion that the maximum number of entries in SolTable is $O(|I|^7)$. $\square$

Theorem 5. PROPORTIONAL ITEM DELETIONS *for three agents can be solved in time* $O(|I|^{9+\omega})$ *where* $\omega < 2.38$ *is the exponent of the best matrix multiplication algorithm.*

Proof. By Lemma 10, we know that algorithm MinDel($T_\emptyset, \emptyset$) returns a solution for $\mathcal{P}$ of minimum size, solving PID. We can use Lemma 11 to bound the running time of MinDel($T_\emptyset, \emptyset$): since SolTable contains $O(|I|^7)$ entries, we know that the number of recursive calls to MinDel is also $O(|I|^7)$. It remains to give a bound on the time necessary for the computations performed by MinDel, when not counting the computations performed in recursive calls. Clearly, Step 0 takes $O(1)$ time. Steps 1 and 2 can be accomplished in $O(|I|^3)$ time, as described in Lemma 1. Using Lemma 8, Step 3 can be performed in $O(|I|^{2+\omega})$ time. Since the cardinality of the branching set found in Step 3 is constant, Steps 4 and 5 can be performed in linear time. This gives us an upper bound of $O(|I|^{9+\omega})$ on the total running time. $\square$

We remark that in order to obtain Theorem 5, it is not crucial to compute a branching set of constant size in Step 3: a polynomial running time would still follow even if we used a branching set of quadratic size. Thus, for our purposes, it would be sufficient to use an extension of Corollary 1 that takes forbidden items into account (an analog of Lemma 8) in Step 3. Therefore, the ideas of Sect. 4.3 – the notion of domination between partial solutions, leading to Lemma 7 – can be thought of as a speed-up that offers a more practical algorithm.

5   Conclusion

In Sect. 4 we have shown that PROPORTIONALITY BY ITEM DELETION is polynomial-time solvable if there are only three agents. On the other hand, if the number of agents is unbounded, then PID becomes NP-hard, and practically intractable already when we want to delete only a small number of items, as shown by the W[3]-hardness result of Theorem 2.

The complexity of PID remains open for the case when the number of agents is a constant greater than 3. Is it true that for any constant n, there exists a polynomial-time algorithm that solves PID in polynomial time for n agents? If the answer is yes, then can we even find an FPT-algorithm with respect to the parameter n? If the answer is no (that is, if PID turns out to be NP-hard for some constant number of agents), then can we at least give an FPT-algorithm with parameter k for a constant number of agents (or maybe with combined parameter (k, n))?

Finally, there is ample space for future research if we consider different control actions (such as adding or replacing items), different notions of fairness, or different models for agents' preferences.

Acknowledgments. This work has been partly supported by COST Action IC1205 on Computational Social Choice, and has been supported by OTKA grants K108383 and K108947 and Austrian Science Fund grant J4047.

References

1. Aziz, H., Gaspers, S., Mackenzie, S., Walsh, T.: Fair assignment of indivisible objects under ordinal preferences. Artif. Intell. **227**, 71–92 (2015)
2. Aziz, H., Schlotter, I., Walsh, T.: Control of fair division. In: IJCAI 2016, Proceedings of the 25th International Joint Conference on Artificial Intelligence, pp. 67–73 (2016)
3. Bartholdi, J.J., Tovey, C.A., Trick, M.A.: How hard is it to control an election? Math. Comput. Model. **16**(8–9), 27–40 (1992)
4. Brams, S.J., Kilgour, D.M., Klamler, C.: Two-person fair division of indivisible items: An efficient, envy-free algorithm. Not. AMS **61**(2), 130–141 (2014)
5. Chen, J., Huang, X., Kanj, I.A., Xia, G.: Strong computational lower bounds via parameterized complexity. J. Comput. Syst. Sci. **72**(8), 1346–1367 (2006)
6. Downey, R.G., Fellows, M.R.: Fundamentals of Parameterized Complexity. Texts in Computer Science. Springer, London (2013)
7. Flum, J., Grohe, M.: Parameterized Complexity Theory. Texts in Theoretical Computer Science. An EATCS Series, vol. XIV. Springer, Berlin (2006)
8. Impagliazzo, R., Paturi, R., Zane, F.: Which problems have strongly exponential complexity? J. Comput. Syst. Sci. **63**(4), 512–530 (2001)
9. Lipton, R.J., Markakis, E., Mossel, E., Saberi, A.: On approximately fair allocations of indivisible goods. In: EC 2004, Proceedings of the 5th ACM Conference on Electronic Commerce, pp. 125–131 (2004)
10. Mucha, M., Sankowski, P.: Maximum matchings via gaussian elimination. In: FOCS 2014, Proceedings of the 45th Annual IEEE Symposium on Foundations of Computer Science, pp. 248–255 (2004)

11. Nguyen, T., Vohra, R.: Near feasible stable matchings. In: EC 2015, Proceedings of the Sixteenth ACM Conference on Economics and Computation, pp. 41–42 (2015)
12. Schlotter, I., Dorn, B., de Haan, R.: Obtaining a proportional allocation by deleting items. CoRR, abs/1705.11060 (2017)
13. Segal-Halevi, E., Hassidim, A., Aumann, Y.: Waste makes haste: Bounded time protocols for envy-free cake cutting with free disposal. In: AAMAS 2014, Proceedings of the 14th International Conference on Autonomous Agents and Multi-Agent Systems, pp. 901–908 (2015)
14. Thulasiraman, K., Arumugam, S., Brandstädt, A., Nishizeki, T.: Handbook of Graph Theory, Combinatorial Optimization, and Algorithms. Chapman & Hall/CRC Computer and Information Science Series. CRC Press, Boca Raton (2015)

Possible and Necessary Allocations Under Serial Dictatorship with Incomplete Preference Lists

Katarína Cechlárová[1(✉)], Tamás Fleiner[2], and Ildikó Schlotter[2(✉)]

[1] Faculty of Science, Institute of Mathematics, P.J. Šafárik University,
Jesenná 5, Košice, Slovakia
katarina.cechlarova@upjs.sk
[2] Budapest University of Technology and Economics,
Magyar tudósok körútja 2, 1117 Budapest, Hungary
{fleiner,ildi}@cs.bme.hu

Abstract. We study assignment problems in a model where agents have strict preferences over objects, allowing preference lists to be incomplete. We investigate the questions whether an agent can obtain or necessarily obtains a given object under serial dictatorship. We prove that both problems are computationally hard even if agents have preference lists of length at most 3; by contrast, we give linear-time algorithms for the case where preference lists are of length at most 2. We also study a capacitated version of these problems where objects come in several copies.

1 Introduction

We study assignment problems that involve a set of agents and a set of objects. Agents have strict ordinal preferences over objects, but not vice versa. We assume that the preference lists can be *incomplete*: an agent might find a given object *unacceptable*. In such situations, there is a very natural and intuitive mechanism called *serial dictatorship* (SD): agents are ordered into a *picking sequence* (which is in our model simply a permutation of the agents) and everybody who has her turn picks her most preferred object out of those that are still available.

Variants of this mechanism are often used in practice. As described by Sönmez and Switzer [26], the United States Military Academy used the following procedure to assign its cadets to branches: after determining a strict priority ranking of the cadets, based on a weighted average of their academic performance, physical fitness test scores, and military performance, the Military Academy applied serial dictatorship with this ranking as the picking sequence to assign the cadets to slots at different specialties. Another example of serial dictatorship is the drafting system used in football, basketball and other professional sports in the

This work was supported by VEGA grants 1/0344/14 and 1/0142/15 from the Slovak Scientific Grant Agency and grant APVV-15-0091 (Cechlárová), by OTKA grants K108383 (Fleiner) and K108947 (Schlotter), and the MTA-ELTE Egerváry Research Group (Fleiner). The authors gratefully acknowledge the support of COST Action IC1205 Computational Social Choice.

J. Rothe (Ed.): ADT 2017, LNAI 10576, pp. 300–314, 2017.
DOI: 10.1007/978-3-319-67504-6_21

United States, where teams pick new players in the draft without the active participation of the players. The first team to pick a new player is the one with the worst win-loss record of the previous season, the one with the second worse record continues, and so on [12]. Serial dictatorship is widely applied in school admission systems as well: the most prominent example may be the centralized university admission system in China [28], where all students take a centralized test and are ordered according to the score they achieve on the test; then the authorities use serial dictatorship with this ordering to determine an allocation. Other examples are the admission system for public schools in Chicago [24] and for specialized high schools in New York City [1]. Further applications include the assignment of students to courses, or allocating rooms at colleges.

We study the following questions for a given agent a and a given object o:

(i) Can agent a receive object o under SD with *some* picking sequence?
(ii) Is it true that agent a receives object o under SD with *any* picking sequence?

Saban and Sethuraman [25] proved that problem (i) is NP-complete and gave a polynomial-time algorithm for (ii). In their model, the number of agents equals the number of objects, and agents find all objects acceptable. The authors expressed their belief that these results hold even if these assumptions are omitted.

Our contributions. We examine a model where agents may consider some objects unacceptable; hence, some agents and also some objects can stay unassigned. This situation arises in many applications: e.g., students may find certain schools unacceptable (in fact, many centralized admission systems set a limit on the number of schools a student can apply to), or certain courses may not be suitable for a student (as a result of missing prerequisites or time-table clashes).

In this setting, NP-completeness of (i) follows from the results of Saban and Sethuraman [25]; we complement this by proving that (ii) is coNP-complete. Then we deal with instances where the length of preference lists is restricted. If each agent finds at most two objects acceptable, we provide polynomial-time algorithms for both problems, based on searching appropriate digraphs. By contrast, we show intractability for the case where preference lists can have length 3.

We also study an extension of our model where objects come in several identical copies. We prove that these capacitated versions of (i) and (ii) are computationally hard already if each object has capacity at most 2 *and* all preference lists are of length at most 2.

2 Related Work

The serial dictatorship mechanism appears under various names in the literature: besides "serial dictatorship" [2,23] it is also dubbed as "queue allocation" [27], "Greedy-POM" [3], "sequential mechanism" [8,9], etc. Its importance is stressed by the fact that the assignment it produces is Pareto-optimal (or, in the economic terminology, efficient); moreover, if each agent may receive at most one object and there is only one copy of each object (called the *one-to-one* case), each Pareto-optimal assignment can be produced by SD with a suitable picking sequence [2,3,11,27].

Recently, two lines of research have emerged. One deals primarily with many-to-many extensions of the basic assignment problem, additionally accompanied by constraints imposed either on the structure of the sets of objects that an individual agent can receive [14,17] or on the whole set of allocated objects [20], or by lower quotas on the number of agents assigned to individual objects [15,19,21] and with possible extensions of serial dictatorship to such settings.

Another line of research explores in detail the properties of serial dictatorship and different types of sequential allocation mechanisms, often with a focus on manipulability [5,9,10]. Asinowski et al. [4] study the sets of objects that can or have to be allocated in some or all Pareto-optimal matchings (without specifying to which agents those objects are allocated).

Saban and Sethuraman [25] consider a randomized setting and prove that computing the proportion of the picking sequences under which agent a receives object o is #P-complete; this was independently obtained also by Aziz, Brandt and Brill [6]. Questions (i) and (ii) are equivalent to asking if the probability of an agent obtaining a given object in the randomized model is greater than 0 or equal to 1, resp. Saban and Sethuraman [25] determine the complexity of these problems assuming that the number of agents equals the number of objects, and agents find all objects acceptable; they find (i) to be NP-complete and (ii) polynomial-time solvable.

Motivated by the intractability results in [25], Aziz and Mestre [7] compute the probability of an agent getting an object in time that is FPT if the parameter is the number of objects, and polynomial if the number of agent types is fixed. Notably, they allow incomplete preferences; up to our knowledge, the only other work considering such a setting is by Asinowski et al. [4].

Aziz, Walsh and Xia [8] examine the complexity of deciding whether an agent can get or necessarily gets some object or set of objects under serial dictatorship with different classes of picking sequences. Their algorithmic results assume that agents have complete preferences and may obtain more than one objects.

3   Definitions and Notation

There is a set A of n agents and a set $\mathcal{O}$ of m objects. Each agent can consume at most one object and each object is available in only one copy. Agents have strict preferences over objects and they are allowed to declare some objects unacceptable. The n-tuple of agents' preferences is called a preference profile and it is denoted by $\mathcal{P}$. The triple $I = (A, \mathcal{O}, \mathcal{P})$ is a matching profile.

We consider serial dictatorship, where agents are ordered into a picking sequence σ, which is a permutation of A. Agents have their turn successively according to σ, and everybody on her turn picks her most preferred object among those that are still available. Obviously, different sequences can lead to different assignments. In this context, we study the following problems associated with a matching profile I, agent a and object o:

Problem POSSIBLE OBJECT POS(I, a, o).
Question: Is it true in I that a receives o under SD with *some* picking sequence?

Problem NECESSARY OBJECT NEC(I, a, o).
Question: Is it true in I that a receives o under SD with *any* picking sequence?

Let us remark that in the one-to-one case, Pareto-optimal matchings (POMs) are exactly those that can be obtained by serial dictatorship, so our questions are equivalent to asking whether a given agent can be allocated a given object in some POM and whether a given agent is allocated a given object in every POM.

4   Incomplete Preference Lists of Unbounded Length

Here we show that if the preference lists are not complete, then both problems are hard. We remark that the NP-completeness of POS(I, a, o) follows from the NP-hardness result of Saban and Sethuraman [25] obtained for the case of complete preference lists. However, the coNP-completeness of NEC(I, a, o) may seem somewhat surprising, sharply contrasting the polynomial-time algorithm given by Saban and Sethuraman [25] for complete preference lists.

Theorem 1. POS(I, a, o) *is* NP-*complete and* NEC(I, a, o) *is* coNP-*complete.*

Proof. POS(I, a, o) belongs to NP and NEC(I, a, o) belongs to coNP, since in both cases it suffices to give a picking sequence σ and check whether a gets (does not get) o under SD with σ.

To prove NP-hardness, we give a polynomial reduction from VERTEX COVER. Let our input be a graph $G = (V, E)$ with $|V| = p$ and $|E| = q$ and some $k \in \mathbb{N}$. We construct a matching profile $I(G)$ involving two agents a, b and object o in a way that G has a vertex cover of size k if and only if agent b gets o in SD under some picking sequence (i.e., POS($I(G), b, o$) is true), which in turn happens exactly if it is not true that a gets o in all picking sequences (i.e., NEC($I(G), a, o$) is false).

We define the set A of agents and $\mathcal{O}$ of objects in $I(G)$ as
$A = \{a(v) \mid v \in V\} \cup \{a(e, u), a(e, v) \mid e = \{u, v\} \in E\} \cup \{a, b\}$ and
$\mathcal{O} = \{s_1, \ldots, s_{p+q-k}\} \cup \{o(v) \mid v \in V\} \cup \{o(e) \mid e \in E\} \cup \{o\}$.

Thus $|A| = p + 2q + 2$, and there are $p + q - k$ *special s-objects*, one *vertex-object* for each vertex, one *edge-object* for each edge, and a distinguished object o. Preferences are as follows:

$$P(a(v)) : s_1, s_2, \ldots, s_{p+q-k}, o(v) \qquad \text{for each } v \in V,$$
$$P(a(e, v)) : s_1, s_2, \ldots, s_{p+q-k}, o(v), o(e) \text{ for each } e \in E \text{ and } v \in e,$$
$$P(b) : o(e_1), \ldots, o(e_q), o,$$
$$P(a) : o.$$

It is easy to see that for any picking sequence, b gets o in SD if and only if a does not get o.

Now suppose that G admits a vertex cover $U \subseteq V$ of size k. Let us orient the edges of G in a way that each edge points toward a vertex of U, and for each edge $e \in E(G)$ let $a_1(e)$ and $a_2(e)$ denote the tail and the head vertex, resp., of e in this orientation. Then order the agents according to sequence σ given in (1) (agents within square brackets are ordered arbitrarily).

$$\underbrace{[a(v) \mid v \notin U]}_{A_1}, \underbrace{[a_1(e) \mid e \in E]}_{A_2}, \underbrace{[a(v) \mid v \in U]}_{A_3}, \underbrace{[a_2(e) \mid e \in E]}_{A_4}, b, a \qquad (1)$$

How are the objects picked under σ? First, the agents in $A_1 \cup A_2$ take all the special s-objects. Then the agents in A_3 pick vertex-objects corresponding to the vertex cover U. Notice that some vertex-objects stay unassigned, but when the agents in A_4 have their turn, all the vertex-objects they are interested in are exhausted. This means that these agents use up all the edge-objects. So when agent b comes to choose, all the edge-objects are gone and b has to pick o. This leaves a with no object assigned. So $\text{POS}(I, b, o)$ is true and $\text{NEC}(I, a, o)$ is false.

Conversely, suppose that there exists a picking sequence σ where b gets o, i.e., $\text{POS}(I, b, o)$ is true and $\text{NEC}(I, a, o)$ is false. This means that when it was b's turn, all the edge-objects were gone. In other words, for each edge $e = \{u, v\}$, one of the pair of agents $a(e, u), a(e, v)$ picked the edge-object $o(e)$. Hence, the remaining q agents in these pairs and all agents in $\{a(v) \mid v \in V\}$ (a total of $p+q$ agents) must have picked all special s-objects and some vertex-objects. As they all prefer the $p + q - k$ special s-objects, exactly k of these agents could have picked vertex-objects. Thus, the k picked vertex-objects define a vertex cover for G of size k. □

5   Preference Lists of Bounded Length

Given the intractability result of Sect. 4, here we shall concentrate on the problems with preference lists of restricted length, and refine the boundary between polynomial-time solvable and intractable cases. Let us call a matching profile a *length-k matching profile*, if each agent finds at most k objects acceptable, and let k-$\text{POS}(I, a, o)$ and k-$\text{NEC}(I, a, o)$ be the restrictions of the studied problems to instances with length-k matching profiles.

5.1   Preference Lists of Length 3

Here we strengthen Theorem 1 for the case where all preference lists have length at most 3.

Given a matching profile with arbitrary preference lists, we eliminate all agents having preference lists longer than 3, while preserving certain crucial properties of I. Our strategy is to eliminate such agents one-by-one. To this end, for any matching profile $I = (A, \mathcal{O}, \mathcal{P})$ we define a matching profile $J(I) = (A', \mathcal{O}', \mathcal{P}')$, called a *substitute for I*, as follows.

Take any agent $x \in A$ whose preference list in I has length $\ell > 3$; let this list be $o_1, o_2, ..., o_\ell$. We introduce $\ell - 3$ *chain objects* $y_1, y_2, \ldots, y_{\ell-3}$, and we replace agent x with new agents $x_1, \ldots, x_{\ell-2}$, called the *brothers* of x. Hence $A' = A \setminus \{x\} \cup \{x_1, \ldots, x_{\ell-2}\}$ and $\mathcal{O}' = \mathcal{O} \cup \{y_1, \ldots, y_{\ell-3}\}$. We set $P'(a) = P(a)$ for each $a \in A \setminus \{x\}$; the preferences of the brothers of x in $J(I)$ are as follows:

$$P'(x_1) : o_1, o_2, y_1,$$
$$P'(x_i) : y_{i-1}, o_{i+1}, y_i \text{ for } i = 2, \ldots, \ell - 3,$$
$$P'(x_{\ell-2}) : y_{\ell-3}, o_{\ell-1}, o_\ell.$$

To state the crucial property of a substitute for each agent-object pair (a, o) in I we define a corresponding agent-object pair $J(a, o)$ in $J(I)$ as follows. If $a = x$ and $o = o_i$ for some i, then we let $J(a, o) = J(x, o_i) = (x_{i-1}, o_i)$; here $x_0 := x_1$ and $x_{\ell-1} := x_{\ell-2}$. Otherwise, we let $J(a, o) = (a, o)$.

Lemma 1. *Let $J(a, o) = (a', o)$ for any $a \in A$ and $o \in \mathcal{O}$. Then agent a can obtain o under some picking sequence in I if and only if a' can obtain o under some picking sequence in J'.*

Proof. For lack of space, we only give a sketch of the proof; for a full version see [16]. For any picking sequence φ for I we can define a picking sequence φ' for $J(I)$ by replacing x in φ with her brothers $x_1, \ldots, x_{\ell-2}$ in that order. It is not hard to show that agent a receives object o under φ if and only if a' receives o under φ'.

For the converse direction, let σ' be a picking sequence in $J(I)$ under which a' receives o.

Case I. If x_i is the brother of x that first picks an object from $\{o_1, \ldots, o_\ell\}$ under σ', then we replace x_i by x and delete all other brothers of x; let σ be the resulting picking sequence in I.

It is easy to see that x_i can receive some object o_{i+1} only if at her turn, all the objects $o_1, \ldots, o_i$ had already been taken, and the brothers $x_1, \ldots, x_{i-1}$ have all picked chain objects before x_i's turn. Hence, agents of A picking before x_i receive the same object under σ in I as under σ' in $J(I)$, and x picks under σ the object picked by x_i under σ'. Further, let us observe that each of the brothers picking after x_i in σ' (that is, brothers $x_{i+1}, \ldots, x_{\ell-2}$) can pick their first choice chain objects under σ'. Hence, all agents of A picking after x in σ receive the same object as they do under σ'.

Case II. If every brother of x receives either a chain object or nothing under σ', then we delete all of them from σ' and append x as the last agent to obtain σ. Observe that the deletion of the brothers of x does not affect what the remaining agents in $A \setminus \{x\}$ obtain. Note also that a' is not a brother of x (as we assume a' to get o), and hence, $a \neq x$. Thus, a gets the same object in I under σ as a' in $J(I)$ under σ'. □

We next apply the above construction iteratively. As the number of agents with preference list longer than 3 is one less in $J(I)$ than in I, repeatedly constructing a substitute for the current matching profile (i.e., taking $J(I)$, then $J(J(I))$, and so on), we finally end up with a length-3 matching profile $J^*(I)$. We also define, for any given agent-object pair (a, o) in I, the corresponding agent-object pair $J^*(a, o)$ obtained in this process (first taking $J(a, o)$, then $J(J(a, o))$, and so on). Applying Lemma 1 repeatedly, we get Corollary 1.

Corollary 1. *Let $J^*(a, o) = (a^*, o)$ for any $a \in A$ and $o \in \mathcal{O}$. Then agent a can obtain o under some picking sequence in I if and only if a^* can obtain o under some picking sequence in $J^*(I)$.*

Theorem 2. *3-POS(I, a, o) is* NP*-complete and 3-*NEC(I, a, o) is* coNP*-complete.*

Proof. We prove our theorem by modifying the reductions and the matching profile I described in the proof of Theorem 1; recall that POS(I, b, o) is true if and only if NEC(I, a, o) is false.

By the definition of a substitute, we can observe that $J^*(b, o) = (b^*, o)$ where b^* is either the last brother of b introduced when eliminating agent b or (if b does not get eliminated) $b^* = b$. By Corollary 1, it is immediate that 3-POS$(J^*(I), b, o)$ is equivalent to POS(I, b^*, o). Furthermore, our construction for $J^*(I)$ ensures that object o is only contained in the preference list of agents b^* and a; the preference list of a remains the same in $J^*(I)$ as in I, containing only o. Therefore, it should be clear that agent a receives o under all picking sequences in $J^*(I)$ if and only if there is no picking sequence in $J^*(I)$ under which agent b receives o. In other words, 3-POS$(J^*(I), b, o)$ is true if and only if 3-NEC$(J^*(I), a, o)$ is false.

Observe that $J^*(I)$ can be computed in time polynomial in $|I|$. Thus, replacing the instance POS(I, b, o) and NEC(I, a, o) constructed in the proof of Theorem 1 with 3-POS$(J^*(I), b, o)$ and 3-NEC$(J^*(I), a, o)$, resp., yields polynomial-time reductions that prove our theorem. □

5.2   Preference Lists of Length 2

Given a length-2 matching profile I and an agent a in I, there are in fact six different questions that can be asked: POSSIBLE OBJECT and NECESSARY OBJECT for the first and second object in the preference list of a (denoted by $f(a)$ and $s(a)$, resp.), and for the special object $\emptyset$ representing the situation when agent a gets nothing. Some of these questions are trivial, see the following assertion.

Lemma 2. *For any $k \in \mathbb{N}$ and $a \in A$, k-*NEC$(I, a, s(a))$ and k-*NEC$(I, a, \emptyset)$ are false, and k-*POS$(I, a, f(a))$ is true. Also, 2-*NEC$(I, a, f(a))$ is true exactly if both 2-*POS$(I, a, s(a))$ and 2-*POS$(I, a, \emptyset)$ are false.*

Hence, for length-2 matching profiles, it suffices to solve the two problems of the form 2-POS$(I, a, s(a))$ and 2-POS$(I, a, \emptyset)$.

How can it happen that agent a does not pick $o = f(a)$? Clearly, if o is the first object in the preference list of some other agent a', then it suffices to place a' before a in the picking sequence. If o is the second object in the preference list of a', then a' may still pick o, if her first choice object was picked before it was her turn, say by an agent a'', etc. To be able to discover such chains of agents, we construct for any length-2 matching profile I the following directed multigraph $G(I)$. Its vertex set is $\mathcal{O}$ and its arc set is A, where each $x \in A$ leads

$$
\begin{aligned}
P(a_1) &: \quad o_1, o_2 \\
P(a_2) &: \quad o_1, o_3 \\
P(a_3) &: \quad o_2, o_1 \\
P(a_4) &: \quad o_3, o_2
\end{aligned}
$$

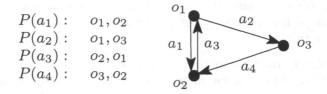

Fig. 1. Example for a length-2 matching profile and its digraph.

from $f(x)$ to $s(x)$ or, if x finds only one object acceptable, to $f(x)$; see Fig. 1 for an illustration. We denote by $\delta_H^+(p)$ the out-degree of any object p in a subgraph H of $G(I)$; for $H = G(I)$ we might omit the subscript.

Theorem 3. *Let $a \in A$ with $|P(a)| = 2$. Then* 2-POS$(I, a, s(a))$ *is true if and only if $G(I)$ contains a directed path starting from a vertex $p \in \mathcal{O}$ with $\delta^+(p) > 1$ and ending with the arc a.*

Proof. Suppose that $a = a_0$ picks her second object $s(a)$ under some picking sequence σ. This means that $f(a)$ was picked before a by some agent a_1. If $f(a) = f(a_1)$, then $\delta^+(f(a)) > 1$ as it is the tail of both a and a_1, so the arc a in itself is a path as required. If $f(a) = s(a_1)$, then the first object of a_1 must have been picked earlier in σ by some other agent a_2. Continuing the same argument, we arrive at a sequence of agents $a_0, a_1, \ldots, a_k, a_{k+1}$ appearing in the order $a_{k+1}, a_k, \ldots, a_1, a_0$ in σ, and picking the objects $f(a_{k+1}) = f(a_k)$, $s(a_k) = f(a_{k-1})$, $\ldots$, $s(a_1) = f(a_0), s(a_0) = s(a)$ in this order. Thus, the arcs $a_k, a_{k-1}, \ldots, a_0$ induce a directed path P in $G(I)$; by $f(a_{k+1}) = f(a_k)$ we get $\delta^+(f(a_{k+1})) > 1$, so P is as required.

Conversely, suppose that $G(I)$ contains a directed path P consisting of arcs $a_k, a_{k-1}, \ldots, a_1$ and vertices $o_k, o_{k-1}, \ldots, o_1, o_0$, appearing on P in this order, with $a_1 = a$ and $\delta^+(o_k) > 1$. By this latter fact, there exists an agent x, $x \notin \{a_1, \ldots, a_k\}$, whose first choice is o_k. Under SD with a picking sequence σ starting with $x, a_k, a_{k-1}, \ldots, a_1 = a$, agent x picks her first object, while each agent a_i with $k \geq i \geq 1$ picks her second object. Hence 2-POS$(I, a, s(a))$ is true. □

To see how to use Theorem 3, let us consider the example shown in Fig. 1. 2-POS$(I, a_4, s(a_4))$=2-POS(I, a_4, o_2) is true, since $G(I)$ contains the path $P = (a_2, a_4)$; note also $\delta^+(f(a_2)) = 2$. This path shows that for a picking sequence starting with a_1, a_2, a_4, agent a_1 picks object o_1, agent a_2 gets object o_3, and agent a_4 picks her second object a_2.

Similarly, in order to decide whether 2-POS$(I, a_3, s(a_3))$=2-POS(I, a_3, o_1) is true, we need to check the existence of a path ending with a_3 and starting from a vertex with out-degree at least 2. However, the only such vertex is o_1, but no path of $G(I)$ can start from o_1 and end with a_3, so 2-POS(I, a_3, o_1) is false.

To decide whether an agent whose preference list contains only one object can end up with nothing, we can use similar arguments as in the proof of Theorem 3. In fact, the following statement can be viewed as a simplified version of Theorem 3.

Theorem 4. *Let $a \in A$ with $|P(a)| = 1$. Then 2-POS$(I, a, \emptyset)$ is true if and only if $G(I)$ contains a directed path (possibly of length 0) leading from a vertex $p \in \mathcal{O}$ with $\delta^+(p) > 1$ to $f(a)$.*

By a similar method, we obtain a condition that can be used to decide whether 2-POS$(I, a, \emptyset)$ is true for an agent a who finds two objects acceptable.

Theorem 5. *Let $a \in A$ with $|P(a)| = 2$. Then 2-POS$(I, a, \emptyset)$ is true if and only if $G(I)$ contains directed paths P_1 and P_2 such that*

(1) P_1 leads from some $p_1 \in \mathcal{O}$ to $f(a)$, and P_2 leads from some $p_2 \in \mathcal{O}$ to $s(a)$, allowing $p_1 = p_2$;
(2) neither P_1 nor P_2 contains the arc a;
(3) $\delta_H^+(p_i) > 0$ for $i = 1, 2$ where H is obtained from $G(I)$ by deleting all arcs of P_1, P_2, and a.

Proof. Suppose a gets nothing under a picking sequence σ. As in the proof of Theorem 3, either $f(a)$ was picked by some agent whose first choice is $f(a)$, or we can find a sequence $a_{k+1}, a_k, \ldots, a_1$ of agents in σ such that a_{k+1} picks $f(a_{k+1}) = f(a_k)$, agent a_i picks $s(a_i) = f(a_{i-1})$ for each $i = k, \ldots, 2$, and agent a_1 picks $s(a_1) = f(a)$ under SD with σ. Let P_1 be the path $(a_k, \ldots, a_1)$; we allow P_1 to contain only the vertex $f(a)$. Similarly, let $b_{\ell+1}, b_\ell, \ldots, b_1$ be the sequence of agents in σ that explains how $s(a)$ was picked before a got her turn under SD with σ, and let P_2 be the path $(b_\ell, \ldots, b_1)$; again, P_2 might only consist of the vertex $s(a)$. Naturally, P_1 and P_2 satisfy (1).

Clearly, neither P_1 nor P_2 contains a, implying (2). Note that P_1 and P_2 may not be arc-disjoint, and $a_{k+1} = b_{\ell+1}$ is possible. However, $a \notin \{a_{k+1}, b_{\ell+1}\}$, and both P_1 and P_2 must be disjoint from $\{a_{k+1}, b_{\ell+1}\}$, because a_{k+1} and $b_{\ell+1}$ obtain their first choice under SD with σ, while all agents on P_1 and P_2 obtain their second choice. Thus a_{k+1} and $b_{\ell+1}$ witness that P_1 and P_2 satisfy (3) too.

Conversely, let P_1 and P_2 be paths in $G(I)$ satisfying conditions (1)–(3), and let $a_k, \ldots, a_1$ and $b_\ell, \ldots, b_1$ be the agents corresponding to the sequence of arcs in P_1 and P_2, resp. By (3), there is a set Q of one or two agents, disjoint from P_1, P_2 and not containing a, for which $\{f(q) \mid q \in Q\} = \{f(a_k), f(b_\ell)\}$. Let us construct a picking sequence σ starting first with the agents in Q, followed by $a_k, \ldots, a_1$ and $b_\ell, \ldots, b_1$; repetitions are ignored (so each agent picks when it first appears in this sequence). Clearly, the agents in Q pick $f(a_k)$ and $f(b_\ell)$ (which may coincide), and then every agent x in $\{a_1, \ldots, a_k\} \cup \{b_1, \ldots, b_\ell\}$ picks either her second choice or, if that is already gone by the time x gets her turn, gets nothing under SD with σ; during this process, both $f(a)$ and $s(a)$ gets picked, at latest by a_1 and b_1, resp., leaving nothing for a to pick.   □

Let us again illustrate Theorem 5 on the instance of Fig. 1. To see that 2-POS$(I, a_4, \emptyset)$ is true, consider the path $P_1 = (a_2)$ leading to $o_3 = f(a_4)$, and the length-0 path P_2 containing only $o_2 = s(a_4)$; note that condition (3) is witnessed by a_1 leaving o_1 and a_3 leaving o_2. A corresponding picking sequence is thus a_1, a_3 followed by a_2, and ending with a_4. The first three agents pick the objects o_1, o_2, o_3, leaving nothing for a_4 at her turn.

Let us now discuss the complexity of the algorithms implied by Theorems 3, 4 and 5. We can construct $G(I)$ in time $O(|A| + |\mathcal{O}|) = O(n)$. Searching for the relevant paths in $G(I)$ can also be performed by, e.g., DFS in $O(n)$ time. This implies that 2-POS$(I, a, s(a))$ and 2-POS$(I, a, \emptyset)$ can be decided in $O(n)$ time for any agent a in I. By Lemma 2, we get Corollary 2.

Corollary 2. *Problems* 2-POS(I, a, o) *and* 2-NEC(I, a, o) *are solvable in* $O(n)$ *time (even for the case* $o = \emptyset$*), where* n *is the number of agents in* I.

6   Multiple Copies of Objects

In this section we allow multiple identical copies for each object. The number of copies available for an object o is its *capacity*, determined by a capacity function $c : \mathcal{O} \rightarrow \mathbb{N}$. Given a *capacitated matching profile* $I = (A, \mathcal{O}, \mathcal{P}, c)$, we refer to the capacitated versions of the studied problems as CPOS(I, a, o) and CNEC(I, a, o).

Since CPOS(I, a, o) and CNEC(I, a, o) are generalizations of POS(I, a, o) and NEC(I, a, o), resp., by Theorem 2 it is immediate that they are NP-complete and coNP-complete, resp., already if the maximum length of preference lists is 3. Hence, we focus on length-2 matching profiles. We write k-CPOS(I, a, o) and k-CNEC(I, a, o) to refer to the corresponding problems restricted to capacitated length-k matching profiles. The following statement is trivial.

Lemma 3. *For any* $k \in \mathbb{N}$, k-CPOS$(I, a, f(a))$ *is true, while* k-CNEC$(I, a, s(a))$ *and* k-CNEC$(I, a, \emptyset)$ *are false for any* $a \in A$.

Next, we show that both 2-CPOS(I, a, o) and 2-CNEC(I, a, o) are computationally intractable in every case not covered by Lemma 3.

Theorem 6. *Problems* 2-CPOS$(I, a, s(a))$ *and* 2-CPOS$(I, a, \emptyset)$ *are* NP-*complete, while* 2-CNEC$(I, a, f(a))$ *is* coNP-*complete.*

Proof. Containment in NP or in coNP for the respective problems is trivial. We first provide a reduction from the EXACT 3-COVER problem to 2-CPOS$(I, a, s(a))$. An instance of EXACT 3-COVER consists of a set $X = \{x_1, x_2, \ldots, x_{3n}\}$ for some $n \in \mathbb{N}$ and a family $\mathcal{T}$ of 3-element subsets of X. The question is whether there exists a subfamily $\mathcal{T}' \subseteq \mathcal{T}$ containing exactly n sets whose union covers X. EXACT 3-COVER is NP-complete also in the case when each element $x \in X$ is contained in at most three sets from $\mathcal{T}$ [18]. We shall denote by $\ell(x)$ the number of sets in $\mathcal{T}$ that contain x.

Given an instance H of EXACT 3-COVER, we define a capacitated length-2 matching profile I as follows. The set A of agents in I contains a special agent a, one agent for each set, and one agent for each element-set pair: $A = \{a\} \cup \{a(T) \mid T \in \mathcal{T}\} \cup \{a(x, T) \mid T \in \mathcal{T}, x \in T\}$. There are four types of objects in I: an object $o(x)$ for each element $x \in X$ with capacity $\ell(x) - 1$, an object $o(T)$ for each set $T \in \mathcal{T}$ with capacity 3, and two special objects: o_1 with capacity n and o_2 with capacity 1. The preferences are as follows:

$P(a(T)) : o(T), o_1$ for each $T \in \mathcal{T}$,

$\quad P(a(x,T)) : o(x), o(T)$ for each $x \in X$ and $T \in \mathcal{T}$ such that $x \in T$,

$\qquad P(a) : o_1, o_2$.

Clearly, the construction is polynomial in the size of H. We claim that agent a can obtain object o_2 in I (that is, 2-CPOS$(I, a, s(a))$ is true) if and only if there is an exact cover in H.

Assume first that there exists an exact cover $\mathcal{T}'$ in H consisting of sets $T_1, T_2, \ldots, T_n$. Let those agents $a(x,T)$ pick first for which $T \notin \mathcal{T}'$. These agents exhaust all the element-objects, i.e., all objects $o(x)$, $x \in X$. Now agents of type $a(x, T_i)$ for $i = 1, 2, \ldots, n$ follow. They all get their second choices and thus completely exhaust all set-objects belonging to $\mathcal{T}'$, i.e. all objects $o(T)$, $T \in \mathcal{T}'$. Next come agents $a(T_i)$ for $i = 1, 2, \ldots, n$. They again get their second choices and exhaust all copies of object o_1. Hence, if agent a gets her turn after this point, she gets her second choice, o_2.

Conversely, assume now that agent a gets o_2 under some picking sequence. This means that o_1 was already exhausted when a got her turn, implying that n set-agents received their second choice. Let these agents be $a(T_1), \ldots, a(T_n)$. To finish the proof, we have to show that sets $T_1, \ldots, T_n$ form an exact cover, or equivalently, that these sets are pairwise disjoint. Assume for the contrary that an element x belongs both to T_r and T_s. As both object $o(T_r)$ and object $o(T_s)$ were exhausted before $a(T_r)$ and $a(T_s)$ pick, this means that both agents $a(x, T_r)$ and $a(x, T_s)$ must have received their second object, $o(T_r)$ and $o(T_s)$, resp. So their first choice, object $o(x)$ was already exhausted by the time they picked. But this could not happen as $o(x)$ has capacity $\ell(x) - 1$ and the number of agents interested in $o(x)$ is only $\ell(x)$, proving our claim.

The above reduction can be modified to show that the problem 2-CPOS$(I, a, \emptyset)$ is NP-complete and 2-CNEC$(I, a, f(a))$ is coNP-complete: we simply need to add a new agent b whose preference list contains only o_2. In this modified instance the following statements are equivalent: (i) a can obtain o_2, (ii) b does not necessarily obtain o_2, and (iii) b might end up with no object assigned to her. Furthermore, from the correctness of the above reduction, these hold exactly if H admits an exact cover, proving our theorem.   □

Observe that in the proof of Theorem 6 each object with capacity $c \geq 3$ has the following property: it is the first choice of a unique agent p, and it is the second choice of several agents $q_1, \ldots, q_k$ for some $k \geq c$; let us call such objects with capacity at least 3 *counter objects*.

Lemma 4. *Given an instance $(I, a, s(a))$ of 2-CPOS where only counter objects have capacity greater than 2 and $s(a)$ is not a counter object, we can in quadratic time construct an equivalent instance $(I', a, s(a))$ of 2-CPOS where all capacities are at most 2.*

Proof. Let o be a counter object with capacity $c \geq 3$ in I, let p be the unique agent whose first choice is o, and let $q_1, \ldots, q_k$ $(k \geq c)$ be those agents whose second choice is o. We describe how to replace o with a gadget containing only objects with capacities at most 2 without changing the answer to our instance.

In what follows, we will denote the modified capacitated matching profile by I', and we will make sure that the agent set A' of I' is a superset of the agent set A of I. We say that a picking sequence φ of I and a picking sequence φ' of I' are A-*equivalent*, if the set of those agents in A that are assigned their second choice is the same under φ and under φ'.

Case I. First assume $k = c$. We start by replacing o in q_i's preference list with a newly introduced object o_i that has capacity 1, for each $i \in \{1, \dots, c\}$. Next, we fix any rooted binary tree T with c leaves. We identify the leaves of T with the objects $o_1, \dots, o_c$, and we identify its root with object o. We add a new object $o(t)$ with capacity 2 for each vertex t of T that is neither a leaf nor root, and we also change the capacity of object o to 2. Furthermore, for any edge $t_1 t_2$ of T where t_1 is the child of t_2, we add a new agent $a(t_1, t_2)$ whose first choice and second choice is $o(t_1)$ and $o(t_2)$, resp. This finishes the construction.

Now we show that $\text{CPOS}(I', a, s(a))$ is true if and only if $\text{CPOS}(I, a, s(a))$ is true. To this end, we prove that for any picking sequence φ in I there is a picking sequence in I' that is A-equivalent to φ, and conversely. By $o \neq s(a)$, this guarantees the equivalence of our two instances.

"$\Leftarrow$": Suppose that φ is a picking sequence in I. Note that if p is assigned its first choice o under φ, then the newly added agents of I' do not "interfere" with the agents of A; simply letting all agents in $A' \setminus A$ pick after agents of A yields a picking sequence A-equivalent to φ. On the other hand, if p is assigned its second choice in I, then by the definition of a counter object and by $k = c$, we know that all agents q_i, $i \in \{1, \dots, c\}$, must also be assigned their second choice (that is, o) under φ. Let us create a picking sequence φ' from φ by inserting the agents $A' \setminus A$ immediately before p in a consecutive, bottom-up way: an agent $a(t_1, t_2)$ corresponding to an edge of T is allowed to pick only after her first choice t_1 is already exhausted. Thus, when p picks in φ', its first choice o is already exhausted (by the two agents corresponding to the two edges connecting o to its children in T). Hence, in the remainder of φ', all agents are assigned the same objects as in φ, showing that φ and φ' are A-equivalent.

"$\Rightarrow$": For the other direction, let φ' be a picking sequence in I'. We prove that the restriction of φ' to A (let us call this picking sequence φ) is A-equivalent to φ'. If p gets o under φ', then this is trivial. If, by contrast, p gets his second choice under φ', then the capacities of the newly introduced objects imply that each agent $a(t_1, t_2) \in A'$ must be assigned its second choice by φ', from which follows also that each agent q_i is assigned her second choice under φ'. This, however, ensures that p gets her second choice in I under φ, proving our claim.

Case II. Assume now $k > c$. First, we create $k - c + 1$ layers of new objects: for each $j \in \{0, \dots, k - c\}$, layer j contains the objects $o_{j,1}, \dots, o_{j,k-j}$; notice that each layer contains one object less than the previous layer. Next, for each $i \in \{1, \dots, k\}$, we replace o in the preference list of agent q_i with $o_{0,i}$. We let all objects in layer $k - c$ have capacity 1. Within some layer j with $0 \leq j < k - c$, we let the two "outermost" objects, that is, $o_{j,1}$ and $o_{j,k-j}$, have capacity 1, and all the remaining objects have capacity 2. Next, we create $k - c$ layers of new agents: for each $j \in \{1, \dots, k - c\}$, layer j contains $2(k - j)$ agents, namely

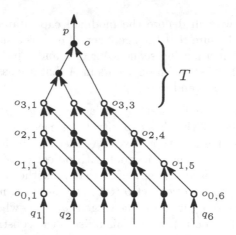

Fig. 2. Illustration for the construction in Lemma 4 for $k = 6$ and $c = 3$, depicting the underlying digraph. Objects with capacity 2 and 1 are black and white circles, resp.

agents $a(j, i, \uparrow)$ and $a(j, i, \searrow)$ for each $i \in \{1, \ldots, k-j\}$. We define the preference list of these agents as follows: both $a(j, i, \uparrow)$ and $a(j, i, \searrow)$ have $o(j, i)$ as their second choice, but the first choice of $a(j, i, \uparrow)$ is $o(j-1, i)$, while the first choice of $a(j, i, \searrow)$ is $o(j-1, i+1)$. We finish the construction by adding the gadget described in Case I, with the only difference that we choose the c objects of layer $k - c$ as the leaves of the binary tree T (and, as in Case I, we again set the capacity of o to 2). See Fig. 2 for an illustration.

Let I' be the obtained instance of 2-POS. We call an agent *active* in a picking sequence, if it is assigned its second choice. Let us now prove the equivalence of $\mathrm{CPOS}(I, a, s(a))$ and $\mathrm{CPOS}(I', a, s(a))$.

"$\Rightarrow$": Let φ' be a picking sequence in I'. By the capacities of the newly added objects, for any $j \in \{1, \ldots, k-c\}$ it holds that layer j contains at most as many active agents as layer $j-1$ (we let layer 0 contain the agents $q_1, \ldots, q_k$). Recalling the properties of the gadget constructed in Case I, it is not hard to verify that p can become active only if layer $k - c$, and hence each of the previous layers too, contains at least c active agents. Hence, at least c agents among $q_1, \ldots, q_k$ are active under φ', implying that the restriction of φ' to A is A-equivalent to φ'.

"$\Leftarrow$": Suppose that φ is a picking sequence in I where p is active, that is, where at least c agents among $q_1, \ldots, q_k$ are active. We can easily construct a picking sequence φ' in I' that is A-equivalent to φ such that under φ' exactly c agents become active in each layer and all the objects in layer $k-c$ get exhausted by agents of layer $k-c$, thus implying that each agent corresponding to an edge of our binary tree T, and therefore also agent p, becomes active in φ'. To determine such a picking sequence, we need to find c object-disjoint paths from the active agents in layer 0 to agents of layer $k - c$ (note that such paths always exist).

The replacement described above takes $O(k^2)$ time; replacing all counter objects therefore takes $O(|I|^2)$ time, proving our lemma.   □

Applying Lemma 4 to the instance constructed in Theorem 6, we get Corollary 3.[1]

Corollary 3. *Problems* 2-CPOS$(I, a, s(a))$ *and* 2-CPOS$(I, a, \emptyset)$ *are* NP-*complete and* 2-CNEC$(I, a, f(a))$ *is* coNP-*complete, even if all capacities are at most* 2.

7   Conclusion

We showed that if we enable agents to declare certain objects unacceptable, both the problems to decide whether a given agent can get a given object or whether a given agent always gets a given object in serial dictatorship are intractable, unless in the very special case when the lengths of preference lists are bounded by 2, and each object comes in a single copy. These results have direct consequences for manipulation possibilities of serial dictatorship: if it is difficult to compute which objects can an agent achieve then it is even the more difficult to compute a successful manipulation.

A possible direction of further research is to investigate a model where preference lists may contain ties. It is known that simply applying serial dictatorship is not enough to find a Pareto-optimal matching (POM) if ties can occur; recently Krysta et al. [22] and Cechlárová et al. [13] provided polynomial-time algorithms combining the greedy approach of serial dictatorship with network flow to find POMs in such situations. Up to our knowledge, the question of possible and necessary allocations has not yet been investigated in the presence of ties.

References

1. Abdulkadiroğlu, A., Pathak, P.A., Roth, A.E.: Strategy-proofness versus efficiency in matching with indifferences: redesigning the NYC high school match. Am. Econ. Rev. **99**(5), 1954–1978 (2009)
2. Abdulkadiroğlu, A., Sönmez, T.: Random serial dictatorship and the core from random endowments in house allocation problems. Econometrica **66**(3), 689–701 (1998)
3. Abraham, D.J., Cechlárová, K., Manlove, D.F., Mehlhorn, K.: Pareto optimality in house allocation problems. In: Deng, X., Du, D.-Z. (eds.) ISAAC 2005. LNCS, vol. 3827, pp. 1163–1175. Springer, Heidelberg (2005). doi:10.1007/11602613_115
4. Asinowski, A., Keszegh, B., Miltzow, T.: Counting houses of Pareto optimal matchings in the house allocation problem. Discrete Math. **339**(12), 2919–2932 (2016)
5. Aziz, H., Bouveret, S., Lang, J., Mackenzie, S.: Complexity of manipulating sequential allocation. CoRR, abs/1602.06940 (2016). arXiv:1602.06940 [cs.GT]
6. Aziz, H., Brandt, F., Brill, M.: The computational complexity of random serial dictatorship. Econ. Lett. **121**(3), 341–345 (2013)
7. Aziz, H., Mestre, J.: Parametrized algorithms for random serial dictatorship. Math. Social Sci. **72**, 1–6 (2014)

---

[1] Note that the modifications in the reduction given in the proof of Theorem 6 that prove the hardness of 2-CPOS$(I, a, \emptyset)$ and 2-CNEC$(I, a, f(a))$ still work after the application of Lemma 4.

8. Aziz, H., Walsh, T., Xia, L.: Possible and necessary allocations via sequential mechanisms. In: IJCAI 2015, pp. 468–474 (2015)
9. Bouveret, S., Lang, J.: A general elicitation-free protocol for allocating indivisible goods. In: IJCAI 2011, pp. 73–78 (2011)
10. Bouveret, S., Lang, J.: Manipulating picking sequences. In: ECAI 2014. Frontiers in Artificial Intelligence and Applications, vol. 263, pp. 141–146 (2014)
11. Brams, S.J., King, D.L.: Efficient fair division: help the worst off or avoid Envy? Rationality Soc. **17**(4), 387–421 (2005)
12. Brams, S.J., Straffin, P.D.: Prisoners' dilemma and professional sports drafts. Amer. Math. Monthly **86**(2), 80–88 (1979)
13. Cechlárová, K., Eirinakis, P., Fleiner, T., Magos, D., Manlove, D., Mourtos, I., Ocelⓐková, E., Rastegari, B.: Pareto optimal matchings in many-to-many markets with ties. Theor. Comput. Syst. **59**(4), 700–721 (2016)
14. Cechlárová, K., Eirinakis, P., Fleiner, T., Magos, D., Mourtos, I., Potpinková, E.: Pareto optimality in many-to-many matching problems. Discrete Optim. **14**, 160–169 (2014)
15. Cechlárová, K., Fleiner, T.: Pareto optimal matchings with lower quotas. Math. Social Sci. **88**, 3–10 (2017)
16. Cechlárová, K., Fleiner, T., Schlotter, I.: Possible and necessary allocations under serial dictatorship with incomplete preference lists. Technical report TR-2017-02, Egerváry Research Group, Budapest (2017)
17. Cechlárová, K., Klaus, B., Manlove, D.F.: Pareto optimal matchings of students to courses in the presence of prerequisites. In: COMSOC 2016 (2016). https://www.irit.fr/COMSOC-2016/proceedings/CechlarovaEtAlCOMSOC2016.pdf
18. Garey, M.R., Johnson, D.S.: Computers and Intractability: A Guide to the Theory of NP-Completeness. W.H. Freeman & Co., New York (1979)
19. Goto, M., Iwasaki, A., Kawasaki, Y., Kurata, R., Yasuda, Y., Yokoo, M.: Strategyproof matching with regional minimum and maximum quotas. Artif. Intell. **235**, 40–57 (2016)
20. Gourvès, L., Martinhon, C.A., Monnot, J.: Object allocation problem under constraints. In: COMSOC 2016 (2016). https://www.irit.fr/COMSOC-2016/proceedings/GourvesEtAlCOMSOC2016.pdf
21. Kamiyama, N.: A note on the serial dictatorship with project closures. Oper. Res. Lett. **41**(5), 559–561 (2013)
22. Krysta, P., Manlove, D., Rastegari, B., Zhang, J.: Size versus truthfulness in the house allocation problem. In: EC 2014, pp. 453–470 (2014)
23. Manea, M.: Serial dictatorship and Pareto optimality. Games Econ. Behav. **61**(2), 316–330 (2007)
24. Pathak, P.A., Sönmez, T.: School admissions reform in Chicago and England: Comparing mechanisms by their vulnerability to manipulation. Am. Econ. Rev. **103**(1), 80–106 (2013)
25. Saban, D., Sethuraman, J.: The complexity of computing the random priority allocation matrix. Math. OR **40**(4), 1005–1014 (2015)
26. Sönmez, T., Switzer, T.B.: Matching with (branch-of-choice) contracts at the United States Military Academy. Econometrica **81**(2), 451–488 (2013)
27. Svensson, L.-G.: Queue allocation of indivisible goods. Soc. Choice Welfare **11**(4), 323–330 (1994)
28. Zhu, M.: College admissions in China: a mechanism design perspective. China Econ. Rev. **30**, 618–631 (2014)

Stable Roommate with Narcissistic, Single-Peaked, and Single-Crossing Preferences

Robert Bredereck[1], Jiehua Chen[1,2(✉)], Ugo Paavo Finnendahl[1],
and Rolf Niedermeier[1]

[1] TU Berlin, Berlin, Germany
{robert.bredereck,rolf.niedermeier}@tu-berlin.de,
ugo.p.finnendahl@campus.tu-berlin.de
[2] Ben-Gurion University of the Negev, Beersheba, Israel
jiehua.chen2@gmail.com

Abstract. The classical STABLE ROOMMATE problem asks whether it is possible to pair up an even number of agents such that no two non-paired agents prefer to be with each other rather than with their assigned partners. We investigate STABLE ROOMMATE with complete (i.e. every agent can be matched with every other agent) or incomplete preferences, with ties (i.e. two agents are considered of equal value to some agent) or without ties. It is known that in general allowing ties makes the problem NP-complete. We provide algorithms for STABLE ROOMMATE that are, compared to those in the literature, more efficient when the input preferences are complete and have some structural property, such as being narcissistic, single-peaked, and single-crossing. However, when the preferences are incomplete and have ties, we show that being single-peaked and single-crossing does not reduce the computational complexity—STABLE ROOMMATE remains NP-complete.

1 Introduction

Given $2 \cdot n$ agents, each having a preference with regard to how suitable the other agents are as potential partners, the STABLE ROOMMATE problem asks whether it is possible to pair up the agents such that no two non-paired agents prefer to be with each other rather than with their assigned partners.

We call such a pairing a *stable matching*. STABLE ROOMMATE was introduced by Gale and Shapley [17] in the 1960's and has been studied extensively since then [21–23,31,32]. While it is quite straightforward to see that stable matchings may not always exist, it is not trivial to see whether an existent stable matching can be found in polynomial time, even when the input preference orders are

---

R. Bredereck was on postdoctoral leave at the University of Oxford (GB) from September 2016 to September 2017, supported by the DFG fellowship BR 5207/2. J. Chen was partly supported by the People Programme (Marie Curie Actions) of the European Union's Seventh Framework Programme (FP7/2007-2013) under REA grant agreement number 631163.11 and Israel Science Foundation (grant no. 551145/14).

J. Rothe (Ed.): ADT 2017, LNAI 10576, pp. 315–330, 2017.
DOI: 10.1007/978-3-319-67504-6_22

complete orders without ties (i.e. each agent can be a potential partner to each other agent, and no two agents are considered to be equally suitable as a partner). For the case without ties, Irving [21] and Gusfield and Irving [19] provided $O(n^2)$-time algorithms to decide the existence of stable matchings and to find one if it exists for complete preferences and for incomplete preferences, respectively. For the case where the given preferences may have ties, deciding whether a given instance admits a stable matching is NP-complete [31].

Solving STABLE ROOMMATE has many applications, such as pairing up students to accomplish a homework project or users in a P2P file sharing network, assigning co-workers to two-person offices, partitioning players in two-player games, or finding receiver-donor pairs for organ transplants [16,24,26,28,33,34]. In such situations, the students, the people, or the players, which we jointly refer to as *agents*, typically have certain *structurally restricted* preferences on which other agents might be their best partners. For instance, when assigning roommates, each agent may have an ideal room temperature and may prefer to be with another agent with the same penchant. Such preferences are called *narcissistic*. Moreover, if we order the agents according to their ideal room temperatures, then it is natural to assume that each agent z prefers to be with an agent x rather than with another agent y if z's ideal temperature is closer to x's than to y's. This kind of preferences is called *single-peaked* [4,7,20]. Single-peakedness is used to model agents' preferences where there is a criterion, e.g. room temperature, that can be used to obtain a linear order of the agents such that each agent's preferences over all agents along this order are strictly increasing until they reach the peak—their ideal partner—and then strictly decreasing. Single-peakedness is a popular concept with prominent applications in voting contexts. It can be tested for in linear time [1,3,8,13] if the input preferences are complete and have no ties. Another possible restriction on the preferences is the *single-crossing* property, which was originally proposed to model individuals' preferences on income taxation [29,30]. It requires a linear order (the so-called single-crossing order) of the agents so that for each two distinct agents x and y, there exists at most one pair of consecutive agents (the crossing point) along the single-crossing order that disagrees on the relative order of x and y. Single-crossingness can be detected in polynomial time [5,8,9] if the input preferences are complete and have no ties. We refer to Bredereck et al. [6] and Elkind et al. [12] for numerous references on single-peakedness and single-crossingness.

Bartholdi III and Trick [3] studied STABLE ROOMMATE with narcissistic and single-peaked preferences. They showed that for the case with linear orders (i.e. complete and without ties), STABLE ROOMMATE *always* admits a unique stable matching and provided an $O(n)$ time algorithm to find this matching. This is remarkable since restricting the preference domain does not only guarantee the existence of stable matchings, but also speeds up finding it to sublinear time. In this specific case, this speed up implies that a stable matching can be found without "reading" the whole input preferences as long as the input is assumed to be narcissistic and single-peaked.

Table 1. Complexity of STABLE ROOMMATE for restricted domains: narcissistic, single-peaked, and single-crossing preferences. Entries marked with $\diamond$ are from Irving [21]. Entries marked with $\spadesuit$ are from Gusfield and Irving [19]. Entries marked with $\triangle$ are from Ronn [31]. Entries marked with $\heartsuit$ are from Bartholdi III and Trick [3]. Entries marked with $*$ and boldfaced are new results shown in this paper. Note that our polynomial-time results also include the existence of a stable matching and that our hardness result even holds for the more restricted tie-sensitive single-crossing property.

	Complete preferences		Incomplete preferences	
	w/o ties	w/ ties	w/o ties	w/ ties
No restriction	$O(n^2)^\diamond$	NP-c$^\triangle$	$O(n^2)^\spadesuit$	NP-c$^\triangle$
Single-peaked & single-crossing	$O(n^2)^\diamond$	?	$O(n^2)^\spadesuit$	**NP-c***
Narcissistic & single-peaked	$O(n)^\heartsuit$	$\boldsymbol{O(n^2)^\star}$	$O(n^2)^\spadesuit$	?
Narcissistic & single-crossing	$\boldsymbol{O(n)^\star}$	$\boldsymbol{O(n^2)^\star}$	$O(n^2)^\spadesuit$	?

In this paper, we first discuss natural generalizations of the well-known single-peaked and single-crossing preferences (that were originally introduced for linear orders) for incomplete preferences with ties. Then, we investigate how some structural preference restrictions can help in guaranteeing the existence of stable matchings and in designing more efficient algorithms for finding one, including the case when the input preferences are not linear orders. We found that for complete preference orders, structurally restricted preferences such as being narcissistic and single-crossing or being narcissistic and single-peaked guarantee the existence of stable matchings. Moreover, we showed that when the preferences are complete, even with ties, narcissistic and single-crossing or narcissistic and single-peaked, then the algorithm of Bartholdi III and Trick [3] always finds a stable matching. The running time for $2\cdot$ agents increases to $O(n^2)$. However, when the preferences are incomplete and ties are allowed, STABLE ROOMMATE becomes NP-complete, even if the given preferences are single-peaked as well as single-crossing. Our results on STABLE ROOMMATE, together with those from related work, are summarized in Table 1. Due to space constraints, some proofs are omitted.

2   Fundamental Concepts and Basic Observations

Let $V = \{1, 2, \ldots, 2\cdot n\}$ be a set of $2\cdot n$ agents. Each agent $i \in V$ has a *preference order* $\succeq_i$ *over a subset* $V_i \subseteq V$ *of agents* that i finds *acceptable* as a partner[1]. We note that although in our stable roommate problem, an agent cannot be matched to itself, it may still make sense to include an agent x in its preference orders, for instance when x represents someone which is very close to its ideal. The set V_i is called the *acceptable set* of i and a *preference order* $\succeq_i$ *over* V_i is a *weak order on* V_i, i.e. a transitive and complete binary relation on V_i. For instance, $x \succeq_i y$ means that i weakly prefers x over y (i.e. x is better than or

---

[1] For technical reasons, an agent may find itself acceptable, which means that $\{i\} \subseteq V_i$.

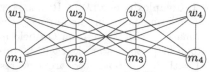

(a) The underlying acceptability graph of a STABLE ROOMMATE instance with complete preferences, where *each* two distinct agents can be matched to each other.

(b) The underlying acceptability graph of a classical STABLE MARRIAGE instance, which is bipartite. In such an instance, each woman from the top row can only be matched with a man from the bottom row, and the converse.

Fig. 1. Acceptability graphs of two special cases of STABLE ROOMMATE.

as good as y). We will use $\succ_i$ to denote the asymmetric part of $\succeq_i$ (i.e. $x \succeq_i y$ and $\neg(y \succeq_i x)$, meaning that i strictly prefers x to y) and $\sim_i$ to denote the symmetric part of $\succ_i$ (i.e. $x \succeq_i y$ and $y \succeq_i x$, meaning that i values x and y equally). We call an agent x a *most acceptable* agent of another agent y if for all $z \in V_y \setminus \{x, y\}$ it holds that $x \succeq_y z$. Note that an agent can have more than one most acceptable agent.

Let $X \subseteq V$ and $Y \subseteq V$ be two disjoint sets of agents and $\succeq$ be a binary relation over V. To simplify notation, we write $X \succeq Y$ to denote that for each two agents x and y with $x \in X$ and $y \in Y$ it holds that $x \succeq y$. (We use $X \succeq y$ as shortcut for $X \succeq \{y\}$ and $X \succ Y$ as well as $X \sim Y$ in an analogous way.)

To model which agent is considered as acceptable in a preference order we introduce the notion of an *acceptability graph* G for V. It is an undirected graph without loops. An edge signifies whether two distinct agents find each other acceptable. We use V to also denote the vertex set of G. There is an edge $\{i, j\}$ in G if $i \in V_j \setminus \{j\}$ and $j \in V_i \setminus \{i\}$. We assume without loss of generality that G does not contain isolated vertices, meaning that each agent could be matched to at least one other agent. We illustrate two prominent special cases of acceptability graphs in Fig. 1.

Blocking pairs and stable matchings. Given a preference profile $\mathcal{P}$ for a set V of agents, a *matching* $M \subseteq E(G)$ is a subset of disjoint pairs of agents $\{x, y\}$ with $x \neq y$ (or edges in $E(G)$), where $E(G)$ is the set of edges in the corresponding acceptability graph G). For a pair $\{x, y\}$ of agents, if $\{x, y\} \in M$, then we denote $M(x)$ as the corresponding partner y; otherwise we call this pair *unmatched*. We write $M(x) = \bot$ if agent x has *no partner*, that is, if agent x is not involved in any pair in M. An unmatched pair $\{x, y\} \in E(G) \setminus M$ *is blocking* M if the pair "prefers" to be matched to each other, i.e. it holds that

$$(M(x) = \bot \vee y \succ_x M(x)) \wedge (M(y) = \bot \vee x \succ_y M(y)).$$

A matching M is *stable* if no unmatched pair is blocking M. Note that this stability concept is called *weak stability* when we allow ties in the preferences. We refer to the textbook by Gusfield and Irving [19], Manlove [27] for two other popular stability concepts for preferences with ties.

We focus on the following stable matching problem.

STABLE ROOMMATE
Input: A preference profile $\mathcal{P}$ for a set $V = \{1, 2, \ldots, 2 \cdot n\}$ of $2 \cdot n$ agents.
Question: Does $\mathcal{P}$ admit a stable matching?

The profile given in Fig. 3 admits a (unique) stable matching: $\{\{1,4\}, \{2,3\}\}$. In fact, as we will see in Sect. 3, a narcissistic and single-peaked preference profile always admits a stable matching. However, if agent 3 changes its preference order to $3 \succ 1 \succ 2 \succ 4$, then the resulting profile is not single-peaked anymore, nor does it admit any stable matching: One can check that any agent i, $1 \leq i \leq 3$ that is matched to agent 4 will form a blocking pair together with the agent that is at the third position of the preference order of i.

Preference profiles and their properties. A preference profile $\mathcal{P}$ for V is a collection $(\succeq_i)_{i \in V}$ of preference orders for each agent $i \in V$. A profile $\mathcal{P}$ may have the following three simple properties:

1. Profile $\mathcal{P}$ is *complete* if for each agent $i \in V$ it holds that $V_i \cup \{i\} = V$; otherwise it is *incomplete*.
2. Profile $\mathcal{P}$ has a *tie* if there is an agent $i \in V$ and there are two distinct agents $x, y \in V_i$ with $x \sim_i y$. Note that linear orders are exactly those orders that are complete and have no ties.
3. Profile $\mathcal{P}$ is *narcissistic* if each agent i *strictly* prefers itself to every other acceptable agent, i.e. for each $j \in V_i$ it holds that $i \succ_i j$.

We note that the completeness concept basically means that each two distinct agents can be matched together. Thus, it does not matter whether $V_i = V$ or $V_i \cup \{i\} = V$ because i cannot be matched to itself anyway. By the same reasoning, the narcissistic property alone, which reflects the fact that each agent prefers to be with someone like itself among all alternatives, does not really restrict the input of our stable roommate problem. However, one can further restrict single-peaked preferences or single-crossing preferences by additionally requiring them to be narcissistic and we show that this affects the existence of stable matchings.

As already discussed in Sect. 1, the single-peaked and the single-crossing properties were originally introduced and studied mainly for linear preference orders (i.e. orders without ties). For preferences with ties, a natural generalization is to think of a possible linear extension of the preferences for which the single-peaked or single-crossing property holds. We consider this variant in our paper. Profile $\mathcal{P}$ is *single-peaked* if there is a linear order $\rhd$ over V such that the preference order of each agent i is *single-peaked with respect to* $\rhd$:

$$\forall x, y, z \in V_i \text{ with } x \rhd y \rhd z \text{ it holds that } (x \succ_i y \text{ implies } y \succeq_i z).$$

Just as for the single-peaked property, the single-crossing property also requires a natural linear order of the agents, the so-called single-crossing order. However, unlike the single-peaked property which assumes that the preferences of an agent i over two agents are measured by their "distance" to the peak along the single-peaked order, the single-crossing property assumes that the agents' preferences over each two distinct agents change (cross) at most once. In fact, for preferences with ties, two natural single-crossing notions are of interest. To define them, we first introduce a notion which denotes a subset of voters that have the same preferences over two distinct agents x and y: Let $V[x \succ y] := \{i \in V \mid x \succ_i y\}$ be the subset of voters i that strictly prefer x to y, and let $V[x \sim y] := \{i \in V \mid x \sim_i y\}$ be the subset of voters i that find x and y to be of equal value. We say that profile $\mathcal{P}$ is *single-crossing* if there is a linear extension of $\mathcal{P}$ to a profile $\mathcal{P}' = (\succ_1', \succ_2', \ldots, \succ_{2 \cdot n}')$ without ties and there is a linear order $\triangleright$ over V such that for each two distinct agents x and y, $\mathcal{P}'$ is *single-crossing with respect to* $\triangleright$, i.e.

$$V[x \sim' y] = \emptyset \text{ and either } V[x \succ' y] \triangleright V[y \succ' x] \text{ or } V[y \succ' x] \triangleright V[x \succ' y].$$

We also consider a more restricted single-crossing concept which compared the single-crossing property introduced above requires that the agents that have ties are ordered in the middle. A profile $\mathcal{P}$ is called *tie-sensitive single-crossing* if there is a linear order $\triangleright$ over V such that each pair $\{x, y\}$ of two distinct agents is *tie-sensitive single-crossing with respect to* $\triangleright$, i.e.

$$\text{either } V[x \succ y] \triangleright V[x \sim y] \triangleright V[y \succ x] \text{ or } V[y \succ x] \triangleright V[x \sim y] \triangleright V[x \succ y].$$

See Fig. 2 for an illustration of the different types of restricted preferences for the case where the preferences are linear orders.

For partial orders, our two single-crossing concepts are *incomparable*. In particular, there are incomplete preferences with ties which are single-crossing but *not* tie-sensitive single-crossing, and the converse also holds. For weak orders and for preferences without ties, however, the following holds. (Notably, a large part of the observation can be found in a long version of Elkind et al. [11].)

Observation 1. *Let $\mathcal{P}$ be an arbitrary preference profile: (i) If $\mathcal{P}$ is complete, then $\mathcal{P}$ is single-crossing if it is tie-sensitive single-crossing. (ii) If $\mathcal{P}$ is without ties, then $\mathcal{P}$ is single-crossing if and only if it is tie-sensitive single-crossing.*

Figure 2(b) demonstrates that the converse of the first statement in Observation 1 does not hold.

There are many slightly different concepts of single-peakedness and single-crossingness for partial orders (a generalization of incomplete preferences with ties) [11,15,25]. It is known that detecting single-peakedness or single-crossingness is NP-hard for partial orders under most of the concepts studied in the literature. For linear orders, all these concepts (including ours) are equivalent to those introduced by Black [4] and Mirrlees [29] and can be detected in polynomial time [1,3,5,8,9,13]. For incomplete preferences with ties, Lackner [25] showed that detecting single-peakedness is NP-complete. For complete preferences with ties, while Elkind et al. [11] showed that detecting single-crossingness is NP-complete, Fitzsimmons [14] and Elkind et al. [11] provided polynomial-time

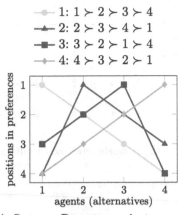

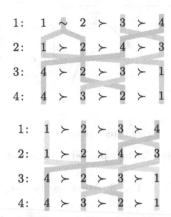

(a) A STABLE ROOMMATE instance with narcissistic and single-peaked preferences. They are *not* single-crossing since $\{1,4\}$ forces the two agents 1 and 3 (resp. 2 and 4) to be ordered next to each other in a single-crossing order whereas $\{2,3\}$ forces the two agents 1 and 2 (resp. 3 and 4) to be ordered next to each other. All these four conditions, however, cannot be satisfied by a linear order.

(b) Top: A STABLE ROOMMATE instance with single-crossing preferences. They are *not* tie-sensitive single-crossing since $\{2,3,4\}$ implies that $1 \rhd 2 \rhd 3 \rhd 4$ and its reverse are the only possible single-crossing orders. But, $\{1,2\}$ is not tie-sensitive single-crossing wrt. either $\rhd$ or its reverse. Bottom: A possible linear extension, showing single-crossingness.

Fig. 2. Visualization of different restricted profiles.

algorithms for detecting single-peakedness and ties-sensitive single-crossingness. All these known hardness results seem to hold only when the preferences have ties. However, we observe that the hardness proof for Corollary 6 by Elkind et al. [11] indeed can be adapted to show NP-completeness for deciding whether an incomplete preference profile without ties is single-peaked or single-crossing.

Observation 2. *Deciding whether an incomplete preference profile without ties is single-crossing (or equivalently tie-sensitive single-crossing) or single-peaked is NP-complete.*

Barberà and Moreno [2] as well as Elkind et al. [10] noted that for complete preferences without ties, narcissistic and single-crossing preferences are also single-peaked. We show that the relation also holds when ties are allowed. We note that Barberà and Moreno [2] also considered complete preferences with ties. However, their single-crossingness for the case with ties only resembles our tie-sensitive single-crossing definition, which is a strict subset of our single-crossingness (Observation 1).

Proposition 1. *If a complete, even with ties, and narcissistic preference profile $\mathcal{P}$ has a single-crossing order $\rhd$, then this order $\rhd$ is also a single-peaked order.*

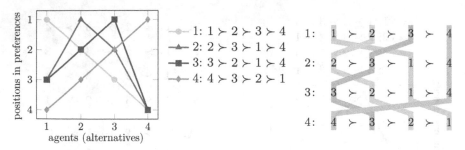

(a) A single-peaked visualization   (b) A single-crossing visualization.

Fig. 3. A narcissistic, single-peaked, and single-crossing profile.

Proof. Suppose for the sake of contradiction that $\rhd$ with $a_1 \rhd a_2 \rhd \cdots \rhd a_{2 \cdot n}$ is not single-peaked. This means that there exists an agent a_i that is not single-peaked wrt. $\rhd$, and there are three agents a_j, a_k, a_ℓ with $j < k < \ell$ such that $a_j \succ_{a_i} a_k$ and $a_\ell \succ_{a_i} a_k$. Together with the narcissistic property, the following holds:

agent a_i: $a_i \succ_{a_i} a_j \succ_{a_i} a_k$ and $a_i \succ_{a_i} a_\ell \succ_{a_i} a_k$,   agent a_j: $a_j \succ_{a_j} a_k$,
agent a_k: $a_k \succ_{a_k} a_j$ and $a_k \succ_{a_k} a_\ell$,   agent a_ℓ: $a_\ell \succ_{a_\ell} a_k$.

On the one hand, the agents' preferences over the pair $\{a_j, a_k\}$ implies that $i < k$. On the other hand, the pair $\{a_k, a_\ell\}$ implies that $i > k$—a contradiction. $\square$

The profile shown in Fig. 3 is narcissistic and single-crossing wrt. the order $1 \rhd 2 \rhd 3 \rhd 4$ and it is also single-peaked with respect to the same order $\rhd$.

3   Complete Preferences

In this section, we consider profiles with complete preferences. It is known that if ties do not exist, then STABLE ROOMMATE can be solved in $O(n^2)$ time [21], while the existence of ties makes the problem NP-hard [31]. For the case of complete, narcissistic, and single-peaked preferences without ties, Bartholdi III and Trick [3] showed that STABLE ROOMMATE is even solvable in $O(n)$ time. Their algorithm is based on the following two facts (referred to as Propositions 2 and 3) that are related to the concept of most acceptable agents. We show that the facts transfer to the case with ties.

Proposition 2. *If the given preference profile $\mathcal{P}$ is complete (even with ties), narcissistic, and single-peaked, then there are two distinct agents i, j that are each other's most acceptable agents.*

Proof. The statement for complete, narcissistic, and single-peaked preferences without ties was shown by Bartholdi III and Trick [3]. It turns out that this also holds for the case when ties are allowed. Let V be the set of all $2 \cdot n$ agents and consider a single-peaked order $\rhd$ of the agents V with $x_1 \rhd x_2 \rhd \cdots \rhd x_n$. For each

---

Algorithm 1. The algorithm of Bartholdi III and Trick for computing a stable matching with input $\mathcal{P}$ being complete, narcissistic, and single-peaked.

$M \leftarrow \emptyset$;
while $\mathcal{P} \neq \emptyset$ **do**
   | Find two agents x, y in $\mathcal{P}$ that consider each other as most acceptable;
   | Delete x and y from profile $\mathcal{P}$;
   | $M \leftarrow M \cup \{x, y\}$;
return M;

---

agent $x \in V$, let M_x be the set of all most acceptable agents of x. Towards a contradiction, suppose that each two distinct agents x and y have $x \notin M_y$ or $y \notin M_x$. By the narcissistic property and single-peakedness, each $M_x \cup \{x\}$ forms an interval in $\triangleright$. This implies that the first agent x_1 and the last agent x_n in the order $\triangleright$ have $x_2 \in M_{x_1}$ and $x_{n-1} \in M_{x_n}$. By our assumption, however $x_2 \in M_{x_1}$ implies that for each $i \in \{2, \ldots, n\}$ the following holds: $x_{i-1} \notin M_x$—a contradiction to $x_{n-1} \in M_{x_n}$. $\qquad\square$

By the stability definition, we have the following for complete preferences.

Proposition 3. *Let $\mathcal{P}$ be a preference profile and let M be a stable matching for $\mathcal{P}$. Let $\mathcal{P}'$ be a preference profile resulting from $\mathcal{P}$ by adding two agents x, y who are each other's most acceptable agents (and the preferences of other agents over x, y are arbitrary but fixed). Then, matching $M \cup \{\{x, y\}\}$ is stable for $\mathcal{P}'$.*

Proof. Suppose for the sake of contradiction that $M \cup \{\{x, y\}\}$ is not stable for $\mathcal{P}'$. This means that $\mathcal{P}'$ has an unmatched blocking pair $\{u, w\} \notin M$. It is obvious that $|\{u, w\} \cap \{x, y\}| = 1$ as otherwise $\{u, w\}$ would also be an unmatched blocking pair for $\mathcal{P}$. Assume without loss of generality that $u = x$. Then, by the definition of blocking pairs, it must hold that $w \succ_x y$—a contradiction to y being one of the most acceptable agents of x. $\qquad\square$

Utilizing Propositions 2 and 3 (in more restricted variants), Bartholdi III and Trick [3] derived an algorithm to construct a *unique* stable matching when the preferences are linear orders (i.e. complete and without ties) and are narcissistic and single-peaked (see Algorithm 1). For $2 \cdot n$ agents their algorithm runs in $O(n)$ time. We will show that Algorithm 1 also works when ties are allowed. The stable matching, however, may not be unique anymore and the running time is $O(n^2)$ since we need to update the preferences of each agent after we match one pair of two agents.

Theorem 1. *Algorithm 1 finds a stable matching for profiles with $2 \cdot n$ agents that are complete, with ties, narcissistic and single-peaked in $O(n^2)$ time.*

Proof. The correctness follows directly from Propositions 2 and 3 and the narcissistic and single-peaked property is preserved when deleting any agent. As for the running time, there are n rounds to build up M, and in each round we find two distinct agents x and y whose most acceptable agent sets M_x and M_y include each other: $x \in M_y$ and $y \in M_x$. Note that Proposition 2 implies that such two agents exist. After each round we need to update the most acceptable agents of at most $2 \cdot n$ agents. Thus, in total the running time is $O(n^2)$.   □

Now, we move on to (tie-sensitive) single-crossingness.

Corollary 1. *Algorithm 1 finds a stable matching for preference profiles with $2 \cdot n$ agents that are complete, with ties, narcissistic and single-crossing (or tie-sensitive single-crossing) in $O(n^2)$ time. The running time for the case without ties is $O(n)$.*

Proof. By Proposition 1 and Observation 1 (i), the stated profiles are single-peaked. The result of Bartholdi III and Trick [3] and Theorem 1 imply the desired statement.   □

4   Incomplete Preferences

Incomplete preferences mean that some agents do not appear in the preferences of an agent, for instance, because two agents are unacceptable to each other or they are not "allowed" to be matched to each other. If in this case no two agents are considered of equal value by any agent (i.e. the preferences are without ties), then STABLE ROOMMATE still remains polynomial-time solvable [19]. However, once ties are involved, STABLE ROOMMATE becomes NP-complete [31] even for complete preferences. In this section, we consider the case where the input preferences may be narcissistic, single-peaked, or single-crossing. First of all, we note that these preference restrictions can no longer guarantee the existence of two consecutive agents that are each other's most acceptable agent. However, this guarantee is crucial for the existence of a stable matching and for why the algorithm by Bartholdi III and Trick [3] can work in time linear in the number of agents. Moreover, for incomplete preferences, even without ties, narcissistic and single-crossing preferences do not imply single-peakedness anymore.

Proposition 4. *For incomplete preferences without ties, the following holds: Narcissistic and single-crossing preferences are not necessarily single-peaked. Narcissistic and single-peaked (resp. single-crossing) preferences guarantee neither the uniqueness nor the existence of stable matchings.*

Proof. Consider the following profile with six agents $1, 2, \ldots, 6$:

agent 1: $1 \succ_1 5 \succ_1 6$, agent 3: $3 \succ_3 5 \succ_3 6$, agent 5: $5 \succ_5 1 \succ_5 2 \succ_5 3 \succ_5 4$,
agent 2: $2 \succ_2 5 \succ_2 6$, agent 4: $4 \succ_4 5 \succ_4 6$, agent 6: $6 \succ_6 4 \succ_6 2 \succ_6 3 \succ_6 1$.

It is single-crossing wrt. the order $1 \rhd 2 \rhd \cdots \rhd 6$, but it is not single-peaked because of the last two agents' preference orders over $1, 2, 3, 4$. It does not admit a stable matching of size three. But it admits a stable matching of size two: $\{\{1, 5\}, \{4, 6\}\}$. The following profile with four agents $1, 2, 3, 4$ is narcissistic and single-peaked wrt. the order $1 \rhd 2 \rhd 3 \rhd 4$, and single-crossing wrt. the order $1 \rhd' 3 \rhd' 2 \rhd' 4$. It admits two different stable matchings $\{\{1,2\},\{3,4\}\}$ and $\{\{1,3\},\{2,4\}\}$.

> agent 1: $1 \succ_1 2 \succ_1 3 \succ_1 4$, agent 2: $2 \succ_2 4 \succ_2 1$,
> agent 3: $3 \succ_3 1 \succ_3 4$,   agent 4: $4 \succ_4 3 \succ_4 2 \succ_4 1$.

The following profile with ten agents $1, 2, \ldots, 10$ is narcissistic and single-peaked wrt. the order $4 \rhd 2 \rhd 1 \rhd 3 \rhd 5 \rhd 9 \rhd 7 \rhd 6 \rhd 8 \rhd 10$. But, no matching M is stable for this profile: First, the agents can be partitioned into two subsets $V_1 = \{1, 2, \ldots, 5\}$ and $V_2 = \{6, 7, \ldots, 10\}$ such that only agents within the same subset can be matched together. Since $|V_1|$ is odd, at least one agent $i \in V_1$ is not matched by M. But, agent i and the agent at the third position of the preference order of i would form a blocking pair.

> agent 1: $1 \succ_1 4 \succ_1 3$,   agent 2: $2 \succ_2 5 \succ_2 4$,   agent 3: $3 \succ_3 1 \succ_3 5$,
> agent 4: $4 \succ_4 2 \succ_4 1$,   agent 5: $5 \succ_5 3 \succ_5 2$,   agent 6: $6 \succ_6 9 \succ_6 8$,
> agent 7: $7 \succ_7 10 \succ_7 9$, agent 8: $8 \succ_8 6 \succ_8 10$, agent 9: $9 \succ_9 7 \succ_9 6$,
> agent 10: $10 \succ_{10} 8 \succ_{10} 7$.   □

For the case with ties allowed, Ronn [31] showed that STABLE ROOMMATE becomes NP-hard even if the preferences are complete. The constructed instances in his hardness proof, however, are not always single-peaked or single-crossing. It is even not clear whether the problem remains NP-hard for this restricted case. If we abandon the completeness of the preferences, then we obtain NP-hardness, by another and simpler reduction. Before we state the corresponding theorem, we prove the following lemma which is heavily used in our preference profile construction to force two agents to be matched together.

Lemma 1. *Let $\mathcal{P}$ be a STABLE ROOMMATE instance for a given voter set V, and let a, b, and c be three distinct agents with the following preferences:*

> *agent a:* $X \succ b \succ c \succ V_a \setminus (X \cup \{b, c\})$,
> *agent b:* $c \succ a \succ V_b \setminus \{a, c\}$,   *agent c:* $a \succ b \succ V_c \setminus \{a, b\}$,

where $X \subseteq (V_a \cap V_b \cap V_c) \setminus \{a, b, c\}$ is a non-empty subset. Then, every stable matching M for $\mathcal{P}$ must fulfill that (i) $M(a) \in X$ and (ii) $\{b, c\} \in M$.

Proof. Assume towards a contradiction to (i) that $\mathcal{P}$ admits a stable matching M with $M(a) \notin X$. There are three cases: (1) $M(a) = b$, implying the blocking pair $\{b, c\}$, (2) $M(a) = c$, implying the blocking pair $\{a, b\}$, and (3) $M(a) \notin \{b, c\}$, implying the blocking pair $\{a, c\}$. Thus, a must be matched with some agent from X. For (ii), statement (i) implies that c cannot be matched with a. Now, if $\{b, c\} \notin M$, then $\{b, c\}$ is a blocking pair.   □

Theorem 2. STABLE ROOMMATE *for incomplete preferences with ties remains NP-complete, even if the preferences are single-peaked and single-crossing or single-peaked and tie-sensitive single-crossing.*

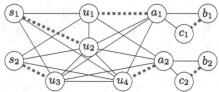

(a) The graph of a VERTEX COVER instance $(G, k = 2)$. The instance is a yes-instance and admits a vertex cover $\{u_2, u_4\}$, marked in light red.

(b) The acceptability graph of the corresponding STABLE ROOMMATE instance. It admits a stable matching, marked by thick dotted lines.

Fig. 4. An illustration of the hardness reduction for Theorem 2. (Color figure online)

Proof. First, the problem is in NP since one can non-deterministically guess a matching and check the stability in polynomial time. To show NP-hardness, we reduce from the NP-complete VERTEX COVER problem [18], which given an undirected graph $G = (U, E)$ and a non-negative integer k, asks whether there is a size-at-most k *vertex cover*, i.e. a subset $U' \subseteq U$ of size at most k such that for each edge $e \in E$, it holds that $e \cap U' \neq \emptyset$. Let $(G = (U, E), k)$ be a VERTEX COVER instance with $p := |U|$. We assume w.l.o.g. that $k < p$. We will construct a STABLE ROOMMATE instance $\mathcal{P}$ with agent set V and show that G has a vertex cover of size at most k if and only if $\mathcal{P}$ admits a stable matching.

Main idea and the constructed agents. To explain the main idea of the reduction, we first describe the agent set V and the corresponding acceptability graph of $\mathcal{P}$ as illustrated through an example in Fig. 4. For each vertex $u_i \in U$, we introduce a *vertex agent* u_i (for the sake of simplicity, we use the same symbol for the vertex and the corresponding agent). Additionally, there is a set of *selector agents* $S := \{s_1, \ldots, s_k\}$ as well as three sets of *collector agents* $A := \{a_1, a_2, \ldots, a_{p-k}\}$, $B := \{b_1, b_2, \ldots, b_{p-k}\}$, and $C := \{c_1, c_2, \ldots, c_{p-k}\}$. The agent set V is defined as $U \cup S \cup A \cup B \cup C$. For the acceptability graph, we have that every vertex agent u_i accepts every selector agent from S, every collector agent from A, and every vertex agent u_j that corresponds to a neighbor of u_j in the input graph G. For each $i \in \{1, 2, \ldots, p - k\}$, the collector agents a_l, b_i, and c_i pairwisely accept each other. We aim at constructing the agents' preferences such that in every stable matching only the selector agents from S and the collector agents from A can be matched to the vertex agents and the vertex agents matched to the selector agents correspond to a vertex cover (of size $|S| = k$). This property is given by the subsequent Claim 1.

Agent preferences. Now, we describe the preferences that realize the idea and the acceptability graph as described above:

$$\text{agent } u_i\colon \ [S] \succ [N(u_i)] \succ a_1 \succ a_2 \succ \ldots \succ a_{p-k} \quad \forall 1 \leq i \leq p,$$
$$\text{agent } s_i\colon \ u_1 \sim u_2 \sim \ldots \sim u_p \qquad\qquad\qquad\quad \forall 1 \leq i \leq k,$$

$$\text{agent } a_i\colon \ [U] \succ b_i \succ c_i,$$
$$\text{agent } b_i\colon \ c_i \succ a_i, \qquad\quad \text{agent } c_i\colon \ a_i \succ b_i \qquad \forall 1 \leq i \leq p-k.$$

Herein, for each subset $X \subset S \cup U$, we denote by $[X]$ some arbitrary but fixed order (e.g. ordered wrt. the names or the indices), called the *canonical order*. This completes the construction and can clearly be performed in polynomial time.

Correctness of the construction. First of all, we claim the following:

Claim 1. *Every stable matching M for $\mathcal{P}$ satisfies the following two properties:*

1. *every vertex agent u_i is matched to either a selector agent from S or a collector agent from A: $M(u_i) \in S \cup A$, and*
2. *no two vertex agents that are both matched to a collector agent are adjacent.*

Proof. (of Claim 1) Let M be a stable matching for $\mathcal{P}$. For the first statement, Lemma 1 immediately implies that for every collector agent $a_i \in A$, it holds that $M(a_i) \in U$. Thus, there are exactly k vertex agents left that are not matched to agents from A. Suppose towards a contradiction that some selector agent s_j is not matched to any vertex agent, implying that at least one vertex agent u_i is left with $M(u_i) \notin A \cup S$. This, however, implies that $\{s_j, u_i\}$ is a blocking pair for M—a contradiction. For the second statement, suppose towards a contradiction that there are two vertex agents u_i, u_j with $\{M(u_i), M(u_j)\} \subseteq A$ as well as $\{u_i, u_j\} \in E$. The preference orders of u_i and u_j immediately imply that agents u_i and u_j form a blocking pair—a contradiction. (of Claim 1) ◇

Now, we show that G has a vertex cover of size at most k if and only if $\mathcal{P}$ admits a stable matching. The "if" part follows immediately from Claim 1. For the "only if" part, suppose that $U' \subseteq U$ is a vertex cover of size k. Without loss of generality, assume that $U' = \{u_1, u_2, \ldots, u_k\}$ and further assume that the canonical order is $u_1 \succ u_2 \succ \cdots \succ u_n$. It is easy to verify that the following matching M is stable:

- for each $i \in \{1, 2, \ldots, k\}$ set $M(u_i) := s_i$;
- for each $i \in \{1, 2, \ldots, p-k\}$ set $M(u_{i+k}) = a_i$;
- for each $i \in \{1, 2, \ldots, p\}$ set $M(b_i) = c_i$.

Single-peakedness and (tie-sensitive) single-crossingness. The constructed profile is single-peaked with respect to the following linear order $\triangleright$:

$$[S] \triangleright [U] \triangleright a_1 \triangleright a_2 \triangleright \cdots \triangleright a_{p-k} \triangleright b_1 \triangleright b_2 \triangleright \cdots \triangleright b_{p-k} \triangleright c_1 \triangleright c_2 \triangleright \cdots \triangleright c_{p-k}.$$

It is also single-crossing, since each preference order (after resolving all ties in favor of the canonical order as discussed when constructing the agent preferences) is a sub-order of one of two different preference orders, and two preference orders are always single-crossing. More specifically, the profile is single-crossing with respect to the order $\triangleright$:

After resolving all ties in the preferences of the selector agents in favor of the canonical order, the preference orders of the agents from $S \cup U \cup A$ are sub-orders of the linear order $[S] \succ [U] \succ a_1 \succ b_1 \succ c_1 \succ a_2 \succ b_2 \succ c_2 \succ \cdots \succ a_{p-k} \succ b_{p-k} \succ c_{p-k}$, and the preference orders of the agents from $B \cup C$ are sub-orders of the linear order

$$[S] \succ [U] \succ c_1 \succ a_1 \succ b_1 \succ c_2 \succ a_2 \succ b_2 \succ \cdots \succ c_{p-k} \succ a_{p-k} \succ b_{p-k}.$$

The tie-sensitive single-crossing property also holds because ties only occur between pairs of agents from U and $\triangleright$ contains first all agents with ties and then the agents with the same canonical order among agents from U.   □

The constructed profile in the proof of Theorem 2 cannot be extended to also satisfy the narcissistic property. However, we conjecture that STABLE ROOMMATE remains NP-complete even if the input preferences are also narcissistic.

5   Conclusion

We investigated STABLE ROOMMATE for preferences with popular structural properties, such as being narcissistic, single-peaked, and single-crossing. We showed the existence of stable matchings and managed to speed up the detection of such matchings when the preferences are complete, narcissistic, and single-peaked (or single-crossing). Some of the speed-up (Corollary 1) is even associated with a sublinear time algorithm. For incomplete preferences with ties, however, single-peakedness combined with single-crossingness does not help to lower the computational complexity—STABLE ROOMMATE remains NP-complete.

We conclude with some challenges for future research. First, considering the NP-completeness result, it would be interesting to study the parameterized complexity with respect to the "degree" of incompleteness of the input preferences, such as the number of ties or the number of agents that are in the same equivalence class of the tie-relation. Second, we were not able to settle the computational complexity for complete preferences that are also single-peaked and single-crossing and for incomplete preferences with ties that are also narcissistic and single-peaked. We conjecture, however, that the NP-hardness reduction by Ronn [31] can be (non-trivially) adjusted to also work for these restricted domains. Third, for incomplete preferences, we extended the concepts of single-peaked and single-crossing preferences. However, there are further relevant extensions in the literature [11,15,25], which deserve study within our framework. Finally, the algorithm of Bartholdi III and Trick [3] strongly relies on the fact that there are always two agents that consider each other most acceptable. It would be interesting to know which generalized structured preferences could guarantee this fact. For instance, the so-called worst-restricted property (i.e. no three agents exist such that each of them is least preferred by any agent) is a generalization of the single-peaked property. We could show that the narcissistic and worst-restricted properties are enough to guarantee this useful property.

References

1. Ballester, M.Á., Haeringer, G.: A characterization of the single-peaked domain. Soc. Choice Welf. **36**(2), 305–322 (2011)
2. Barberà, S., Moreno, B.: Top monotonicity: a common root for single peakedness, single crossing and the median voter result. Game Econ. Behav. **73**(2), 345–359 (2011)
3. Bartholdi, J.J., Trick, M.: Stable matching with preferences derived from a psychological model. Oper. Res. Lett. **5**(4), 165–169 (1986)
4. Black, D.: The Theory of Committees and Elections. Cambridge University Press, Cambridge (1958)
5. Bredereck, R., Chen, J., Woeginger, G.J.: A characterization of the single-crossing domain. Soc. Choice Welf. **41**(4), 989–998 (2013)
6. Bredereck, R., Chen, J., Woeginger, G.J.: Are there any nicely structured preference profiles nearby? Math. Soc. Sci. **79**, 61–73 (2016)
7. Coombs, C.H.: A Theory of Data. Wiley, New York (1964)
8. Doignon, J., Falmagne, J.: A polynomial time algorithm for unidimensional unfolding representations. J. Algorithms **16**(2), 218–233 (1994)
9. Elkind, E., Faliszewski, P., Slinko, A.: Clone structures in voters' preferences. In: Proceedings of EC 2012, pp. 496–513 (2012)
10. Elkind, E., Faliszewski, P., Skowron, P.: A characterization of the single-peaked single-crossing domain. In: Proceedings of AAAI 2014, pp. 654–660 (2014)
11. Elkind, E., Faliszewski, P., Lackner, M., Obraztsova, S.: The complexity of recognizing incomplete single-crossing preferences. In: Proceedings of AAAI 2015, pp. 865–871, Long version available as manuscript on M. Lackers homepage (2015)
12. Elkind, E., Lackner, M., Peters, D.: Structured preferences. In: Endriss, U. (ed.) Trends in Computational Social Choice. AI Access (2017). Upcoming. Draft online available
13. Escoffier, B., Lang, J., Öztürk, M.: Single-peaked consistency and its complexity. In: Proceedings of ECAI 2008, pp. 366–370 (2008)
14. Fitzsimmons, Z.: Single-peaked consistency for weak orders is easy. Technical report. arXiv:1406.4829v3 [cs.GT] (2016)
15. Fitzsimmons, Z., Hemaspaandra, E.: Modeling single-peakedness for votes with ties. In: Proceedings of STAIRS 2016, vol. 284. Frontiers in Artificial Intelligence and Applications, pp. 63–74 (2016)
16. Gai, A.-T., Lebedev, D., Mathieu, F., Montgolfier, F., Reynier, J., Viennot, L.: Acyclic preference systems in P2P networks. In: Kermarrec, A.-M., Bougé, L., Priol, T. (eds.) Euro-Par 2007. LNCS, vol. 4641, pp. 825–834. Springer, Heidelberg (2007). doi:10.1007/978-3-540-74466-5_88
17. Gale, D., Shapley, L.S.: College admissions and the stability of marriage. Am. Math. Mon. **120**(5), 386–391 (2013)
18. Garey, M.R., Johnson, D.S.: Computers and Intractability–A Guide to the Theory of NP-Completeness. W. H. Freeman and Company, New York (1979)
19. Gusfield, D., Irving, R.W.: The Stable Marriage Problem-Structure and Algorithms. Foundations of Computing Series. MIT Press, Cambridge (1989)
20. Hotelling, H.: Stability in competition. Econ. J. **39**(153), 41–57 (1929)
21. Irving, R.W.: An efficient algorithm for the "stable roommates" problem. J. Algorithms **6**(4), 577–595 (1985)
22. Irving, R.W., Manlove, D.: The stable roommates problem with ties. J. Algorithms **43**(1), 85–105 (2002)

23. Knuth, D.E.: Stable marriage and its relation to other combinatorial problems. CRM Proceedings & Lecture Notes, vol. 10. AMS (1997)
24. Kujansuu, E., Lindberg, T., Mäkinen, E.: The stable roommates problem and chess tournament pairings. Divulgaciones Matemáticas **7**, 19–28 (1999)
25. Lackner, M.: Incomplete preferences in single-peaked electorates. In: Proceedings of AAAI 2014, pp. 742–748 (2014)
26. Lebedev, D., Mathieu, F., Viennot, L., Gai, A., Reynier, J., de Montgolfier, F.: On using matching theory to understand P2P network design. Technical report. arXiv:cs/0612108v1 [cs.NI] (2006)
27. Manlove, D.F.: Algorithmics of Matching Under Preferences. Series on Theoretical Computer Science, vol. 2. WorldScientific, Singapore (2013)
28. Manlove, D.F., O'Malley, G.: Paired and altruistic kidney donation in the UK: algorithms and experimentation. ACM J. Exp. Algorithmics **19**(1), 2.6:1–2.6:21 (2014)
29. Mirrlees, J.A.: An exploration in the theory of optimal income taxation. Rev. Econ. Stud. **38**, 175–208 (1971)
30. Roberts, K.W.: Voting over income tax schedules. J. Public Econ. **8**(3), 329–340 (1977)
31. Ronn, E.: NP-complete stable matching problems. J. Algorithms **11**(2), 285–304 (1990)
32. Roth, A.E., Sotomayor, M., Matching, T.-S.: A Study in Game-Theoretic Modeling and Analysis. Cambridge University Press, Cambridge (1990)
33. Roth, A.E., Sönmez, T., Ünver, M.U.: Pairwise kidney exchange. J. Econ. Theor. **125**(2), 151–188 (2005)
34. Roth, A.E., Sönmez, T., Ünver, M.U.: Efficient kidney exchange: coincidence of wants in markets with compatibility-based preferences. Am. Econ. Rev. **97**(3), 828–851 (2007)

Short Papers (Poster Presentations)

Compact Preference Representation via Fuzzy Constraints in Stable Matching Problems

Maria Silvia Pini[1]([✉]), Francesca Rossi[2], and Kristen Brent Venable[3]

[1] University of Padova, Padua, Italy
pini@dei.unipd.it
[2] IBM T.J. Watson Research Center, Yorktown Heights, NY, USA
frossi@math.unipd.it
[3] IHMC, Tulane University, New Orleans, USA
kvenabl@tulane.edu

Abstract. We define a framework for stable matching problems where agents are allowed to express their preferences in a compact way, via fuzzy constraints over the features describing the agents of the other group. We provide a solving engine for this new kind of stable matching problems that does not increase the time complexity of the classical GS algorithm, while maintaining stability of the matching returned. We then evaluate the approach experimentally.

1 Introduction

The stable matching (SM) problem has two sets of agents (men and women) that need to be matched in a stable way, that is, no man and woman, who are not matched to each other, both prefer each other to their current partner [6]. SM problems arise, for example, in assigning junior doctors to hospitals, children to schools, students to campus housing, and kidney transplant patients to donors.

The most well-known and used algorithm to find a stable matching is the GS algorithm [5], that runs in polynomial time. It assumes that both men and women express a preference ordering over all members of the other gender. However, this can be unfeasible, since the number of men and women can be very large. However, if men and women possess some features, this allows for expressing preferences in a compact way by referring to features rather than men or women. In this paper we study how to adapt the GS algorithm to work with preferences expressed by fuzzy constraints over features.

Fuzzy constraints are a special case of soft constraints [7], where quantitative preferences over features are between 0 and 1, and the objective is to maximize the minimal preference. To use such formalism within the GS stable matching procedure, we describe a fuzzy constraint solver to perform the operations needed in the GS algorithm.

The use of fuzzy constraints reduces the time and space needed by each agent to specify its preference ordering. Moreover, the stable matching procedure

---

Francesca Rossi—(on leave from University of Padova, Italy).

© Springer International Publishing AG 2017
J. Rothe (Ed.): ADT 2017, LNAI 10576, pp. 333–338, 2017.
DOI: 10.1007/978-3-319-67504-6_23

receives an input which is much smaller. If men's preferences can be modelled by a constraint graph with a bounded tree-width [4], the time complexity of the stable matching procedure does not increase, as shown by our theoretical and experimental evaluation.

A study similar to the one in this paper was done using CP-nets instead of fuzzy constraints [8]. As different techniques are needed to reason with CP-nets [1] and soft constraints in SMs, the linearization studied for CP-net in [8] is very different than those we define in this paper for fuzzy constraints.

2   Background

Stable matching problems. A *stable matching problem* (SM) [6] of size n includes n men and n women that each have a strict preference ordering over the members of the other gender. A *matching* is a one-to-one correspondence between men and women. Given a matching M, a man m, and a woman w, the pair (m, w) is a *blocking pair* for M if m prefers w to his partner in M and w prefers m to her partner in M. A matching is said to be *stable* if it does not contain blocking pairs. Given a SM P, there is at least one stable matching for P.

The *Gale-Shapley algorithm* (GS) [5] is widely used to solve SMs. The algorithm takes $O(n^2)$ steps and constructs a stable matching. It consists of a number of rounds in which each un-engaged man proposes to his most preferred woman to whom he has not yet proposed. Each woman receiving a proposal becomes "engaged", provisionally accepting the proposal from her most preferred man. In subsequent rounds, an already engaged woman can "trade up", becoming engaged to a more preferred man and rejecting a previous proposal, or if she prefers him, she can stick with her current partner.

Given a matching M, is $M(w)$ (resp., $M(m)$) the man (resp., woman) associated to the woman w (resp., man m) in M. Also, $pref(x)$ is the preference list of a man or a woman x. The GS algorithm includes the following operations: $Opt(pref(m))$: computes the optimal woman for m (i.e., m's first proposal); $Next(pref(m), w)$: computes the next best woman after w for man m (i.e., a new proposal for m); $Compare(pref(w), m, m')$: returns true if woman w prefers man m to m'. This is needed when woman w, currently matched with m', must decide whether to accept or decline a proposal from m.

Soft and fuzzy constraints. A *soft constraint* [7] associates a preference value from a (totally or partially ordered) set to each instantiation of its variables. A *Soft Constraint Satisfaction Problem* (SCSP) is a set of variables and a set of soft constraints (each one involving a subset of variables. An instance of the SCSP framework is obtained by choosing a specific preference structure. For instance, in fuzzy CSPs (FCSPs) [7] preference values are in $[0, 1]$ and we want to maximize the minimum preference. This is the instance that we consider in the paper.

A solution of an SCSP P is an assignment s to all its variables and its preference value, written $pref(P, s)$, is obtained by combining the preference values associated by each constraint to the part of s related to the variables of the constraint. In fuzzy CSPs the preference value of a complete assignment is the minimum preference value given by the constraints. An optimal solution is a solution s such that there is no other solution s' with $pref(P, s) < pref(P, s')$, where $<$ is the preference ordering of the considered preference structure. We denote with opt the preference value of an optimal solution.

Finding an optimal solution for a soft CSP is computationally hard in general, but it is polynomial, for example, for tree-shaped fuzzy CSPs, where a technique called directional arc-consistency, applied bottom-up on the tree shape of the problem, is enough to make the search for an optimal solution backtrack-free and thus polynomial. A tree-shaped fuzzy CSP is a fuzzy CSP whose constraint graph (where nodes represent variables and arcs connect variables involved in the same constraint) is a tree.

3   Stable Matching Problems with Fuzzy Constraints

We consider stable marriage problems with n men and n women, where each man and each woman specify their preferences over the members of the other gender via a set of fuzzy constraints. We call this a fuzzy CSP based SM (FSM).

Each man and woman is described by a set of features, that are represented by the variables of the fuzzy constraint problems. If each variable has d possible values, the number of variables, say f, of each fuzzy constraint problem is $O(log_d n)$. We have $2f$ features, of which f describe men and f describe women.

GS operations. $Opt(pref(m))$ must return the optimal solution of a fuzzy CSP defining the preferences of man m over the women. We recall that, in general, finding the optimal solution of a soft CSP is a computationally difficult problem. However, if the SCSP has a tree-like shape, or bounded tree-width, it can be done in polynomial time [3]. Thus this operation takes polynomial time if the constraint graph is a tree (or has a bounded tree-width).

$Compare(pref(w), m_1, m_2)$ compares two complete assignments m_1 and m_2 and checks if m_1 is strictly more preferred to m_2 for w. In fuzzy constraint problems, this is computationally easy to do, if there is a polynomial number of constraints. In fact, m_1 is strictly preferred to m_2 when the preference of m_1 for w is strictly greater than that of m_2 for w. Notice that women need only to perform Compare operations. Thus we do not need any restriction on the shape of the constraint graph for women's preferences to make Compare polynomial.

For the $Next(pref(m), w)$ operation, we need to linearize the solution ordering of a fuzzy CSP. In fact, this operation is used to find the next most preferred woman in a man's preference ordering, so when two or more women are tied, we need to put an order over them to understand who to propose first.

Linearizations. In fuzzy constraints, the solution ordering is in general a total order with ties. In this context, linearizing the solution ordering means giving an order over the elements in each tie.

We aim to define linearizations where finding the next best solution (that is, applying operation $Next$) is tractable and where solutions which are less distant from optimal ones appear earlier in the linearization. We define three linearizations L_1, L_2, and L_3 which break ties by taking into account the distance of a solution preference from the optimal preference (L_1), or also the minimum number of preference values for parts of the solutions to be changed to make the solution optimal (L_2), or also the amount of change required (L_3). Among the solutions which are still in a tie, we put first those that are lexicographically earlier, according to an ordering over variables and domain values.

We will now see how to perform operation $Next$ on such three linearizations. We will call these operations $Next_i$, for $L_i = 1, 2, 3$.

From results in [2], we know that performing $Next_1$ can be accomplished in polynomial time when we have a tree-shaped fuzzy CSPs. To perform $Next_2$ and $Next_3$ for tree-shaped fuzzy CSPs, when we already have the top $k-1$ solutions, we find the top k solution according L_2 and L_3 by computing the top k solutions of a set of weighted CSPs with a bounded tree-width. This is polynomial [4].

Our algorithm, $KCheapest$, takes in input a tree-shaped fuzzy CSP P, an integer k, and a linearization L (either L_2 or L_3), and returns the top k solutions of P according to L_2 and L_3 in polynomial time, when k is polynomial in f and d. We know from [8] that on average only 2% of all proposals are made, so the idea is to call $KCheapest$ with $k = 2\%$ of the maximum number of proposals and cache the returned set of solutions. Only when all cached solutions have already been returned, we need to call $KCheapest$ again.

If we run the GS algorithm on any of the linearizations we defined, by definition we obtain a matching which is stable w.r.t. this linearization. With n men and n women, and preference lists given explicitly, finding a SM needs $O(n^2)$ space and $O(n^2)$ time [6]. When we use fuzzy constraints, each man and woman needs $O(d^k m)$ space and time to state their preferences, where d is the size of domains, k is the number of features involved in the largest soft constraint, and m is the number of soft constraints. When we assume a bounded tree-width constraint graph for men, each proposal in the GS algorithm takes $O(poly(f))$ time, where f is the number of features and each Compare operation takes $O(poly(m))$. So, overall, the GS algorithm may need up to $O(n^2 \times poly(f) \times poly(m))$ time, although the number of proposals have been shown to be much lower in practice [8].

Experimental setting and results. The described linearizations can be used in two different ways within the GS algorithm. One way is to compute the whole preference lists for each man and then run GS. The other is to use the linearization only when GS requires a new proposal. We ran experiments to compare the running time of these two scenarios on a 2.4 Ghz Intel Core i5 machine with 8 GB of RAM, and averaged values over 100 executions, setting a time limit of 10 min. For each man, a tree-shaped fuzzy CSP over f features of domain size d is generated by randomizing the preference values. Each woman is represented by a randomly generated fuzzy CSP with a generic topology over the same number of f features and a constraint density of 50%. Thus, the whole generated FSM problem consists of $n = d^f$ individuals on each side.

In the first test we fixed d and measured the execution time needed to find a stable matching while increasing the number of features f. Results for $Next_1$

are shown in Fig. 1(a), where GS-next1-lists is GS run on precomputed lists obtained running $Next_1$ exactly n^2 times, whereas SoftGS1 is GS which calls $Next_1$ on demand. For $f = 15$ GS-next1-lists didn't complete within the time limit. SoftGS1 substantially outperforms GS-next1-lists both in space and time. As expected from [8], GS makes on average only 2% of all possible proposals. This justifies the advantage of an on demand implementation.

We ran the same tests on $Next_{2-3}$, that is, the algorithm that calls $KCheapest$, only when needed, for either L_2 or L_3. The results are plotted in Fig. 1(c) where GS-next23-lists is again the implementation that runs over full precomputed lists while SoftGS23 only precomputes the first 2% of every preference list and then eventually asks for more. SoftGS23 is considerably better than its equivalent that runs on precomputed lists, with an average running time of $4.875s$ versus $107.202s$. In Fig. 1(d) we compared the performance of the different linearizations L_1 and L_{2-3}. As expected, solving the Weighted CSP cost minimization problem brings additional overhead to the Next operation, resulting in a worse performance compared to $Next_1$.

In the second test setting, we fixed f to 5, and measured execution time as a function of the domains cardinality d. As shown in Fig. 1(b), SoftGS1 behaves better than GS with precomputed lists, despite the fact that $n = d^f$. The performance of SoftGS1 is very promising as for a setting of $d = 2$ and $f = 12$ (which means 8192 individuals to be matched), the average computation time is $1.83s$ versus $137.43s$ of the version with precomputed lists. In a real world scenario of this size, ranking 8192 individuals of the opposite sex may be impractical, while the compact preference representation makes it feasible, as each individual only needs to express his/her preferences over 12 features.

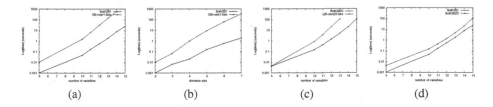

(a)	(b)	(c)	(d)

Fig. 1. Computation time varying the number of variables with $d = 2$, and varying d with 5 variables ($f = 5$).

References

1. Boutilier, C., Brafman, R.I., Domshlak, C., Hoos, H.H., Poole, D.: CP-nets: a tool for representing and reasoning with conditional ceteris paribus preference statements. JAIR **21**, 135–191 (2004)
2. Brafman, R.I., Rossi, F., Salvagnin, D., Venable, K.B., Walsh, T.: Finding the next solution in constraint-and preference-based knowledge representation formalisms. In: Proceedings KR 2010 (2010)

3. Dechter, R.: Tractable structures for CSPs. In: Rossi, F., Van Beek, P., Walsh, T. (eds.) Handbook of Constraint Programming. Elsevier (2005)
4. Dechter, R., Flerova, N., Marinescu, R.: Search algorithms for m best solutions for graphical models. In: Proceedings AAAI 2012. AAAI Press (2012)
5. Gale, D., Shapley, L.S.: College admissions and the stability of marriage. Amer. Math. Monthly **69**, 9–14 (1962)
6. Gusfield, D., Irving, R.W.: The Stable Marriage Problem: Structure and Algorithms. MIT Press, Boston (1989)
7. Meseguer, P., Rossi, F., Schiex, T.: Soft constraints. In: Rossi, F., Van Beek, P., Walsh, T. (eds.) Handbook of Constraint Programming. Elsevier (2005)
8. Pilotto, E., Rossi, F., Venable, K.B., Walsh, T.: Compact preference representation in stable marriage problems. In: Rossi, F., Tsoukias, A. (eds.) ADT 2009. LNCS, vol. 5783, pp. 390–401. Springer, Heidelberg (2009). doi:10.1007/978-3-642-04428-1_34

Determining Nash Equilibria for Stochastic Positional Games with Discounted Payoffs

Dmitrii Lozovanu[1](✉) and Stefan Pickl[2]

[1] Institute of Mathematics and Computer Science, Academy of Sciences, Moldova,
Academy str., 5, 2028 Chisinau, Moldova
lozovanu@math.md

[2] Institute for Theoretical Computer Science, Mathematics and Operations Research,
Universität der Bundeswehr München, 85577 Neubiberg-münchen, Germany
stefan.pickl@unibw.de

Abstract. A class of discounted stochastic games is formulated and studied by applying the concept of positional games to Markov decision processes with expected total discounted reward criteria. Existence results of pure and mixed stationary Nash equilibria for the considered class of discounted stochastic positional games are presented and an approach for determining the optimal strategies of the players is proposed.

Keywords: Stochastic positional game · Discounted payoffs · Stationary strategy · Nash equilibrium

1 Introduction

In this paper we formulate and study a class of stochastic games by applying the concept of positional games to discounted Markov decision processes with finite state and action spaces. We consider Markov decision processes that may be controlled by several actors (players) as follows. The set of states of the system in a Markov process is divided into several disjoint subsets that represent the position sets for the corresponding players. Each player controls the process only in his position set via the feasible actions in the corresponding states. The aim of each player is to determine which action should be taken in each state of his position set in order to maximize his own discounted sum of step rewards. The step rewards in the states with respect to each player are known for an arbitrary feasible action in the corresponding states of the position sets. We consider the infinite horizon stochastic games and assume that players use stationary strategies of a selection of the action in the states, i.e. each player in his arbitrary position uses the same action for an arbitrary discrete moment of time. For the considered class of games we are seeking for a Nash equilibrium. We show that for stochastic positional games with discounted payoffs stationary Nash equilibria exist. Moreover, we show that for the considered class of games there exist stationary Nash equilibria in pure strategies. Based on these results we propose an approach for determining the optimal strategies of the players.

J. Rothe (Ed.): ADT 2017, LNAI 10576, pp. 339–343, 2017.
DOI: 10.1007/978-3-319-67504-6_24

2 Formulation of a Discounted Stochastic Positional Game

An n-player stochastic positional game with discounted payoffs is determined by the following elements:

- a finite set of states X;
- a partition $X = X_1 \cup X_2 \cup \cdots \cup X_n$ of X, where X_i represents the position set of player $i \in \{1, 2, \ldots, n\}$, $X_i \cap X_j = \emptyset$ for $i \neq j$;
- a finite set of actions $A(x)$ for an arbitrary state $x \in X$;
- a step reward $r^i(x, a)$ with respect to each player $i \in \{1, 2, \ldots, n\}$ for an arbitrary state $x \in X$ and an arbitrary action $a \in A(x)$;
- a transition probability function $p : X \times \prod_{x \in X} A(x) \times X \to [0, 1]$ that gives the transition probabilities $p^a_{x,y}$ from an arbitrary $x \in X$ to an arbitrary $y \in X$ for a fixed action $a \in A(x)$, where $\sum_{y \in X} p^a_{x,y} = 1$, $\forall x \in X$, $a \in A(x)$;
- a discount factor λ, $0 < \lambda < 1$;
- a starting state x_0.

The game starts in the state (position) x_0 at the moment of time $t = 0$. If this position belongs to the set of positions of player $k \in \{1, 2, \ldots, n\}$, i.e. $x_0 \in X_k$, then player k selects an action $a_0 \in A(x_0)$. After that players $1, 2, \ldots, n$ receive the corresponding rewards $r^1(x_0, a_0), r^2(x_0, a_0), \ldots, r^n(x_0, a_0)$ and the dynamical system passes randomly to a state $y = x_1 \in X$ according to probability distributions $p^{a_0}_{x_0,y}$ where x_1 is reached at the moment of time $t = 1$. In general, if at the moment of time t the state of the dynamical system is $x_t \in X$ and this state belongs to the set of positions of player $i \in \{1, 2, \ldots, n\}$, i.e. $x_t \in X_i$, then player i selects an action $a_t \in X_i$ and players $1, 2, \ldots, n$ receive the corresponding rewards $r^1(x_t, a_t), r^2(x_t, a_t), \ldots, r^n(x_t, a_t)$. After that the system passes randomly to the a state $y = x_{t+1}$ according to probability distributions $p_{x_t,y}$ and so on. Such a play induces a sequence of states and actions $x_0, a_0, x_1, a_1, \ldots, x_t, a_t, \ldots$ that defines a stream of rewards $r^i_0, r^i_1, r^i_2, \ldots$, where $r^i_t = r^i(x_t, a_t)$, $i = \overline{1, n}$, $t = 0, 1, 2, \ldots$. The infinite stochastic positional game with discounted payoffs is the game with the following payoffs of the players

$$\sigma^i_{x_0} = \lim_{t \to \infty} \inf \mathsf{E}\left(\sum_{\tau=0}^{t} \lambda^\tau r^i(x_\tau, a_\tau) \right), \quad i = \overline{1, n}.$$

Each player in this game selects actions from the feasible set of actions in the state of his position set in order to maximize his own discounted sum of rewards.

Note that in a standard formulation of a stochastic game it is assumed that to each player $i \in \{1, 2, \ldots, n\}$ in each state $x \in X$ it is associated a feasible action set $A^i(x)$ and for a fixed set of actions $a^1, a^2, \ldots, a^n$ of the players from the corresponding feasible action sets $A^1(x), A^2(x), \ldots, A^n(x)$ the corresponding rewards $r^i = r^i(x, a^1, a^2, \ldots, a^n), i = \overline{1, n}$ are determined uniquely.

So, the rewards in a state depend on actions of all players. Additionally the transition probability distribution in a state $x \in X$ also depends on the profile $a = (a^1, a^2, \ldots, a^n)$ of actions chosen by all players, i.e. $p_{x,y}^a = p_{x,y}^{(a^1, a^2, \ldots, a^n)}$. We can see that a stochastic game becomes a stochastic positional game if for a given partition $X = \cup_{k=1}^n X_k$ the rewards $r^i(x, a^1, a^2, \ldots, a^n)$ and the probability distributions $p_{x,y}^{(a^1, a^2, \ldots, a^n)}$ are defined as follows:

$$r^i(x, a^1, a^2, \ldots, a^k, \ldots, a^n) = r^i(x, a^k), \forall a^k \in A^k(x), \; i = \overline{1, n} \; \text{if} \; x \in X_k.$$

$$p_{x,y}^{(a^1, a^2, \ldots, a^k, \ldots, a^n)} = p_{x,y}^{a^k}, \; \forall a^k \in A^k(x), \; \text{if} \; x \in X_k.$$

3 Stochastic Positional Games in Stationary Strategies

A *strategy of player* $i \in \{1, 2, \ldots, n\}$ in a stochastic positional game game is a mapping s^i that provides for every state $x_t \in X_i$ a probability distribution over the set of actions $A^i(x_t)$. If these probabilities take only values 0 and 1, then s^i is called *a pure strategy*, otherwise s^i is called *a mixed strategy*. If these probabilities depend only on the state $x_t = x \in X$ (i.e. s^i does not depend on t), then s^i is called *a stationary strategy*, otherwise s^i is called a *non-stationary strategy*. So, we can identify the set of stationary strategies S^i of player $i \in \{1, 2, \ldots, n\}$ with the set of solutions of the system

$$\begin{cases} \sum_{a^i \in A^i(x)} s_{x,a}^i = 1, & \forall x \in X_i; \\ s_{x,a}^i \geq 0, & \forall x \in X_i, \; \forall a \in A^i(x). \end{cases} \tag{1}$$

A basic solution of this system corresponds to a pure strategy for player i. If s^k is a stationary strategy of player k then the probability transitions from a state $x \in X_k$ to the states $y \in X$ can be calculated as follows $p_{x,y}^{s^k} = \sum_{a \in A(x)} s_{x,a}^k p_{x,y}^a$ and the corresponding step rewards $r^i(x, s^k), i = \overline{1, n}$ in a state $x \in X_k$ for a given strategy s^k of player k are calculated as $r^i(x, s^k) = \sum_{a \in A(x)} s_{x,a}^k r_{x,a}^i$, $i = \overline{1, n}$. If $s = (s^1, s^2, \ldots, s^n)$ is a profile of strategies of the players then we can determine the matrix of transition probabilities $P^s = (p^s)$ and the matrix $Q^s(\lambda) = (I - \lambda P^s)^{-1}$ induced by the profile s. Therefore, for the given starting state x_0 and a given profile $s = (s^1, s^2, \ldots, s^m)$ of stationary strategies we can determine the corresponding discounted sum of rewards for the players $\sigma_{x_0}^i(s) = \sum_{k=1}^m \sum_{y \in X_i} q_{x_0, y}(\lambda) r^i(y, s^k), \; i = \overline{1, n}$, where $q_{x,y}(\lambda)$ represent the elements of the matrix $Q^s(\lambda)$. The functions $\sigma_{x_0}^1(s), \sigma_{x_0}^2(s), \ldots, \sigma_{x_0}^n(s)$ on $S = S^1 \times S^2 \times \cdots \times S^n$ define a game in normal form that we denote by $\langle \{S^i\}_{i=\overline{1,n}}, \{\sigma_{x_0}^i(s)\}_{i=\overline{1,n}} \rangle$. This game corresponds to a *discounted stochastic game in stationary strategies with a fixed starting state*. In the case when the starting state is chosen randomly according to a given distribution $\{\theta_x\}$ on X we obtain a *discounted stochastic positional game with random starting state*. For such a game we assume that the play starts in the states $x \in X$ with probabilities $\theta_x > 0$ where $\sum_{x \in X} \theta_x = 1$.

If the players use mixed stationary strategies of a selection of the actions in the states then the payoff functions $\psi_\theta^i(s^1, s^2, \ldots, s^n) = \sum_{x \in X} \theta_x \sigma_x^i(s^1, s^2, \ldots, s^n)$, $i = \overline{1, n}$ on S define a game in normal form. In the case $\theta_x = 0$, $\forall x \in X \setminus \{x_0\}$, $\theta_{x_0} = 1$ the considered game becomes a stochastic game with fixed starting state x_0.

In a more detailed form the considered class of games in stationary strategies can be formulated as follows: Let $S^1, S^2, \ldots, S^n$ be the corresponding sets of stationary strategies for players $1, 2, \ldots$, where each S^i, $i \in \{1, 2, \ldots, n\}$ corresponds to the set of solutions of system (1). On S we consider n payoff functions $\varphi_{x_0}^i(s^1, s^2, \ldots, s^n) = \sigma_{x_0}^i$, $i = \overline{1, n}$, where σ_x^i for $x \in X$ satisfy the conditions

$$\sigma_x^i - \lambda \sum_{y \in X} \sum_{a \in A(x)} s_{x,a}^k p_{x,y}^a \sigma_y^i = \sum_{a \in A(x)} s_{x,a}^k r^i(x,a)^i, \quad \forall x \in X_k; \ i, k = \overline{1, n}.$$

These functions on S define a game in normal form that corresponds to the discounted stochastic positional game with starting state x_0. In the case when the starting state of the game is chosen randomly according to a given distribution $\{\theta_y\}$ on X we have the payoffs $\varphi^i(s^1, s^2, \ldots, s^m) = \sum_{y \in X} \theta_y \sigma_y^i$, $i = \overline{1, n}$.

4   Pure and Mixed Stationary Nash Equilibria

In Sect. 2 we have shown that the discounted stochastic positional game represents a special case of the general discounted stochastic game. Therefore based on results from [1,2,5] we obtain the following theorem.

Theorem 1. *An arbitrary discounted stochastic positional game possesses of a stationary Nash equilibrium.*

In the following we show that for an arbitrary discounted stochastic game there exist a pure stationary Nash equilibrium. Such a result we obtain on the bases of the following theorem.

Theorem 2. *Let a discounted stochastic positional game with a discount factor $0 < \lambda < 1$ be given. Then there exist the values σ_x^i, $i = \overline{1, n}$ for $x \in X$ that satisfy the following conditions:*

(1) $r^i(x,a) + \lambda \sum_{y \in X} p_{x,y}^a \sigma_y^i - \sigma_x^i \leq 0$, $\forall x \in X_i$, $\forall a \in A(x)$, $i = \overline{1, n}$;

(2) $\max\limits_{a \in A(x)} \left\{ r^i(x,a) + \lambda \sum_{y \in X} p_{x,y}^a \sigma_y^i - \sigma_x^i \right\} = 0$, $\forall x \in X_i$, $i = \overline{1, n}$;

(3) on each position set X_i, $i \in \{1, 2, \ldots, n\}$ there exists a map $s^{i^*} : X_i \to A$

such that $s^{i^*}(x) = a^* \in \arg\max\limits_{a \in A(x)} \left\{ r^i(x,a) + \lambda \sum_{y \in X} p_{x,y}^a \sigma_y^i - \sigma_x^i \right\}$, $\forall x \in X_i$

and $r^j(x, a^*) + \lambda \sum_{y \in X} p_{x,y}^{a^*} \sigma_y^j - \sigma_x^j = 0$, $\forall x \in X_i$, $j = 1, 2, \ldots, n$.

The set of maps $s^{1^*}, s^{2^*}, \ldots, s^{n^*}$ determines a Nash equilibrium for the discounted stochastic positional game and $\varphi_{\overline{x}}^i(s^{1^*}, s^{2^*}, \ldots, s^{n^*}) = \sigma_{\overline{x}}^i$, $\forall \overline{x} \in X$, $i = \overline{1, n}$. Moreover $s^* = (s^{1^*}, s^{2^*}, \ldots, s^{n^*})$ is a Nash equilibrium for an arbitrary starting position $\overline{x} \in X$. Taking into account that such a property holds for each player we obtain the proof of the theorem.

Proof. According to Theorem 1 for a discounted stochastic positional game there exists a Nash equilibrium $s^* = (s^{1^*}, s^{2^*}, \dots, s^{n^*})$. Let us fix the strategies $s^{1^*}, s^{2^*}, \dots, s^{i-1^*}, s^{i+1^*}, \dots, s^{n^*}$ of the players $1, 2, \dots, i-1, i+1, \dots, n$ and consider the problem of determining the expected total discounted cost with respect to player i. Then the optimal stationary strategy for this Markov decision problem is s^{i^*} and based on results from [3,4] there exist the values σ_x^i for $x \in X$ that satisfy the conditions $r^i(x,a) + \lambda \sum_{y \in X} p_{x,y}^a \sigma_y^i - \sigma_x^i \leq 0,$ for

$$x \in X_i, \quad \forall a \in A(x), \text{ where } \max_{a \in A(x)} \left\{ r^i(x,a) + \lambda \sum_{y \in X} p_{x,y}^a \sigma_y^i - \sigma_x^i \right\} = 0, \quad \forall x \in X_i$$

and $s^{i^*}(x) = a^* \in \arg \max_{a \in A(x)} \left\{ r^i(x,a) + \lambda \sum_{y \in X} p_{x,y}^a \sigma_y^i - \sigma_x^i \right\}, \quad \forall x \in X_i.$

So, the theorem holds. □

Remark 1. In the considered stochastic positional game it is assumed that the discount factor λ is the same for all players. In general, such a game can be formulated for the case when for different players the discount factors may be different. So, we may assume that each player $i \in \{1, 2, \dots, m\}$ has his own discount factor λ_i and the payoffs $\varphi_{x_0}^i(s^1, s^2, \dots, s^n)$, $i = \overline{1, n}$ for a fixed starting state x_0 are determined taking into account the corresponding discount factors λ_i, $i = \overline{1, n}$. In this case Theorem 2 holds if in conditions (1)–(3) we replace λ by λ_i. So, for the considered game there exists a pure Nash equilibrium.

5   Conclusion

The discounted stochastic positional games represents a special class of discounted stochastic games for which pure stationary Nash equilibria exist. The optimal pure stationary strategies of the players can be found on the basis of Theorem 2.

References

1. Fink, A.M.: Equilibrium in a stochastic n-person game. J. Sci. Hiroshima Univ. Ser. A-1 **28**, 89–93 (1964)
2. Lozovanu, D., Pickl, S.: On Nash equilibria for stochastic games and determining the optimal strategies of the players. Contrib. Game Theory Manage. **8**, 187–198 (2015). St. Petersburg University
3. Lozovanu, D., Pickl, S.: Optimization of Stochastic Discrete Systems and Control on Complex Networks. Springer, New York (2015)
4. Lozovanu, D., Pickl, S.: Determining the optimal strategies for discrete control problems on stochastic networks with discounted costs. Discrete Appl. Math. **182**, 169–180 (2015)
5. Takahashi, M.: Equilibrium points of stochastic noncooperative n-person stochastic games. J. Sci. Hiroshima Univ. Ser. A-1 **28**, 95–99 (1964)

A Qualitative Decision-Making Approach Overlapping Argumentation and Social Choice

Pierre Bisquert[1(✉)], Madalina Croitoru[2], and Nikos Karanikolas[1]

[1] IATE, INRA, Montpellier, France
pierre.bisquert@inra.fr
[2] GraphIK, LIRMM, University of Montpellier, Montpellier, France

Abstract. Collective decision making is classically done via social choice theory with each individual expressing preferences as a (total) order over a given set of alternatives, and the group's aggregated preference is computed using a voting rule. However, such methods do not take into account the rationale behind preferences. Our research hypothesis is that a decision made by participants understanding the qualitative rationale (i.e., arguments) behind each other's preferences has better chances to be accepted and used in practice. To this end, we propose a novel qualitative decision process which combines argumentation with computational social choice. We show that a qualitative approach based on argumentation can overcome some of the social choice deficiencies.

1 Introduction

Taking decisions is a part of our everyday life. From the simplest ones, e.g., choosing which movie we are going to watch in the theater, to the most complex ones, e.g., selecting a government, a decision has to be made. The way to achieve a decision though can be a very complex task. Usually decision makers make their decisions based on different criteria and aspects that they consider to be important. One should wonder then what happens when we want to take a justified and fair collective decision, which leads us to the following questions. *How do agents form their thoughts and reason their preferences?* How should we aggregate them in order to have a democratic collective decision? That is the problem we are dealing with in this paper.

The commonly used way of making such a collective decision is using social choice theory. Each agent of the group expresses her preference as a total order over a set of alternatives, and then the group's preference is computed from the individual preferences using a voting rule. In the classical voting, the collective decision is computed from quantitative methods by taking into account only the agents' preferences without knowing why the agents have these preferences and what is the rationality behind it. Thus, classical social choice presents a framework where the justifications for the agents' preferences are not considered.

We believe that qualitative methods where humans can understand the reasoning behind the preferences have more chances to be accepted. This gives us

© Springer International Publishing AG 2017
J. Rothe (Ed.): ADT 2017, LNAI 10576, pp. 344–349, 2017.
DOI: 10.1007/978-3-319-67504-6_25

the motivation to combine Argumentation with Computational Social Choice: we believe that enriching the collective decision making procedure with an argumentation framework will provide the explanation behind the decision. To this end, we are placing the decision problem within the boundaries of an *abstract argumentation framework*. Abstract argumentation provides a flexible and robust tool for non-monotonic reasoning. It was introduced by Dung [4] and is based on the evaluation of interacting entities called arguments. The argumentation systems are represented by graphs, where the nodes represent the arguments and the edges represent the attacks, or conflicts, between them. Various semantics defined by Dung and other researchers have been proposed to identify coherent sets of arguments, which are based on the attack relations between them.

In our problem, the decision to be recommended lies on a set of *alternatives*. The decision will be derived from the justified preferences of the set of *agents* over the alternatives. The justified preferences are the outcome of a deliberation phase where each agent reveals her preferences and their justifications. The collection of agents' rankings is known as *preference profile*. The preference profile of the agents and the justifications are used to build an argumentation framework that will help us build the *justified preference profile*, which includes the preferences produced after the deliberation and corresponds to the different collective viewpoints of the agents. The objective is to fairly aggregate the justified viewpoints by using a *voting rule*.

The classical problem in social choice theory is which voting rule is the most appropriate for aggregating the preference profile. Unfortunately, due to the impossibility results by Arrow [1] and Gibbard-Satterthwaite [5,6] there is no hope of finding a voting rule that can be "perfect". Despite that, social choice theory has enhanced our perception among proposed voting rules, where each of them has different characteristics, qualities and weaknesses. One of the most prominent rules in the history of social choice, and which is generally acclaimed as a founding method of the field, is the one proposed by the Marquis de Condorcet. The *Condorcet method* [3] relies on comparisons between each pair of alternatives. An alternative x is said to beat alternative y in a *pairwise election (comparison)* if a majority of agents prefer x to y, i.e. rank x above y. The alternative who beats every other alternative in a pairwise comparison is the winner. Unfortunately, there are preference profiles where the collective preferences are cyclic, i.e., not transitive. For example if we have 3 alternatives x, y, z and the results of the pairwise comparisons are: x beats y, y beats z and z beats x then we say that a *voting cycle* occurs. This contradictory phenomenon is known as the *Condorcet paradox* [2]. Despite this paradox, the Condorcet criterion is widely acclaimed as the most intuitive way of voting and it is the aim of this paper to provide an approach that always avoids the Condorcet paradox thanks to the construction of justified preference profiles.

2 Preliminaries

Social Choice Theory. We consider a set of $N = \{1, \ldots, n\}$ *agents* and a set of *alternatives* A, $|A| = m$. Each agent $i \in N$ has preference relations ($\succ$)

over the alternatives denoted with $x \succ_i y$ which means that agent i *prefers* alternative x to y. We define that each preference relation satisfies transitivity and hence, the set of all the preference relations for agent i produces a linear (total) order $\succ_i$ on A, i.e., the ranking of agent i over the alternatives. Let $\mathcal{L}_A$ be the set of linear orders over A. A *preference profile* $\succ_{PP} = \langle \succ_1, \ldots, \succ_n \rangle \in \mathcal{L}_A^n$ is a collection of the linear orders for all the agents. A *voting rule* is a mapping $f : \mathcal{L}_A^n \to 2^A \setminus \{\emptyset\}$ from preference profiles to nonempty subsets of alternatives, which designates the winner(s) of the election. For two candidates $x, y \in A$, and $\succ_{PP} \in \mathcal{L}_A^n$, alternative x *beats* y in a *pairwise comparison* if $|\{i \in N : x \succ_i y\}| > n/2$, that is, if a (strict) majority of agents prefer x to y. A well-known voting rule is the *Condorcet method*: the *Condorcet winner* is an alternative that beats every other alternative in a pairwise comparison. The *Condorcet paradox* is a situation in which collective preferences can be cyclic (i.e. not transitive), even if the preferences of individual agents are not cyclic. A *voting cycle* occurs when we have 3 alternatives x, y, z such that $|\{i \in N : x \succ_i y\}| > n/2$, $|\{i \in N : y \succ_i z\}| > n/2$, and $|\{i \in N : z \succ_i x\}| > n/2$.

Argumentation. In order to be general with regards to the deliberation step, we are using the abstract argumentation framework proposed in [4]:

Definition 1 (Argumentation framework). *An argumentation framework (AF) is a pair* $(\mathbf{A}, \mathbf{R})$*, where* $\mathbf{A}$ *is a finite nonempty set of arguments and* $\mathbf{R}$ *is a binary relation on* $\mathbf{A}$*, called attack relation. Let* $a, b \in \mathbf{A}$*,* $a\mathbf{R}b$ *means that* a *attacks* b *.*

The coherent sets of arguments (called "extensions") are determined according to a given semantics whose definition is usually based on the following concepts:

Definition 2 (Conflict-free set, defense and admissibility). *Given an AF* $(\mathbf{A}, \mathbf{R})$*, let* $a \in \mathbf{A}$ *and* $\mathcal{S} \subseteq \mathbf{A}$*,*

- $\mathcal{S}$ *is* conflict-free *iff there does not exist* $a, b \in \mathcal{S}$ *such that* $a\mathbf{R}b$*.*
- $\mathcal{S}$ defends *an argument* a *iff each attacker of* a *is attacked by an argument of* $\mathcal{S}$*.*
- $\mathcal{S}$ *is an* admissible set *iff it is conflict-free and it defends all its elements.*

Definition 3 (Semantics). *Given an AF* $(\mathbf{A}, \mathbf{R})$*, let* $\mathcal{E} \subseteq \mathbf{A}$*.* $\mathcal{E}$ *is*

- *a* complete *extension iff* $\mathcal{E}$ *is an admissible set and every argument which is defended by* $\mathcal{E}$ *belongs to* $\mathcal{E}$*.*
- *a* preferred *extension iff* $\mathcal{E}$ *is a maximal admissible set (wrt set inclusion* $\subseteq$*).*
- *the* grounded *extension iff* $\mathcal{E}$ *is a minimal (wrt* $\subseteq$*) complete extension.*
- *a* stable *extension iff* $\mathcal{E}$ *is conflict-free and attacks any argument* $a \notin \mathcal{E}$*.*

Given a semantics, the set of extensions of $(\mathbf{A}, \mathbf{R})$ *is denoted by* $\mathbf{E}$*.*

It should be noted that in this paper we focus on the preferred semantics since it ensures the existence of at least one extension, which is needed since extensions will be used as voters, and their maximality, which ensures that each extension represents a full ranking over the alternatives. Other semantics will be considered in future work.

3   A Decision Model Based on Justified Preferences

In the proposed model we are considering the case of taking a decision using a qualitative argumentative approach and voting theory. Observe that the suggested process is an argumentative approach that relies on combining the qualitative preferences and not a voting rule whose role is to aggregate the individual preferences with quantitative methods. In our problem we have a set of alternatives and the agents whose justified preferences over the alternatives will determine the decision to be taken. Each agent provides a justification for each of the preference relations on the alternatives and we demand the preference relations to be transitive so a ranking with the preferences of the agent is built. We use this information to formulate arguments which express the agents' preferences. More precisely, we are going to distinguish between three types of arguments: *preference relation* arguments, *ranking* arguments and *generic* arguments.

Preference relation arguments. A preference relation argument a_{xy} represents a *justification* given by an agent to consider the preference $x \succ y$. Note that we may have multiple a_{xy} arguments, in the case where some agents have different justifications for the preference $x \succ y$. The set of preference relation arguments is denoted $\mathbf{A}_P$.

It should be noted that due to what they represent, the arguments a_{xy} and a_{yx} cannot be considered together in a coherent view point since they are "opposed". Consequently, we assume that those arguments attack each other.

Ranking arguments. A ranking argument represents one of the possible ranking over the considered alternatives. It is important to note that in our setting, we always consider all the possible ranking arguments; it will be the agents' prerogative to justify why a ranking should not be considered as we will see below. We denote by $\mathbf{A}_R$ the set of all the possible ranking arguments and by $\mathbf{A}_{R_{x\cdots y}}$ the set of ranking arguments where the preference $x \succ \cdots \succ y$ is satisfied. Moreover, we define a special ranking argument $a_\emptyset$ that represents a ranking without preference; it can be seen as the blank vote resulting from either non-transitive preference relations or no justified preferences.

Like preference relation arguments, we consider ranking arguments as mutually inconsistent. For this reason, we assume that ranking arguments attack each other, with the exception of $a_\emptyset$ that attacks no argument. In this way, we represent the fact that having a reason to consider a ranking forbids the possibility of considering blank voting. Furthermore, ranking arguments can be attacked by preference relation arguments. Indeed, giving a justification for $x \succ y$ (i.e. giving an argument a_{xy}) is a reason for ignoring the rankings with $y \succ x$ (i.e. $\mathbf{A}_{R_{yx}}$): a_{xy} is attacking the elements of $\mathbf{A}_{R_{yx}}$.

Generic arguments. Generic arguments regroup all the other possible arguments that can arise during a debate. In particular, those arguments are only able to attack other generic arguments and preference relation arguments (for instance if the reason given for considering $x \succ y$ is itself justified to be wrong). It is important to note that while the flexibility offered by the abstract argumentation setting is convenient for its generality, it can also lead to undesirable behaviors. Hence, we propose the following restriction.

Axiom 1 (Independence of preference justifications). *Given two preference relation arguments a_{xy} and a_{uv}, such that $\{x, y\} \neq \{u, v\}$, then there is no generic argument a_g such that both paths of attacks from a_g to a_{xy} and from a_g to a_{uv} exist.*

The intuition is that the discussions about each pairwise preference are independent, i.e. a generic argument cannot have an impact on preferences over different alternatives.

Computing the justified profile. Using the arguments and attacks shown above in an argumentation framework, it is possible to compute the sets of "coherent preferences", represented by the extensions. Hence, it is important to remark that this process allows to move from the direct aggregation of agents' preferences to the aggregation of rational and justified preferences (and their corresponding rankings).

More precisely, multiple extensions are computed (unless there is a consensus among the agents). Each extension contains preference relation arguments and a single ranking argument which corresponds to a coherent aggregation of possible preference relations with their justifications. Hence, it is now possible to consider the extensions as (virtual) voters and aggregate their rankings. We consider the ranking of each extension (except if the extension contains the blank vote) as a *justified preference* $\mathcal{JP}_k$. The set of justified preferences is denoted by $\mathcal{JP}$; hence, $|\mathcal{JP}| = |\boldsymbol{E} \setminus \{\mathcal{E} \in \boldsymbol{E} : a_\emptyset \in \mathcal{E}\}|$. Each justified preference has preferences over the alternatives denoted with $x \succ_{\mathcal{JP}_k} y$ which means that justified preference $\mathcal{JP}_k$ *prefers* alternative x to y. Informally, the collective justified preference profile is the set of all the justified preferences.

Definition 4 (Justified preference profile). *A justified preference profile $\succ_{\mathcal{JP}} = \langle \succ_{\mathcal{JP}_1}, \ldots, \succ_{\mathcal{JP}_{|\mathcal{JP}|}} \rangle \in \mathcal{L}_A^{|\mathcal{JP}|}$ is a collection of linear orders for all the justified preferences.*

As noted before, the justified preference profile can have multiple justified preferences (extensions) so we refer to classical social theory for aggregating them. The construction of the justified preference profile allows to avoid the Condorcet paradox.

Theorem 1. *There are no voting cycles in any justified preference profile under the preferred semantics.*

4   Conclusion and Future Work

In this paper, we have proposed a framework for decision-making using qualitative preferences instead of social choice methods which rely only on the quantitative aggregation of the individual preferences. The method allows to take into account the justifications behind these preferences, and compute the collective justified preferences which allows to overcome the Condorcet paradox.

As future work, we want to extend our research on Argumentation and Computational Social Choice towards multi-criteria decision-aiding. The combination of these two fields will allow to explain the decisions rationally, which may allow for decisions procedures that will have more chances to be accepted by the society. To strengthen this view we plan to propose quantitative methods that can evaluate the different decision-making procedures, in particular in the context of real world examples.

References

1. Arrow, K.J.: A difficulty in the concept of social welfare. J. Polit. Econ. **58**(4), 328–346 (1950)
2. Black, D.: Theory of Committees and Elections. Cambridge University Press, Cambridge (1958)
3. Condorcet, M.D.: Essai sur l'application de l'analyse à la probabilité de décisions rendues à la pluralité de voix. Imprimerie Royal, Stockholm (1785). Facsimile published in 1972 by Chelsea Publishing Company, New York
4. Dung, P.M.: On the acceptability of arguments and its fundamental role in non-monotonic reasoning, logic programming and n-person games. Artif. Intell. **77**(2), 321–357 (1995)
5. Gibbard, A.: Manipulation of voting schemes: a general result. Econometrica **41**(4), 587–601 (1973)
6. Satterthwaite, M.A.: Strategy-proofness and arrow's conditions: existence and correspondence theorems for voting procedures and social welfare functions. J. Econ. Theor. **10**(2), 187–217 (1975)

Efficient Satisfiability Verification
for Conditional Importance Networks

Zachary J. Oster[✉]

University of Wisconsin-Whitewater, 800 West Main Street,
Whitewater, WI 53190, USA
osterz@uww.edu

Abstract. Conditional importance networks (CI-nets) provide a formal framework for modeling and reasoning with qualitative preferences over sets of many variables. Existing approaches for verifying the satisfiability of a CI-net operate on a complete model of the CI-net's semantics, but the time required to construct this model limits the practical usefulness of CI-nets. We present an algorithm that decides a CI-net's satisfiability by analyzing a sufficient partial model of its semantics, and we show how to efficiently construct such a model. Our method significantly reduces the average time needed to verify the satisfiability of CI-nets.

1 Introduction

Conditional importance networks, or CI-nets [1], provide an expressive language for specifying and reasoning over qualitative conditional preferences among sets of items. To ensure that preference reasoning over a CI-net is sound, one must verify that the CI-net is *satisfiable*, i.e., it contains no cycles of preferences. Existing methods for this, including [1,4], construct a *preference graph* that explicitly represents the (partial) preference ordering defined by a CI-net. A CI-net is satisfiable if and only if its preference graph is acyclic [1], but the time cost to construct the graph dominates the time cost to verify satisfiability.

This paper presents a new algorithm for verifying CI-net satisfiability without the full preference graph, which significantly reduces the average time needed to verify a CI-net's satisfiability. We describe and empirically evaluate our implementation of this algorithm, then discuss future applications of this work.

2 Overview of CI-Nets

We use the definitions for conditional importance statements (CI-statements) and conditional importance networks (CI-nets) from [1], given here as Definition 1. Let V be a finite set of binary variables, each of which indicates whether a given proposition is true or false. We assume that each proposition is preferred to be true and that propositions do not directly contradict each other.

Definition 1. *A CI-statement on V is a quadruple (S^+, S^-, S_1, S_2) of pairwise disjoint subsets of V. This CI-statement can be written as $S^+, S^- : S_1 \rhd S_2$. A CI-net on V is a set C of CI-statements on V.*

© Springer International Publishing AG 2017
J. Rothe (Ed.): ADT 2017, LNAI 10576, pp. 350–354, 2017.
DOI: 10.1007/978-3-319-67504-6_26

Informally, a CI-statement can be interpreted to mean: "Given two sets of items chosen from $\mathbf{V}$, if both sets include the items in S^+ and neither set includes the items in S^-, then I would rather have the set that has all items in S_1 instead of the set that has all items in S_2, *ceteris paribus* (all else being equal)."

The semantics of CI-nets are defined by [1] in terms of *worsening flips*, which are pairs of sets of variables (V_1, V_2) such that V_1 is preferred to V_2 ($V_1 \triangleright V_2$). In contrast, [4] defines CI-net semantics in terms of *improving flips*, where (V_1, V_2) means that V_2 is preferred to V_1 ($V_2 \triangleright V_1$). Our proofs are based on improving flips, but it is simple to rewrite them using worsening flips. From this point on, the word "flip" denotes either a worsening or an improving flip unless noted.

Given variable sets V_1 and V_2, V_1 is preferred to V_2 under a CI-net $\mathbf{C}$ ($\mathbf{C} \models V_1 \triangleright V_2$) if and only if $\mathbf{C}$ defines an improving flipping sequence from V_2 to V_1.

Definition 2 *([4], after [1]). A sequence of sets of variables $V_1, V_2, \cdots, V_n$ is an improving flipping sequence w.r.t. a CI-net if and only if, for $1 \leq i < n$, either*

1. *(Monotonicity Flip) $V_{i+1} \supset V_i$; or*
2. *(Importance Flip) a CI-statement $S^+, S^- : S_1 \triangleright S_2$ satisfies the following:*
   (a) *$V_{i+1} \supseteq S^+$, $V_i \supseteq S^+$, and $V_{i+1} \cap S^- = V_i \cap S^- = \varnothing$;*
   (b) *$V_{i+1} \supseteq S_1$, $V_i \supseteq S_2$, and $V_{i+1} \cap S_2 = V_i \cap S_1 = \varnothing$; and*
   (c) *if $\bar{V} = V \setminus (S^+ \cup S^- \cup S_1 \cup S_2)$, then $\bar{V} \cap V_i = \bar{V} \cap V_{i+1}$.*

Condition 1 states that a set with more variables is always preferred to a set with fewer variables. Condition 2 states that if the variables in S^+ are included and the variables in S^- are not included, then including the variables in S_1 is more important than including the variables in S_2, all else being equal (*ceteris paribus*). The same flip may be both a monotonicity flip and an importance flip, and monotonicity and importance flips may directly oppose each other.

3   Efficiently Verifying CI-Net Satisfiability

A CI-net is *satisfiable* if and only if it does not define any cycle of flips (Proposition 2 in [1]). We begin this section by proving several results that lead to a new necessary and sufficient condition for satisfiability of a CI-net, which is equivalent to the condition defined in [1] but can be verified for a given CI-net using only its importance flips. We then use these results to define our new algorithm for verifying a CI-net's satisfiability without constructing the full preference graph.

Theorem 1. *Let an* empty *CI-net be a CI-net that contains no CI-statements. Every empty CI-net is satisfiable.*

Proof. Suppose an empty CI-net $\mathbf{C}$ over a set of preference variables $\mathbf{V}$ is not satisfiable. Then $\mathbf{C}$ has a cycle of k flips $V_1 \triangleright \cdots \triangleright V_k \triangleright V_1$, where $k \geq 1$ and $V_1, \ldots, V_k$ are unique subsets of $\mathbf{V}$. Since $\mathbf{C}$ is empty, all flips of $\mathbf{C}$ are monotonicity flips. Therefore $V_1 \supset \cdots \supset V_k \supset V_1$, which is not possible. $\square$

It follows directly from Theorem 1 that monotonicity edges alone cannot form a cycle. Therefore, it is sufficient for a model of a CI-net's semantics to encode only importance flips. The following results show that we can focus our efforts on a subset of the importance flips without compromising correctness.

Definition 3. *An importance flip (or edge)* $V_1 \triangleright V_2$ *of a CI-net* $\boldsymbol{C}$ *is* widening *if* $|V_1| > |V_2|$, steady *if* $|V_1| = |V_2|$, *or* narrowing *if* $|V_1| < |V_2|$. *These classes form a partition of the set of importance flips (or edges) of* $\boldsymbol{C}$.

The following proofs use improving-flip semantics, i.e., widening flips move in the same direction as monotonicity flips. If worsening-flip semantics are used, then narrowing flips must be used instead of widening flips (and vice versa).

Theorem 2. *Every CI-net with zero or more widening importance flips, but no other (steady or narrowing) importance flips, is satisfiable.*

Proof. By induction on the number n of widening flips. When $n = 0$, Theorem 1 holds. For $n > 0$, let $\boldsymbol{C}$ be a satisfiable CI-net with $n - 1$ widening flips and no steady or narrowing flips. Add one widening flip $V_1 \triangleright V_2$ to $\boldsymbol{C}$. Since $|V_1| > |V_2|$ (Definition 3), there is no monotonicity flip from V_1 to any subset of V_2. There is also no importance flip from V_1 to any subset of V_2, since $\boldsymbol{C}$ has no narrowing flips. Thus V_2 is not reachable from V_1, so $\boldsymbol{C}$ has no cycles and is satisfiable. □

Theorem 3. *A CI-net* $\boldsymbol{C}$ *is satisfiable if and only if it has no steady or narrowing importance flips that are part of a cycle.*

Proof. Immediate from Proposition 2 in [1] and from Theorem 2. □

Theorem 3 suggests an effective algorithm for verifying CI-net satisfiability, shown here as Algorithm 1. The IsSATISFIABLE function first constructs the set $\boldsymbol{C}_{Imp}$ of all importance flips induced by a given CI-net $\boldsymbol{C}$. Each pair (V_i, V_j) in $\boldsymbol{C}_{Imp}$ denotes an importance flip $V_j \triangleright V_i$ specified by $\boldsymbol{C}$. IsSATISFIABLE then calls INCYCLE to begin analyzing the CI-net semantics. With each recursive call, INCYCLE checks whether each flip from a given variable set V_j leads to a "safe" set of variables, i.e., one that has been verified to not be in a cycle. If all recursive calls to INCYCLE return false, then V_j is safe, so the original call to INCYCLE returns false. If INCYCLE is called twice on the same variable set during the recursion, then V_j leads to a cycle; in this case, INCYCLE returns true.

To avoid repeatedly exploring flips, IsSATISFIABLE and INCYCLE share a mapping M that stores the current status of each variable set: either *New* (not explored yet), *OnPath* (currently being explored), or *Safe* (already explored and found to be safe). Initially, $M(V_i)$ is set to *New* for every $V_i \subseteq \boldsymbol{V}$. The mapping is updated by INCYCLE as it explores the CI-net's semantics. After a variable set's status is set to *Safe*, its outgoing flips will not be explored again.

4   Evaluation and Future Work

Theorem 4. *A given CI-net* $\boldsymbol{C}$ *is satisfiable if and only if* IsSATISFIABLE *returns true when invoked with* $\boldsymbol{C}$ *as its argument.*

Algorithm 1. Decide if CI-net is satisfiable

function IsSATISFIABLE($\mathbf{C}$)
    let $\mathbf{V}$ be the set of CI-variables over which $\mathbf{C}$ is defined
    let $\mathbf{C}_{Imp} = \varnothing$ $\triangleright$ $\mathbf{C}_{Imp}$ is the set of importance flips
    for all statements $S^+, S^- : S_1 \triangleright S_2 \in \mathbf{C}$ **do**
        for all $\gamma \subseteq \mathbf{V} \setminus (S^+ \cup S^- \cup S_1 \cup S_2)$ **do**
            $\mathbf{C}_{Imp} \leftarrow \mathbf{C}_{Imp} \cup \{(\gamma \cup S^+ \cup S_2, \gamma \cup S^+ \cup S_1)\}$
    let $M : 2^{\mathbf{V}} \rightarrow \{New, OnPath, Safe\}$, where initially $\forall V_i \subseteq \mathbf{V} : M(V_i) = New$
    for all CI-variable sets $V_i \subseteq \mathbf{V}$ in descending order from $\mathbf{V}$ to $\varnothing$ **do**
        for all importance flips $V_j \triangleright V_i \in \mathbf{C}_{Imp}$ where $|V_j| \leq |V_i|$ **do**
            if INCYCLE($\mathbf{C}_{Imp}, M, V_j, |V_i|$) **then**
                return false
    return true $\triangleright$ if no statement is in a cycle

function INCYCLE($\mathbf{C}_{Imp}, M, V_j, n$)
    if $M(V_j) = Safe$ **then**
        return false $\triangleright$ V_j is not part of any cycle
    else if $M(V_j) = OnPath$ **then**
        return true $\triangleright$ cycle found: V_j already visited on this path
    else if $M(V_j) = New$ **then**
        $M(V_j) \leftarrow OnPath$
        if $\exists V_k \subseteq \mathbf{V} : V_k \triangleright V_j \in \mathbf{C}_{Imp}$ and INCYCLE($\mathbf{C}_{Imp}, M, V_k, n$) **then**
            return true
        if $\exists V_k \subseteq \mathbf{V} : V_k \supset V_j, |V_k| \leq n$, and INCYCLE($\mathbf{C}_{Imp}, M, V_k, n$) **then**
            return true
        $M(V_j) \leftarrow Safe$ $\triangleright$ no path from V_j is in a cycle
        return false

Proof. IMPORTANCEFLIPS returns $\mathbf{C}_{Imp}$, which is the set of all importance flips induced by $\mathbf{C}$. For any given set of variables V_i, INCYCLE returns true if and only if it finds a cycle of improving flips that is reachable from V_i according to the CI-net semantics. Since IsSATISFIABLE invokes INCYCLE on every steady or narrowing importance flip of $\mathbf{C}$ and returns true if any one call to INCYCLE returns true, the satisfiability condition of Theorem 3 is fulfilled. $\qquad\square$

It is PSPACE-complete to check the satisfiability of CI-nets and similar models [3]. The theoretical time complexity of Algorithm 1 is $O(N^2)$, where N is the number of possible outcomes (sets of variables). This is not an apparent improvement over [4], but we suspect that it is not a tight upper bound for our algorithm's complexity; space constraints prevent a detailed discussion here.

Figure 1 compares the mean time and memory used by our method and the method of [4] (using version 2.6.0 of the NuSMV model checker) to verify 780 randomly generated CI-nets, defined over 5 to 17 binary variables with either 5 or 10 CI-statements. The test set contained 30 CI-nets with each combination of n variables and m statements. This figure shows that our algorithm significantly reduces the average time and memory needed to check the consistency of CI-nets, compared to an approach that constructs the entire preference gra

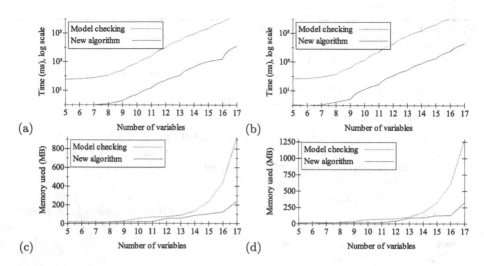

Fig. 1. Empirical evaluation results: mean time usage with (a) 5 statements and (b) 10 statements, and mean memory usage with (c) 5 statements and (d) 10 statements.

We are exploring how to further improve this algorithm's efficiency while preserving its correctness. One possibility might be a hybrid approach that uses our method to construct reduced models of parts of the CI-net semantics, which are then verified using model checking as in [4]. We are also exploring the feasibility of a Datalog approach to CI-net satisfiability verification, similar to the Datalog implementation of CP-nets as Logical Conditional Preferences (LCP) in [2]. This algorithm is a step toward our long-term goal of developing a "preference workbench", which will include facilities for editing, visualizing, comprehending, and tracing preferences within a CI-net (or related) preference model.

References

1. Bouveret, S., Endriss, U., Lang, J.: Conditional importance networks: a graphical language for representing ordinal, monotonic preferences over sets of goods. In: Boutilier, C. (ed.) Proceedings of the 21st International Joint Conference on Artificial Intelligence, IJCAI 2009, Pasadena, California, USA, 11–17 July 2009, pp. 67–72 (2009). http://ijcai.org/papers09/Papers/IJCAI09-022.pdf
2. Cornelio, C., Loreggia, A., Saraswat, V.A.: Logical conditional preference theories. CoRR abs/1504.06374 (2015). http://arxiv.org/abs/1504.06374
3. Goldsmith, J., Lang, J., Truszczynski, M., Wilson, N.: The computational complexity of dominance and consistency in CP-nets. JAIR **33**, 403–432 (2008)
4. Santhanam, G.R., Basu, S., Honavar, V.: Dominance testing via model checking. In: Fox, M., Poole, D. (eds.) Proceedings of the 24th AAAI Conference on Artificial Intelligence, AAAI 2010, Atlanta, Georgia, USA, 11–15 July 2010. AAAI Press (2010). http://www.aaai.org/ocs/index.php/AAAI/AAAI10/paper/view/1844

Discovery of Energy Network Topology from Uncertain Flow Measurements

Wilfried Joseph Ehounou[1,2]($\boxtimes$), Dominique Barth[1]($\boxtimes$),
and Arnaud De Moissac[2]($\boxtimes$)

[1] DAVID - Lab, Université de Versailles - St Quentin, 55 Avenue des états unis,
78000 Versailles, France
dominique.barth@uvsq.fr
[2] DCbrain, 23 Avenue d'italie, 75013 Paris, France
{ehounou_willy,arnaud}@dcbrain.com
http://www.dcbrain.com

Abstract. The objective we focus on here consists in discovering the topology of an energy distribution network modeled by a flow digraph from which we only know the set of arcs without identification of their extremities. We also have as inputs a set of temporal series of flow measures on these arcs and the correlation matrix of the arcs, with possible errors. From these inputs, we consider the graph which incidence matrix is the correlation one. If the correlation matrix contains no errors, this graph is the line graph of the network to be discovered. Thus, given this graph, we then propose here algorithms determining the graph with the same vertex set being a line graph and maximizing the set of similar edges with the initial correlation graph. We then evaluate the performances of this approach by simulation on 50 networks, randomly generated.

Keywords: Flow network · Line graph · Topology discovery

1 Introduction

The aim of our work is to predict a probable topology of the network, considered as a DAG, knowing only the links (without knowing their extremities) and matching to the flow measures on them. Indeed, the contribution of this paper consists in algorithms to discover the topology of a flow network from a binary adjacency matrix called $matE$ obtained by the correlation of the flow measures. Let's remember that 1 means two measures are correlated and 0 means two measures are not correlated. The method used to correlate measures is Symbolic Aggregate approXimation (SAX) [2]. These correlations, if they are correct, induce the line graph [3] of the underlying undirected graph of the network under consideration. In fact, a line graph $L(G)$ of an undirected graph G is a graph such that each vertex of $L(G)$ represents an edge of G and two vertices of $L(G)$ are adjacents iff their corresponding edges share a common endpoint in G. Unfortunately, measure errors often appear and the topology deduced from the

J. Rothe (Ed.): ADT 2017, LNAI 10576, pp. 355–360, 2017.
DOI: 10.1007/978-3-319-67504-6_27

correlations is not that of the desired line graph. The objective is to correct this graph in order to obtain a line graph as close as possible to that of the DAG.

Various works have been devoted to the discovery of a graph from its line graph [1,3,4]. In particular, initial works [4] show that a graph is a *line graph* iff it admits a decomposition of its set of edges in cliques such that each vertex is covered by one or two cliques. Line graphs can also be characterized by a set of nine forbidden induced subgraphs [3]. A concept of root graph was introduced by Whitney [5] and he proved that if two line graphs are isomorphic and connected, then their root graphs are isomorphic except for the triangle graph.

Algorithms, we present here, are two main features. First, they propose the best clique decomposition whether there are no errors in the correlation matrix. We use the Kirchhoff law of flows to find out a cover of vertex in the ambiguity situation (Fig. 1). Second, for cases in which the correlation matrix contains errors i.e. some 1 in the matrix should be 0 and vice versa, algorithms suggest a flow network topology which is as near as possible to the initial topology of the network considered. The predicted topology is close to the real topology if few false negative and false positive correlations are corrected.

This paper is structured as follows. Section 2 relates to the definition of the problem. Section 3 defines the algorithms for resolving our problem and finally Sect. 4 presents the results obtained from the simulation of algorithms on 50 networks of 25 vertices generated randomly.

2   Problem

We consider that we only know the set of arcs on the flow network, i.e. the extremities of each arc are unknown. To each arc, we have a serie of measures with possible errors. Our goal is to find the topology of the initial network from these measures. We suppose that the binary adjacency matrix $matE$ of arcs has already be computed through a SAX method [2]. Columns and rows of the matrix $matE$ are the arcs of the flow network we want to predict. Let G be the undirected graph from which $matE$ is the adjacency matrix. If $matE$ is correct then G is the line graph of the unknown target DAG G_R. The aim is then to determine G_R from G, using line graph properties. If not, we look for a line graph $G' = (V, E')$ the closest one to G. To be done, we define the Hamming distance $dH(G, G')$ between two graphs like the Hamming distance between these adjacency matrices i.e. the number of different edges between two matrices. We denote by **line-distance** of G, noted $DL(G)$, the smallest Hamming distance between G and a predicted line graph, assuming that they have the same vertices. If $matE$ is incorrect then the graph G is not a line graph and we use **proxy-line** to get this property. Let the following problem be:

Problem: Proxi-Line
Data: A graph $G = (V, E)$ and an integer k.
Question: Is there $DL(G) \leq k$?

We are interested in a minimisation problem combined with Proxi-Line i.e. the resolution of $DL(G)$.

The problem is clearly *NP*. We conjecture that it is *NP-complete*.

3   Algorithms

To solve this problem, we propose to use two consecutive algorithms in order to resolve Proxi-Line problem. Consider G the undirected graph from which $matE$ is the adjacency matrix. It is known that a graph $G = (V, E)$ is a **line graph** iff it exists a set of cliques C of G such as each vertex of V belongs to one or two cliques of C and each edge of E belongs to a subgraph induced by only one clique of C. In this case, the set C is called a **correlation cover** of G.

3.1   Cover Algorithm

This algorithm is similar to Lehot's algorithm [1] since it looks for the clique decompositions of the line graphs. In fact, at each step, the algorithm selects first the vertices not treated enough covered by one clique and afterwards the vertices not treated covered by two cliques. At the end of the algorithm, the *correction algorithm* is launched if it exists the vertices belonging to any cliques.

We have demonstrated that the only cases, where a line graph has two correlation covers, are in the Fig. 1 (framed vertex). These ambiguities are resolved using the measure correlations.

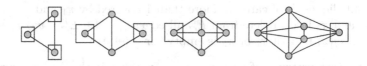

Fig. 1. Possible graphs with two covers and framed ambiguity points

3.2   Correction Algorithm

We suggest an algorithm which applies to the vertices covered by any cliques and modifies the initial set E_C by adding or deleting edges in order to get a *line graph*. Let E_C^i be the edge set of G after the (i-1)-th vertices had been already treated in the order $z_1, z_2, ..., z_t$. In the following, we denote by $z = z_i$ a vertex and by C^i the set of cliques of G_C at the i-th step. Thus $E_C^1 = E_C$ et $C = C^1$.

The idea of this algorithm is to cover the neighborhood of the vertex z into three partitions π_1, π_2, π_s such as:

- $\pi_1 \cap \pi_2 = \{z\}$
- π_1 (resp. π_2) is the union of two cliques C_1 and $C_2 \in C$ which any edge $[u, v]$ of E_C such that $u \in C_1$ and $v \in C_2$ is not covered by a clique (other that $\{u, v\}$) in C. Edges, missing to the partition π_1 (resp π_2), are added to the edge set E_C.
- π_s is the set of vertices such as the edges $[v, z] \in E_C, v \in \pi_s$ have to be removed.

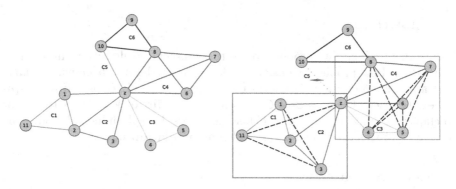

Fig. 2. A partition example (Color figure online)

Figure 2 is an example of the partition π_1, π_2, π_s in which the vertex z belongs to two cliques π_1 (in red) and π_2 (in green). The dashed lines are the added edges and the dotted lines are the removed edges.

A cost $c(T)$ of partition $T = \pi_1, \pi_2, \pi_S$ is minimal and takes into account the edges to add (linked to π_1 and π_2) and to remove (linked to π_S). The following update procedures allow to obtain $\mathcal{C}^{i+1}$ and E_C^{i+1} by applying this partition:

- Delete all cliques $\mathcal{C}_z$ of cardinal more than 1 covered by π_1 and π_2 in $\mathcal{C}^i$.
- Add π_1 and π_2 in $\mathcal{C}^i$, delete from E_C^i all edges $\{[z,v] \in E_C^i : v \in \pi_S\}$.
- Assign $Cliq(z)$ to 1 (if π_1 or π_2 is empty) or 2 (else).

Thus, for each vertex z_i of a list $z_1, z_2, \ldots, z_t$ taken in that order, we consider a minimum partition cost c_m^i and we apply it. At the end of process, we obtain a correlation graph $G_C^t = (V, E_C^t)$ whose the modified set $\mathcal{C}$ is a correlation cover. The line-distance verifies $DL(G_C^0, G_C^t) \leq \sum_{1 \leq i \leq t} c_m^i$. We notice that during a $j > 1$ step, the vertex z_j and its neighborhood are covered by one or two cliques after the processing of $j - 1$ previous vertices.

The two algorithms process once each vertex in the graph. The complexity of vertex processing is exponential depending of each vertex degree and the cliques to which the vertex belongs, here again depends on the size and the number of cliques. Global algorithm is thus pseudo-polynomial depending on the degree of the graph.

4   Result Analysis

We consider $n = 50$ flow networks having 25 vertices and the degree of each vertex is between 2 and 5. The line graphs of the flow network are generated and we deleted $k = [1..5]$ edges on these graphs. We introduce two metrics: line-distance mean (noted moy_DL) and Hamming distance mean (noted moy_DH) corresponding respectively to the means of the line-distance and the Hamming distance when k different edges are deleted α times randomly. If this number (moy_DL or moy_DH) is 0, there are no edge differences between the predicted

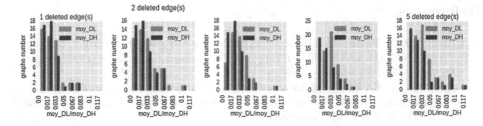

Fig. 3. Mean line/Hamming distances

and initial line graphs. Whereas, these line graphs have no common edge when the number is 1.

Figure 3 shows the distribution of the number of graphs in relation to moy_DL or moy_DH. In the x-axis, there are values of the line-distance mean and of the Hamming distance mean. In the y-axis, there is the number of line graphs associated to a specific mean.

We find that:

- More than 50% of the predicted line graphs are close to line graphs with deleted edges. They have the line-distance mean less than 0.033.
- If the initial line graphs have a lot of deleted edges, most predicted line graphs are not near from the latter. This is illustrated by the case of 5 deleted edges where there is no graph having moy_DL <0.017.
- There is a gap between the line-distance mean and Hamming distance mean. This is explained by the discovery of some deleted edges in a proportion of 25%. However, it is difficult to find out deleted edges when there are removed many edges. This is illustrated in the 4 deleted edges and 5 deleted edges cases (Fig. 3) where the histogram points of Hamming distance mean are greater than those of line-distance mean.

We conclude that our algorithms provide line graphs very close to the original line graphs. However, they tend to add missing edges (i.e. missing correlations) in the case of weak errors in the correlation matrix and imaginary edges when the adjacency matrix has too much correlation errors. Furthermore, tests are realized on real flow networks. It was not conclusive because of the number of errors in the correlation matrix and the high degree (i.e. around 10) of each vertex.

The evaluation of simulations of these algorithms shows that they predict corrected graphs close to the real ones when the number of measure correlation errors is weak. In the future, we improve the correction algorithm in order to treat a graph in which each vertex has a high degree and also look for directing edges i.e. which sets on edges are incoming or outgoing of a vertex.

References

1. Degiorgi, D.G., Simon, K.: A dynamic algorithm for line graph recognition. In: Nagl, M. (ed.) WG 1995. LNCS, vol. 1017, pp. 37–48. Springer, Heidelberg (1995). doi:10. 1007/3-540-60618-1_64
2. Lin, J., Keogh, E., Fu, A.: Hot sax: efficiently finding the most unusual time series subsequence. In: Fifth IEEE International Conference on Data Mining, ICDM 2005, pp. 226–233. IEEE, 27–30 November 2005
3. Harary, F.: Line graphs. In: Graph Theory, pp. 71–83. Addison-Wesley, Massachusetts (1972). Chap. 8
4. Lehot, P.G.H.: An optimal algorithm to detect a line graph and output its root graph. J. ACM **21**, 569–575 (1974). Association for Computing Machinery (ACM)
5. Whitney, H.: Congruent graphs and the connectivity of graphs. In: Hassler Whitney Collected Papers, pp. 61–79. Birkhäuser Boston (1992)

Measuring Border Security for Resource Allocation

Paul B. Kantor[✉]

Rutgers/CCICADA, 96 Frelinghuysen Road, Piscataway, NJ 08854-8018, USA
paul.kantor@rutgers.edu

Abstract. Effective management of border security requires effective measurement of the impact of alternative policies on unwanted cross-border flows. There are no agreed-upon ways to measure the amount of any particular inflow; some indicators are believed to rise and fall with each in-flow, but they do not measure the total flow. To combine multiple measures proportional to an unmeasurable flow, we introduce Principal Ray Analysis. The resulting estimates of fractional change in unseen flows, combined with cost information, support optimal incremental resource allocation, for any single type of flow. Extension to multiple flows requires agreement on the relative importance of reducing each of the flows. A common data store is recommended to support rational debate on cross-flow allocations and overall "security."

Keywords: Decision making · Resource allocation · Principal Ray Analysis

1 Introduction

Effective management of border security requires measuring the impacts of alternative policies. Despite considerable research [1, 4, 5, 8], and principles from other domains (Espenshade [2]), there are no accepted ways to measure any particular inflow, in a given week, month or year. Some indicators may rise and fall with each inflow, but none provides an accurate measure of the total flow. Thus counterfeit purses detected (with constant effort; not with a change in practices) likely represents some fraction of all counterfeit purses, at least on the average. If so, a 25% increase in detected purses suggests a 25% increase in the (unknown) number of undetected purses. Similar arguments will apply to other inflows.

For any specific flow of goods or persons, the key "measure of harm" is the part of the flow that is neither deterred nor detected. Stakeholders in one state may be more concerned to reduce flows into their own state, and less concerned if the flow is merely diverted to another state, rather than being reduced. This very brief note considers the problem of a single flow, as it is assessed by a single state or stakeholder.

For a single flow, and a single stakeholder we ask: "which of several alternative policies should receive the next increment of funding?" A radical restructuring of the entire program might yield lower flow than can be found by incremental search among policy alternatives, but seems politically unrealistic.

With limitation to a single flow the decision problem is: which of several increments in resource allocation yields the greatest reduction in that flow? We do not know how

© Springer International Publishing AG 2017
J. Rothe (Ed.): ADT 2017, LNAI 10576, pp. 361–365, 2017.
DOI: 10.1007/978-3-319-67504-6_28

to measure the volume of flow and propose using a collection of parallel measures to estimate the fractional reduction, but not the absolute reduction. The flow, and actions to reduce it, form a temporal and conceptual sequence. Flow begins with "motivating factors," leading actors to initiate a cross-border activity. Their actions next reflect any "deterrent factors," which are perceived to increase the cost or risks of cross-border activity. The next factors are "detection/apprehension." Finally, experts suggest that "consequences" of apprehension, although coming last in the chain, affect the motivating factors. They are only in play if apprehension occurs.

The key ideas are:

1. Assess the fractional decrease in flow per unit cost for each alternative.
2. Fractional measure of improvement supports allocation if the proportionality of measure to unseen variable is stable over the data analysis and policy implementation.
3. With several imperfect indicators of the unmeasured flow: (1) apply "ray clustering," [not illustrated here] to find a cluster of imperfect indicators moving together over time and believed (by experts) to "move in proportion" to the unseen flow; and (2) use a relative of orthogonal projection which we call Principal Ray Analysis to extract an indicator that is, in a well-defined sense, proportional to the unseen flow.
4. Use this common indicator to assess the fractional impact of a policy change, during a trial period.
5. After a set of trial periods, rank alternate policies, based on the ratio of the fractional improvement to the cost of that policy.

2   Context of the Problem

These suggestions flow from several years of research, interviews within the US Department of Homeland Security (DHS), including operational personnel and Washington decision makers, and review of relevant literature. That research sought a "single measure for the security of the border." While some agencies and decision makers might offer such measures, for some flows, none are directly observable. An example is "total dollar value of counterfeit luxury items." Similarly, one U.S. Coast Guard officer noted "what really counts is not how much cocaine is caught or dumped. What counts is reducing the number of deaths from imported narcotics, isn't it."

Stakeholders for the problem of measuring border security include: lawful manufacturers and retailers of proprietary goods, employers with undocumented workers; property owners along the border, etc. Local, tribal, state and federal agencies are tasked with enforcing laws that constrain these inflows. Finally, the citizens of each nation, have concerns from, on the one hand, stemming flow of synthetic drugs to, on the other hand, owning an inexpensive counterfeit handbag or wristwatch.

No policy change will jointly increase the welfare of these diverse stakeholders. We suggest that a common collection of objective data will serve them all, permitting rational and democratic discussion of the tensions among goals, and of the effectiveness of various plans and policies. Despite recent public attention to the bizarre notion of "alternative facts," we hold, with the late US Senator Moynihan [9], that each party to a debate is entitled to his own opinions, but not to "his own facts."

3 Sequential Causal Factors

A flow occurs when a member of some potential population is motivated to attempt a border crossing, is not deterred and is not detected/apprehended. This chain may be represented as a product: ($F_x(t)$ = undetected attempts of type x in time period t. $G_x(t)$ = undeterred attempts of type x in time period t; $H_x(t)$ = potential attempts of type x in time period t; $q_x(t)$ = the fraction of attempts that are deterred; $c_x(t)$ = fraction of (undeterred) attempts that are detected and apprehended). These are related by:

$$F_x(t) = \left(1 - c_x(t)\right) G_x(t)$$
$$G_x(t) = \left(1 - q_x(t)\right) H_x(t)$$

If we find indicators approximately proportional to these inflows at any stage, we can infer useful information about the effect of factors "to the right" in these equations, even while the magnitude of the overall inflow remains undetermined.

4 Measuring Flow Reduction

Suppose for the moment, that there is a single preferred indicator, $I_x(t)$ for each of the flows $F_x(t)$. Information collected during trials of alternative policies lets us divide the percentage reduction by the added cost of the practice (see Table 1). We must be careful. First, many techniques from statistics or engineering use interval scale data; cost and the fractional improvement are ratio scale data [6, 7]. Second, detection or search exhibits diminishing returns, making it important to test similar effort levels.

Table 1. Notional comparison of several practices. If experiments are done at the same border point, to improve comparison; they can't be done at once. The prior value for the indicator will not be the same in each trial. Therefore improvement is measured as the fractional reduction in the flow. Added cost is measured, e.g., in Euros. The Impact is the fractional improvement divided by the incremental cost of the tested practice, rescaled to O(1). See text.

Added practice	Prior indicator	Effect indicator	Improvement	Cost of practice	Impact (×10,000)
Practice 1	243	216	11% (2)	€1,800	0.62 (1)
Practice 2	195	180	8%	€1,600	0.48
Practice 3	302	285	6%	€800	0.70

Notes: (1) Impact is measured in "percentage improvement per €10,000." (2) For a large change, such as 11%, the relation between effort and Indicator may not be linear.

Table 1 shows that Practice 3 has the greatest impact, suggesting that it is the best way to spend the next increment to reduce this flow. But efforts at deterrence or detection (like search efforts generally) improve more slowly as they approach 100%. Practice 1, with nearly twice the improvement, at more than twice the cost, might show higher impact than Practice 3, if funded at only €800. Whether to change policy now, or do another trial is best left to the agencies. Note, our analysis applies only if the amount of funding for controlling this (or any other) specific flow is fully pre-allocated, which is in principle sub-optimal. To compare policies affecting different or multiple flows, one must weight the relative importance of distinct improvements.

5   Inferring Single Measures: Principal Ray Analysis (PRA)

Our procedure is akin to orthogonal regression, used in econometrics [3]. We seek a single ray (normalized Euclidean vector) most nearly aligned with *all* of the rays representing the several indicator time series. We minimize the summed squared distance from the ray sought, to the indicator rays. The indicator rays are: x_t^k, (k labels the K indicators; t labels the time periods). Using Lagrange multipliers, one finds that the closest ray, itself a time series, is the non-negative eigenvector of the (T-by-T, where T is the number of time periods) matrix:

$$M_{st} = \sum_{1 \leq k \leq K} u_s^k u_t^k.$$

This eigenvector time series is shown as column "PRA" of Table 2 (data simulated).

Table 2. Principal component analysis contrasted to principal ray analysis (see text)

X_1	X_2	X_3	X_4	PCA factor 1	PRA	True level	Fract change
2.90	3.47	10.58	14.61	−0.80	0.23	3	NA
2.89	3.15	9.65	15.69	−0.95	0.24	3	
2.85	3.07	11.62	14.35	−0.90	0.24	3	
4.01	4.26	15.45	21.11	1.09	0.34	4	41%
4.03	4.13	19.06	20.98	1.42	0.36	4	
4.41	3.53	15.50	16.87	0.62	0.30	4	
2.76	1.92	5.87	7.63	−2.59	0.13	2	−53%
2.23	1.93	11.21	8.15	−2.20	0.17	2	
1.90	1.73	10.47	8.73	−2.46	0.17	2	
5.31	5.84	19.54	24.20	3.01	0.40	5	138%
5.09	4.56	16.99	25.03	2.19	0.39	5	
4.49	4.76	19.11	17.43	1.59	0.33	5	

An unwary reader might seek the hidden driver using Factor Analysis, or Principal Component Analysis (PCA). We show the results of applying PCA in Table 2. While PCA (similarly, Factor Analysis) immediately shows the four time periods to be quite different, what it cannot do is to tell us the relative size of the latent driving variable, denoted by "True Level." PRA yields a unique eigenvector whose elements are all different from zero and of the same sign. This vector is shown as PRA in Table 2. Since the interventions are constant in each set of three time periods we average the corresponding values of, PRA, to support estimates of fractional change in the (hidden) variable of interest. The reader may verify that the ratio of this mean, to the "Truth" is very nearly constant. We then compute the fractional change in the mean. This estimator of proportional variation is in pretty good agreement with the actual change from period to period (e.g., from 3 to 4, 33% estimated as 41%). These estimators have larger errors than the estimates for the sub-periods, as they use differences of two independently noisy estimators. Increasing their precision requires more measurement per period, increasing the duration of the trial, decreasing the interval used for measurement, or increasing the number of indicator variables entering the analysis.

6   Translational Considerations

In a real application the set of indicators $(X_1,...X_K)$ will be determined by a combination of mathematical and expert analyses. Details will be presented elsewhere. For multi-flow comparison we recommend that the Impact ratios be reported objectively, to support open discussion of their relative importance (weight) to stakeholders.

Excel is widely accepted and serves for data entry. The PRA eigen-vector finding is done in freely available Octave, and if invoked by Excel, completes a usable package.

Acknowledgments.   The author thanks Isaac Maya, Henry Willis, and Ali Abbas, of the CREATE Center at USC for valuable discussions, and experts at the US Department of Homeland Security (DHS) who contributed time and expertise to educating us; DHS/OUP program officers Joseph Kielman and Gia Harrington, head DHS Office of University Programs (OUP), Matt Clark who saw the potential of rigorous management science in managing border security. Crucial thanks to Fred Roberts, and colleagues at CCICADA; and to Vicki Bier and Jeff Linderoth at the University of Wisconsin Dept. of Industrial and Systems Engineering, for support. Supported by DHS Contract 2009-ST-061-CCI002-07 and NSF Grant #1247696. Opinions expressed are those of the author and not of CCICADA, the DHS or the NSF.

References

1. Cfir, D.: A model of Border Patrol to support optimal operation of border surveillance sensors. Thesis, Naval Postgraduate School (2005). http://calhoun.nps.edu/handle/10945/1837
2. Espenshade, T.J.: Undocumented migration to the United States: evidence from a repeated trials model. In: Bean, F.D., Edmonston, B., Passel, J.S. (eds.) Undocumented Migration to the United States: IRCA and the Experience of the 1980s, pp. 159–181 (1990)
3. Malinvaud, E.: Statistical Methods of Econometrics. North-Holland Pub., American Elsevier Pub., Amsterdam, New York (1970)
4. Morral, A.R., et al.: Measuring Illegal Border Crossing Between Ports of Entry: An Assessment of Four Promising Methods. Rand Corporation, Santa Monica (2014)
5. Roberts, B., Hanson, G.: An Analysis of Migrant Smuggling Costs Along the Southwest Border. Office of Immigration Statistics Working Paper (2010)
6. Roberts, F.S.: Limitations on conclusions using scales of measurement. In: Handbooks in Operations Research and Management Science, vol. 6, pp. 621–671 (1994)
7. Stevens, S.S.: On the theory of scales of measurement. Science **103**, 677–680 (1946)
8. Wein, L.M., et al.: Analyzing the Homeland security of the US-Mexico Border. Risk Anal. **29**(5), 699–713 (2009)
9. Moynihan, D.: Wikiquote. https://en.wikiquote.org/wiki/Daniel_Patrick_Moynihan

Doctoral Consortium

Doctoral Consortium

Incremental Preference Elicitation for Collective Decision Making

Margot Calbrix[✉]

Sorbonne Universités, UPMC Univ Paris 06, UMR 7606, LIP6 CNRS,
UMR 7606, LIP6, 75005 Paris, France
margot.calbrix@lip6.fr

1 Introduction

Most of collective decision procedures used in group decision making require a full ranking or rating of the alternatives from the agents. However, the complete specification of preferences can be a cognitively difficult task when the number of alternatives increases. It is particularly the case on combinatorial domains, as alternatives are very numerous and implicitly defined. Moreover, the determination of a winner does not necessary requires the agents to express their preferences over the entire set of alternatives. Even when preference information is too poor to determine a winner, it may be sufficient to rule out some alternatives. Additionally, a partial specification of preferences would allow them to keep a certain privacy. This shows the importance of designing computer-aided elicitation procedures aiming to focus the elicitation burden on the useful part of preference information and motivates the topic of my PhD.

2 Related Works

Different ways of dealing with incompletely specified preferences are being developed. A first approach consists of checking if available preferences are sufficient to determine a necessary winner (i.e. an alternative that is optimal whatever the missing preferences are). If no such alternative exists, we may focus our attention on possible winners (i.e. alternatives that are optimal for at least one complete extension of preferences), see [4,11,17]. However, it may be the case that the available preference information is too poor for a necessary winner to exist, and that possible winners are too numerous. Incremental elicitation enables to overcome this problem by gathering new preference information to progressively reduce the set of possible winners and potentially determine a necessary winner.

Several methods are being developed to minimize the communication and cognitive burden imposed on the agents by the preference elicitation procedure. Some methods [6,7] rely on the progressive updating of a probability distribution on the set of preferences: preference queries are selected according to the expected value of information provided by the possible answers. Some other methods [5,16] adopt a more pessimistic approach based on max-regret minimization. In this

© Springer International Publishing AG 2017
J. Rothe (Ed.): ADT 2017, LNAI 10576, pp. 369–373, 2017.
DOI: 10.1007/978-3-319-67504-6_29

case, preference queries are selected according to their ability to reduce regrets in the worst case scenario of answer.

Incremental elicitation is of particular interest on combinatorial domains, as the agents cannot be expected to provide preferences over the entire set of feasible alternatives. However the determination of a necessary winner as well as the active selection of informative preference queries are much more difficult due to the implicit definition of the set of alternatives. Some recent works tackle this problem in specific contexts such as constraint satisfaction [9], planning [14], state space search [1], spanning trees [2]. This approach has also been extended to multi-agents decision problem, e.g. in stable matching problems [8] and multiwinner social choice [3,12]. Following this line, we investigate in this thesis the parallel incremental elicitation of individual preferences in new collective decision contexts. For the sake of illustration, we will consider below the incremental elicitation of cardinal utilities in the Multi-agents Knapsack Problem (MKP).

3   The Example of the Multi-agent Knapsack Problem

Consider a set of items $N = \{1, \ldots, n\}$, and a set of agents $A = \{1, \ldots, p\}$. Each item i is characterized by its positive weight w_i, and p positive utilities $u_i^1, \ldots, u_i^p$ that represent the subjective values attached to this item by the agents. A knapsack is a subset of N modeled by a vector $x = (x_1, \ldots, x_n)$ of binary variables where $x_i = 1$ if and only if item i belongs to the subset of selected items. The goal is to maximize the agents satisfaction under a weight constraint: $\sum_{i=1}^{n} w_i x_i \leq W$. Applications of the MKP cover multiple concrete issues, such as multiwinner elections, projects selection, and capital budgeting [12,13,15]. As usual in the standard knapsack problem (KP), agent's preferences over subsets are represented by additive utilities of the form: $f_u^j(x) = \sum_{i=1}^{n} u_i^j x_i$, $j \in A$. For purposes of simplicity, we consider here the utilitarian criterion to represent social preferences, measuring the overall utility of a solution by: $f_u(x) = \sum_{j=1}^{p} f_u^j(x)$. In this problem, the preference information is completely specified by the matrix u of general term u_i^j, $i = 1, \ldots, n$, $j = 1, \ldots, p$.

When the matrix u is known, maximizing the overall utility $f_u(x)$ clearly boils down to a standard single agent KP. However, when individual utilities are unknown, the multi-agents version of the problem cannot be reduced straightforwardly to a standard KP. Incremental elicitation takes place in this context, and consists in interleaving preference elicitation with a combinatorial optimization algorithm to progressively focus the search on the preferred alternatives until determining a necessary optimal knapsack.

Combining Elicitation and Dynamic Programming. The standard KP is usually solved by dynamic programming as a sequential decision problem. Starting with the empty set, the algorithm considers the decision of selecting or not selecting item k at step k. Let X^k be the set of all sequences of decisions of type $(x_1, \ldots, x_k)$ that fulfill the weight constraint $\sum_{i=1}^{k} w_i x_i \leq W$. Let U be the set of possible utility matrices (uncertainty set) constrained by some knowledge of the agents' preferences. Given the uncertainty set U, we consider the following dominance relation over X^k:

$$(x_1, \ldots, x_k) \succsim_U^k (y_1, \ldots, y_k) \iff \begin{cases} \forall u \in U, \ \sum_{i=1}^k (x_i - y_i) \sum_{j=1}^p u_i^j \geq 0 \\ \sum_{i=1}^k (x_i - y_i) w_i \leq 0 \end{cases}$$

The asymmetric part of relation $\succsim_U^k$ (denoted $\succ_U^k$ hereafter) is useful to prune partial solutions that cannot be extended into an optimal solution. If $(x_1, \ldots, x_k) \succ_U^k (y_1, \ldots, y_k)$, then any extension of solution $(y_1, \ldots, y_k)$ is dominated by the same extension of $(x_1, \ldots, x_k)$, whatever the utility matrix $u \in U$. Remark that any new preference statement of type "I prefer x to y" from an agent j in A translates into a linear constraint of the form $\sum_{i=1}^n (x_i - y_i) u_i^j \geq 0$ restricting U. Hence, assuming that some preference statements have been obtained from the agents, U is nothing else but the convex polyhedron defined by the linear constraints derived from these statements. Interestingly, relation $\succsim_U^k$ can be tested using linear programming by checking that:

$$\min_{u \in U} \sum_{i=1}^k (x_i - y_i) \sum_{j=1}^p u_i^j \geq 0 \ \text{ and } \ \sum_{i=1}^k (x_i - y_i) w_i \leq 0$$

These remarks lead us to propose a dynamic programming algorithm to construct the possibly optimal knapsacks for a given uncertainty set U, using at any step $k \in \{1, \ldots, n\}$ the relation $\succ_U^k$ to prune dominated solutions in X^k. This algorithm is actually a baseline to design incremental elicitation algorithms. If at any step k, we obtain a new preference statement reducing the uncertainty set U, relations $\succ_U^k$ will be refined, allowing a better discrimination between partial solutions under consideration. So we propose inserting preference queries at every step of the dynamic programming algorithm. In order to generate useful preference queries, we introduce now a new regret minimization strategy specially tailored for the KP.

The fact that, at a given step k, partial solutions of different weights are present in X^k prevents us to implement a standard max-regret minimization process. In order to overcome this problem, we define symmetrical pairwise max regret (SPMR) of a pair of solutions $x, y \in X^k$, as the minimum between $\mathrm{PMR}(x, y)$ and $\mathrm{PMR}(y, x)$ where $\mathrm{PMR}(x, y) = \max_{u \in U} \{f_u(y) - f_u(x)\}$. The pair of solutions (x^*, y^*) that maximizes the SPMR is used to generate the queries we ask to the agents (one per agent), and we keep asking preference queries until the max-SPMR drops below a given positive threshold. The preference queries are generated as follows: we identify for each agent the item having the most imprecise utility in those items belonging to x^* XOR y^*; then we use a standard gamble query per agent to reduce the uncertainty attached to the items under consideration. We also test another approach based on the direct rating of item k at step k.

Numerical Tests. We have implemented and tested both strategies on randomly generated instances of different size (the number of items varies from 10 to 20 and the number of agents varies from 10 to 30) and we evaluated their performances in terms of computation time and number of questions. The results obtained by averaging over 30 instances can be seen in Fig. 1 for the approach based on gamble queries, and in Fig. 2 for the approach based on direct rating. We can see that both strategies are efficient. The average computational time never exceed 10 min. If the average number of questions is greater than the number of items in the approach based on the gamble queries, it is nonetheless much smaller than the number of feasible knapsacks, and of course smaller than the number of queries necessary to obtain a precise definition of individual utility functions. Numbers of queries are significantly reduced with the strategy based on direct rating but the direct assessment of items may be cognitively difficult for users, especially when the range of the proposed utility scale is important.

n	10	15	20
10 agents	3.1	16.2	52.1
20 agents	4.9	42.5	248.2
30 agents	9.9	62.4	363.1
(a) computation time (in s)			

n	10	15	20
10 agents	12.3	24.5	36.0
20 agents	13.9	28.2	40.6
30 agents	14.0	29.0	43.3
(b) queries per agent			

Fig. 1. Computation time and number of gamble queries

n	10	15	20
10 agents	5.9	42.2	130.3
20 agents	11.4	104.2	407.7
30 agents	19.1	147.3	572.1
(a) computation time (in s)			

n	10	15	20
10 agents	6.1	10.5	15.1
20 agents	6.0	10.3	15.0
30 agents	5.9	10.5	14.5
(b) queries per agent			

Fig. 2. Computation time and number of direct rating queries

4   Perspectives

The model we studied (utilitarian model with additive preferences) is simple: we chose to address difficulties one at the time, and, this way, we have been able to tackle the problem of building an appropriate incremental elicitation procedure. The simplicity of this model also makes it improvable. A first way to improve it would be to resort to more sophisticated social welfare functions. It would be interesting to extend the study by considering Ordered Weighted Average (OWA) for f_u. However, the OWA is not a linear criterion, it is therefore difficult to compute the PMR values. Moreover, OWA-optimal solutions may include subsolutions at step k that are OWA-dominated, which prevents resorting to standard dynamic programming algorithm.

The representation of agents preferences can be improved too. In practical problems that can be modeled as a KP, items often interact and we may want to model positive or negative synergies between items. Preferences involving synergies can be represented by Choquet Integrals, or k-additives utilities [10].

Of course, these representations would require a greater number of questions to be elicited. There is a subtle balance to keep between expressiveness of the model used and the elicitation burden imposed by the model.

Besides, the incremental approach is worth investigating in other combinatorial optimization problems involving multiple agents such as multiagent assignment problems or multiagent scheduling problems.

References

1. Benabbou, N., Perny, P.: Combining preference elicitation and search in multiobjective state-space graphs. In: IJCAI, pp. 297–303 (2015)
2. Benabbou, N., Perny, P.: On possibly optimal tradeoffs in multicriteria spanning tree problems. In: International Conference on Algorithmic DecisionTheory, pp. 322–337 (2015)
3. Benabbou, N., Perny, P.: Solving multi-agent knapsack problems using incremental approval voting. In: 22nd European Conference on Artificial Intelligence (ECAI 2016) (2016)
4. Betzler, N., Dorn, B.: Towards a dichotomy for the possible winner problem in elections based on scoring rules. J. Comput. Syst. Sci. **76**(8), 812–836 (2010)
5. Boutilier, C., Patrascu, R., Poupart, P., Schuurmans, D.: Constraint-based optimization and utility elicitation using the minimax decision criterion. Artif. Intell. **170**(8–9), 686–713 (2006)
6. Chajewska, U., Koller, D., Parr, R.: Making rational decisions using adaptive utility elicitation. In: Proceedings of AAAI 2000, pp. 363–369 (2000)
7. Dery, L.N., Kalech, M., Rokach, L., Shapira, B.: Reaching a joint decision with minimal elicitation of voter preferences. Inf. Sci. **278**, 466–487 (2014)
8. Drummond, J., Boutilier, C.: Elicitation and approximately stable matching with partial preferences. In: IJCAI, pp. 97–105 (2013)
9. Gelain, M., Pini, M.S., Rossi, F., Venable, K.B., Walsh, T.: Elicitation strategies for soft constraint problems with missing preferences: Properties, algorithms and experimental studies. Artif. Intell. J. **174**(3–4), 270–294 (2010)
10. Grabisch, M., Labreuche, C.: A decade of application of the choquet and sugeno integrals in multi-criteria decision aid. Ann. Oper. Res. **175**(1), 247–286 (2010)
11. Konczak, K., Lang, J.: Voting procedures with incomplete preferences. In: Proceedings of IJCAI-05 Multidisciplinary Workshop on Advances in Preference Handling, vol. 20 (2005)
12. Lu, T., Boutilier, C.: Multi-winner social choice with incomplete preferences. In: Proceedings of IJCAI 2013, pp. 263–270 (2013)
13. Oren, J., Lucier, B.: Online (budgeted) social choice. In: AAAI, pp. 1456–1462 (2014)
14. Regan, K., Boutilier, C.: Regret-based reward elicitation for Markov decision processes. In: UAI, pp. 444–451 (2009)
15. Skowron, P., Faliszewski, P., Lang, J.: Finding a collective set of items: from proportional multirepresentation to group recommendation. In: Proceedings of AAAI 2015, pp. 2131–2137 (2015)
16. Wang, T., Boutilier, C.: Incremental utility elicitation with the minimax regret decision criterion. In: Proceedings of IJCAI 2003, pp. 309–316 (2003)
17. Xia, L., Conitzer, V.: Determining possible and necessary winners given partial orders. J. Artif. Intell. Res. (JAIR) **41**, 25–67 (2011)

Logic-Based Merging in Fragments of Classical Logic with Inputs from Social Choice Theory

Adrian Haret$^{(\boxtimes)}$

Databases and Artificial Intelligence Group (DBAI),
Institute of Information Systems (E184/2), TU Wien, Vienna, Austria
haret@dbai.tuwien.ac.at

1 Background

Aggregation. One does not need to look far to find problems which require harmonization of different, possibly conflicting, preferences. Problems of this sort occur, on a grand scale, in national elections. But they also occur on smaller scales, in day-to-day decision processes such as friends settling on a place for lunch, or co-workers choosing available timeslots for a meeting. And, of course, they are encountered in the technical domain: an automated recommender system designed to suggest items to a group of agents, will first have to get an idea of the group's preferences. To do that it should take into account the individual preferences, typically by looking at how they rate various products, and distil them into a portrait of the group's desires [15,16]. Elections, hungry friends and target groups are all situations in which a collection of agents needs to be thought of and modelled as a single agent. And one wonders, naturally, about the best, most efficient way to do this.

Combinatorial domains. Aggregation of preferences has been studied for some time under the umbrella of Social Choice Theory. In the standard Social Choice model agents collectively choose from a given set of alternatives, each agent ranking the alternatives on a list. An aggregation rule then combines the lists and selects a winner.

But we can easily imagine that, instead of having to pick one winner, agents are required to vote on several issues at the same time; or, that they may have to choose a *combination* of alternatives, such as a committee. We can also imagine that preferences have a structure which is richer than what a simple list of the alternatives can represent. An agent may say to itself: "I want A only if B is there as well; if not I would rather have C and D, but then A and C is better than B and C," and so on. In other words, agents can have preferences over bundles of alternatives, and it might not be straightforward to infer these complex preferences from rankings on individual alternatives. My father, for example, likes to smoke and drink coffee. And his tastes are such that if he cannot do both activities at the same time, he would rather do none of them than either in isolation.

Known as *combinatorial voting*, this topic has lately been subject to intense scrutiny [1,6,14]. Most work is dedicated to reconciling two extremes, both of

© Springer International Publishing AG 2017
J. Rothe (Ed.): ADT 2017, LNAI 10576, pp. 374–378, 2017.
DOI: 10.1007/978-3-319-67504-6_30

which are untenable: either we simplify the aggregation procedure to a point where results threaten to become meaningless, or we require things from voters which are prohibitively costly. A workable solution needs to be found somewhere in between, and this frames the underlying question addressed by my research:

> What are the means available for artificial agents to express complex preferences? And how shall (inevitable) conflicts be resolved?

What we need is a language in which voters can express preferences in a compact form, a way in which one can say very much with very little. Preferably, such a language is easy to handle (by machines) and intuitive (for us), an agreeable middle ground between the formal rigour of computers and the cognitive sensibilities of human reasoners.

2 Research Topics

The approach I take in the thesis is to start from a known and beloved formal language which has already proven its worth in representing knowledge, namely propositional logic. To handle aggregation problems in propositional logic we design an operator tasked with performing exactly this kind of reasoning: given multiple sources of information, the operator analyses them, resolves conflicts and returns a consistent compromise view. 'Design' is an appropriate word here, since it involves formulating suitable properties which such an operator is expected to satisfy. The sub-field of Artificial Intelligence called *belief merging* [13] studies the variety of ways in which the aggregated perspective is obtained and gives a systematic picture of the entire process.

Nonetheless, belief merging still does not go all the way towards solving the issues raised in the previous section. Below I elaborate on the main shortcomings, which constitute the subject of my thesis.

Topic 1. Reasoning in fragments. The price to be paid is that, from the point of view of computational complexity, the reasoning involved in merging propositional knowledge bases can be costly [12]. Often, though, it seems that propositional logic actually provides *more* than enough resources, and for practical purposes we may find that we can get along with a restricted version—a *fragment*—of the language. Merging in fragments has, except for some notable exceptions [5,7], not been investigated to a great extent.

Topic 2. More than one way of being fair. Aggregation, in the social sphere, is supposed to be fair towards the parties involved. Since merging was mainly motivated by the idea of obtaining a consistent result, the basic properties expected of merging operators tend to be light on the social aspect, and only consider the most basic of fairness criteria. On the other hand the Social Choice literature, in its analysis of voting rules, has developed a sophisticated arsenal of properties relating to aggregation procedures. With some exceptions [11,18], the relation between these properties and merging operators has until quite recently been largely unexplored.

Topic 3. Branching out. We know how to do merging in propositional logic and, from work on Topic 2, we would be able to understand merging in fragments. But human knowledge comes in other forms as well. Representation results show us that merging operators behave *as if* they have preferences over the possible outcomes described by the language, but these preferences are not encoded in the language itself. However, there do exist formalisms designed specifically to handle preferences [3,4] and it would be a significant achievement to extend our results to these languages.

Research questions. Building on the topics elaborated above, the main issues engaged with in the thesis can be summarized as follows.

1. How apt are well-known fragments of propositional logic at supporting merging tasks?
2. How do merging operators (in the case of both propositional logic and fragments thereof) behave with respect to properties studied in Social Choice Theory?
3. Can we export insights gained in (1) and (2) to other notable formalisms?

Progress so far. My research on Topic 1 has consisted in work on merging in the Horn fragment [19] and on the deviation of standard operators from known fragments [17]. Deviation has been introduced as a measure for quantifying a problem often encountered when working in fragments, which is that the output cannot always be expressed in the fragment we are working with. Research on Topic 2 has gone towards adapting properties from Social Choice Theory to the context of (propositional) merging, and on studying their relationship with existing operators [18]. It remains to be seen how these properties behave in fragments.

One outcome of Topic 1 is that we have developed a toolbox for working with fragments, and we also have an idea of what fragment works for what kind of task. I believe, though, that these insights can be applied more broadly. A first confirmation was obtained by results on Argumentation Frameworks [8,10], another prominent formalism in AI. What would give significant weight to the thesis, now, would be an understanding of merging in languages conceived to handle preferences. Such an application would be striking and useful, and it is also the subject of ongoing work.

DC presentation. Voting on multiple issues is tricky because, as described in Sect. 1, we have to strike a balance between two extremes: requiring full rankings over all possible outcomes places too much of a burden on the agent, but using too little information can lead to universally loathed results.

Using a language like propositional logic allows one to package combinations of outcomes in the form of formulas, which one can then combine with a belief merging operator. But viewed from a Social Choice perspective, merging works by effectively asking the agents to state only their most preferred options: the knowledge bases they provide are taken to encode the interpretations which they believe (or want) to be true, and these are taken to be the top choices in a set of rankings over all possible worlds. The remaining possible worlds are then ranked as less plausible, though under the condition that each ranking is completely fixed by the top choices (known as the principle of irrelevance of syntax). It is an interesting question if aggregating rankings over possible worlds under such restrictions is still subject to celebrated results from Social Choice Theory, such as Arrow's theorem [2]. There exists some recent work in this direction [9], but it requires giving up some key merging properties, like the irrelevance of syntax mentioned above. Part of the presentation will focus on obtaining an Arrow-like impossibility result in the merging framework, if such properties are kept.

An attractive alternative to merging is to use formalisms designed especially for representing and reasoning with preferences [3]. This approach has its advantages, because preference languages are very good at expressing the kind of structures on the space of alternatives that often characterize the preferences of real world agents—which should come as no surprise, since preference languages are designed with exactly this kind of thing in mind.

However, preference languages are confronted with the same kind of problems that appear in reasoning with propositional logic: expressive formalisms make it hard to do efficient computation. So the idea arises, here as elsewhere, of restricting the languages in some systematic way—that is to say, of using fragments—in order to make the reasoning simpler.

Another part of the presentation will talk about merging applied to specialized formalisms for representing knowledge and preferences, such as the preference logic *PL* and its conjunctive fragment. The focus here is on formulating appropriate rationality constraints for a merging operator, and then exploiting the semantic structure of the logic to obtain representation results.

References

1. Ahn, D.S., Oliveros, S.: Combinatorial voting. Econometrica **80**(1), 89–141 (2012)
2. Arrow, K.: Social Choice and Individual Values, 2nd edn. Yale University Press, New Haven (1963)
3. Bienvenu, M., Lang, J., Wilson, N.: From preference logics to preference languages, and back. In: Proceedings of KR 2010 (2010)
4. Boutilier, C., Brafman, R.I., Domshlak, C., Hoos, H.H., Poole, D.: CP-nets: a tool for representing and reasoning with conditional ceteris paribus preference statements. J. Artif. Intell. Res. (JAIR) **21**, 135–191 (2004)
5. Brewka, G., Mailly, J., Woltran, S.: Translation-based revision and merging for minimal horn reasoning. In: Proceedings of ECAI 2016, pp. 734–742 (2016)
6. Chevaleyre, Y., Endriss, U., Lang, J., Maudet, N.: Preference handling in combinatorial domains: from AI to social choice. AI Mag. **29**(4), 37–46 (2008)
7. Creignou, N., Papini, O., Rümmele, S., Woltran, S.: Belief merging within fragments of propositional logic. ACM Trans. Comput. Log. **17**(3), 20:1–20:28 (2016)
8. Delobelle, J., Haret, A., Konieczny, S., Mailly, J., Rossit, J., Woltran, S.: Merging of abstract argumentation frameworks. In: Proceedings of KR 2016, pp. 33–42 (2016)
9. Díaz, A.M., Pérez, R.P.: Impossibility in Belief Merging. CoRR abs/1606.04589 (2016)

10. Diller, M., Haret, A., Linsbichler, T., Rümmele, S., Woltran, S.: An extension-based approach to belief revision in abstract argumentation. In: Proceedings of IJCAI 2015, pp. 2926–2932 (2015)
11. Everaere, P., Konieczny, S., Marquis, P.: The strategy-proofness landscape of merging. J. Artif. Intell. Res. (JAIR) **28**, 49–105 (2007)
12. Konieczny, S., Lang, J., Marquis, P.: DA^2 merging operators. Artif. Intell. **157**(1–2), 49–79 (2004)
13. Konieczny, S., Pérez, R.P.: Logic based merging. J. Philos. Logic **40**(2), 239–270 (2011)
14. Lang, J., Xia, L.: Voting in combinatorial domains. In: Brandt, F., Conitzer, V., Endriss, U., Lang, J., Procaccia, A.D. (eds.) Handbook of Computational Social Choice, pp. 197–222. Cambridge University Press, New York (2016). Chapter 9
15. Masthoff, J.: Group modeling: selecting a sequence of television items to suit a group of viewers. User Model. User-Adapt. Interact. **14**(1), 37–85 (2004)
16. Senot, C., Kostadinov, D., Bouzid, M., Picault, J., Aghasaryan, A., Bernier, C.: Analysis of strategies for building group profiles. In: De Bra, P., Kobsa, A., Chin, D. (eds.) UMAP 2010. LNCS, vol. 6075, pp. 40–51. Springer, Heidelberg (2010). doi:10.1007/978-3-642-13470-8_6
17. Haret, A., Mailly, J., Woltran, S.: Distributing knowledge into simple bases. In: Proceedings of IJCAI 2016, pp. 1109–1115 (2016)
18. Haret, A., Pfandler, A., Woltran, S.: Beyond IC postulates: classification criteria for merging operators. In: Proceedings ECAI 2016, pp. 372–380 (2016)
19. Haret, A., Rümmele, S., Woltran, S.: Merging in the horn fragment. In: Proceedings of IJCAI 2015, pp. 3041–3047 (2015)

Distances in Voting and Committee Election Rules for General Preference Types

Lisa Rey[✉]

Heinrich-Heine-Universität Düsseldorf, Düsseldorf, Germany
lrey@cs.uni-duesseldorf.de

Distances have a major role in voting theory. They can determine the outcome of an election or ways to influence elections via manipulation or bribery. In the first part of this abstract, previotus results concerning distance-based committee election rules are summarized, while current research questions are introduced in the second part.

There is detailed work on elections where the desirable outcome of an election is a single winner. However, in many situations a group of candidates is supposed to be elected, such as a group of people for a certain project, a delegation of politicians, or a selection of food for a party. Here we use a setting in which the size of the committee is fixed, such that a committee election is defined by a tuple (C, V, k), where $C = \{c_1, \ldots, c_m\}$ is the set of candidates, $V = (v_1, \ldots v_n)$ a list of all votes given by the voters in the set $\{1, \ldots, n\}$, and k the committee size. As election rules the minisum- and minimax approaches are used. Brams et al. [5] studied these procedures regarding approval voting and Baumeister and Dennisen [3] enlarged on this matter by also considering trichotomous votes, incomplete-, and complete linear orders. The complexity of winner determination and manipulation of these different variations was studied by Baumeister et al. [4].

While casting votes in these forms may apply to many situations, it may not always be sufficient or possible to rank all voters to different positions. As an intermediate approach Baumeister et al. [2], inspired by Elkind et al.'s work on Doodle poll games [8], introduced ℓ-group rules.

In ℓ-group rules, voters assort the candidates into ℓ groups. Here a group may be empty. The concluding vote is a list $(G_1, \ldots, G_\ell)$ with disjoint sets $G(i), 1 \leq i \leq \ell$, where $G(i)$ contains all candidates allocated to group i by the voter in question. As committee elections with a fixed size are considered, and as, in order to calculate the election outcome, votes are compared to possible election winners, only committees of size k are of relevance. We write $W(i) = 1$ for a candidate c_i in the committee W of size k and $W(i) = \ell$ for a candidate $c_i \notin W$. Similarly, let $v(i)$ denote the group number of $v(i)$ in v.

To define minisum- and minimax ℓ-group rules the distance between a ballot v and a committee W of size k is introduced as

$$\delta_\ell(v, W) = \sum_{i=1}^{m} |v(i) - W(i)|,$$

© Springer International Publishing AG 2017
J. Rothe (Ed.): ADT 2017, LNAI 10576, pp. 379–382, 2017.
DOI: 10.1007/978-3-319-67504-6_31

where m is the number of candidates in the election. When denoting a committee accordingly this distance corresponds to the Hamming distance between the vote and the committee. In a minisum approach the committee (or the committees) which minimizes the sum of the distances between all votes and the committee are the election winners.

Definition 1 (minisum ℓ-group rule). *The minisum ℓ-group rule is a function f_{sum}^ℓ so that $f_{sum}^\ell(C, V, k) = \arg\min_{W \text{ of size } k} \sum_{v \in V} \delta_\ell(v, W)$, i. e., f_{sum}^ℓ minimizes the sum of disagreements of voters with the winning committees.*

In the minimax approach the committee (or the committees) that minimizes the maximal distance between a vote and the committee wins the election. The minimax ℓ-group rule is defined as follows:

Definition 2 (minimax ℓ-group rule). *The minimax ℓ-group rule is a function f_{max}^ℓ so that $f_{max}^\ell(C, V, k) = \arg\min_{W \text{ of size } k} \max_{v \in V} \delta_\ell(v, W)$, i. e., f_{max}^ℓ minimizes the disagreement of the worst satisfied voter with the winning committees.*

The following example illustrates these voting rules.

Example 1. Let $E = (\{c_1, c_2, c_3\}, \{v_1, v_2, v_3\}, 2)$ be an election with the following votes:

$$v_1 : \qquad \{c_1\} >_1 \{c_2\} >_1 \{c_3\}$$
$$v_2 : \qquad \{c_3\} >_2 \{\} >_2 \{c_1, c_2\}$$
$$v_3 : \qquad \{c_2, c_3\} >_3 \{\} >_3 \{c_1\}$$

Table 1 shows each voter's dissatisfaction with each possible committee of size 2 as well as the voters' sum of (respectively, the maximal) dissatisfaction with the respective committee. For example, the dissatisfaction of the first voter with the committee $\{c_2, c_3\}$ is $\delta_\ell(v_1, \{c_2, c_3\}) = 2 + 1 + 2 = 5$.

The winners under the minisum/minimax 3-group rule are highlighted in Table 1. Under the minisum 3-group rule $\{c_2, c_3\}$ is the unique winner of the election ($f_{sum}^\ell(E) = \{c_2, c_3\}$), whereas under the minimax 3-group rule candidate c_2 is replaced by candidate c_1 ($f_{max}^\ell(E) = \{\{c_1, c_3\}\}$).

A central aspect of the work regards axiomatic properties. A summary of the results by Baumeister et al. [2] is given in Table 2, while we refer to Baumeister et al. [2] for the definitions of the properties.

However distances are not only used to determine election winners. One can also introduce a distance to limit changes that can be made to a vote or to several votes when elections are manipulated or bribed. In this setting we stick to single winner elections, although an extension to committee elections is of great interest. Manipulation in single winner elections, motivated by the famous results by Gibbard [6] and Satterthwaite [9], was formally introduced by Bartholdi et al. [1] and describes the problem of as to whether it is possible to obtain a more preferred result by voting insincerely. My current research, which is joint work with

Table 1. Computing the winners of the committee election under the minisum/minimax 3-group rule

	$\{c_1, c_2\}$	$\{c_1, c_3\}$	$\{c_2, c_3\}$
v_1	1	3	5
v_2	6	2	2
v_3	4	4	0
Sum	11	9	**7**
Max	6	**4**	5

Table 2. Property results for minisum and minimax ℓ-group rules

Properties	ℓ-group rules	
	Minisum	Minimax
Non-imposition, Homogeneity	✓	✓
Consistency	✓	✗
Independence of clones	✓	✗
Committee monotonicity	✓	✗
(Candidate) monotonicity	✓	✓
Positive responsiveness	✓	✗
Pareto criterion	✓	✓
(Committee) Condorcet consistency	✗	✗
Solid coalitions, Consensus committee	✗	✗
Unanimity	strong	strong

my supervisor Dorothea Baumeister and Tobias Hogrebe from Heinrich-Heine-Universität Düsseldorf, is based on a model of including distances in manipulation problems introduced by Obraztsova and Elkind [7], who included distances into manipulation by stating that it might be natural in some situations, that a voter can only change his vote to a certain extent. In the case of constructive manipulation in the unique winner model, which is of interest here, the manipulator tries to make a distinguished candidate the unique winner of an election.

A single winner election is a tuple (C, V) with the set of candidates C and the profile V, as above. The set of voters is $\{1, \ldots n\}$, such that the vote v_i of voter $i, 1 \leq i \leq n$ is given in V. The election outcome under a voting rule $\mathcal{E}$ (we write $\mathcal{E}(C, V)$ for short) is a set, containing the winner or the winners of the election. A distance function is denoted by d. When exchanging the vote v_i in the profile V by another, here dishonest, vote v_j the resulting new profile is denoted by (V_{-i}, v_j). With these notations, the general form of the optimal manipulation problem, where $\mathcal{D}$ is a family of integer-valued distances and $\mathcal{E}$ is again a voting rule, can be stated as follows.

$(\mathcal{D},\mathcal{E})$-OptManipulation [7]
Given: An election (C,V) with $C = \{c_1, c_2, \ldots, c_m\}$ and $V = (v_1, v_2, \ldots, v_n)$, a voter $i \in \{1, \ldots, n\}$, a distinguished candidate $p \in C$, and a non-negative integer b.
Question: Does there exist a vote s over C, such that $\mathcal{E}(C, (V_{-i}, s)) = \{p\}$ and $d(v_i, s) \le b$?

One of the used distance measures, which we also consider here, is the swap-distance. When writing $x \succ_i y$ if voter i strictly preferes candidate x over candidate y, the swap distance between two votes v_i and v_j is given by

$$d_{swap}(v_i, v_j) = |\{(x,y)| x \succ_i y \text{ and } y \succ_j x\}|.$$

A central task for further research is the adaption of this model to coalitional manipulation, where several manipulators try to achieve their mutual goal by giving possibly insincere votes, and its computational complexity.

References

1. Bartholdi, J., Tovey, C., Trick, M.: The computational difficulty of manipulating an election. Soc. Choice Welf. **6**(3), 227–241 (1989)
2. Baumeister, D., Böhnlein, T., Rey, L., Schaudt, O., Selker, A.: Minisum and minimax committee election rules for genral preference types. In: Proceedings of the 22nd European Conference on Artificial Intelligence, pp. 1656–1657. IOS press (2016)
3. Baumeister, D., Dennisen, S.: Voter dissatisfaction in committee elections. In: Proceedings of the 14th International Conference on Autonomous Agents and Multiagent Systems, pp. 1707–1708. IFAAMAS (2015)
4. Baumeister, D., Dennisen, S., Rey, L.: Winner determination and manipulation in minisum and minimax committee elections. In: Walsh, T. (ed.) ADT 2015. LNCS, vol. 9346, pp. 469–485. Springer, Cham (2015). doi:10.1007/978-3-319-23114-3_28
5. Brams, S., Kilgour, D., Sanver, R.: A minimax procedure for electing committees. Public Choice **132**, 401–420 (2007)
6. Gibbard, A.: Manipulation of voting schemes: a general result. Econometrica **41**(4), 587–601 (1973)
7. Obraztsova, S., Elkind, E.: Optimal manipulation of voting rules. In: Proceedings of the 11th International Conference on Autonomous Agents and Multiagent Systems, vol. 2, pp. 619–626. IFAAMAS (2012)
8. Obraztsova, S., Elkind, E., Polukarov, M., Rabinovich, Z.: Doodle poll games. In: Proceedings of the First IJCAI-Workshop on Algorithmic Game Theory (2015)
9. Satterthwaite, M.: Strategy-proofness and arrow's conditions: existence and correspondence theorems for voting procedures and social welfare functions. J. Econ. Theor. **10**(2), 187–217 (1975)

Profile Distances and Optimal Manipulation of Voting Rules

Tobias Alexander Hogrebe[(✉)]

Institut für Informatik, Heinrich-Heine-Universität Düsseldorf,
40225 Düsseldorf, Germany
`tobias.hogrebe@uni-duesseldorf.de`

Profile Distances

In recent years, there has been increasing interest in defining and applying various variants of distances for measuring the difference between two votes. Classic examples are Dodgson and Kemeny voting [7] which already include vote distances. Another important example for this is the optimal manipulation of voting rules [2] which defines manipulation problems, where the distance between the truthful and nontruthful vote is limited. And finally the distance rationalization of voting rules which already introduces the idea of measuring the distance between two elections or profiles.

We introduce a new, easy to apply, technique of extending well-known vote distances to profile distances with respect to the anonymity of every voter. Consider the following well-known distances between votes.

Given a set of alternatives $A = \{a_1, a_2, ..., a_m\}$ and two votes v, v' as linear orders over A. Let $pos_v(a)$ be the position of a in vote v.

Spearman Distance [4]: $d_{Spear.}(v, v') = \sum_{a \in A} |pos_v(a) - pos_{v'}(a)|$

Swap Distance [3]: $d_{Swap}(v, v') = |\{(a_i, a_j) \in A \times A \mid a_i >_v a_j \wedge a_j >_{v'} a_i\}|$

One way of extending the previous vote distances to distances between profiles is to implement them in the following approach, based on the Bottleneck distance [6].

Given two profiles P, P' over the set of alternatives A and a vote distance d. The *matching distance* is given by:

$$D_{Mat.,d}(P, P') = \left[\min_{M \in \mathcal{M}(P,P')} \sum_{(v,v') \in M} d(v, v') \right] + ||P| - |P'|| \cdot M_d$$

where $\mathcal{M}_d(P, P')$ is the set of all possible maximum matchings between the votes in P and P' and M_d is the maximum value of d between two votes over A.

By maintaining the anonymity of the voters, it is not possible to compare the two voices of a voter directly to each other. Because of this, the most natural way of measuring the distance between two profiles seems to be to match the votes in the minimal way. Considering two profiles, where one profile emerged from the other through changed votes, a matching with minimal overall distance between the votes appears to be the most likely assumption.

© Springer International Publishing AG 2017
J. Rothe (Ed.): ADT 2017, LNAI 10576, pp. 383–384, 2017.
DOI: 10.1007/978-3-319-67504-6

Optimal Manipulation of Voting Rules
As mentioned by Obraztsova and Elkind [2], a natural direction for working with the optimal manipulation of voting rules is to extend the study to more vote distances and more general (e.g. weighted) distances.

Let D be some vote distance and F some voting rule. We extend the original definition of *optimal manipulation* by Obraztsova and Elkind [2] with an additional matrix which can be used as input to D.

(D,F)-**OptManipulation**: Given a set of alternatives $A = \{a_1, a_2, ..., a_m\}$, profile $P = (v_1, v_2, ..., v_n)$, a voter $i \in \{1, ..., n\}$, a preferred alternative $p \in C$, a distance limit $k \in \mathbb{N}$ and some D-specific parameter matrix M.

The above instance is a yes-instance if and only if there exists a vote v_i' with $F((v_1, ..., v_{i-1}, v_i', v_{i+1}, ..., v_n)) = \{p\}$ and $D_M(v_i, v_i') \leq k$.

The natural direction of extending the swap distance is to add the possibility of setting non-uniform costs [5]. This version is indirectly used in swap bribery [1].

Let $A = \{a_1, a_2, ..., a_m\}$ be the set of alternatives and $\pi : A \times A \to \mathbb{R}_{\geq 0}$ a symmetric swap price function. The swap (pseudo) distance with non-uniform costs is given by:

$$d_\pi(v, v') = \sum_{(a_i, a_j) \in inv(v,v')} \pi(a_i, a_j)$$

where $inv(v, v') = \{(a_i, a_j) \in A \times A \mid a_i >_v a_j \wedge a_j >_{v'} a_i\}$ is the set of inversions. π can be expressed through a symmetric matrix $M \in \mathbb{R}_{\geq 0}^{m \times m}$.

In terms of optimal manipulation this distance takes care of votes in a biased election, e.g. where swapping some very conflicting alternatives seems to be a very unnatural behaviour.

Our goal is to find new distances for measuring differences between votes and profiles with respect to wishful properties. We are planning to use them in the study of optimal manipulation and related topics.

References

1. Elkind, E., Faliszewski, P., Slinko, A.: Swap bribery. In: Mavronicolas, M., Papadopoulou, V.G. (eds.) SAGT 2009. LNCS, vol. 5814, pp. 299–310. Springer, Heidelberg (2009). doi:10.1007/978-3-642-04645-2_27
2. Obraztsova, S., Elkind, E.: Optimal manipulation of voting rules. In: Proceedings of AAMAS 2012, vol. 2, pp. 619–626. Richland, SC (2012)
3. Kendall, M.: A new measure of rank correlation. Biometrika **30**(1/2), 81–93 (1938)
4. Spearman, C.: The proof and measurement of association between two things. Am. J. Psychol. **15**(1), 72–101 (1904)
5. Kumar, R., Vassilvitskii, S.: Generalized distances between rankings. In: Proceedings of the 19th International Conference on World Wide Web, pp. 571–580. ACM (2010)
6. Deza, M.M., Deza, E.: Encyclopedia of Distances. Springer, Heidelberg (2009)
7. Bartholdi, J., Tovey, C.A., Trick, M.A.: Voting schemes for which it can be difficult to tell who won the election. Soc. Choice Welfare **6**(2), 157–165 (1989)

Strategy-Proofness of Scoring Allocation Correspondences under Social Welfare Maximization

Anna Maria Kerkmann[(⊠)]

Heinrich-Heine-Universität Düsseldorf,
Universitätsstraße 1, 40225 Düsseldorf, Germany
anna.kerkmann@uni-duesseldorf.de

Introduction

A central theme in fair division is to study attacks on allocation procedures and to discover the restrictions that make them strategy-proof. I concentrate on scoring allocation correspondences, which are based on a model due to Brams and King [1] and were further developed by Baumeister $et\ al.$ [2]. They are concerned with allocating resources to agents with the goal of social welfare maximization. Further investigations were made by Nguyen $et\ al.$ [3]. They considered determination of optimal allocations according to utilitarian social welfare, while I intent to concentrate on egalitarian social welfare and social welfare by the Nash product. Furthermore, I plan to focus on some other relevant restrictions, which are mentioned below.

Set-Up

Consider a set $A = \{1, ..., n\}$ of agents and a set $R = \{r_1, .., r_m\}$ of indivisible, nonshareable resources. All agents have ordinal preferences over all subsets of resources, but for practical reasons each agent i only gives a linear order $>_i$ over R. This leads to a preference profile $P = (>_1, ..., >_n)$. An allocation π of the resources to the agents is a partition $\pi = (\pi_1, ..., \pi_n)$ of the resources, where $\pi_i \subset R$ is agent i's share.

Scoring Allocation Correspondence

Under this set-up, the scoring allocation procedure pursues the goal to provide a set of allocations, which maximize the social welfare according to a given welfare function.

First, a scoring vector $s = (s_1, ..., s_m) \in \mathbb{R}^m$ is chosen, where $s_1 \geq ... \geq s_m \geq 0$ and $s_1 > 0$. Let $rank(r, >) \in \{1, ..., m\}$ be the rank of resource $r \in R$ under preference order $>$. Then the utility function $u_{>,s} : 2^R \rightarrow \mathbb{R}$ for order $>$ and scoring vector s is defined by $u_{>,s}(B) = \sum_{r \in B} s_{rank(r,>)}$, for a subset $B \subseteq R$. Furthermore, a social welfare function, which provides a measure of overall welfare under an allocation $\pi = (\pi_1, ..., \pi_n)$, is chosen by $sw(\pi) = \star_{i=1}^n u_{>_i,s}(\pi_i)$, where $\star$ is a symmetric, nondecreasing mapping from $\mathbb{R}^n$ to $\mathbb{R}$.

Finally, a scoring allocation correspondence F_s is a function mapping a given preference profile P to a nonempty set of allocations, and is defined by

© Springer International Publishing AG 2017
J. Rothe (Ed.): ADT 2017, LNAI 10576, pp. 385–386, 2017.
DOI: 10.1007/978-3-319-67504-6

$F_s(P) = \text{argmax}_\pi\, sw(\pi)$. Hence, $F_s(P)$ maximizes social welfare by using the welfare function sw.

Considered Welfare Functions

Unlike Nguyen et al. [3], who observed scoring allocation correspondences under utilitarian social welfare, I plan to consider the following welfare functions, leading to different scoring allocation correspondences:

- egalitarian social welfare: $sw_e(\pi) = \min\{u_{>_i,s}(\pi_i)|i \in A\}$ and
- Nash social welfare: $sw_N(\pi) = \prod_{i \in A} u_{>_i,s}(\pi_i)$.

Strategy-Proofness

The question I want to study is whether a single agent i can benefit from submitting an insincere order $>_i$ over R. Since scoring allocation correspondences determine a set of allocations, agents have to value their set of possible shares instead of only single shares. Therefore, we use set-extension principles, which lift the agents' preferences. Consider the Kelly extension: Given two sets of shares Π_1 and Π_2, we say that agent i weakly Kelly-prefers Π_1 to Π_2 ($\Pi_1 \succsim_i^K \Pi_2$) if for every $A \in \Pi_1$ and $B \in \Pi_2$, it holds that $A \succ_i B$ or $A = B$. Another considered principle is the Gärdenfors extension (given by $\succsim_i^G$), whose definition we omit due to lack of space. For both extensions $e \in \{K, G\}$ we define $\Pi_1 \succ_i^e \Pi_2 \Leftrightarrow \Pi_1 \succsim_i^e \Pi_2 \wedge \neg\Pi_2 \succsim_i^e \Pi_1$.

We say that a scoring allocation correspondence F is e-manipulable by an agent i if there exists a profile $P = (>_1, ..., >_n)$ and an order $>_i'$ such that $F(P')_i \succ_i^e F(P)_i$, where $F(P)_i$ is the set of possible shares for agent i when submitting $>_i$ and $F(P')_i$ is the analogous set for $>_i'$. F is said to be e-strategy-proof if F is not e-manipulable by any agent.

Intended Investigations

My plans include to identify restrictions of the chosen scoring vector and the number of agents, which guarantee strategy-proofness under different welfare models. A good starting point could be fixing the number of agents. As results from Nguyen et al. [3] suggest, it might be promising to investigate the impact of limiting the number of different values in the scoring vector.

Finally, as further work it would also be interesting to consider the impact of changing the underlying extension principle.

References

1. Brams, S.J., King, D.L.: Efficient fair division. Ration. Soc. **17**(4), 387–421 (2005)
2. Baumeister, D., Bouveret, S., Lang, J., Nguyen, N.-T., Nguyen, T.T., Rothe, J., Saffidine, A.: Positional scoring-based allocation of indivisible goods. Auton. Agents Multi-Agent Syst. **31**(3), 628–655 (2017)
3. Nguyen, N.-T., Baumeister, D., Rothe, J.: Strategyproofness of scoring allocation correspondences for indivisible goods. In: Proceedings of the Twenty-Fourth International Joint Conference on Artificial Intelligence, pp. 1127–1133 (2015)

Expanding the Fair Division Framework by Agents with Altruistic Influence

Dominique Christine Komander[✉]

Institut für Informatik, Heinrich-Heine-Universität Düsseldorf,
40225 Düsseldorf, Germany
dominique.komander@uni-duesseldorf.de

This abstract can be classified into the field of fair division and more precisely into multi-agent resource allocation of indivisible goods, where the aim is to distribute some resources to the agents and meanwhile getting a fair allocation. We consider the framework from *Chevaleyre et al.*'s work "Allocating Goods on a Graph to Eliminate Envy" [1], where interactions between the agents are limited to a negotiation topology, represented as a graph. After introducing the details of this work I want to present an idea for expanding the environment by agents that are not only utility maximizers but that can have altruistic influences.

There is a finite set of agents $\sigma = \{1, \ldots, n\}$ and a finite set of indivisible, nonshareable goods $R = \{R_1, \ldots, R_m\}$ that should be completely distributed to the agents. An allocation is a function $A : \sigma \to 2^R$ that allocates disjoint bundles of goods to each agent. The negotiation topology is an undirected graph $G = (\sigma, E)$, where each node represents an agent. The nodes are connected if the represented agents can see each other. Only if that is the case, the agents are enabled to envy each other or to negotiate with each other as explained below.

A deal $\delta = (A, A')$ is an exchange of two allocations, from A to A'. The admitted deals are *clique-deals*, which are deals that involve only the agents of a *clique*. A clique is defined as known from graph theory - it is a subset of the nodes, in which each node is connected to every other node inside the clique. The supermodular preferences of every agent i are expressed by a *valuation function* $v_i : 2^R \to \mathbb{Q}$ that is normalised. In order to reach envy-freeness, a monetary sidepayment is connected to the deals. The payment function is $p : \sigma \to \mathbb{Q}$ with the restriction $\sum_{i \in \sigma} p(i) = 0$. In the negotiation process only individually rational deals are allowed. A deal is individually rational (IR) when there exists a payment function that satisfies the following condition: $v_i(A') - v_i(A) > p(i)$ for all $i \in \sigma$. If $A(i) = A'(i)$, it is allowed that $p(i) = 0$, so that there does not have to be a payment, if nothing in the bundle changes from i's point of view. Nevertheless, it is possible that i receives money, although he is not directly affected by the deal.

The payment function that is used is the globally uniform payment function (*GUPF*): $(v_i(A') - v_i(A)) - (sw(A') - sw(A))/n$ where $sw(A) = \sum_{i \in \sigma} v_i(A)$ is the *utilitarian social welfare* of allocation A as defined by Moulin [2]. The payment balance $\pi(i)$ sums up all the payments agent i made so far. A negotiation state (A, π) combines an allocation with the associated payment balance. The definition of envy-freeness inside the graph relies on a utility function $u_i(R, x) = v_i(R) - x$, where x is the payment that is subtracted from the valu-

© Springer International Publishing AG 2017
J. Rothe (Ed.): ADT 2017, LNAI 10576, pp. 387–388, 2017.
DOI: 10.1007/978-3-319-67504-6

ation of agent i's bundle. A state (A, π) is graph-envy-free (GEF) based on the graph $G = (\sigma, E)$ iff $u_i(A(i), \pi(i)) \geq u_i(A(j), \pi(j))$ holds for all agents i and j that are connected in the graph G. That means that agents are only jealous of bundles combined with payments of another agent. At least initial equitability payments are provided, before the negotiations begin when the initial allocation A_0 is given. This initial payments must be of the form $\pi_0(i) = v_i(A_0) - sw(A_0)/n$.

From these definitions Chevaleyre et al. [1] conclude the following result.

Theorem 1. (Chevaleyre et al. [1]). *If all valuations are supermodular and if initial equitability payments have been made, then any sequence of IR clique-deals using the GUPF will eventually result in a GEF state.*

This theorem holds only by using exactly this payment function, which ignores the negotiation topology of the graph and distributes the wealth globally all over the agents.

Until here only selfish agents are considered. But it seems to be very natural that humans do not only care about their own wealth. The following example will give a better comprehension of the dilemma: There is a scholarship and my sister and me both have preferences about getting it. In the assumption that we are utility maximizers, the problem is easy to describe. But what if my utility does not only depend on getting the scholarship for myself? I might also be just as happy when my sister will have it. And on top of that, what if my sister is even more altruistic so that she would prefer me to getting the scholarship instead of her, provided the chance that one of us can have it at all?

For allocating the goods with regard to agents with this altruistic influence, the preferences of the agents should be modified, so that some agents profit by the fact that one special other agent gets a good in the allocation. It is obvious, that agents can only care about the wealth of other agents, that they can see. Transferred to the negotiation framework these are agents, that have to be connected to each other in the graph.

To find the right form for describing the preferences of agents that care about the wealth of other agents seems to be technically difficult, because now the profit of an agent is not only depending on the allocation he gets for himself. Before caring about envy-freeness in this modified enviroment, there must be a reasonable representation of the altruistic influence of each agent. At this point I am currently researching and trying to find a method that is intuitively natural and that nevertheless does not complicate the fair allocation problem too much.

References

1. Chevaleyre, Y., Endriss, U., Maudet, N.: Allocating goods on a graph to eliminate envy. In: Proceedings of AAAI, pp. 700–705 (2007)
2. Moulin, H.: Axioms of Cooperative Decision Making. Cambridge University Press (1998)

Author Index

Printed in the United States
by Bookmasters

Printed in the United States
By Bookmasters